The Casting Powders Book

Kenneth C. Mills · Carl-Åke Däcker

The Casting Powders Book

Springer

Kenneth C. Mills
Royal School of Mines
Imperial College London
London
UK

Carl-Åke Däcker
Swerea KIMAB AB
Kista
Sweden

ISBN 978-3-030-09607-6 ISBN 978-3-319-53616-3 (eBook)
https://doi.org/10.1007/978-3-319-53616-3

Softcover re-print of the Hardcover 1st edition 2017

This Springer imprint is published by the registered company Springer Nature Switzerland AG
The registered company address is: Gewerbestrasse 11, 6330 Cham, Switzerland

Foreword

Swedish steelmakers have a long-standing commitment to product development. This demands that improved skills must be developed in the steel shop, in order to manufacture steel products free from defects and process problems. This is particularly important in the casting operation where it is necessary to optimise the process parameters especially, for the new products under development. The mould powder is the key factor affecting casting process control and product quality and has attracted considerable research effort in Sweden and other parts of the world over the last 15–20 years. However, many of the parameters affecting continuous casting are interactive; thus, changing one casting parameter can have a knock-on effect on other parameters. Furthermore, there is an absence of collated and structured data for casting powders; this has proved a major obstacle to researchers in steel plants, research institutes and universities. This is particularly apparent when new researchers enter the complex world of mould powders and continuous casting.

The Swedish Steel Producers identified the need for a comprehensive book on continuous casting powders. This was discussed in 2012 at a meeting of the Swedish Steel Producers' Association by its CEO, Bo-Erik Pers, the Chairman of the Technical Area on Casting and Solidification, Bo Rogberg, the research manager, Lars-Henrik Österholm and Carl-Åke Däcker, Manager of Process & Material Department, Swerea KIMAB. Following this meeting, Carl-Åke Däcker was tasked with contacting Prof. Ken Mills (Imperial College) with a view to writing a book on casting powders.

The project of writing a book on mould powders and slags was discussed at a meeting in London in August 2012 between Ken Mills and Carl-Åke Däcker, who have both participated in a significant number of ESCS and RFCS projects involving mould powders. The idea of a combined effort was persuasive since it would bring together Ken Mills' knowledge of academic research on mould powders with Carl-Åke Däcker's more practical experience in Swedish steel plants. The Swedish Steel Producers' Association (Hugo Carlsson Foundation) provided the necessary funding to take on the project and to provide supervision of the progress by a reference group.

The book is not meant to be read from binder to binder and each chapter is stand-alone with its own abstract, introduction and reference list. For that reason, the authors have tried to minimise the amount of searching the reader must do, through the various chapters, for figures, tables, etc. However, providing all the relevant information comes at the expense of some repetition.

The original version of the book was revised: For detailed information please see correction. The correction to the book is available at https://doi.org/10.1007/978-3-319-53616-3_12

Acknowledgements

We wish to thank the Swedish steelmakers and Swerea KIMAB for their continual support. We also acknowledge the additional funding and enthusiastic support of the Technical Area on Casting and Solidification (TO 24).

We also thank Swerea KIMAB, and its CEO, Staffan Söderberg, for funding and support and the Materials Department, Imperial College, London for their support.

We would like to give special thanks to the reference group that was formed at the beginning of the book project. They attended a number of meetings and provided, support, critical discussion and important technical input throughout the project:

- Tomas Sohlgren and Anders Lagerstedt, (SSAB Special Steels in Oxelösund).
- Arashk Memarpour, (Sandvik Materials Technology).
- Fatemeh Shahbazian, (Swerea KIMAB).

We also wish to thank the following for their help:

- Klaus Schultz and Dirk Eckhard (Imerys) and Claudio Valadares and Omar Afrange (Carbox, Brazil) for their inputs on Chap. 8 "Manufacture of mould fluxes".
- KCM would like to thank the late, Dr. Adrian Normanton, Vince Ludlow, Dr. Shahid Riaz and Dr. Bridget Steward (Tata Steel/Corus) for many informative discussions.
- Pavel Ramirez-Lopez (Swerea MEFOS), Peter Andersson (Swerea KIMAB), Brian Thomas (University of Illinois), Masahiro Susa and Miyuki Hayashi (Tokyo Institute of Technology), Qian Wang and Bing Xie (Chonquing University), Masayuki Kawamoto and M. Hanao (NSSMC), Koichi Tsutsumi (JFE Steel Corporation), Dr. S. Ogibayashi (Nippon Steel, NSSM Corp.), J.W. Cho (Gift, POSTECH) and Dr. Begona Santillana (Tata Steel, IJmuiden) for providing important information and valuable discussions on various points arised during the project.
- Mariana Ursu Däcker for valuable discussions, information and literature regarding silicate chemistry.

Finally, Ken Mills would like to thank his wife Margaret for her constant love, support and understanding throughout the project.

Kenneth C. Mills
Carl-Åke Däcker

Contents

Chapter 1
Introduction and Overview

Abstract This chapter provides an introduction into the use of casting powders in the continuous casting of steel. It includes brief descriptions of the following:

(i) The continuous casting process for steel;
(ii) The various changes that the mould powder undergoes as it transforms into a liquid slag and then into a slag film;
(iii) The various tasks carried out by the mould powder and the slag formed from it;
(iv) The factors affecting the formation and performance of slag films and slag rims;
(v) The effect of changes in casting conditions (a) casting speed, (b) metal flow, (c) mould dimensions, (d) oscillation characteristics, (e) steel grades, (f) Ar flow rate on powder consumption, heat flux, etc. and explains how mould powder can be modified to deal with these changes;
(vi) The important physical properties of mould slags and
(vii) It points out how the continuous casting mould is in a continual state of fluctuation and how this affects casting performance.

Finally, definitions are given of various terms used in this book.

Symbols, Abbreviations and Units

A	Area (m^2)
d_l	Thickness of liquid slag film (m)
d_s	Thickness of solid slag film (m)
F	Friction force (N)
f	Frequency (Hz or cycles min^{-1})
f^*	Fraction casting powder forming slag

The original version of this chapter was revised: For detailed information please see correction. The correction to this chapter is available at https://doi.org/10.1007/978-3-319-53616-3_12

Within this book minor changes to some of the figure permissions and the readability of the chapters have been made and several figures have been replaced with new versions.

K.C. Mills and C.-Å. Däcker, *The Casting Powders Book*,
DOI 10.1007/978-3-319-53616-3_1

q_{hor} Horizontal heat flux (Wm^{-2})
Q_{MR} Melting rate (kg min^{-1} or kgs^{-1})
Q_s Powder consumption (kgm^{-2})
R^* (surface area/volume) of mould
s Stroke length (m)
T Temperature (°C)
T_{br} Break (or solidification) temperature
t Thickness (mould, m) or time (sec, min)
t_{cycle} Time for one cycle (s)
t_n Negative strip time (s)
t_p Positive strip time (s)
V_c Casting speed (m min^{-1})
V_m Velocity of mould (m min^{-1} or ms^{-1})
w Width of mould (m)
η Slag viscosity (dPas)
ρ Density (kg m^{-3})
C/S Basicity = %CaO/%SiO_2
SEN Submerged entry nozzle

Superscripts

^{powd} Refers to powder
^{slag} Slag formed from powder

1.1 Introduction

The continuous casting process for steel is a highly successful process; more than 90% of the world steel production (>1 billion tonne p.a.) is cast using this process. It will be seen below that the performance of the casting powder has been one of the key factors in the success of the continuous casting process.

1.2 The Continuous Casting Process for Steel

The continuous casting process is shown in Fig. 1.1. Molten steel from the ladle pours through a submerged nozzle into the tundish; the tundish provides a constant head for molten steel to flow into the mould. The steel passes into the mould via a submerged entry nozzle (SEN). The copper mould is water-cooled and the bottom of the mould is initially sealed by a steel *dummy bar*. Thus, steel freezes against the cold, mould wall and forms a shell of solidified steel, the name, *shell*, comes from

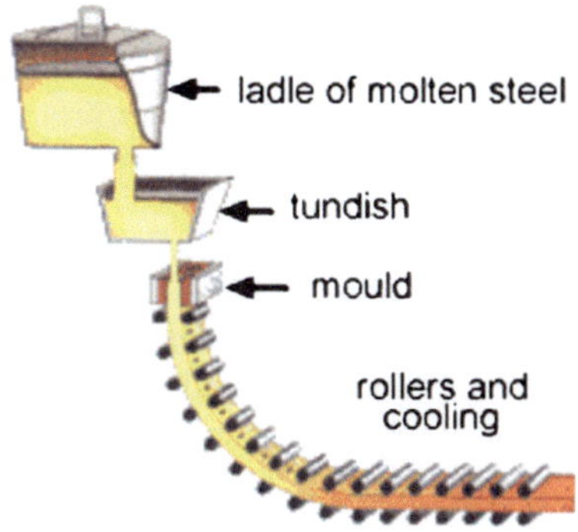

Fig. 1.1 Schematic diagram showing the continuous casting process (*courtesy* of AB Fox)

its relation to an *egg shell* since it consists of a thin, solid containing a liquid inside. The dummy bar is then removed and the shell is pulled gradually through the mould (where it continually thickens as it moves gradually down the mould) and then passes into a spray chamber (where it thickens even more). When the shell is thick enough, the slab is bent. Later, when the steel is completely solidified, the slab is cut off by an oxy-acetylene torch. The casting speed (V_c) (i.e. the rate of withdrawal of the steel strand) is a key factor affecting the performance of casting powders.

However, the shell would stick to the mould unless action is taken to prevent it sticking. Two measures are taken to prevent the shell from sticking to the mould. The first measure involves oscillating the mould vertically, and typical values for the stroke length (s) are 5–8 mm and frequencies (f) are ca. 2 Hz (or 120 cpm) [1, 2]. However, in practice, values of V_c, f and s are selected using Eq. 1.1 [3] so as to maintain a specific *negative strip time* (t_n) (i.e. the time when the mould is descending faster than the shell (V_c)). The *positive strip time* (t_p) constitutes the remaining portion of the cycle time (t_{cycle}) as shown in Eq. 1.2. This action is due to the fact that oscillation marks are formed on the steel surface and the depths of these marks (d_{OM}) are linearly related to t_n and transverse cracking tends to increase with increasing d_{OM}:

$$t_n = \left(\frac{60}{\pi f}\right) \cdot \arccos\left(\frac{V_c}{\pi s f}\right) \quad (1.1)$$

$$t_p + t_n = t_{cycle}. \quad (1.2)$$

The second measure taken to prevent sticking of the shell to the mould was to feed oil, or casting powders, to the top of the mould. Liquid (oil or mould slag) infiltrates between the shell and the mould and provides some lubrication to the moving shell.

The oil and mould powder are expected to perform the following tasks:

1. To prevent the steel surface from oxidising;
2. To provide thermal insulation to the top of the steel meniscus and thereby prevent it from freezing;

3. To infiltrate between the shell and the mould and provide lubrication to the shell and
4. To control the heat extraction from the steel shell.

Initially, rape seed oil was fed continuously onto the top of the mould; the oil helps to create a reducing atmosphere and prevents the oxidation of the molten steel. The heat given out during the combustion of the oil provides some thermal insulation of the steel surface. Furthermore, a thin oil layer infiltrates between the steel shell and the mould; this helps to prevent the shell from sticking to the mould. This oil film evaporates and forms a thin gaseous layer between shell and mould which provides a thermal barrier to heat extraction from the shell to the mould.

1.3 The Introduction of Casting Powders

Mould powders were first used in bottom-pouring, ingot casting in Belgium in 1958 and were, subsequently, applied to continuous casting in 1963 (see Sect. 7.4). These mould powders were based on *fly ash*, a waste product from power stations, which contained high contents of SiO_2, Al_2O_3 and unburnt carbon to which CaO and various fluxes (e.g. Na_2O and CaF_2) were added. The powders heat up as they descend the mould and eventually melt to form a slag pool (Fig. 1.2). This slag pool provides a reservoir of liquid slag which infiltrates between the shell and the mould during the period when the mould is descending (which occurs throughout t_n and into early t_p). This liquid slag film partially solidifies to form a slag film, consisting

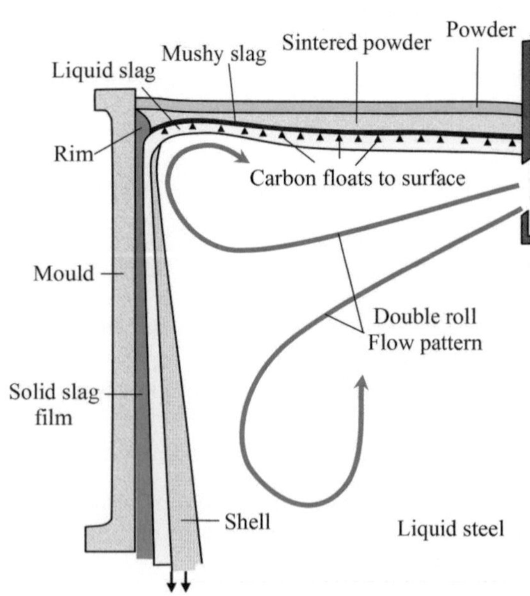

Fig. 1.2 Schematic diagram showing various phases and their location in the mould; note the solid layer of the slag film is shown as *grey* and the liquid layer (and the slag pool) as *white* (*courtesy* of Ramirez-Lopez [5])

of a solid layer (1–2 mm thick) and a liquid layer (ca. 0.1 mm thick) [4]. The initial slag film is predominantly glassy (because of the high cooling rate) but over time, the fraction of crystalline phases (f_{crys}) gradually increases until it reaches a steady-state value. The thickness of the liquid slag film (d_l) determines the lubrication supplied to the shell and the thickness of the solid layer (d_s) and f_{crys} control the amount of heat extracted from the shell [4].

Casting powders were preferred over oil because they provided better thermal insulation of the steel surface and allowed steelmakers to operate with significantly lower levels of superheat (ΔT) for the steel. The slag films formed between shell and mould also provided better control of the horizontal heat extracted from the shell (cf. the gaseous layer formed in oil casting). Furthermore, the liquid slag pool sealed off the steel from the atmosphere and prevented oxidation. The slag pool also provided an extra benefit (cf. oil casting) by absorbing some of the non-metallic inclusions (e.g. Al_2O_3) in the molten steel.

The level of lubrication provided by the liquid slag layer is usually described in terms of the powder consumption, Q_s in kg (slag) m^{-2} (of mould), which is equal to $\rho.d_l$ where ρ is the density of liquid slag. The required powder consumption also increases with increasing surface area of the shell (or mould). Thus, Q_s must be adjusted for the mould surface area; this adjustment is achieved by control of the melting rate (Q_{MR}, which is related to Q_s). In practice, the melting rate is adjusted by the content (%) and size of the free carbon in the powder; thus, typical C_{free} values for slabs and billets (with large and small surface areas) are 4% and 25%, respectively.

There is considerable serendipity associated with the use of *fly ash,* since not only were the powder manufacturers paid to remove the fly ash but the unburnt carbon in the fly ash aided the control of the melting rate and the formation of a reducing atmosphere.

The constituents of mould powders can be divided into four classes: (i) *network formers* (SiO_2, Al_2O_3), (ii) *network breakers* (CaO, MgO), (iii) *Fluxes* (Na_2O, K_2O, CaF_2, B_2O_3) and (iv) *Carbon* particles with different sources and size (e.g. lamp-black, coke breeze, etc.). In addition, mould powders contain (i) some minerals which are present as carbonates and these decompose on heating giving off CO_2(g) and (ii) some impurities present in the minerals (e.g. TiO_2, FeO). The minerals and carbon particles are mixed into a slurry and are then spray-dried to form granules. More details on powder manufacture are given in Chap. 8.

Some physical properties (such as viscosity and thermal conductivity) are very dependent upon the degree of polymerisation of the slag which increases with increasing SiO_2 and Al_2O_3 contents and decreasing amounts of CaO, MgO, CaF_2 and Na_2O (see Chap. 9). The basicity (%CaO/%SiO_2) is a useful measure of the de-polymerisation. Glassy slags are highly polymerised but increasing basicity results in both decreasing polymerisation and increasing crystallisation of the slag (i.e. f_{crys} increases). In recent years, there has been a movement to significantly reduce fluoride emissions and this has led to the development of F-free fluxes. Furthermore, novel mould slags have been developed to cast certain steel grades (such as the calcium aluminate slags developed to cast high-Al (TRIP) steels) [6–9]. Further details of different types of mould powders are given in Chap. 6.

The performance of the mould fluxes has improved gradually over the last 50 years. *Fly ash* powders were gradually replaced by synthetic mould powders to ensure better quality control. However, *fly ash* powders are still used in ingot casting (described in Chap. 7). These improvements were brought about mostly through ad hoc research. Scientific understanding has tended to trail behind the developments resulting from this empirical research. However, some recent innovations have been based on our scientific knowledge, although some empirical (ad hoc) research was needed in their development (e.g. the development of (i) non-Newtonian slags [10, 11] and (ii) the calcium aluminate slags for casting high-Al steels mentioned above.

1.4 Mould Powder Behaviour in the Mould

Mould powders are dispensed onto the top of the steel meniscus in the mould and form a *powder bed.* As the mould powders work their way down the powder bed, they heat up and undergo various reactions. The bed contains three layers, namely, *powder*, *sintered* and *liquid* layers (Fig. 1.2). The various reactions and events occur in the following sequence (i.e. as the mould powder heats up):

(i) Any moisture in the flux evaporates at the top of the bed.
(ii) Carbonates (such as $CaCO_3$ and Na_2CO_3) decompose to form oxide and CO_2(g) at a temperature of around 500 °C; these are endothermic reactions.
(iii) Carbon in the powder combusts with oxygen and CO_2 (g) to form CO (g) in those regions of the bed lying between 500 and 900 °C; these reactions are exothermic.
(iv) In this temperature range the mineral particles also start to sinter.
(v) Finally, the solid slag begins to melt at temperatures above 900 °C and forms a liquid slag pool; any remaining carbon particles float in this pool (see Fig. 1.2). This slag pool serves as a reservoir of molten slag for subsequent infiltration into the shell/mould channel; the liquid pool should have a depth of at least 10 mm to ensure satisfactory infiltration of liquid slag.

The slag pool is very important to the process since it carries out the following tasks:

(a) It seals off the steel meniscus from the atmosphere and prevents oxidation of the steel.
(b) It provides a constant stream of liquid slag to lubricate the shell throughout the length of the mould.
(c) It keeps the floating carbon particles away from the shell and thereby minimises C-pick-up by the steel.

1.5 Slag Film and Slag Rim Characteristics

The slag film and slag rim are shown in Fig. 1.2.

1.5.1 Slag Film

The shell and its characteristics are central to the success of the continuous casting process. The slag film is important because it is our main tool to control the shell. The slag film controls both the lubrication and the heat transfer from the shell, and so it determines the characteristics of the shell (e.g. shell length and thickness). Hence, the characteristics of the slag film are the key to the success of the casting process.

The lubricating properties of the slag increase with increasing liquid slag thickness (d_l) (this term is related to the powder consumption, $Q_{s} = \rho\ d_l$). Lubrication is inversely related to the liquid friction force (F_l) and the parameters affecting liquid friction are given in Eq. 1.3, where A is the surface area of the shell (or mould), η is the slag viscosity and V_m is the velocity of the mould:

$$F_l = A\eta(V_m - V_c)/d_l. \quad (1.3)$$

Thus, liquid friction decreases with decreasing values for mould surface area, viscosity and ($V_m - V_c$) and increasing slag film thickness. The plot of F_l versus V_c exhibits a minimum with ($V_m - V_c$) dominating at low casting speeds and V_c at higher speeds. In practice, the slag liquid thickness (d_l) can be increased by reducing the solidification temperature (T_{sol} or T_{br}) of the mould slag.

It was mentioned above that the heat extraction from the shell was controlled by both the solid layer thickness (d_s) and the fraction crystalline phase (f_{crys}) in the slag film. The crystalline phases reflect and scatter radiant energy leaving the shell; thus q_{hor} decreases with increasing f_{crys} and with increasing crystal size [12, 13]. The initial slag film formed is probably glassy because of the high cooling rate involved, but it crystallises over time. Crystals have higher densities than glasses and thus crystallisation is accompanied by shrinkage. This shrinkage, in turn, results in the creation of (i) porosity in the slag film and (ii) an interfacial thermal resistance (associated with an "*air or gas gap*") at the Cu/slag interface, $R_{Cu/sl}$, which is sometimes denoted R_{int} (see Chap. 3). Increasing porosity and $R_{Cu/sl}$ both lead to decreases in q_{hor} [14–19]. The heat flux also decreases with increasing solid layer thickness (d_s) which, in turn, increases with increasing solidification temperature (T_{sol} or T_{br}) [19, 20]. In practice, d_s, f_{crys} and $R_{Cu/sl}$ all increase with increasing basicity (= %CaO/%SiO_2, denoted (C/S)) so these parameters can be increased by increasing the basicity of the mould powder.

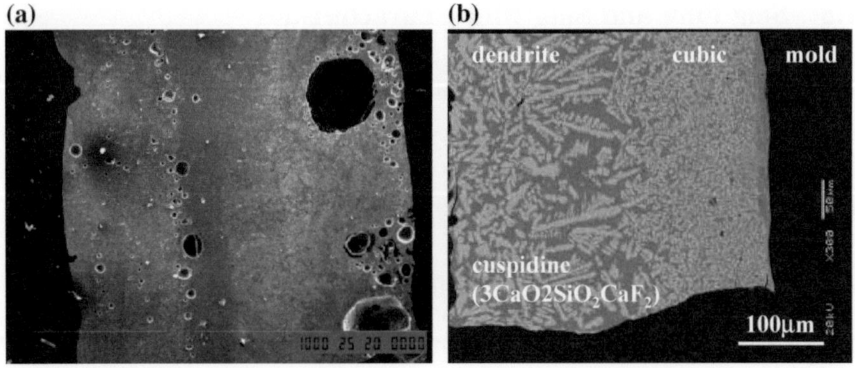

Fig. 1.3 Photographs of slag films taken from the mould, showing **a** pores due to crystallisation at one-third position from the mould (*left*) side and large pores in glassy phase due to CO (g) formation on the shell (*right*) side (*courtesy* of Carboox [22]) and **b** large dendritic crystals formed near the shell side and fine crystals formed on the mould side (permission granted, ISIJ [24])

In conventional, mould powders (i.e. containing fluorides) *cuspidine* ($3CaO{\cdot}2SiO_2{\cdot}CaF_2$ denoted as, C_3S_2Fl) crystals are usually the first to precipitate (i.e. at high temperatures). Thus crystals are precipitated at the mould (cold) side of the slag film (Fig. 1.3a) due to the undercooling of the liquid [21–23]. However, it can be seen from Fig. 1.3b) that the large dendritic crystals can form on the shell side of the slag film; this occurs because crystal growth is promoted by high temperatures. Thus, small, cubic crystals form on the mould (cold) side [24] but crystals grow in high-temperature region grow to form dendritic crystals.

The crystals in the slag film scatter IR radiation and thereby reduce the rate of heat extraction from the shell. Thus the degree of crystallisation in the slag film [12, 13] is a key factor in the control of the horizontal heat flux. The slag film does change over time by either slow downward movement of the slag film [25, 26] or by fracture of the slag film and repair with liquid slag [27]. However, any such changes are slow and the original slag film formed may persist throughout the cast for those regions of the film [27] adjacent to the copper mould [28].

The movement of both the shell and the mould impose stresses on the slag film and crystalline slag films have a tendency to fracture [25, 27]. Glassy slag films form a super-cooled liquid (*scl*) above 600 °C and this *scl* will move when stress is applied. Thus glassy slags, despite high viscosities, do provide some measure of lubrication. Crystalline slags tend to fracture in the upper mould and glassy slags tend not to fracture, but if they do, it is usually near the bottom of the mould [25]. The fractured slag film is repaired by liquid slag filling the gap and solidifying but this may take time in the lower mould where the liquid slag flow is low [28–30].

1.5.2 Slag Rim

Slag films are formed by the infiltration of molten slag into the shell/mould channel. Slag rims are their equivalent but are formed higher up in the mould (Fig. 1.2) and are formed by the "*painting*" mechanism (see Fig. 11.92a, b) [31] as the mould travels through the powder, sinter and liquid slag layers. The contents of the slag film reflect this and contain unmelted mould powder, sinter, glassy carbon and molten slag. An examination of a rim revealed it contained a series of layers bounded by Na_2O-rich frozen liquid which provided evidence for the painting mechanism [27, 32]. The low melting, Na_2O-rich liquid was considered to act as a glue [27, 32]. The carbon comes from both the powder bed [33] and from the C-rich layer formed by carbon particles floating at the top of the slag pool (see Fig. 1.2).

A slag rim is shown in Fig. 1.4a; typically the bulge of the slag rim is 3–5 mm wide. The slag rim is thought to act as a piston and pushes liquid slag into the mould/strand channel. No study relating the size of the slag rim to powder consumption has been identified by the authors.

However, excessively large slag films, like those shown in Fig. 1.4b, can be formed; these are frequently referred to as "*rope*s" or "*bear*s". The *ropes* can have girths up to 60 mm [27]. Large rims are promoted by

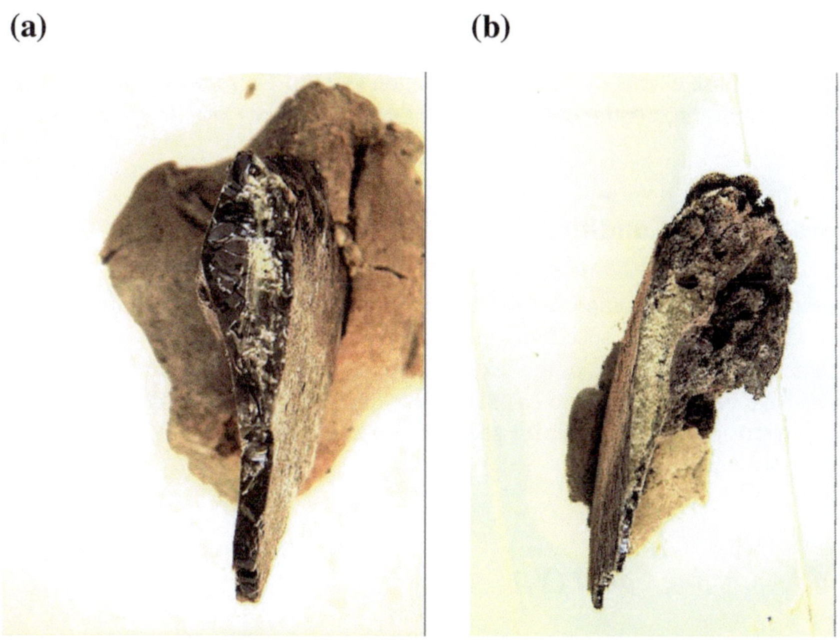

Fig. 1.4 Photographs of slag rims formed with **a** a glassy rim formed by a low-basicity mould slag and **b** a crystalline rim formed by high-basicity mould slag (*courtesy* of C-Å Däcker, Swerea KIMAB)

- Large mould level variations;
- Using a mould slag with a high solidification (or break) temperature (e.g. high-basicity slags);
- Mould powders with high, free carbon contents (for the given casting conditions) which lead to low melting rates [27];
- Low meniscus temperatures, aided by low casting speeds, low superheat and low melting steels (e.g. austenitic stainless steels or high-Al steels);
- Large Al_2O_3 pick-up by the slag which increases the melting temperature, the viscosity and the fraction of crystalline phase (f_{crys}) (see Fig. 6.22 [34]); this is a particular problem when casting high-Al steels where Al_2O_3 pick-up can reach 35%;
- Low metal flow velocities, which decrease the horizontal heat flux; frequently, slag rims are not formed on the narrow wall because of the high metal flow velocities [35]; *single-roll* flow systems reduce the meniscus temperature and thus will favour slag rim growth. (cf. *double-roll* flow);
- Low-argon flow rates, since increased Ar flow increases both the vertical heat flux and the meniscus temperatures; and
- Poor thermal insulation in the powder bed; thermal insulation can be improved using (i) a thicker bed, (ii) smaller powder granules or (iii) exothermic agents in the powder.

Highly crystalline slag rims tend to fracture easily [25]. Fractured slag rims can (i) cause depressions in the surface of the steel [36–38], (ii) get trapped in the shell/mould channel and block off slag infiltration which results in sticker breakouts and (iii) create false alarms in the sticker detection system. Thus the formation of ropes can be harmful to continuous casting.

1.6 Casting Conditions

The mould powder composition is selected for the given casting conditions; these include (i) the casting speed, (ii) the mould dimensions, (iii) the oscillation characteristics, (iv) the nature of the steel being cast and (v) the Ar flow rate. Empirical rules have been developed to calculate the required values of powder consumption (Q_s), viscosity (η) and break (or solidification) temperature (T_{br}) [39]. For further detail, see Chap. 10.

1.6.1 Casting Speed (V_c)

The demands for ever-increasing productivity have resulted in the use of higher casting speeds. This has imposed increasing demands on the mould powders which have always been expected to be "*forgiving*" or "*flexible*", i.e. that they will readily accommodate changes in casting conditions.

Increases in casting speed result in the following, simultaneous, responses:

(i) Decreased powder consumption (i.e. less lubrication of the shell);
(ii) Increased heat flux density (q_{hor}) from the shell;
(iii) Shorter residence time in the mould leading to both a decrease in the total heat (q_{tot}) leaving the shell and a thinner shell;
(iv) Increased metal flow turbulence, including increased metal velocities and the formation of standing waves and vortices, leading to increased slag and gas entrapment; and
(v) Shorter negative strip time (t_n) leading to shallower oscillation marks.

Mould powders are usually selected to minimise the incidences of both "longitudinal cracking" and "sticker breakout" and involves careful selection of viscosity, T_{br} and f_{crys} for the slag [39]. Slag entrapment can be reduced by increasing slag viscosity but this adversely affects the powder consumption (see Chap. 11).

1.6.2 Metal Flow

The metal flow pattern (Fig. 1.5) is an important factor in continuous casting. The *double-roll* pattern is the preferred pattern and provides satisfactory casting conditions. The *single-roll* flow pattern tends to cause both a decrease in meniscus temperature (which can lead to the freezing of the steel meniscus) and high levels of inclusions in the steel. Asymmetric flows pose many problems and can lead to high levels of slag entrapment.

At high-casting speeds, turbulent metal flow occurs. This involves high metal velocities and the formation of both standing waves on the metal/slag interface and

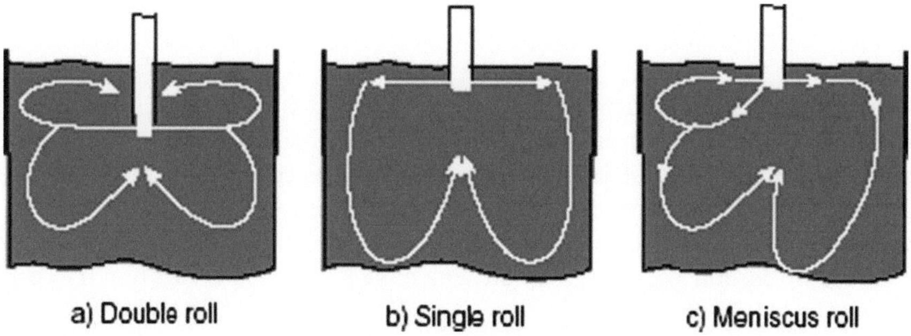

Fig. 1.5 Various metal flow patterns established in the mould **a** *Double roll* **b** *single roll* and **c** *Meniscus or Asymmetric roll*; (permission granted, UNESID [47])

vortices [40, 41]. All of these lead to slag entrapment and the incidence of longitudinal cracking has been linked with the height of the standing waves [42]. Electromagnetic braking (EMBr, see Sect. 5.8.4 for more details) has proved useful in dealing with turbulent flow. The upper coils (see Fig. 5.14) retard the velocity of the metal leaving the SEN and the lower coils reduce the penetration of the metal into the mould [43–45].

1.6.3 Mould Dimensions

It was mentioned in Sect. 1.3 that the lubrication supplied by the liquid slag increases with increasing surface area of the mould and the *required* powder consumption is in the hierarchy, $Q_s^{slabs} > Q_s^{blooms} > Q_s^{billets}$. The demand for slag is regulated through the melting rate which is mainly controlled through the free carbon content (typically, 4% C for slabs and 25% C for billets). For more details, see Chap. 2. Mould dimensions also affect the flow patterns and slag entrapment tends to be worse with very wide slabs [48] (see Sect. 11.9.2.8).

1.6.4 Oscillation Characteristics

Most steel works tend to operate with a fixed negative strip time. Thus when the casting speed is increased, the frequency and stroke are adjusted (Eq. 1.1) to maintain the same value for negative strip. Although there is some dispute about the effects of oscillation parameters on powder consumption, most plant trial data indicate that Q_s increases with decreasing frequency and stroke. The depths of oscillation marks increase with increasing values of stroke and negative strip times.

1.6.5 Steel Grade

Longitudinal cracks occur when the horizontal heat flux exceeds a certain critical value (q_{hor}^{crit}) resulting in an overly thick shell. The shells of steels undergoing the peritectic reaction (Eq. 1.4) are especially vulnerable to cracking because of the 4% differences in thermal shrinkage coefficients for *austenite* and *ferrite*. Consequently, to avoid longitudinal cracking, it is necessary to form a thin, uniform shell; this is achieved by ensuring q_{hor} is low by creating a thick, crystalline, slag film (i.e. with high values of T_{br} and f_{crys}):

$$\mathrm{Fe(liq)} + \mathrm{Fe}_{\delta}(\mathrm{ferrite}) \rightarrow \mathrm{Fe}_{\gamma}(\mathrm{austenite}). \quad (1.4)$$

Sticker breakouts are associated with the formation of a thin, weak shell. Thus, these breakouts can be minimised by creating a thick, strong shell and this is achieved with a high q_{hor} by creating a glassy, thin, slag film (i.e. with low values of T_{br} and f_{crys}). Thus, the nature of the steel affects the required properties of the mould slag.

In recent years, the mould slags have had to respond to the challenges of casting new steel grades, such as new high (>1% Al) steels where the slag has to absorb large amounts of Al_2O_3 formed by metal/slag reactions (Eq. 1.5) where the underline denotes in the metal. New families of mould slags have been developed to meet this challenge but further development is still required:

$$2\underline{\mathrm{Al}} + 1.5\mathrm{SiO}_{2\,\mathrm{sl}} = \underline{\mathrm{Si}} + \mathrm{Al}_2\mathrm{O}_{3\,\mathrm{sl}}. \quad (1.5)$$

1.6.6 Ar Flow Rate

Argon is flowed through the SEN to reduce SEN clogging (i.e. the accumulation of Al_2O_3 in the ports and the interior of the SEN). However, the argon is also the source of "pinholes" in the steel. Furthermore, Ar bubbling is helpful in removing inclusions from the steel to the slag pool. Argon flows also tend to oppose the effects of casting speed and to *cushion* the metal flow. However, excessive Ar flow rates (>5 l min^{-1}) can transform a *double-roll* flow pattern in the steel to *single-roll* flow pattern [49, 50], which provides less satisfactory casting conditions.

1.7 Physical Properties of Mould Slags

The mould slag must provide good lubrication (i.e. powder consumption) and the key properties here are the fluidity (or reciprocal viscosity) and the break temperature ($Q_s\uparrow$ as $\eta\downarrow$ and $T_{br}\downarrow$) [4]. The key properties for the control of the heat flux (q_{hor}) are T_{br} (which determines the thickness of the solid layer of the slag film) and f_{crys}, the fraction crystalline phase in the slag film [4]. However, the thermal conductivities of the glass and crystalline phases and the optical properties of the slags are also important but not as important as T_{br} and f_{crys}. In slag and gas entrapment, the key properties are the metal/slag interfacial tension (γ_{msl}) and the viscosity (η). The principal factor affecting γ_{msl} is the S content of the steel but reducing the B_2O_3, Na_2O and CaF_2 contents of the slag would help to increase γ_{msl}. Further details of the physical properties of mould slags are given in Chap. 9.

1.8 Fluctuations in the Process

When you look at the top of the mould it looks very peaceful. However, the powder bed frequently hides a mass of turbulent flow in the mould; surface waves and vortices are formed on the steel meniscus and these are in a state of permanent movement. The turbulence increases with increasing casting speed. The conditions are continually fluctuating and are continually reacting to the current conditions. Under such conditions the mould slag needs help. This can be achieved by (i) optimising parameters, like the immersion depth of the SEN and SEN port design and (ii) using electromagnetic devices to reduce metal flow turbulence and to homogenise temperatures and steel composition. Nevertheless, the variation in the depths of oscillation marks is an indication of the fluctuations. If all the bad casting conditions were to coincide, there is the possibility of a disaster occurring (the "*butterfly effect*") [51].

1.9 Definitions

1.9.1 Powders, Slags, Fluxes

A number of terms are used to describe mould powders (e.g. casting powders, mould powders, mould fluxes, etc.) which can lead to some confusion. In this work we use the following terms to define the casting powder:

Powder implies that the flux is in its "*as-received*" state from the manufacturers, i.e. it contains carbon and carbonates in pulverised or granule form; for instance, it could be referred to as either casting powder or mould powder.

Slag (e.g. mould slag, casting slag or slag film) denotes the composition of the slag pool (i.e. where carbon and carbonates have been removed during heating but Al_2O_3, ZrO_2, TiO_2, TiN etc.) have been picked up by the slag.

The term, ***flux***, is not extensively used here, where it is used, its meaning is synonymous with "powder".

1.9.2 Powder Consumption Terms

The lubrication supplied by the slag to the shell is usually cited in terms of powder consumption. Not all of the powder forms (lubricating) slag since carbon particles are oxidised and carbonates undergo gaseous decomposition. Consequently, when defining specific powder consumption terms, it is necessary to specify whether this refers to *powder* or *slag* (e.g. Q_s^{slag} (kg.m^{-2}) or Q_t^{powd} (kg tonne^{-1})). The key factor here is the fraction (f^*) of the powder forming slag (Eq. 1.6); typical values of f^* are >0.9 and ca. 0.75 for casting powders used to cast slabs and billets, respectively:

$$Q_s^{slag} = f * Q_s^{powd}. \tag{1.6}$$

1.9.3 Temperature

In this work, temperature (*T*) is normally expressed in °C since this is the scale normally used in the steel industry and is denoted by the symbol, T. However, it is necessary to use the thermodynamic temperature scale (K) when describing thermodynamic functions and using Arrhenius relationships for viscosity and diffusion coefficient. In these cases, the thermodynamic temperature will be used and is denoted as T_K.

Various temperatures associated with the slag film and the slag pool are used here:

Glass transition temperature (T_g) is the temperature where a frozen glass transforms into a super-cooled liquid (*scl*); the viscosity is considered to be $10^{13.4}$ (dPas) at T_g.

Deformation temperature (T_d) is the temperature where the sample collapses when measuring, for instance, the thermal expansion.

Critical temperature (T_{crit}) is the temperature where the thermal conductivity (*k*) of partially glassy samples decreases dramatically with increasing temperature which occurs at ca. 770 °C for most mould slags; T_{crit} is probably identical or linked to the deformation temperature, T_d.

Liquidus temperature (T_{liq}) is a thermodynamic entity and is the equilibrium temperature where the mould slag becomes completely liquid.

Solidification or Break temperature (T_{sol} *or* T_{br}) is the temperature where crystals are first precipitated during a cooling cycle; the *break temperature* refers specifically to the temperature where there is sudden increase in viscosity during cooling; T_{sol} and T_{br} are not equilibrium values and tend to decrease with increasing cooling rate, and thus it is customary to associate them with the cooling rate (e.g. 10 °C min^{-1}).

Crystallisation temperature (T_{crys}) is not used in this text because it has been used in the literature to describe different entities, e.g. for temperatures where (i) crystallisation occurs on heating cycles [52] and (ii) where solidification occurs in viscosity measurements during cooling (i.e. T_{br} or T_{sol}) [53].

1.9.4 Viscosity

In the continuous casting fraternity, the dynamic viscosity is usually cited in dPas (which is the equivalent of the old unit, *poise*) but some workers cite values in Pas; 1 dPas (= 0.1 Pas) is used here.

References

1. P. Andrzejewski, A. Drastik, K. U. Köhler, W. Pluschkell, Proc. 9th Process Technol. Conf., 1990, p. 173.
2. M. Wolf, Discussion Group on Continuous casting of mould fluxes, London, 1984, (Inst. Of Metals, London, 1984).
3. ES Szerkeres, Iron and Steel Engineer, **73** (7), 29, (1996).
4. KC Mills, AB Fox, ISIJ Intl., **43**, 1479, (2003).
5. P. Ramirez- Lopez, "*Modelling shell and oscillation mark formation during continuous casting via Explicit incorporation of slag infiltration*" PhD Thesis, Imperial College, London, 2010.
6. JW Cho, KE Blazek, MJ Frazee, HB Yin, JE Park, SW Moon, ISIJ Intl., **53**, 62, (2013).
7. K Blazek, HB Yin, G Skoczylas, M McClymonds, M Frazee, Iron and Steel Technology, **8** (3), 232, (2011).
8. TS Kim, JH Park, ISIJ Intl., **54**, 2031, (2014).
9. Q Liu, GH Wen, JZ Li, XJ Fu, P Tang, W Li, Ironmaking and Steelmaking, **41**, 292, (2014).
10. K Tsutsumi, K Watanabe, J Kubota, S Hatori, Y Miki, T Suzuki, T Omoto, Proc. 7th Europ. Cont. Casting Conf., Dusseldorf, 2011, (VDEh, Dusseldorf, 2011) Session p. 1.
11. K Watanabe, K Tsutsumi, M Suzuki, H Fujita, S Hatori, T Omoto, ISIJ Intl., **54**, 865, (2014).
12. M Susa, A Kushimoto, H Toyota, M Hayashi, R Endo, Y Kobayashi, ISIJ Intl., **49**, 1722, (2009).
13. M Susa, A Kushimoto, R Endo, Y Kobayashi, ISIJ Intl., **51**, 1587, (2011).
14. J. W. Cho, H. Shibata, T. Emi,M. Suzuki, ISIJ Intl., **38** (5), 440 (1998).
15. K Watanabe, H. Okamoto, M. Suzuki, H. Kondo, T. Shiomi,. Proc. 79th Steelmaking Conf., 1996, (ISS, Warrendale, PA, 1996) p. 265.
16. H Shibata, JW Cho, T Emi, M Suzuki, Proc. 5th Intl. Conf. Molten slags, fluxes and salts, Sydney, 1997, (ISI–AIME,Warrendale, PA, 1997) p. 771.
17. H Nakada, K Nagata, ISIJ Intl., **46**, 441, (2006).
18. M Hanao, M Kawamoto, ISIJ Intl., **48**, 180, (2008).
19. M Hanao, M Kawamoto, A Yamanaka, ISIJ Intl., **52**, 1310, (2012).
20. A Yamauchi, K Sorimachi, T Yamauchi, Ironmaking and Steelmaking, **29**, 203, (2002).
21. M Hanao M Kawamoto, M Hara, T Murakami, H Kikuchi, K Hanazaki, Tetsu-to-Hagane, **88**, 23, (2002).
22. MV Fonseca, ODC Afrange, A Lavinas, AA Ramos, CA Valadares, Proc. 5th Intl. Conf. Molten slags, fluxes and salts, Sydney, 1997 (ISS, Warrendale, PA, 1997) p 851.
23. Z Li, KC Mills, MC Bezerra, Proc. XXXV Semin. De Fusao Refino e Solidifacao Metals, Salvador, Brazil (2004) p. 281.
24. H Nakada, K Nagata, ISIJ Intl., **46**, 441, (2006).
25. YA Meng, BG Thomas,. Met. Mater. Trans. B, **34B**, 707, (2003).
26. C-Å Däcker, A Salwén, P Andersson, C Eggertsson, Proc. 7th Europ. Conf. Continuous Casting, Düsseldorf, 2011, (VDEh, Dusseldorf, 2011) Session 12.

27. JA Kromhout, PhD Thesis, "*Mould powders for the high speed continuous casting of steel*", Univ of Delft, (2011).
28. E Lainez, Proc. 3rd Europ. Conf. Continuous Casting, Madrid, 1998, (UNESID, Madrid, 1998) p. 155.
29. RJ O'Malley, J Neal, Proc. Intl. Conf. New Developments in Metallurgical Process Technol., Dusseldorf, 1999, METEC Congress (VDEh, Dusseldorf, 1999) p. 73.
30. TJ Billany AS Normanton, KC Mills, P Grieveson, Ironmaking and Steelmaking, **18,** (1991) 403.
31. C Perrot, JN Pontoire, C Marchionni, MR Ridolfi, LF Sancho, Proc. 5th Europ. Conf. Continuous Casting, Nice, 2005, (La Rev. Metall., Paris, 2005,) p. 36.
32. J Kromhout, RS Schimmel, Proc 8th Europ. Conf. Continuous casting, Graz, 2014, (Austrian Metals Soc., Vienna, 2014).
33. JJ Macho, G Hecko, B Golinmowski, M Frazee, Preprints 33rd McMaster Symp. Iron and Steelmaking, Hamilton, Ont, 2005, (McMaster Univ. Hamilton, 2003), p. 131.
34. M Hanao, Y Tsukaguchi, M Kawamoto, Proc. 4th Intl. Cong. Sci. Technol. Steelmaking, Gifu, Japan, 2008 (ISIJ, Tokyo, 2008), p. 94.
35. CA Dacker, M Glaes, SP Andersson, A Salwen, C Eggertsson, Proc. 6th Europ. Conf. Continuous Casting, Riccone, Italy, 2008, (AIM, Milan, 2008), 8 pdf.
36. JW Kim, S. K. Kim, D. S. Kim, Y. D. Lee, J. I. Eum, E. S. Lee; Proc. 78th Steelmaking Conf.1995 (ISS, Warrendale, PA,1995) p. 333
37. M Jenkins, BG Thomas, WC Chen, RB Mahapatra, Proc. 77th Steelmaking Conf. 1994, (ISS, Warrendale, PA., 1994) p. 337.
38. M S Jenkins, BG Thomas, Proc. 80th Steelmaking Conf., Chicago, IL, 1997, (ISS, Warrendale, PA, 1997) p. 285.
39. KC Mills, AB Fox, PD Lee, S Sridhar, Proc. Sci. Technol. Steelmaking, Swansea, 2001, (IOM, London, 2001) p. 445.
40. LC Hibbeler, BG Thomas, Proc. AISTech Conf 2010, Pittsburgh (ISS, Warrendale. PA, 2010) p. 1215.
41. LC Hibbeler, BG Thomas, Iron and Steel Technology, **2013** (Oct), 121, (2013).
42. M Hanao, M Kawamoto, M Hara, T Murakami, H Kikuchi, A Yamanaka, Proc. 5th Europ. Conf. Continuous Casting, Nice, 2005 (La Revue Metall., Paris, 2005), see also Tetsu-to Hagane, **88**, (1), 23, (2002).
43. E Takeuchi, JOM, **1995** (May), 42, (1995).
44. E Takeuchi, H Harada, H Tanaka,T Ishii, T Toh, M Zeze, M Hojo, K Shigematsu, Nippon Steel Technical Report, **61**, 29, (1994).
45. M Washio, M Sugizawa, S Moriwaki, K Kariya, S Idogawa, S Takeuchi, Revue de Metallurgie, CIT, **90** (April), 507, (1993).
46. SG Kollberg, HR Hackl, PJ Hanley, Iron and Steel Engineer, **73** (7), 24, (1996).
47. D Gotthelf, P Andrezjewski, E Julius, H Haubrich, Proc. 3rd Europ.. Conf. Continuous casting, Madrid, 1998, (UNESID, Madrid, 1998), vol. 2, p. 825.
48. W Emling, T A Waugaman, SL Feldbauer, AW Cramb, Proc. 77th Steelmaking Conf. (ISS, Warrendale, PA., 1994) p. 371.
49. PE Ramirez- Lopez, PN Jalali, J Bjorkvall, U Sjostrom, C Nilsson, ISIJ Intl. 54, 342. (2014) and Proc. 8th Europ. Conf Continuous Casting, Graz, Austria, 2014 (ASMET, Vienna, 2014).
50. E van Vliet, DW van der Plas, SP Carless, A A Kamperman, AE Westendorp, Proc.. 7th Europ. Conf. Cont. Casting, Dusseldorf, 2011, (VDEh, Dusseldorf, 2011) Session 4.
51. PD Lee, PE Ramirez-Lopez, KC Mills, B Santillana, Ironmaking and Steelmaking **39**(4), 244, (2012).
52. JV Dubrawski, JM Camplin, J Thermal Analysis, **40**, 329, (1993).
53. JW Kim, J Choi, OD Kwon, IR Lee YK Shin and JS Park, Proc. 4th Intl. Conf. Molten slags and fluxes, Sendai,1992,. (ISIJ, Tokyo, 1992) p. 468.

Chapter 2
Slag Infiltration, Lubrication and Frictional Forces

Abstract It is essential to lubricate the shell since inadequate lubrication leads to defects in the steel product (e.g. longitudinal cracks, sticker breakouts and star cracks). The liquid layer of the slag film, formed between the shell and the mould, lubricates the newly formed shell; the lubrication increases with increasing liquid slag thickness (d_l). Lubrication is usually represented by the powder consumption (Q_s in units of kg slag (or powder) m^{-2}) which is related to liquid film thickness (d_l). However, there are several terms used for powder consumption and these terms are interrelated (e.g. Q_s, Q_t and Q_{MR}). The frictional forces acting on the shell are highest in the centre of slabs and thus slabs need more lubrication. The required powder consumption, Q_s increases with increasing distance from the corner and thus $Q_s^{slab} > Q_s^{bloom} > Q_s^{billet}$. The *required* powder consumption can be calculated from the relation, $Q_s^{req} = 2/(R^* - 5)$ where $R^* = \{2(w + t)/w \cdot t\}$ = (surface area/volume) of the mould. However, the powder consumption, Q_s, is also affected by other parameters, namely, the casting speed (V_c), slag viscosity (η), the break temperature of the slag and the oscillation frequency (f) and stroke (s). There is general agreement that Q_s decreases with increasing casting speed and viscosity (e.g. empirical rules, $Q_{sreq}^{slag} = 0.55/\eta^{0.5} \cdot V_c$). There is some dispute with regard to the effect of f, s and T_{br} but most plant studies indicate that Q_{sreq}^{slag} decreases as f, s and T_{br} increase. The *required* values of powder consumption and viscosity can be calculated for the given casting conditions using empirical rules. The predictions of a mathematical model indicate that slag infiltration into the model/ strand channel occurs when the mould and slag rim are descending but little powder consumption occurs when the mould is ascending. The changes in mould direction are accompanied by periods of *confused flow* in the mouth of the channel and little slag infiltration occurs in these periods. Frictional forces and the factors affecting them are also discussed; it was found that liquid friction increased with increasing mould dimensions, slag viscosity, casting speed and ($V_m - V_c$). Plots of liquid friction (F_l) versus casting speed exhibit a minimum since F_l increases with increasing V_c but decreases with decreasing ($V_m - V_c$).

The original version of this chapter was revised: For detailed information please see correction. The correction to this chapter is available at https://doi.org/10.1007/978-3-319-53616-3_12

Within this book minor changes to some of the figure permissions and the readability of the chapters have been made and several figures have been replaced with new versions.

K.C. Mills and C. Däcker, *The Casting Powders Book*,
DOI 10.1007/978-3-319-53616-3_2

Symbols, Abbreviations and Units

A	Area (m^2)
$\%C_{free}$	Percentage of free carbon
$\%C_{total}$	Percentage of total carbon
%LOI	Percentage of loss on ignition
D_C	Mean particle size of the carbon
D_{corn}	Distance mould corner to centre (m)
D_l	Thickness of liquid slag film (m)
F_l	Frictional force (N)
f	Frequency (Hz or cycles min^{-1})
f^*	Fraction of powder forming slag
Q_{cycle}	Powder consumption (kg m^{-1} $cycle^{-1}$)
Q_{MR}	Melting rate (kg/min or kg/s)
Q_s	Powder consumption (kg/m^2)
Q_t	Powder consumption (kg/$tonne^{-1}$)
R^*	Mould (surface area/volume) (m^{-1})
s	Stroke length (m)
T	Temperature (^oC)
T_{br}	Break (or solidification) temperature
t	Time (s) or thickness of mould (m)
t_{cycle}	Time for one cycle (s or min)
t_n	Negative strip time (s)
t_p	Positive strip time (s)
V_c	Casting speed (m/min)
V_m	Velocity of mould (m/min or m/s)
w	Width of mould
η	Slag viscosity (dPas)
ρ	Density (kg/m^3)

Superscripts

^{powd}	Refers to powder
^{slag}	Slag formed from powder

2.1 Introduction

The newly formed steel shell is lubricated as it progresses down the mould by a flow of liquid slag from the slag pool (Fig. 2.1). This flow of slag helps to reduce the frictional forces acting on the shell. The liquid frictional force (F_l) operating on the shell can be calculated using Eq. 2.1 where V_m and V_c are the velocities of the mould and the casting speed, respectively, A is the active surface area of the mould, η = slag viscosity and d_l is the liquid slag film thickness.

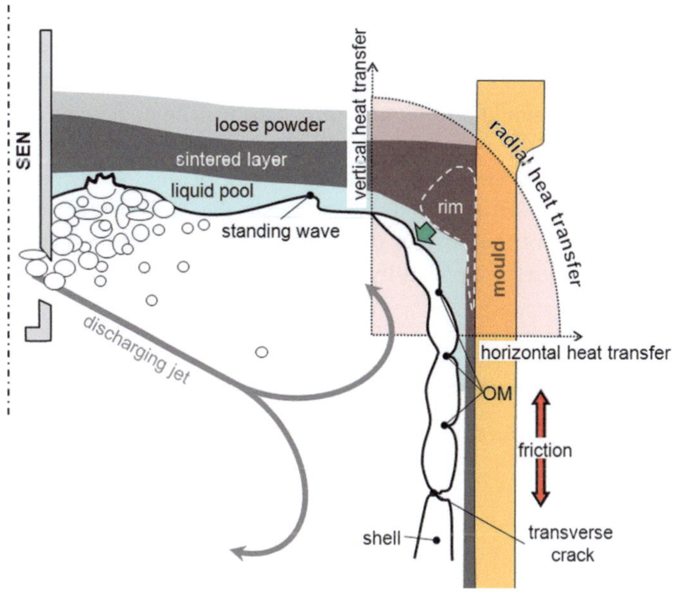

Fig. 2.1 Schematic drawing showing the lubrication of the sausage-shaped shell by liquid slag, which is shown in light blue (permission granted, ISS/AIST, [1])

$$F_l = A\eta(V_m - V_c)/d_l \tag{2.1}$$

Thus the liquid frictional force decreases as the:

- Thickness of the liquid film increases.
- Surface area of the mould decreases.
- Viscosity decreases.
- With a decrease in the difference between the velocities of mould and steel strand.

It will be seen below that the thickness of the liquid slag film (d_l) is related to the powder consumption Q_s (with units of kg/m^2). In addition to the liquid friction forces there are also solid friction forces which tend to occur in the lower mould.

The powder consumption, Q_s (in kg/m^2 of mould) is often used as a measure of the lubrication supplied by the liquid slag since it is linearly related to d_l (see Eq. 2.4).

2.2 Powder Consumption (Q)

The powder consumption (Q) can be expressed in several ways, each term having different units. The most common form for powder consumption is Q_t (in kg/t steel) which is the mass of powder consumed per tonne of steel cast. This provides a measure of the cost of the casting powders in continuous casting.

It should be pointed out that the mould powder contains carbon, carbonate and other volatiles which burn off and, hence, do not contribute to lubrication by the slag. Some billet fluxes contain up to 25% free carbon and consequently, it is necessary to differentiate between mould *powder* and *slag*. This is denoted here by attaching the superscripts, ^{powd} and ^{slag}, respectively, to the various consumption terms. It is possible to calculate one term from the other by calculating the fraction of powder forming slag (f^*) by Eq. 2.2 and using either the free carbon ($\%C_{free}$) and total carbon contents of the powder ($\%C_{total}$) or alternatively, the loss on ignition (%LOI) which are usually supplied by the powder manufacturer.

$$f^* = [100 - (\%C_{free}) - \{(44/12)(\%C_{total} - \%C_{free})\}]/100 = (100 - \%\mathrm{LOI})/100 \tag{2.2}$$

The powder consumption of slag can be calculated from the relation:

$$Q_t^{slag} = f^* Q_t^{powd}. \tag{2.3}$$

2.2.1 Various Powder Consumption Terms

As mentioned above, the powder consumption can be expressed in various ways, each with different units; the various terms can be calculated from one another, e.g. Q_s and Q_t by Eq. 2.4.

$$Q_s^{slag}\,(\mathrm{kg/m^2}) = \left(f^*\,7.6\,Q_t^{powd}\right)/R^* = \rho\, d_l \approx 2550\, d_l. \tag{2.4}$$

where R^* = (surface area/volume) of mould = $\{2(w + t)/\, w{\cdot}t$ (and has units = of m^{-1}) w = width of the mould (m), t = thickness of the mould (m) and ρ = the density (kg/m^3) of the molten slag, 7.6 is the density of steel in t/m^3. Equation 2.4 also shows the link between Q_s and the thickness of the liquid slag film (d_l).

Furthermore, the melting rate of the mould powder, Q_{MR}^{powd} (in units of kg/min or kg/s) can be calculated from Q_s^{slag}, by Eq. 2.5 where V_c is the casting speed (m/min).

$$Q_{MR}^{slag}\,(\mathrm{kg/s}) = f^* Q_{MR}^{powd} = 2(w + t) Q_s V_c/60 \tag{2.5}$$

The melting rate should match the powder consumption (Q_s) needed to provide good lubrication of the shell. There are a number of variables which affect the melting rate but it is primarily controlled through the free carbon content ($\%C_{free}$) and the mean particle size of the carbon ($D_{C_{free}}$) of the powder (i.e. $Q_{MR}^{slag}\uparrow$ as $\%C_{free}\downarrow$ and $D_{C_{free}}\uparrow$).

The powder consumption per oscillation cycle, Q_{cycle}^{slag} (kg/cycle) can be calculated from Q_s *via* Eq. 2.6 where f = oscillation frequency (in Hz)

$$Q_{\text{cycle}}^{\text{slag}}\left(\text{kg cycle}^{-1}\right) = 2(w+t)Q_s V_c/60f \tag{2.6}$$

All of the above parameters can be derived by considering a liquid slag film of uniform thickness (d_l) distributed around the mould and by assuming $A_{\text{shell}} = A_{\text{mould}}$.

2.2.2 Measurement of Powder Consumption

The powder consumption, Q_t^{powd}, is frequently determined as the number of bags (N) of known mass of casting powder (m) used in the entire cast, for which the total mass of steel cast (m_{steel}) is known. Thus Q_t^{powd} can be calculated ($Q_t^{\text{powd}} = Nm/m_{\text{steel}}$).

Powder consumption rates for different periods during a cast can be determined by measuring the mass of powder dispensed from the hoppers in a known time period. However, these measurements are affected by the height of the powder in the hopper and thus measured rates are affected by recharging the hopper; for true significance, the measurements should refer to the same height of powder in the hopper. There is some variability in the consumption values. Some of this variability probably arises from small variations in the casting conditions (e.g. changes in casting speed) through the cast.

The most precise method for measurement of mould powder consumption, when, for example, performing plant trials with new mould powders, is to use a bucket with known amount of mould powder and count the number of buckets needed for the casting.

Powder consumption data for 32 casts of the same steel under the same conditions (where any casting speed variations were $<\pm 10\%$) is shown in Fig. 2.2. The mean Q_t value was 0.63 kg/t, the maximum range in values was 0.24 kg/t and the uncertainty was calculated as $\pm 12\%$ [1, 2].

Studies where the powder supply was monitored at nine positions between the centre and the edge of the mould showed that consumption (i) was highest at the edge (presumably because it has to feed the narrow face too) and (ii) went through a minimum located midway between the edge and the centre [3].

The powder consumption at different periods of the cast is shown in Fig. 2.3, it can be seen that the powder consumption appears to be higher at the beginning of the cast and gradually decreases to a steady value.

2.2.3 Methods Used to Understand Slag Infiltration Mechanisms

Several techniques have been used to gain an understanding of the mechanisms responsible for slag infiltration, they are:

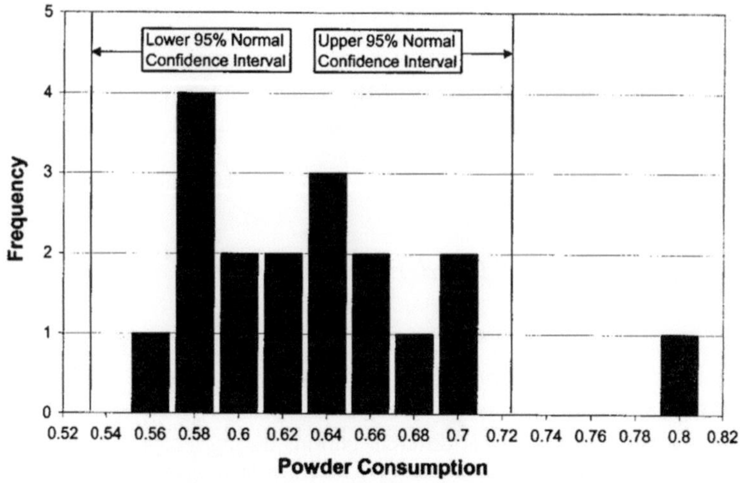

Fig. 2.2 Histogram of powder consumption (kg/t) in similar trials where a variation of $< \pm 10\%$ in casting speed was allowed (courtesy of A.B. Fox [2])

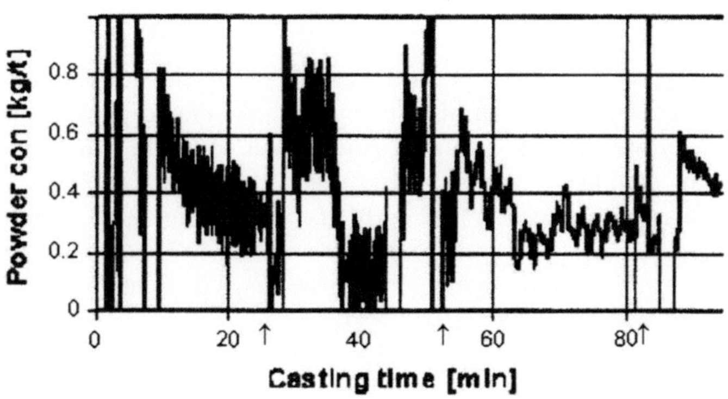

Fig. 2.3 Powder consumption at different periods of the cast (*courtesy* of A.B. Fox [2])

2.2.3.1 Analysis of Plant Data

Plant trials are carried out where powder consumption is measured as one variable (for instance, as a function of casting speed or slag viscosity) is changed while all the other variables are held constant. Most of the empirical relations were derived in this manner (see Table 2.3). One major problem with this approach lies in determining the effects of oscillation characteristics where it is common practice to operate with a fixed negative strip ratio. Thus if the frequency is increased, it is common practice to alter the stroke (s) or casting speed to ensure that negative strip time remains the same. Thus there are few plant trial data carried out where only one oscillation parameter (say frequency, f) is varied systematically.

An alternative approach was taken by Saraswat et al. [4]; they carried out a statistical analysis for a large database of powder consumption data from various steel plants in order to identify those variables (V_c; η, f, s, T_{br}) which had a statistically significant effect on Q_s.

2.2.3.2 Physical Modelling Studies

In physical models, the effect of changing one variable is studied by observation of the slag flow. The physical models can be divided into three types, given below, and details of the various studies are given in Table 2.1:

(a) *Cold (or water) models* which are usually carried out at near-ambient temperatures. In some cases the slag rim is represented by a solid ledge [5] and other cases by freezing of the water representing the slag [6]. The main problem with these tests is that they tend to be isothermal whereas the conditions in the mould are not isothermal.

Table 2.1 Details of some physical modelling studies of slag infiltration and factors affecting powder consumption

Reference	Type	Metal	Slag	Comments, findings
Anzai [12]	CM	Rubber belt	Paraffin	Pressure measured in mould/shell gap; mould = oscillating plate, shell = moving belt
Jenkins [6]	CM	Ice	Gycerol/water	Rim = solid ledge; Measures pressure (p) in gap. Predicts (a) $Q_s\uparrow$ as $f\downarrow$ as $s\uparrow$ and $\eta\downarrow$. (b) P versus η plot goes through maximum at 2 dPas
Tsutsumi [7]	HM	Sn/Pb	Stearic acid	Slag infiltration occurs in both t_n and t_p but more in t_p—slag inflow occurs at end of t_n. $Q_s\uparrow$ as $V_c\downarrow$ $\eta\downarrow$ $f\downarrow$ $s\,f\uparrow$ $T_{sol}\downarrow$ Friction measured
Itoh [7], Nebeshima [7]	HM	Cu	Mould flux	Q_s dependent upon ($t_n + 0.5t_p$). No η dependence for Q_s.
Kajitani [5]	CM	Moving polyester belt	Silicone oil	Mould = oscillating acrylic plate; Mould/ shell gap monitored. Gap profile is important. If gap narrows (as go downward) $Q_s\uparrow$ as $V_c\uparrow$, $\eta\uparrow$; if gap widens $Q_s\downarrow$ as $V_c\uparrow$, $\eta\uparrow$ suggests gap widens downward
Badri [9, 10]	M Sim	Steel	Mould slag	Oscillating mould. Principally focused on shell profile, heat flux and OM formation
Ko [11]	M Sim	Steel	Mould slag	Oscillating mould. focused on shell profile, heat flux and OM formation

CM cold model; *HM* hot model; *M Sim* mould simulator

(b) *Hot models* use low-melting metals, like copper or tin, to represent the steel shell and the effects of changing variables are observed visually [7, 8].
(c) *Miniature continuous casters* [9–11] are, bench-top versions of a continuous casting machine; they are fully instrumented to provide information about powder consumption, heat flux, friction, etc. The apparatus (see Sect. 3.2.4.2) consists of (i) a water-cooled copper mould which is inserted in a bath of steel covered with mould powder (the mould is in *inverse mode,* i.e. where the mould is surrounded by a slag film and a steel shell) (ii) an extraction system to provide continuous casting (iii) mould level control (iv) sinusoidal oscillation of the mould and (v) sensors to provide mould and steel temperatures, shell and mould displacements [9–11].

2.2.3.3 Mathematical Modelling Studies

The importance of the slag infiltration process is reflected in the number of mathematical models developed to predict the powder consumption; these models mostly make use of the Navier–Stokes equation [1, 13–25]. Billany et al. [26] and Gray [27] compared the predictions of the earlier models with plant data and observations. They found that:

- The models predicted the correct trends with casting speed, viscosity, etc.
- The models tended to slightly underestimate powder consumption.

However, one common problem with the model predictions is that they predict that Q_s increases with increasing frequency whereas most experimental data indicate the reverse relation, i.e. Q_s and Q_t decrease with increasing frequency (see Sect. 2.2.6.4) (Fig. 2.4).

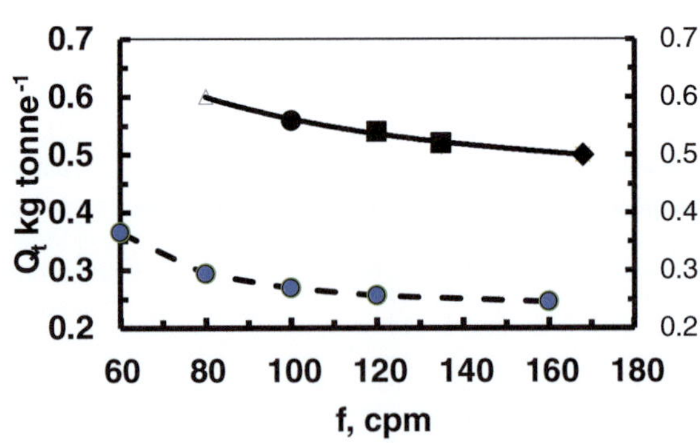

Fig. 2.4 Powder consumption, Q_t, as a function of frequency showing decreasing consumption with increasing frequency; measured values for V_c = 1m min^{-1}, η = 2.7 dPas, [28] (solid line, ●,◆, ■) and values calculated using Tsutsumi Equn [8] (– –,●) see Table 2.3

There are two approaches to modelling powder consumption. The first considers the powder consumption in its entirety. In the second approach, the various contributions to powder consumption are calculated individually [7, 19, 20] and collated. For instance, Meng and Thomas [19] consider that the powder consumption (Q_s) is made up of the following contributions:

- Liquid slag consumption (Q_l).
- Solid slag consumption (Q_{sol}) caused by any downward movement of the solid slag film.
- Powder consumption (Q_{OM}) caused by liquid slag trapped in the oscillation mark.

In some cases, $Q_{sol} + Q_l = Q_{lub}$ is assumed [7]. The powder consumption which is trapped in the oscillation mark (Q_{OM}) has been derived from calculations of the mean depth and the pitch of the oscillation mark [19] or from the relation, $Q_{OM} = 0.001 f\, t_n \rho_l\, (t_n + 0.5 t_p)$ [20].

Itoyama et al. [20] divided the powder consumption (Q_s) into the following contributions:

$$Q_s = Q_m + Q_g + Q_f + Q_{OM}. \quad (2.7)$$

where Q_m ($=\rho d_l/2$) and Q_g ($=g\rho^2 d_l^3/12\eta V_c$) are associated with the pressure drop for parallel plates and the effect of gravity on the flow, respectively, Q_f is the flow associated with mould oscillation ($Q_f = (L_u/L_e)\rho s f\, L_e \sin(\beta/V_c)$ and $Q_{OM} = A_{OM}\, \rho\, f/\, V_c$ where β is the mould taper, L_e = effective length of mould) and L_u = relative, decreasing, distance when mould is ascending and A_{OM} = area of oscillation mark.

Saraswat et al. [4] calculated the values of Q_{OM} (denoted Q_{OM}^{calc}) and compared them with a database of powder consumption values recorded in different steelworks (Q_s^{total}). It was found that all of these models tended to seriously over-calculate Q_{OM}, i.e. ($Q_{OM}^{calc}/Q_s^{total}$) > 1 and frequently with values of 2 or more for this ratio.

Details of the various mathematical models are summarised briefly in Table 2.2.

2.2.3.4 Empirical Rules

A number of empirical equations have been deduced form analysis of plant data or from observations made in simulation tests. These are discussed in detail in Sect. 2.2.6. The equations and the effects of changes in the various factors are summarised in Table 2.3.

Table 2.2 Details of some published mathematical models covering slag infiltration

References	Model details	Comments, findings
Niggel [13]	Initially oscillation excluded. Assumes solid slag film adheres to mould; liquid moves down. Thermo-mechanical approach	Two regimes (i) $Q_{MR} \leq Q_{MR}^{req}$; (ii) $Q_{MR} \geq Q_{MR}^{req}$ Model predicts Q_s versus V_c exhibits a maximum
Schwerdtfeger [18]	Based on N–S eqn. for slag flowing between parallel plates. Needs data for d_{s+l} at meniscus and parallel plate regions Calculates velocity profile and P in gap	Shear stress and Friction calculated from velocity profile. Predicts shell bent away from mould during descent of mould and towards mould during ascent of the mould
Kor [14]	Based on N–S eqn. for slag flowing between two parallel plates—(1) moving at constant velocity and (2) oscillating	Must provide value of d_{s+l} and no account of thermal effects in the slag film
Hill [29]	Model to calculate slag flow, d-$_{OM}$. Max.vol flow rate, $Q_s^{max} = (2/3) \{\eta / \Delta\rho g\}^{0.5} V^{1.5}$–N–S Eqn.	Predictions for d_{OM}: $d_{OM}\downarrow$ as $V_c\uparrow$ as $s\downarrow$ as $\eta\downarrow$
Nippon Steel [15, 16, 30]	Values of d_s and d_l not needed. Assumes mould oscillation affects viscosity. Rheological approach to calculate slag film thickness and velocity profile	Predicts (i) $d_l\uparrow$ as $\eta\uparrow$ and as $f\downarrow$ (ii) $Q_s\uparrow$ as $V_c\downarrow$ as $\eta\downarrow$ [26]. In agreement with plant data
Meng [19]	Couples transient slag flow model with solid slag stress and FD model of heat transfer. Uses Slag fracture and TTT data. Validated via Friction measurements on plant	Models used to check effects of V_c and η. F_l gives negligible stresses. Slag film Fracture occurs near meniscus for crystalline and near mould exit for glass phase
Okazawa [17]	Uses Reynolds eqn, not N–S eqn. Steady state Unsteady state	Mould/shell gap widen as mould ascends; gap narrows during the ascent Gap opened during ascent and closed during descent of mould
Ojeda [22]	Couples, FF and HT. Values assumed for d_s, d_l; men shape = Bikerman	Steel overflows at end of t_p. Rim exerts positive pressure. Positive Q_s during (t_n to early t_p) and negative Q_s during t_p. when V_m is high
Ramirez–Lopez [21]	Couples FF, HT and solidification. No assumptions re: d_s, d_l or men. shape	Follows Q_s, q, $d_{\underline{shell}}$, d_s, d_l through osc.cycle, Q_s increases as mould/rim descends with max. in early t_p. Related to direction of flow in slag pool- downward-descending; outward-ascending
Jonayat [25]	Couples transient flow of metal and slag phases with heat transfer and solidification models Validated against plant data for Q_s	Parametric data: $Q_s\uparrow$ as $V_c\downarrow$, as $s\uparrow$ as $\alpha'\uparrow$ relation with f complex;-Q (kg m^{-1}cycle^{-1})$\downarrow$ as $f\uparrow$

FF fluid flow in slag and metal; *HT* heat transfer; *men* meniscus; *osc* oscillation cycle; *N–S* Navier–Stokes; *P* pressure; Q_{MR}^{req} is *required* Q_s value; α' modification ratio

Table 2.3 The predicted effect (increase ↑ or decrease, ↓) on Q_s with an increase in parameter (e.g. f)

References		Equation, Q_s	V_c	η	f	T_{sol} T_{liq}	s	t_p	t_n	t_{cycle}	ρ_{slag}	d_l	d_{OM}
Wolf [62]	P	$Q_s = 0.1+ 0.55\ (60/f\eta^{0.5}\ V_c)$	↓	↓									
Ogibayashi [42, 43]	P	$Q_s^{powd} = 0.60/\eta V_c$	↓	↓									
Wolf [41, 42], Fox [1]	P	$Q_s^{powd} = 0.70/\eta^{0.5} V_c$ *modified* $Q_s^{pslag} = 0.55/\eta^{0.5}\ V_c$	↓	↓									
Kwon [51]	Ps	$Q_s = 0.22 + 0.4\{60/(0.5\ s)^{0.3} f\eta^{0.5} V_c\}$	↓	↓	↓		↓						
Maeda [49]	P	$Q_s = 0.15\ f \cdot t_p/\ (\eta \cdot V_c)$	↓	↓	↑	↓		↑					
Nakajima [52]	P	$Q_s = 205400 V_c^{-0.628} T_{liq}^{-0.866} \cdot s^{0.341}\ t_{cycle}^{-0.076}\ t_p^{0.116}$	↓			↓	↑	↑		↓			
Sridhar [45]	P	$Q_s = 0.3/\ (\eta^{0.045} V_c)$	↓	↓									
Wolf 5 [63]		$Q_s = 0.5\ (s{\cdot}f\eta\ V_c/1000)^{-0.5}$	↓	↓	↓		↓						
Jenkins [6] Fox-modified [2]	Sim M	$Q_s = (0.433/V_c^{0.5})\ \{1 + (0.0283/\eta V_c)\}$ $Q_s = (0.369/v_c^{0.5})(1 + 0.1564/\eta V_c^2) - 0.123$	↓	↓									
Nebeshima [7]	Sim	$Q_s = (\rho_l{\cdot}d_{OM} d_x f/2V_c) + (\rho_l d_l)$	↓		↑						↑	↑	↑
Tsutsumi [53, 64]	Sim	$Q_s = (k\beta s^{0.4}/T_{sol}\eta^{0.5} V_c)\ \cos^{-1}\ [1000 V_c/2\pi f s]$	↓	↓	↓	↓	↑						
Emi [65]		$Q_t = 0.6\eta^{-0.15}$		↓									
Nakato [66, 67]		$Q_s = \rho_{slag}(0.143–0.003\eta)$		↓	↓						↑		
Kitagawa [57]		$Q_s = 0.0085\ t_p{\cdot}f/\ Vc$	↓		↑			↑					
Noguchi [68]		$Q_s = 10^{-3}(1952 - 246V_c - 44\eta - 1.07T_{liq}.)$	↓	↓		↓							
Kobayashi [69]		$Q_s = 0.003(t_{cycle} + t_p)^{1.5}{\cdot}f/V_c$	↓		↑			↑		↑			
Shimizu [70]		$Q_s = f\ t_p\{0.0184(T_s – T_{liq}) – 2.58\}/10V_c$	↓		↑			↑					
Itoyama [20]		$Q_s = Q_m + Q_g + Q_f + Q_{OM}$											
Shin [23]	P	$Q_s = [0.025\ \rho_l{\cdot}k^{1.43}\{(2\Delta\gamma/\ \Delta\rho g)^{0.5}\}^{0.556} \cdot\ t_n^{0.389}$ $\cdot\ v_s^{-1.49}\ + 0.507e^{3.59tp}]\ (f/v_s)$ where $v_s = V_c 10^3/60$											
Saraswat [4]	StA	$Q_s = e^{28.81}/V_c\eta^{0.46} \cdot\ f^{0.49} \cdot s^{1.37} \cdot T_{br}^{3.48}$	↓	↓	↓	↓	↓						

when casting a LC steel where *f* frequency; *s* stroke; t_n negative strip time; t_p positive strip time; Q_m, Q_f, Q_g, Q_s defined in Sect. 2.2.3.1 (iii); *P* analysis of plant trial data; *Sim* simulation expts; *StA* statistical analysis of plant data

2.2.4 Problems Arising from Poor Powder Consumption

Inadequate powder consumption has been reported to cause:

- Longitudinal cracking [31].
- Sticker breakouts (which are associated with a thin shell and poor lubrication).
- Deep oscillation marks [32, 33].
- Transverse cracking [34].
- Off-corner cracking [35, 36].
- The formation of depressions (when poor mould level control leads to the capture of the slag rim and the simultaneous cutting off of slag infiltration) [37–39].

Furthermore, heat flux (or mould temperature) variations were found to increase with decreasing powder consumption (Fig. 2.5) [40]. One of the principal causes of heat flux variations is the creation of a corrugated shell which is prevalent when casting peritectic, MC steels. The powders used to cast these steels are designed to give a thick, crystalline slag film to reduce the horizontal heat transfer. This is achieved by increasing T_{br} for the slag. However, a high T_{br} slag tends to create in a thin liquid slag film (d_l) which results in a low Q_s value. Corrugated shells are less common in other steel grades and so slags tend to have lower T_{br} values and hence higher Q_s values. Thus mould temperature fluctuations tend to exhibit an inverse relation with powder consumption (Fig. 2.5).

2.2.5 Optimum Casting Conditions

Wolf [41, 42] proposed that there was an optimum range for casting (i.e. the minimum friction and the optimum horizontal heat flux) which was defined in terms of the parameter (ηV_c^2). This is shown in Fig. 2.6. The optimum conditions were found to occur when $(\eta V_c^2) = 5 \pm 3$ dPas (m min^{-1})^2.

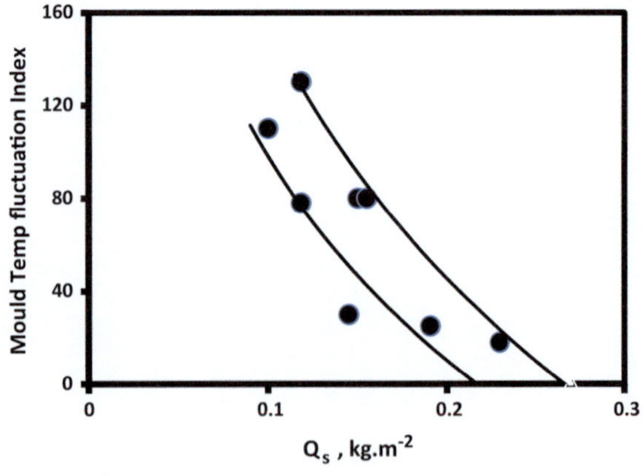

Fig. 2.5 Fluctuations in mould temperature as a function of powder consumption when using different casting powders (permission granted, ISS/AIST, [40])

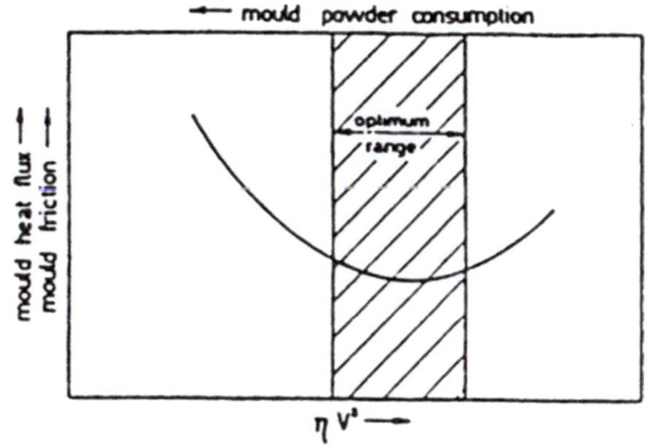

Fig. 2.6 Schematic drawing showing mould friction and horizontal heat flux as function of the parameter (ηV_c^2) (Reproduced with permission from Verlag Stahleisen GmbH [42])

Ogibayashi et al. [43] reported that *fluctuations* in molten slag infiltration and mould temperature (heat flux) were at a minimum when the parameter, ηV_c, had a value in the range 1–3 dPas m min^{-1} (Fig. 2.7a, b). A similar figure, involving fluctuations in the frictional force, also exhibits a minimum in this region (Fig. 2.7c) [44]. Thus the minimum fluctuations (i.e. optimum casting) occurs when $\eta V_c = 2 \pm 1$ dPas m min^{-1}.

It can be seen from Figs. 2.5 and 2.6 that the equations, $\eta V_c^2 = 5 \pm 3$ dPas (m min^{-1})^2 and $\eta V_c = 2 \pm 1$ dPas (m min^{-1}) that powder consumption increases with decreasing viscosity. A plot of the viscosity of the casting slag as a function of the casting speed for mould fluxes used in a large number of steel plants revealed that the bulk of the data fell between the curves representing the Wolf [42] and Ogibayashi [43, 44] relations (Fig. 2.8 [45]). The outliers in the figure all relate to high-viscosity powders used in high-speed billet casting and this behaviour will be discussed in Sect. 2.2.6.10.

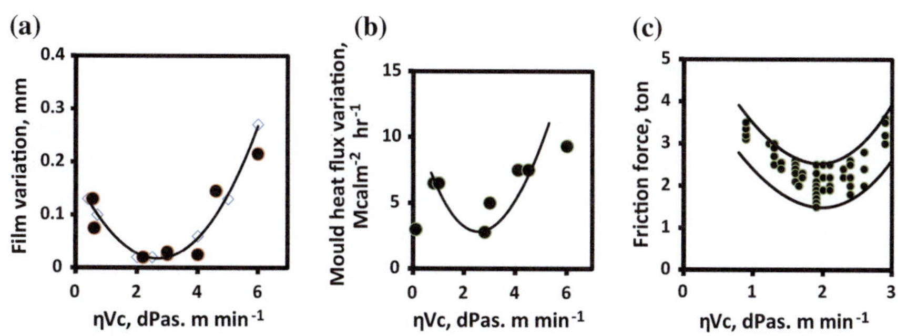

Fig. 2.7 Fluctuations in **a** Liquid slag film thickness on broad face (d_l) [43] **b** mould heat transfer (i.e., horizontal heat flux) [43] [and **c** frictional force [44] as functions of the parameter, ηV_c (permission granted **a, b** Nippon Steel Sumitomo Metal Corp. [43] and **c** ISIJ [44])

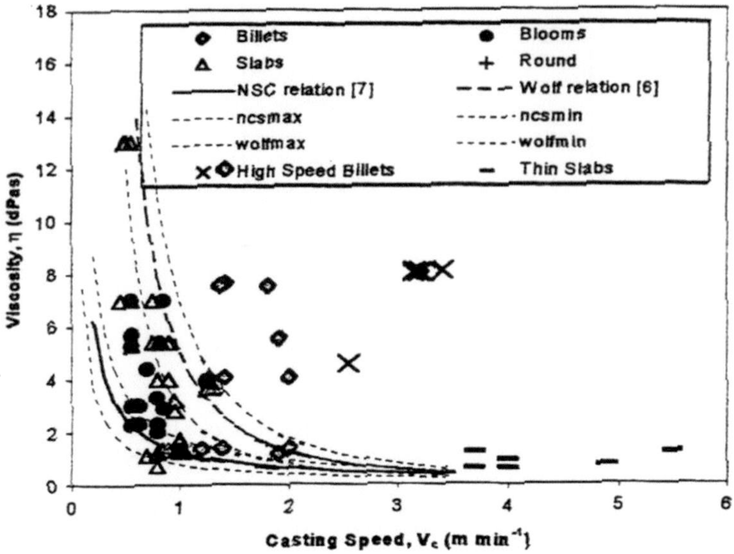

Fig. 2.8 Viscosity of mould slag at 1300 °C as a function of casting speed used in casting at a number of steel plants; *dashed line* Wolf relation ($\eta V_c^2 = 5 \pm 3$ dPas (m min^{-1})^2); *solid line* Ogibayashi relation, ($\eta V_c = 2 \pm 1$ dPas (m min^{-1})); *dotted lines* bounds of above relations; Δ = slabs; ▬ = thin slabs; o = blooms; ◇ = billets; + = rounds; X, ◇ = high-speed billets (permission granted, UNESID, [45])

2.2.6 *Factors Affecting Powder Consumption*

There are a number of factors affecting powder consumption, these are outlined below. These factors can be classified as following: (i) casting conditions, (ii) characteristics of the mould powders and slags and (iii) conditions arising from process problems.

2.2.6.1 Mould Dimensions

It was noted above in Eq. 2.1 that the frictional force increased as the surface area of the mould increased. Neumann et al. [46] proposed the powder consumption (Q_s) could be represented as function of the parameter, R^* (Fig. 2.9a, Eq. 2.8). This term represents the ratio of the (surface area/volume) for the mould. Alternatively, it can be regarded as the ratio of (the surface area of shell/volume of steel in the mould). It can be calculated from Eq. 2.9 where w and t are the width and thickness of the mould. The parameter, R^*, was used in Eq. 2.4 to calculate Q_s^{slag} from Q_t^{powd}.

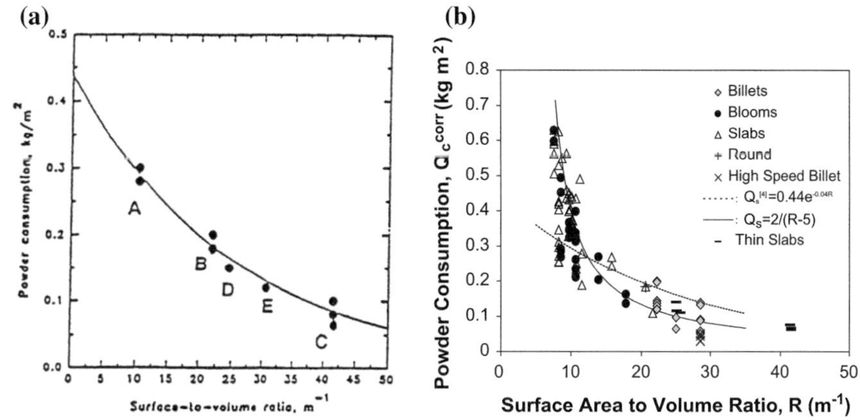

Fig. 2.9 Powder consumption, Q_s, as a function of the parameter, R^* **a** used in moulds with different dimensions; ($w \cdot x \ t$ in m) A (1.5 × 0.22) B (1.0 × 0.1); C (1.3 × 0.050); D (0.16 × 0.16); E (0.13 × 0.13) [46] and **b** ◊ = billets; o = blooms; Δ = slabs; ▬ = thin slabs; + = Rounds; X = high-speed billets; dotted line = Eq. 2.8 [46]; solid line = ($Q_s = 2/(R - 5)$) [45, 47] (permission granted), ISS/AIST, [46] and **b** UNESID, [45]

$$Q_s = 0.44 \exp^{-0.44R*} \tag{2.8}$$

$$R^* = 2(w + t)/w \cdot t \tag{2.9}$$

Subsequently, a similar study based on a much larger database and found that the data shown in Fig. 2.9b could be represented better by Eq. 2.10 [45, 47].

$$Q_s\left(\mathrm{kg\,m^{-2}}\right) = 2/(R^* - 5) \tag{2.10}$$

Typical R^* values are for Slabs 9–15; for Blooms 10–18; for Billets 22–30, Thin slabs 40. It can be seen from Fig. 2.9b that the *required* powder consumption is much greater for slabs than that for billets. Ogibayashi [34] pointed out that frictional forces on the broad face tended to increase as the distance from the corner increased. Consequently, Fox [2] derived an alternative relation for powder consumption, ($Q_t f^*$), in terms of distance from the corner, D_{corn} (Fig. 2.10 and Eq. 2.11,); note ($Q_t f^*$) was calculated to avoid using R^* in the conversion of Q_t^{powd} to Q_s^{slag}.

$$(Q_t f^*) = 0.615\, D_{corn}^{0.586} \tag{2.11}$$

The overall conclusion is that mould dimensions have a marked influence on the required powder consumption, regardless of whether the parameters R^* or D_{corn} is used. However, with thin slabs the high value of R^* suggests thin slabs require little powder consumption whereas the fact the D_{corn} is quite large suggests that a high powder consumption is needed. It is also true that powder consumption values tend to be low in thin-slab casting. The shell in thin-slab casting tends to be very thin

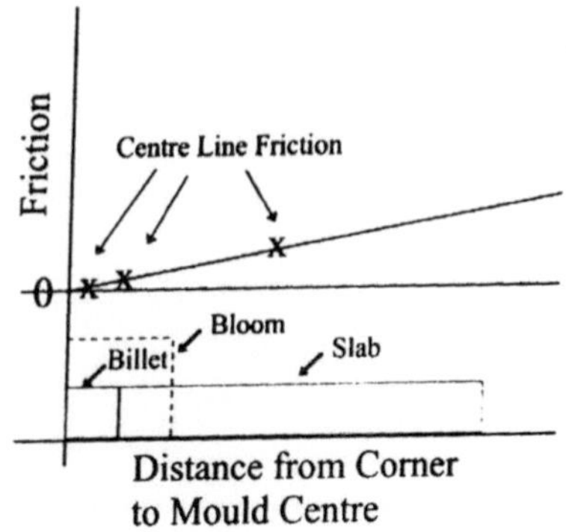

Fig. 2.10 Schematic diagram showing friction increasing with increasing distance of corner to centre; X = friction at centre line for 3 mould geometries; solid line = billet; dotted line = bloom and solid line = slab; (courtesy of A.B. Fox [2])

because of the short residence time in the mould and thus shell shrinkage and the size of the gap tend to be low also. Some support for this view is that 400 mm thick slabs at Dillinger (where the shrinkage is high) tend to have high powder consumption (Figs. 2.9b and 2.10).

2.2.6.2 Effect of Casting Speed

The effect of casting speed on powder consumption has been investigated by a large number of workers [1, 3, 4, 20, 21, 23, 25, 32, 42, 46–52].

In Sect. 1.4 above, it was pointed out that Wolf [41, 42] proposed that optimum casting was obtained when $\eta V_c^2 = 5 \pm 3$ dPas (m min^{-1})^2 and that Ogibayashi et al. [43] found the fluctuations were at a minimum, (for both slag film thickness and heat flux) when $\eta V_c = 2 \pm 1$ dPas (m min^{-1}). Subsequently, Wolf [42] converted these relations into optimum powder consumption equations, i.e. Eqs. 2.12 and 2.13, respectively.

Wolf:

$$Q_{\mathrm{s\,req}}^{\mathrm{powd}} = 0.7/\eta^{0.5} \cdot V_{\mathrm{c}} \tag{2.12}$$

Ogibayashi:

$$Q_{\mathrm{sreq}}^{\mathrm{powd}} = 0.6/\eta \cdot V_{\mathrm{c}} \tag{2.13}$$

An investigation using a much larger database of Q_s values, subsequently indicated a modified version of Eq. 2.12 (given in Eq. 2.14) provided a better fit with plant data [1, 48].

$$Q_{\mathrm{sreq}}^{\mathrm{slag}} = 0.55/\eta^{0.5} \cdot V_{\mathrm{c}} \tag{2.14}$$

These equations all indicate that powder consumption decreases as the casting speed increases.

2.2.6.3 Slag Viscosity

The effect of viscosity on powder consumption has been studied by a number of investigators [1, 15, 20, 21, 25, 31–33, 38, 45, 49–54].

It can be seen from Eqs. 2.12 to 2.14 that powder consumption increases with decreasing viscosity. The powder consumption values obtained on plant with high-viscosity fluxes used in high-speed billet casting are significantly lower than *required* Q_s values calculated with Eqs. 2.12–2.14. This is discussed in Sect. 2.2.6.10. Low consumption values are also recorded on plants casting Ti-stabilised stainless and LC steels. This is thought to be due to the blocking off of slag infiltration by accretions of Ti(CN) or *perovskite* ($CaO{\cdot}TiO_2$) at the mouth of the shell/mould channel (see Sect. 2.2.6.9). The minimum powder consumption providing acceptable levels of lubrication has been discussed [42]. Recommended values of Q_s for casting sticker sensitive steel grade are 0.3 and 0.4 kg m^{-2} for round billets and HC steel grades, and for crack-sensitive grades, values of 0.25 and 0.4 for round billets and heavy plate, respectively [42]. For high-speed casting, minimum values of $Q_s = 0.1$ kg m^{-2} have been recommended [55]. The minimum levels of powder consumption required to avoid solid friction (which causes fracture of the slag film) has also been discussed [19].

Values of Q_s for non-Newtonian mould slags have been compared with those for conventional slags (Q_s c_{onv} < $Q_{s\ non\text{-}Newt}$) since the high shear forces in the infiltration region result in a reduction of their viscosity [56].

2.2.6.4 Oscillation Parameters

The effects of the various oscillation parameters have been investigated by a number of workers [1, 4, 20, 21, 23, 31, 32, 42, 49, 51–53, 57–61]. Various equations have been reported for the calculation of powder consumption; these involve a number of variables (e.g. casting speed, frequency, etc.). A number of the reported equations are given in Table 2.3; the effect of an increase in the variable on the powder consumption is also given. Inspection of Table 2.3 indicates:

- There is reasonable agreement that increases in casting speed, viscosity and break temperature (T_{br}) all result in a decrease in powder consumption.
- There is little agreement on the effect of other variables controlling powder consumption
- There is no agreement on the effects of the oscillation characteristics; it should be pointed out that some workers considered slag infiltration to occur during negative strip time t_n, (i.e. when the mould is descending faster than the shell) which results in increased pressure produced by a descending slag rim. In the other school of thought, slag inflow is considered to be difficult in t_n (since the channel gets blocked by the bending of the shell in this period) and thus slag infiltration largely occurs in positive strip time, t_p, (where $t_p + t_n = t_{cycle}$). It has been reported that Q_s exhibited a stronger correlation with t_p than with t_n [25].

- Some of the experimental studies indicate that Q_s decreases with increasing frequency (Fig. 2.4) but as can be seen from Table 2.3 other studies show the reverse relation with frequency (i.e. $Q_s\uparrow$ as $f\uparrow$)
- Wolf [42] examined plant data where all casting variables were kept constant except the stroke length which was altered (i.e. t_n and t_p will change) and found that as the stroke increased Q_s increased ($Q_s\uparrow$ as $s\uparrow$) as shown in Fig. 2.11. This is supported by plant data from Oxelösund Steelworks which exhibited a strong relationship between stroke length and mould powder consumption. However, a statistical analysis of plant data showed the reverse trend ($Q_s\uparrow$ as $s\downarrow$) and a parametric study suggested that increasing stroke only had a slight effect on Q_s [25].

2.2.6.5 Non-sinusoidal Oscillation

Several workers have reported that non-sinusoidal oscillation increases powder consumption [25, 49, 53, 60, 64, 71]. It can be seen from Fig. 2.12 that the use of non-sinusoidal oscillation leads to a 10% increase in powder consumption.

2.2.6.6 Solidification (or Break) Temperature (T_{br})

The effect of break (or solidification) temperature on powder consumption has been studied by a number of investigators [4, 16, 20, 51–53, 64].

It can be seen from Table 2.1 that those equations which involve T_{br} (or T_{sol}, or T_{liq}) all agree that an increase in T_{br} results in a decrease in powder consumption ($Q_s\downarrow$). This can be seen in Fig. 2.13. This is consistent with intuition since it can be

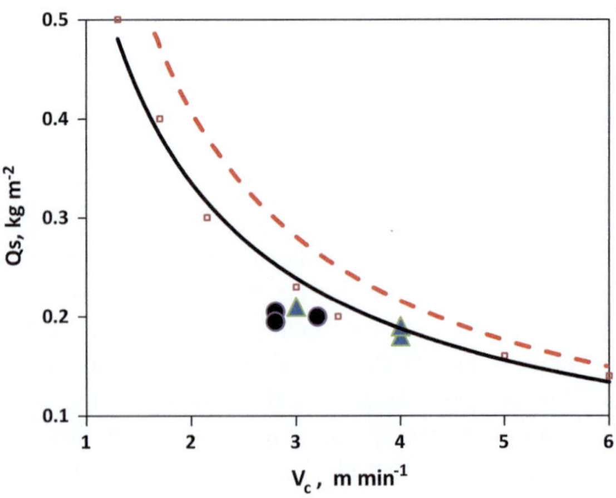

Fig. 2.11 Powder consumption as a function of casting speed showing Q_s increasing with increasing stroke length; - - -, s = 8 mm; — = 6 mm; •,▲ = experimental data (permission granted, VDEh/ Stahl Eisen, [42])

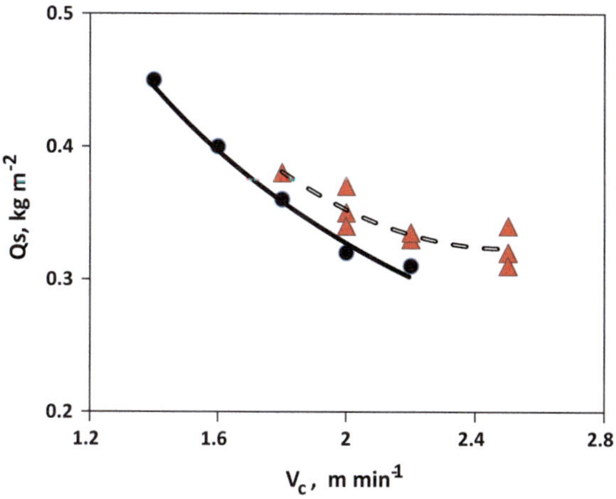

Fig. 2.12 Powder consumption as a function of casting speed showing non-sinusoidal oscillation - -▲- -) provides a 10% higher Q_s than sinusoidal oscillation (—•—) (permission granted, ISIJ, [49])

Fig. 2.13 Powder consumption, Q_s, as a function of casting speed showing Q_s decreasing as T_{sol} or T_{br} increases as T_{sol} or T_{br} increases; ▲,● = experimental values; T_{br}; • • •, = 952 °C; — = 970 °C, - - - = 1027 °C; (permission granted, VDEh/ Stahl Eisen, [42], re-drawn)

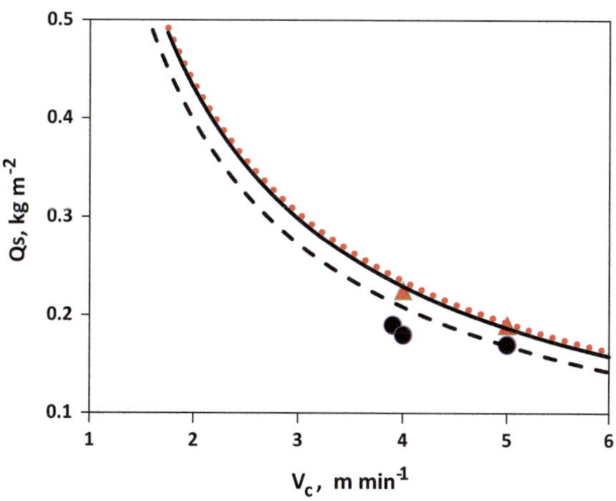

seen from Fig. 2.14 that an increase in T_{br} leads to a decrease in thickness of the liquid film (d_l↓) and an increase in solid film thickness (d_s↑).

The mathematical model reported by Ramirez–Lopez [1, 21, 54] indicated that the horizontal heat flux increased gradually in the period of the oscillation cycle where the mould was descending. The increased heat flux, in turn, caused solid slag to melt, resulting in an increase in liquid film thickness (d_l↑) and a decrease in the solid film thickness (d_s↓). This process was reversed when the mould ascended. Thus, the liquid layer increased at the expense of the solid layer when the heat flux was increasing and *vice versa* when the heat flux was decreasing.

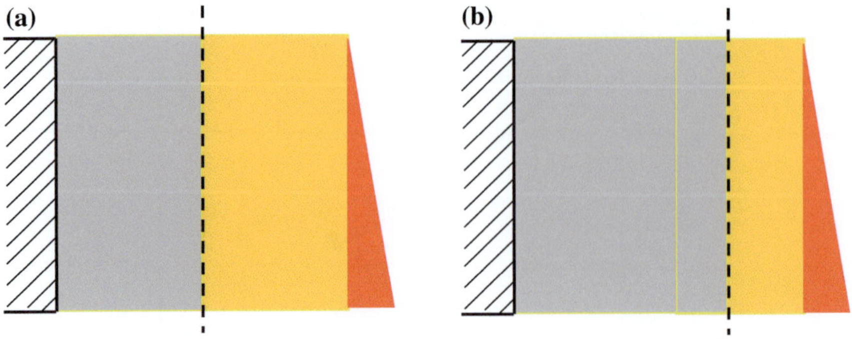

Fig. 2.14 Schematic drawing showing effect of increasing T_{br} on the thicknesses of solid and liquid films, where $T_{br(b)} > T_{br(a)}$ (note relative thickness values are not to scale)

2.2.6.7 Melting Rate

It was pointed out in Sect. 2.2.1 that powder consumption (in kg min^{-1} or kg s^{-1}) in Eq. 2.5 (Q_{MR}^{slag}) can be viewed as the melting rate. The powder consumption has been found to increase with increasing slag pool depth (d_{pool}) [72] and d_{pool} is affected by the melting rate. In practice, the melting rate is controlled by the amount of free carbon and, to a lesser extent, by the size of the carbon particles (see Sect. 4.3). Consequently, it is important that powder consumption is close to the *required* Q_s for the given casting conditions (mould dimensions, casting speed, etc.), i.e. it is not restricted by an excessively high carbon content of the mould powder. A few casting powders have a carbon content which is too high and thus leads to a restricted slag infiltration.

2.2.6.8 Superheat (ΔT)

The effect of superheat on powder consumption has been studied by several investigators [1, 3, 16]. Increasing superheat results in increased powder consumption. On the basis of Eq. 2.12 or Eq. 2.13 increases in superheat will lead, sequentially, to an increase in slag temperature, a lower viscosity and a higher value of Q_s.

2.2.6.9 Argon Flow

It has been reported that powder consumption increases with increasing argon flow rate [73] (Fig. 2.15). One possible reason for this behaviour is that an increased Ar flow rate causes more convection and hence, a higher vertical heat flux which, in turn, increases the melting rate. A high Ar flow rate is known to affect the metal flow patterns in the mould.

Fig. 2.15 Plant measurements of powder consumption, Q_t, as a function of Ar flow rate; (courtesy of A.B. Fox [2])

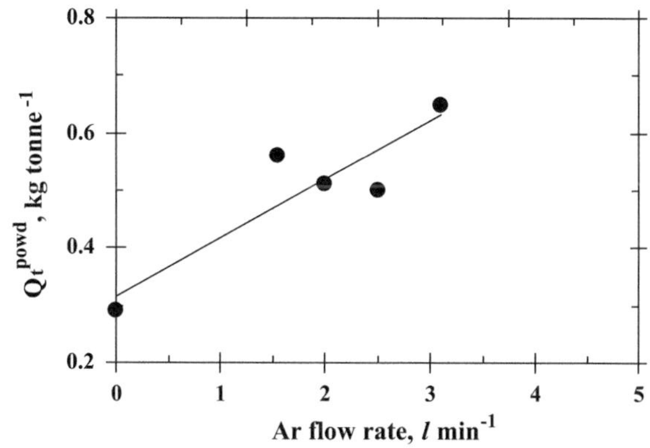

2.2.6.10 Continuous Casting of Steels Containing Ti

It has been observed that powder consumption is frequently lower than predicted when casting steel grades containing Ti. This is thought to be due to the formation of TiN or Ti(C, N) which has a low solubility in the slag pool and thus, tends to exist as solid particles [74, 75]. These particles agglomerate through turbulent collisions and the agglomerates restrict the slag flow when they are sited in the mouth of slag/mould channel (Fig. 2.16) [74, 76]. The solid particles also increase the slag viscosity and thus, decrease Q_s. Alternatively, TiO_2 particles can form *perovskite* ($CaO.TiO_2$) which has a high melting temperature and thus reduces both the thickness of the liquid slag film (d_l) and Q_s. It is necessary to keep the basicity, (C/S) < 1.0 to avoid *perovskite* formation [77].

2.2.6.11 High-Viscosity Powders

The powder consumption data for most powders follow the empirical rules based on the viscosity and casting speed [2, 42, 43] the only exceptions are the high-viscosity (η_{1300} = 10–30 dPas) powders used for high-speed billet casting [2] and those used in casting steel grades containing Ti. In high-speed billet casting, considerable turbulence is generated which results in significant levels of slag entrapment. One way of reducing slag entrapment is to increase the slag viscosity. However, the reduction in slag entrapment levels comes at the expense of a significant decrease in powder consumption. Fortunately, the powder consumption required for billets is low (since the distance from centre line to corner is low and, furthermore, R* has values > 22) which means this practice is widely used to minimise slag entrapment.

It has been suggested that these high-viscosity slags will form super-cooled liquids (*scl*) rather than crystallites during cooling. The *scl*, although viscous, will move in response to any stress applied by the ferro static pressure, the downward movement of the shell or the oscillating motion of the mould; hence, the slag (*scl*)

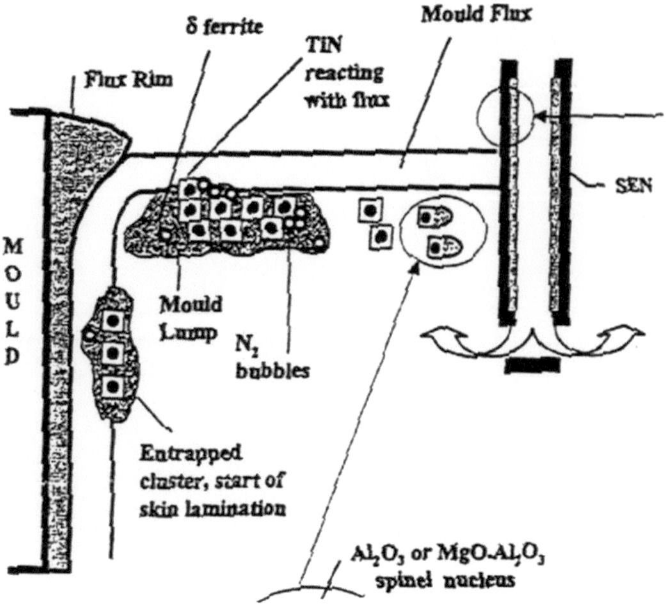

Fig. 2.16 Schematic drawing showing formation of agglomerates of Ti(C, N) when casting steels containing Ti; (permission granted, ISS/AIST [74])

will supply some lubrication to the shell. The importance of retaining some glass phase in the slag film has been pointed out by Hanao et al. [78].

2.2.6.12 Electromagnetic Braking (EMBr) and Casting (EMC)

Electromagnetic devices are reported to increase powder consumption. The application of electromagnetic braking (EMBr) results in a 30% decrease in vertical heat transfer [79] from the steel, resulting in a *ca.* 10 °C increase in meniscus temperature [80–83]. This increased meniscus temperature results in a lower slag viscosity and higher powder consumption. EMBr is widely used in high-speed, thin-slab casting where powder consumption tends to be low [84].

Increases of 20% in Q_s, have been reported when using pulsative, Electromagnetic casting (EMC) [85]; it is known that the pinch force in EMC reduces the horizontal heat transfer, which, in turn, results in increases in both meniscus temperature and Q_s.

2.2.6.13 Liquid Slag Feeding to the Mould

Some high Al-and Mn-steels have low-melting temperatures, so the vertical heat flux is insufficient to melt the mould powder. Consequently, liquid slag feeding technology has been developed to provide liquid slag to the steel surface. It is reported that powder consumption is increased with liquid slag feeding [86].

2.3 Slag Infiltration During the Oscillation Cycle

A number of empirical rules have been proposed to calculate the powder consumption. These are given in Table 2.3 and are based on plant observations and physical modelling results. It was mentioned above that there is general agreement that the powder consumption increases as the casting speed (V_c), slag viscosity (η) and break temperature (T_{br}) all decrease. However, there is no consensus as to (i) which of the various oscillation parameters affect Q_s or (ii) the way in which they affect Q_S. Furthermore, there are two schools of thought concerning the period of the oscillation cycle where slag infiltration occurs;

- The first identifies t_n as the primary period of slag infiltration, where the descending slag rim increases the pressure on the molten slag which responds by infiltrating into the shell/mould channel and
- The second considers t_p as the principal period of infiltration, since infiltration is restricted in t_n because the bending-back of the shell is considered to block off the slag flow during this period; thus slag infiltration is restricted to t_p where the shell does not interfere with the infiltration.

The infiltration mechanism has been studied in (i) plant trials [3, 55, 87–89] (ii) cold modelling studies [5, 6, 12, 58, 90] (iii) hot modelling studies [7, 53, 91] and (iv) mathematical modelling of the heat and fluid flow [1, 12, 16, 21, 22, 25, 36, 92, 93].

Mathematical models based on Navier–Stokes equations do provide a reasonable description of the effects of casting speed, slag viscosity and T_{br}, but, for the most part, they also predict that Q_s increases with increasing frequency which disagrees with most experimental observations, e.g. Fig. 2.4. In an attempt to explain these discrepancies, mathematical models have been developed to explore in which part of the oscillation cycle the slag infiltration takes place [1, 21, 22, 25, 54].

It is customary to characterise oscillation in terms of negative and positive strip times (t_n and t_p, respectively) where t_n represents the time when the mould is descending faster than the shell and t_p constitutes the remainder of the cycle (i.e. $t_n + t_p = t_{cycle} = 60/f$ where f is in cpm). However, the oscillation cycle can also be characterised in terms of the position of the mould. The mould and the slag rim will be at their highest position in late t_p (denoted t_p^{late}). It descends throughout t_n and reaches its lowest position in early t_p (t_p^{early}). The mould will then ascend steadily through t_p.

The findings of the two studies due to Ojeda [22] and Ramirez–Lopez [1, 21, 54] are in good agreement with both proposing that slag infiltration occurs during the descent of the mould/rim covering the period (t_n–t_p^{early}) and that there is little slag infiltration during the ascent of the mould (in t_p). This can be seen in Fig. 2.17c and it was also noted that the rate of slag flow into the shell/ mould channel is at its highest when the mould/slag rim was at its lowest position (i.e. between t_n^{late} and t_p^{early}).

The directions of flow in the slag pool at different periods of the oscillation cycle have also been studied in the several investigations [1, 21, 22, 54] and are shown in Fig. 2.18. It can be seen when the mould and slag rim are at their highest position (Fig. 2.18a) that the slag flow into the mouth of the channel is radially outward and upward. As the mould/rim descends the slag direction changes to downward and there is evidence of a vortex in the region of the mouth (Fig. 2.18b).When the mould and slag rim reach their lowest positions (in t_p^{early}) the flow is strongly downward into the channel (Fig. 2.18c). Finally, halfway through t_p the flow changes direction to radially—outward and upward (Fig. 2.18d). Thus the direction of the flow in the slag pool plays a significant role in the slag infiltration into the shell/ mould channel and this is affected by the direction of movement of the mould and slag rim.

The slag rim acts like a piston and helps to inject slag into the channel. However, the movement of the mould alone will cause some downward flow of slag but the slag rim certainly accentuates the downward slag flow.

Just after the mould reaches its highest position there is a *tide—change* in the slag flow (radially outward and upward to downward) which results in a period of "*confused flow*" (the remnants of which are shown in Fig. 2.18b). There is a similar period of confused flow after the mould and rim reach their lowest points. There is very little slag infiltration during these periods of confused flow. It has been suggested that the lack of slag infiltration in the periods following a tide-change is responsible for the failure of the models based on the Navier–Stokes equation to predict the correct Q_s dependency on frequency (namely, $Q_s\downarrow$ as $f\uparrow$) observed on plant [1]. For example, if $f = 60$ cpm and is increased to 120 cpm, there will twice as many tide-changes per unit time. Since little powder consumption occurs during these tide-changes, the increased number of tide-changes per minute will result in an overall decrease in Q_s. A parametric study showed [25] indicated that a 60% increase in frequency resulted in only 2% change in Q_s (where both increases and decreases in Q_s were recorded for different stroke lengths). However, it was found Q_{cycle} (kg $m^{-1}cycle^{-1}$) decreased by 35% with a 60% increase in f.

In summary, slag infiltration occurs through the downward flow of slag resulting from the downward movement of the mould; the slag rim accentuates this slag flow. The size of the slag rim is dependent upon the steel grade being cast, with high basicity (C/S) slags (used for MC steels) forming large rims and low-(C/S) slags forming smaller rims (see Fig. 1.4).

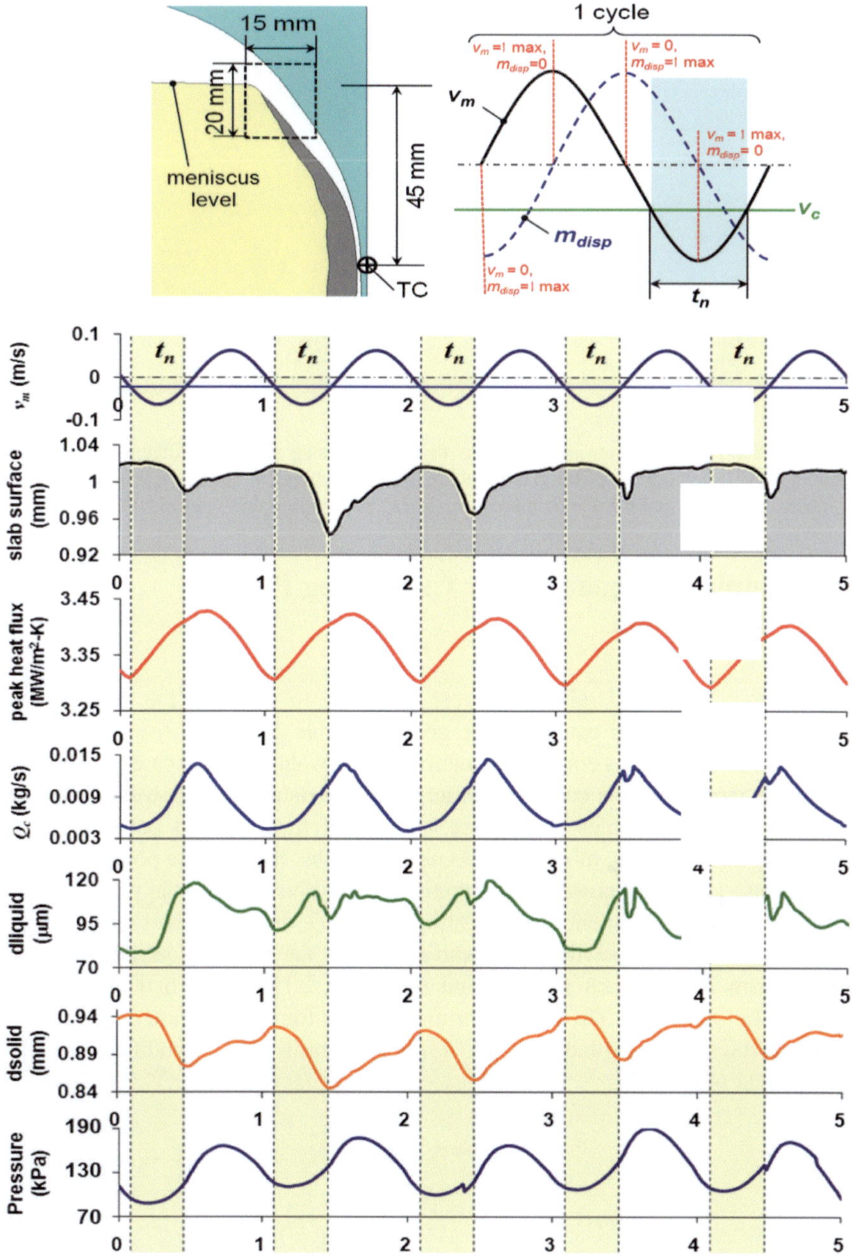

Fig. 2.17 Mathematical model predictions of **a** profile of strand surface, **b** heat flux, **c** powder consumption in kg s^{-1}, **d** liquid film thickness $d_{l,}$ **e** solid slag film thickness, d_s and **f** pressure during five oscillation cycles [1, 21, 54] the dotted, *vertical lines* indicate the onset (*left*) and end (*right*) of negative strip periods of each cycle (0–5) Note that peaks in Q and d_l lie in early t_p; (permission granted, ISIJ, [21])

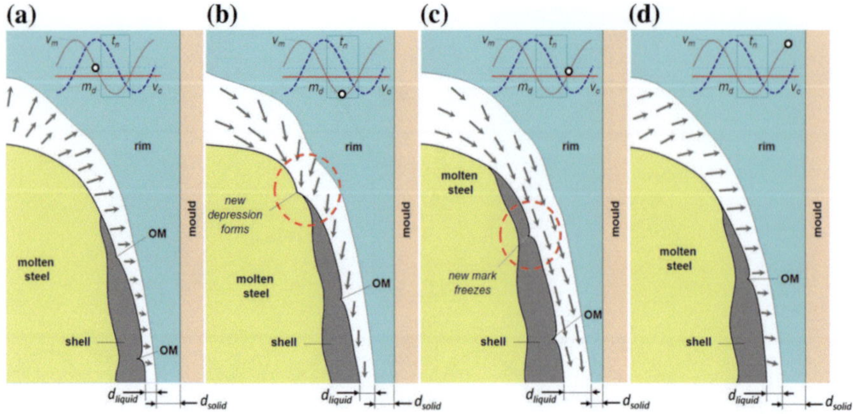

Fig. 2.18 Schematic diagram showing the direction of slag flow at different parts of the oscillation cycle **a** at highest position of mould in late t_p, **b** halfway through t_n, **c** at lowest position of mould in early t_p and **d** midway through t_p [1, 21, 54]; (permission granted, ISIJ, [21])

2.4 Empirical Equations for Calculating Powder Consumption

A number of empirical rules have been proposed to calculate the powder consumption; the proposed equations are given in Table 2.3. Fox [2] carried out an evaluation of the various empirical equations. This evaluation compared predictions with plant measurements contained in an extensive database of powder consumption and casting variables for a large number of trials carried out at different steelworks casting slabs, blooms, billets and thin slabs. It should be pointed out that the database contained powder consumption for high-viscosity billet powders used to minimise slag entrapment at high casting speeds; Q_s data for these slags are much lower than for other powders and tend to distort the fit. The performance was judged from Δ_{RMS} which is calculated from Eqs. 2.15 and 2.16 where N = the number of mould slags. The best performing models for this database were found to be in the hierarchy, Ogibayashi [42, 43] > Kobayashi [69] > modified Wolf [1, 2] > Maeda [49]

$$\delta = 100\left(Q_{\text{meas}} - Q_{\text{pred}}\right)/Q_{\text{meas}} \tag{2.15}$$

$$\Delta_{\text{RMS}} = \left\{\Sigma\left(\delta_1^2 + \delta_2^2 + \delta_3^2 + \cdots\right)\right\}^{0.5}/N \tag{2.16}$$

Analysis of plant data for powder consumption for slab-, bloom- and billet-casting indicated that Q_t^{slag} ($= f^* \cdot Q_t^{powd}$) is reasonably constant at 0.48 kg (tonne steel)$^{-1}$ except for the "*high-viscosity billet powders*" mentioned above in Sect. 2.2.6.10.

2.4.1 Frictional Forces

Frictional forces (F) acting on the shell contain contributions from the liquid frictional (F_l) and solid frictional (F_s) forces (Eq. 2.17).

$$F = F_l + F_s \tag{2.17}$$

It was pointed out in Eq. 2.1 that the liquid frictional force was inversely, dependent on the thickness of the liquid layer (d_l). Since d_l = (Q_s/ρ_l) Eq. 2.1 can be re-written as

$$F_l = A \cdot \eta(V_m - V_c)\rho_l/Q_s. \tag{2.18}$$

It can be seen that that the liquid friction force is inversely dependent upon the powder consumption and directly related to the slag viscosity and the difference between the velocities of the mould and the strand.

The frictional forces tend to be highest when casting MC steels because the corrugated shell formed increases both A and F_l and d_l tends to low because high T_{br} slags are used to cast these steels. The friction forces measured when casting with mould powders are lower than those for oil casting [94]. Longitudinal cracking has been correlated with high frictional forces [95]. Frictional forces tend to decrease as the cast proceeds [95].

Solid friction (between the shell and the solid slag) tends to occur in the lower half of the mould. The formation of *star and spongy cracks* (Sect. 11.7) is associated with solid/solid friction and the consequent *spalling* of the solid slag film which can even result in the pick-up of copper by the strand. However, solid friction may also occur in the upper mould in the corner regions if the corners are overcooled [4].

Since solid friction can occur in the bottom of the mould it is important to ensure that all of the mould receives liquid lubrication [96]. A lubrication index (LI) was proposed by Billany et al. [96] which is a measure of the fraction of the mould enjoying liquid lubrication; this index is defined in Eq. 2.19. Ideally, the parameter, LI, should have a value of 1.0.

$$\begin{aligned} \text{LI} = &(\text{Distance from meniscus to point where } T = T_{br})/ \\ &(\text{Distance} - \text{meniscus to mould exit}) \end{aligned} \tag{2.19}$$

If solid friction is a problem in the lower half of the mould, probably the best measure is to increase the casting speed. In theory, a decrease in the flow rate of the cooling water would also be beneficial but in practice, the remedial effect is small.

Sorimachi et al. [97] have pointed out that:

- The frictional forces refer to the entire mould wall and not to the local frictional forces in the meniscus region which is of key importance in the formation of sticker breakouts.
- The measured friction is that acting on the mould wall and is not that acting on the shell.

2.4.1.1 Measurement of Frictional Forces

Measurements on frictional forces have been derived using plant trials, simulation experiments and mathematical modelling of the frictional forces.

Plant Measurements of Friction

In the past, a number of investigators have measured frictional forces by using load cells attached to the oscillating mechanism and then applying Fourier analysis of the signals produced [27, 71, 98]. Alternatively, frictional forces can be measured using the MLTEKTOR system [99]. These devices have been used for the detection of defects and longitudinal cracks [44, 100, 101].

Friction Measurements in Simulation Tests

Several tests have been devised to simulate the frictional forces acting on the shell when using different mould fluxes. Short descriptions of these tests are given below.

Rotating Cold Finger Test

In this test, a water-cooled, copper finger (representing the mould) is rotated in a steel crucible (representing the strand) containing the molten mould flux [98]. The copper finger becomes covered with a solid slag film of ca. 3 mm thickness and a thin liquid layer. The frictional forces are measured by determining the torque developed on the steel crucible as the finger is rotated at constant velocity (10–50 rpm). The apparatus is shown in Fig. 2.19a.

Oscillating Cold Finger Test

This test resembles the rotating cold finger test but the cold finger is oscillated vertically instead of being rotated [9, 11, 53, 102].

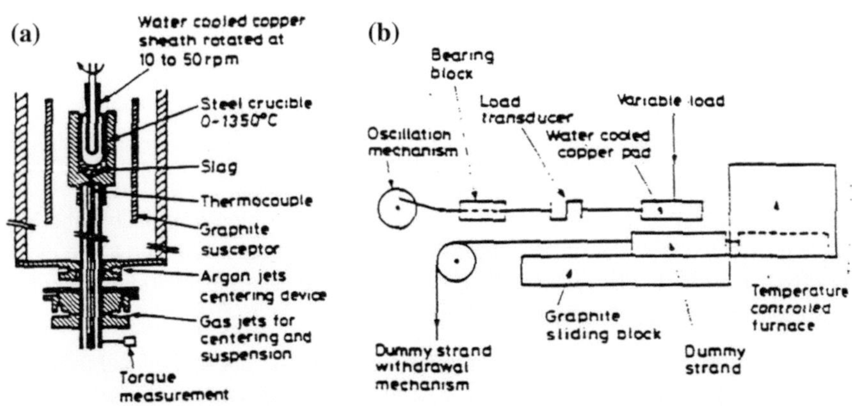

Fig. 2.19 Schematic drawings showing the apparatus used in **a** the rotating cold finger test [98] and **b** the oscillating pad method [27, 103] (permission granted, Europ. Comm. Sci. and Tech. Publ. [27])

Oscillating Pad Test

In these tests, a water-cooled copper pad (mould), which can be oscillated at different frequencies, is lowered onto a heated steel block (strand) covered with molten mould slag (Fig. 2.19b). The block is then withdrawn at a fixed speed and the frictional force exerted by the pad is measured as it bears down on the strand by using a load cell mounted on the oscillation arm. The thickness of the molten slag layer (ca. 0.3 mm) was monitored using a displacement transducer [27, 103].

Rotating and Oscillating Pads Test

A simulation test was reported by Sorimachi et al. [97] and is shown in Fig. 2.20. In this test a graphite disc (representing the strand is rotated unidirectionally). A second, lower, graphite disc (representing the mould) contains a 2 mm deep liquid, mould slag of known viscosity; and this disc is oscillated sinusoidally. The torque is measured continuously.

Miniature Continuous Caster

Friction measurements have been carried out in a miniature continuous caster [91]. In order to view the solidification process of the shell the steel was replaced by Sn–5%Pb, the mould slag by stearic acid with Al_2O_3 particles to act as tracers and one side of the oscillating, 50 mm^2, Cu mould was replaced by silica to facilitate viewing [91]. The friction between mould and shell was measured by load cells sited

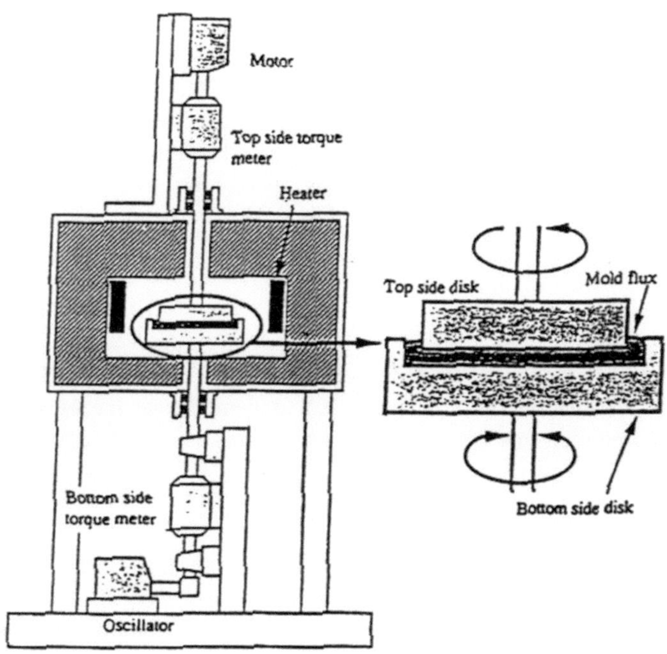

Fig. 2.20 Schematic drawing showing the apparatus used by Sorimachi et al. [97] (permission granted, ISS/AIST [97])

below the mould. The frictional force per unit area between mould and shell was taken as $\Delta F/A = (F_{max} - F_{min})/A$ where A = surface area and the subscripts max and min represent the maximum and minimum load in any one cycle, respectively. Values of $\Delta F/A$ were found to decrease as liquid film thickness (d_l) increased and $(V_m - V_c)/ d_l$ decreased (i.e. $\Delta F/A\downarrow$ *as* $d_l\uparrow$ and as $\{(V_m - V_c)/ d_l\}\downarrow$) [91]. Friction measurements can be made in a similar manner in mould simulators [9–11].

2.4.1.2 Mathematical Modelling of Friction in Mould

Mathematical modelling of the frictional forces in the mould has been reported [12, 18, 53, 97]. Schwerdtfeger and Tacke [18] derived a relation for the shear stress in the liquid slag based on computations of the velocity. The frictional force was calculated by multiplying the calculated stress by the area wetted by the slag.

2.4.2 *Factors Affecting Frictional Forces in the Mould*

It can be seen from Eq. 2.18 ($F_l = A\eta\ (V_m - V_c)\ \rho_l/Q_s$) that the liquid friction force (F_l) is inversely dependent upon the powder consumption (Q_s); thus it follows F_l will increase as Q_s decreases ($F_l\uparrow$ as $Q_s\downarrow$).However, Q_s is dependent upon other factors, e.g. casting speed; the effect of the various parameters are given below.

However, high friction measurements recorded on plant (i.e. 10–20 kPa) have been attributed to (i) movement of the solid slag layer (ii) excessive taper and (iii) mould misalignment [19]. At low casting speeds the critical consumption is high so variations in consumption, Q_s, can lead to slag film fracture and high, solid friction forces. In contrast, at high casting speeds the principal causes of high frictional forces are excessive taper (see Fig. 1.48) and mould misalignment [19].

2.4.2.1 Casting Speed (V_c)

The powder consumption, $Q_{s,}$ is dependent upon $(V_c)^{-1}$ (Eqs. 2.12–2.14). Using the modified Wolf relation ($Q_s = 0.55/\eta^{0.5}\ V_c$) to demonstrate the effect of casting speed, it can be seen that Eq. 2.18 can be re-written as

$$F_l = A\eta(V_m - V_c)\rho_l\eta^{0.5}V_c/0.55 \qquad (2.20)$$

Similar relations could be derived with other relationships for Q_S. It can be seen from Eq. 2.20 that an increase in casting speed causes a decrease in the $(V_m - V_c)$ term, in addition, to the increase in the V_c term. These conflicting responses to a V_c increase, result in a minimum (V_c^{min}) in the F_l–V_c plot shown in Fig. 2.21 reported by D'Haeyer [99]. It can be seen that the $(V_m - V_c)$ is dominant at low speeds and the V_c term tends to dominate at higher casting speeds.

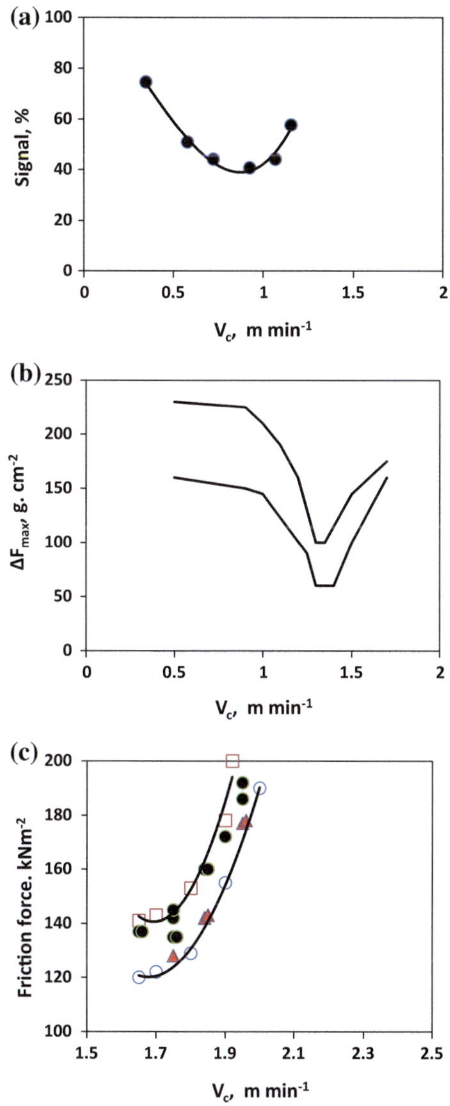

Fig. 2.21 Frictional force as a function of casting speed **a** friction signal from ML Tektor, **b** ΔFmax and **c** frictional force using two mould powders, *Q* (η_{1300} = 0.9 dPas) = , ●; R (η_{1300} = 0.8 dPas) = ▲) [99]; (permission granted, Europ. Comm. Sci. and Tech. Publ., [99])

Tsutsumi et al. [91] carried out simulation experiments and reported that F_l decreased:

- As casting speed increased ($F_l\downarrow$ as $V_c\uparrow$) indicating, $V_c < V_c^{min}$ in their experiments.
- As the liquid slag film thickness (d_l) increased ($F_l\downarrow$ as $d_l\uparrow$).
- As the velocity gradient (($V_m - V_c$)/ d_l) decreases.
- With the introduction of non-sinusoidal oscillation.

2.4.2.2 Viscosity (η)

According to Eq. 2.20 the liquid frictional force is a function of ($\eta^{1.5}$) and the Ogibayashi [42, 43] relation (Eq. 2.13) leads to F_l exhibiting a dependence on (η^2). Thus the frictional forces increase with increasing viscosity. Wolf [42] reported a minimum in the F_l- ($\eta^{0.5}$ V_c) plot (Fig. 2.6) at 5 ± 2 (dPas)$^{0.5}$ · m min^{-1}; the equivalent plot for the Ogibayashi relation for Q_s leads to a minimum at 2 ± 1 dPas. m min^{-1} as shown in Fig. 2.7c [44].

Hering et al. [104] reported that the liquid frictional force increased with increasing Al_2O_3 content in the casting slag (Fig. 2.22a). This is presumably due to the increase in viscosity with increasing Al_2O_3 content. However, Hering et al. [104] found that this was not always the case, (as can be seen from Fig. 2.22a) and proposed that the friction was affected the nature of the mineralogical phase formed. It is possible that with the *wollastonite/gehlenite* curve in Fig. 2.22b could be explained by a lowering of T_{br} with increasing Al_2O_3 which offsets the effect of increasing viscosity.

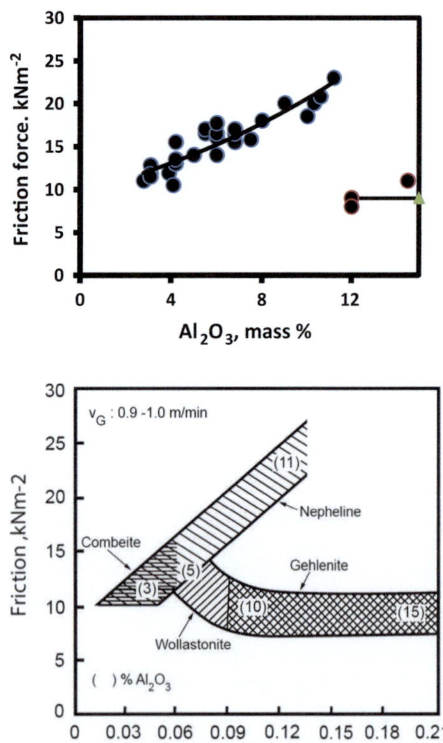

Fig. 2.22 Frictional force as function of **a** Al_2O_3 content and **b** viscosity (cited in Pas, thus multiply by 10 for dPas [104]); (permission granted, Stahl Eisen/VDEh, [104], re-drawn)

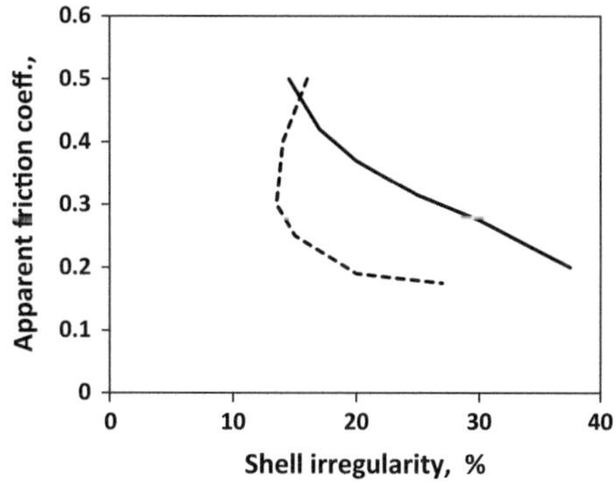

Fig. 2.23 The apparent frictional coefficient as a function of the unevenness (or irregularity) of the shell for $V_c = 2.0$ (—) and 4.0 m min^{-1} (- - -) (permission granted, ISS/AIST, [105])

2.4.2.3 Mould Dimensions and Surface Area (A)

It can be seen from Eq. 2.1 and Fig. 2.20 that the liquid frictional force (F_l) increases as the surface area of the mould (or shell) increases. Ogibayashi et al. [105] pointed out that the shrinkage of the steel will be greatest at the centreline, the point where the shell is at its thinnest. Ogibayashi et al. [105] also pointed out that in *peritectic*, MC steels the shell (in the meniscus region) tends to become uneven or corrugated; this unevenness increases the surface area of the shell. Consequently, the friction coefficient tends to increase as the unevenness of the shell increases (Fig. 2.23).

2.4.2.4 Break (or Solidification) Temperature (T_{br})

Increases in break temperature would be expected to reduce the thickness of the liquid slag film (d_l), as shown in Fig. 2.14; this would result in higher frictional forces ($F_l\uparrow$ as $T_{br}\uparrow$). Furthermore, increasing T_{br} will also enhance the amount of solid friction. Thus on both counts the friction will tend to increase as T_{br} increases.

Measurements of friction and friction coefficient are shown in Fig. 2.24a, b, respectively. These figures show that there is a sharp increase in friction at a temperature slightly below the break temperature, T_{br}; this may imply that the cooling rates in the friction experiments were slightly higher than those used in the viscosity experiments, since T_{br} decreases with increasing cooling rate. These figures show that relatively small amounts of solid friction can have a significant effect on the total friction. The rate of friction rise was much greater in some cases, denoted. Type A, e.g. Powder J with sharp T_{br} temperatures) than in others (Type B i.e. more "*glassy*" slags, e.g. Powder A).

2.4.2.5 Frequency (f)

Since most plant observations indicate that powder consumption Q_s decreases as f increases, it is expected that an increase in frequency would increase the liquid

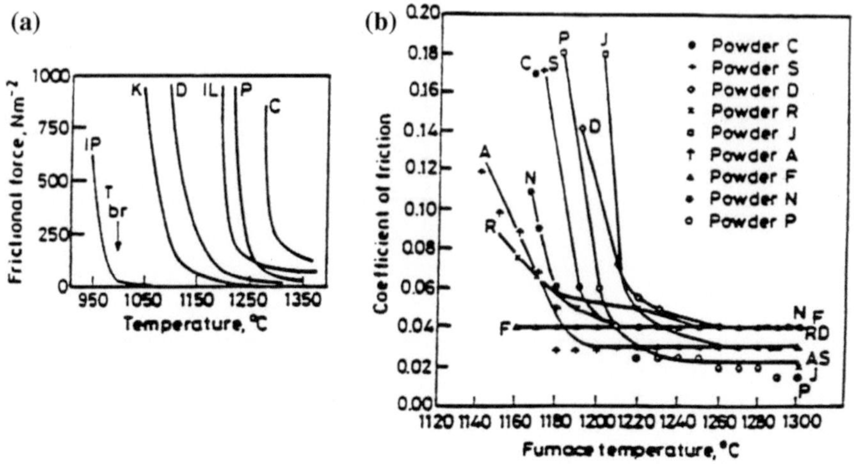

Fig. 2.24 **a** Frictional force [98] and **b** Coefficient of friction [27, 103] as functions of temperature derived in rotating cold finger and oscillating pad tests, respectively (permission granted, Europ. Comm. Sci. and Tech. Publ. [27, 98])

friction ($F_l\uparrow$ as $f\uparrow$). This relationship ($F_l\uparrow$ as $f\uparrow$) has been confirmed by several investigators [53, 64, 106]. It should be noted that an increase in frequency also increases the velocity of the mould (V_m) and it can be seen from Eq. 2.18 that an increase in V_m will result in an increase in friction.

2.4.2.6 Stroke Length (S)

It can be seen from Table 2.3 that there is no consensus on the effect of the stroke length (s) on the powder consumption, Q_s. The statistical analysis of plant data due to Saraswat et al. [4] indicates that Q_s decreases as the stroke increases ($Q_s\downarrow$ as $s\uparrow$); on this basis, F_l would be expected to increase (i.e. $F_l\uparrow$ as $s\uparrow$). Other workers have reported that Q_s increases with increasing stroke ($Q_s\uparrow$ as $s\uparrow$) [53, 64] which would result in ($F_l\uparrow$ as $s\downarrow$).However, Q_s (or d_l) is not the only factor affecting friction forces and an increased stroke would lead to increased values for V_m and ($V_m - V_c$) and F_l. Thus no relation between F_l and s can be recommended at this stage.

2.4.2.7 Negative and Positive Strip Time (T_n and T_p)

As mentioned above, there has been considerable debate as to whether powder consumption occurs in negative strip time or positive strip time. Mathematical models [1, 21, 22] indicate that slag infiltration occurs predominantly in the period when the mould (plus rim) are descending, with the infiltration rate being at its highest in late t_n and early t_p. Thus it may be concluded that increased negative strip would increase Q_s and hence decrease F_l.

However, Tsutsumi et al. [53, 64] reported that in their simulation experiments that increased positive strip resulted in a decrease in F_l. It has also been reported that the correlation of powder consumption with t_p is stronger than that with t_n [25].

2.4.2.8 Steel Temperature

It has been reported that a decrease in steel temperature results in increased friction measurements, presumably due to the effect of the slag viscosities (which increase at lower temperatures) on the friction.

2.4.2.9 Non-sinusoidal Oscillation

Non-sinusoidal oscillation has been reported to reduce liquid friction [60, 71, 99, 106–108], however, the reverse relation ($F_l\uparrow$ as NSO↑) has been found in trials in Sweden [109]. Mizukami et al. [60] reported that non- sinusoidal oscillation can

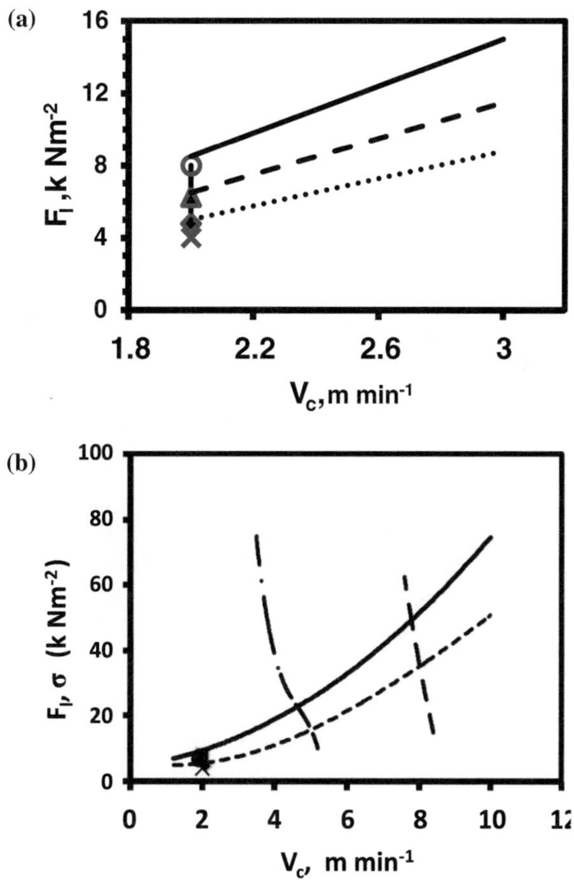

Fig. 2.25 **a** Frictional force, F_l, as a function of casting speed, showing decreasing F_l, with increasing non-sinusoidal component (α' %) (α' = 0 %, sinusoidal, solid line [60] o [61]); α' = 15 %, ▲ [61]; α' = 18 %, dashed line [60]; α' = 28 %, ◆, [61]; α' = 40 %, dotted line [60], x, [61] and **b** Comparison of F_l in relation to tensile strength ($\sigma_{B-\text{ aver shell T}}$ = – • –; and $\sigma_{\text{sutf T}}$ = – –; [60]) as functions of casting speed showing (i) maximum casting speed (i.e. $\sigma > F_l$,) lies between 5 and 8m min^{-1} and (ii) the benefits of increasing the non-sinusoidal component (α') on F_l,(—, [60], ■, [61] α', = 0%; ●, α' , = 15% [61]; ◆, α' = 28% [61]; - - - [60], x = [61] α' = 40%))

result in a 40% decrease in liquid friction compared with conventional, sinusoidal oscillation (Fig. 2.25a) but only resulted in a 5% decrease in solid friction (F_s). Non-sinusoidal oscillation results in a decrease in ($V_m - V_c$) which leads to a concomitant decrease in the liquid friction (F_l) as shown in Eqs. 2.18 and 2.20. The frictional forces were compared with the tensile strength of the steel (Fig. 2.25a) from which it was concluded that the upper limit for the casting speed when casting with an oscillating mould, lay between 5 and 8 m min^{-1} [60].

2.5 Summary

The following observations were made concerning the lubrication of the shell and the frictional forces acting on it:

(i) Inadequate lubrication of the shell leads to various defects in the steel product, such as, longitudinal cracks, sticker breakouts and star cracks.
(ii) The lubrication is supplied by the liquid slag infiltrating into the channel between mould and shell; this occurs principally in the period where the mould (and slag rim) are descending and Q_s increases gradually through this period with maximum infiltration corresponding with the lowest position of the mould.
(iii) Changes in mould direction are accompanied by periods of *confused flow* where little slag infiltration occurs.
(iv) The powder consumption, Q_s (in kg m^{-2}) provides a good measure of the lubrication supplied and is related to the thickness of the liquid slag layer in the slag film ($Q_s^{req} = \rho \cdot d_l$) where ρ is the density of the liquid slag.
(v) Several powder consumption terms are used and these terms are interrelated; the melting (Q_{MR}) rate must match the required powder consumption.
(vi) Since mould powders contain carbon and volatile materials it is necessary to distinguish between *powder* and *slag* ($Q_s^{slag} = f^* \, Q_s^{powd}$) where f^* is the mass fraction of the powder forming slag
(vii) Analyses of plant data for powder consumption revealed that Q_s^{req} increases:

- with increasing mould surface area, ($Q_s^{req} = 2/ (R^* -5)$)
- with decreasing casting speed and slag viscosity
- with decreasing oscillation frequency and stroke (although these relations are disputed by some workers); these effects are smaller than those above.
- with increasing Argon flow rate.

(viii) The required values of viscosity, break temperature and Q_s^{req} can be calculated for the given casting conditions using empirical rules
(ix) High frictional forces in the mould arise from (i) fracture of slag films at low casting speeds and (ii) excessive taper and mould misalignment at higher casting speeds
(x) Liquid friction increases with the:

- Increasing surface area of the shell (including any shell corrugations)
- Increasing viscosity, casting speed and ($V_m - V_c$); the plot of F_l versus V_c exhibits a minimum due to the conflicting responses to increasing V_c on ($V_m - V_c$) and F_l with ($V_m - V_c$) dominant at low casting speeds and the V_c effect being dominant at high speeds.
- Increasing oscillation frequency but there is no consensus regarding the effect of stroke length.
- With decreasing non-sinusoidal oscillation.

References

1. P.E. Ramirez-Lopez, K.C. Mills, P.D. Lee, B. Santilanna, Met. Mater. Trans. **43,** 109, (2011).
2. A.B. Fox, PhD Thesis, *Mould fluxes their properties and performance.* Dept. of Materials, Imperial College, London, (2003).
3. Y Nuri, T Ohashi, N Miyasaka, K Shima, Y Uchida, Trans. ISIJ, **20**, B170, (1980).
4. R. Saraswat, A.B. Fox, K.C. Mills, P.D. Lee, B. Deo, Scand. J. Met., **33**, 85, (2004).
5. T. Kajitani, K. Okazawa, W. Yamada, H. Nakamura, ISIJ Intl. **46**, 250, (2006).
6. M.S. Jenkins, PhD Thesis "*Heat transfer in the continuous casting mould*", Monash Univ., Clayton, Vic., Australia, (1999).
7. Y. Itoh, S Nebeshima, K Sorimachi, Proc. 6th Intl. Conf. Molten slags, fluxes and salts, Stockholm, Paper 152 (2000) see also S Nebeshima, Y Itoh, H Tozawa, H Nakato, K Sorimachi, Proc. 4th Intl. Conf. Solidification Processing, Sheffield, 1997 (Sheffield Univ., 2997) p 10.
8. K. Tsutsumi, Tetsu- to Hagane, **84**, 617, (1998).
9. A. Badri, B.T. Natarajan, CC Snyder, K.D. Powers, F.J. Byrne, M. Byrne, A.W. Cramb, Met. Mater. Trans. B, **36B,** 355, (2005).
10. A. Badri, B.T. Natarajan, C.C. Snyder, K.D. Powers, F.J. Byrne, M. Byrne, A.W. Cramb, Met. Mater. Trans B, **36B,** 373, (2005).
11. EY Ko, J Choi, JY Park, I Sohn, Met. Mater. Intl. **20**, 141 and 1103 (2014).
12. E Anzai, T Shigezumi, T Nakano, T Ando, M Ikeda, Nippon Steel Technical Report, **34**, 35, (1987).
13. C. Niggel, F. Felder. *Lubrication by slag of continuous casting of steel.* Report 9339, ECSC, (Europ. Comm. Sci and Tech. Publ., Luxembourg, 1985).
14. J. Kor, "An analysis of the fluid flow of liquid mould flux into space between continuous casting mold and steel shell" US Steel Report.
15. Y. Nuri, T. Ohashi, Trans. ISIJ, **20**, B172, (1980).
16. S. Ogibayashi, Proc. 85th Steelmaking Conf. (2002) (ISS/AIME, Warrendale PA.) p. 175.
17. K. Okazawa,T. Kajitani, W. Yamada, H. Nakamura, ISIJ Intl., **46**, 226 and 234 (2006).
18. K. Schwerdtfeger, K.H. Tacke, *Fundamental study of behaviour of casting powders*; Report EUR 9560, (Europ. Comm.Sci. and Tech Publ., Luxembourg, 1985).
19. YA Meng, BG Thomas, Met. Mater. Trans. B, **34B**, 707, (2003).
20. S. Itoyama, CAMP- ISIJ, **14**, 893, (2001).
21. P.E. Ramirez-Lopez, K.C. Mills, P.D. Lee, B. Santilanna, ISIJ Intl., **30**, 1797, (2010).
22. C. Ojeda, J. Sengupta, B.G. Thomas, J. Barco, J.L. Aruna, Proc. AIST Tech., 2006 vol 1 (ISS, Warrendale, PA) p. 1017.
23. H.J. Shin, S.H. Kim, B.G. Thomas, G.G. Lee, J.M. Park, J. Sengupta, ISIJ Intl., **46**, 1635, (2006).

24. XN Meng, MY Zhu, Can. Metall. Q, **50**, 45, (2011).
25. ASM Jonayat, BG Thomas, Met. Mater. Trans. B, **45B**, 1862, (2014).
26. T.H. Billany, K.C. Mills, *Mould flux performance during continuous casting*. Final Report ECSC Contract 7210 CA 820, 1987, (Europ. Comm. Sci. and Tech. Publ. Luxembourg, 1987).
27. R.J. Gray, *Behaviour of mould fluxes during continuous casting*. Report EUR 9495 EN 1985) (Europ. Comm. Sci. and Tech. Publ., Luxembourg.1985.
28. T. Okazaki et al., Tetsu-to Hagane, **65** (10), 265, (1985).
29. JM Hill, YH Wu, B Witwatanapataphee, J Eng. Math., **36**, 311, (1999).
30. S Ogibayashi, CAMP- ISIJ, **18**, 126 and 127 (2003).
31. T Emi, H Nakato, K Suzuki, Y Iida, T Ueda, Tetsu- to Hagane, **60** (7), 981, (1974) Henry Brutcher Translation HB9357.
32. T Fastner, C Furst, HP Narzt, G Xia, G Zuba, Proc. 3rd Europ. Conf. Continuous Casting Madrid, (UNESID, Madrid,1998) p. 791.
33. Y Fukuda, H Kawai, M Okimori, M Hojo, S Tanaka, Proc. 5th Intl. Conf. Slags, Fluxes and molten salts, Sydney,1997, (ISS, Warrendale, PA, 1997) p. 791.
34. S Ogibayashi, T Mizoguchi, T Kajatani, Intl. Workshop on Thermophys. Data for the Development of Mathematical models of solidification, Gifu City, Japan (1995).
35. BG Thomas, A Moitra, R McDavid, Iron and Steelmaker, **23** (4), 51 (1996).
36. Y. Meng, B.G. Thomas, Proc. ISS Tech., Indianapolis, 2003 (ISS, Warrendale, PA, 2003) p. 589.
37. MS Jenkins Proc. 78th Steelmaking Conf.,1995, (ISS, Warrendale, PA, 1995) p. 669.
38. MS Jenkins, BG Thomas, WC Chen, RB Mahapatra, Proc. 77th Steelmaking Conf., 1994, (ISS, Warrendale, PA 1994) p. 337.
39. JW Kim, S. K. Kim, D. S. Kim, Y. D. Lee, J. I. Eum, E. S. Lee., Proc. 78th Steelmaking Conf. 1995, (ISS, Warrendale, PA,1994) p. 333.
40. K Suzuki, C Matsumura, H Yamamoto, Y Kanrda, Proc. 73rd Steelmaking Conf., 1990, (ISS, Warrendale, PA, 1990) p 197.
41. M. Wolf, AIME Elect. Furn. Proc., **40,** 335, (1982).
42. M. Wolf, Proc. 2nd Europ. Conf. Continuous Casting, Dusseldorf, 1994 (VDEh, Dusseldorf, 1994) vol 1, p 78.
43. S. Ogibayashi, K Yamaguchi, T Mukat, T Takahashi, Y Mimura, K Koyama. Y Nagano, T Nagano. Nippon Steel Technical Report, **34**, 1, (1987).
44. Y. Nakamori, Y Fujikake, K Tokiwa, T Kataoka, S Tsuneoka, H Misumi, Proc. 10th Conf. IMEKO TC3 on Measurement and Mass held Kobe, Japan Sept. (1984) and Tetsu-to Hagane, **70**(8),1282, (1984).
45. S. Sridhar, K.C. Mills, V. Ludlow, S.T. Mallaband, Proc. 3rd Europ. Conf. Continuous Casting, Madrid, 1998, (UNESID, Madrid,1998) p. 807.
46. F Neumann, J Neal, MA Pedroza, AH Castillejos, FA Acosta, Proc. 79th Steelmaking Conf. 1996. (ISS, Warrendale, PA,1996) p. 249.
47. K.C. Mills, S. Sridhar, A.S. Normanton, S.T. Mallaband, Proc. Brimacombe Conf., Vancouver, BC, 2000,p 781.
48. K.C.Mills, A.B. Fox, ISIJ Intl.,**43,** 1479, (2003).
49. H. Maeda, T. Hirose, CAMP-ISIJ, **6**, 280, (1993).
50. K Koyama, K Nagano, Y Nagano, T Nakano, Nippon Steel Technical. Report, **34**, 41, (1987).
51. OD Kwon, J Choi, IR Lee, JW Kim, KH Moon, YK Shin, Proc.74th Steelmaking Conf.,1991, (ISS, Warrendale, PA,1991) p. 561.
52. K. Nakajima, S Hiraki, T Kanazawa, T Murakami, CAMP-ISIJ, **5,** 1221, (1992).
53. K. Tsutsumi, H Murakami, S Nishioka, M Tada, M Nakada, M Komatsu, Tetsu- to- Hagane**,** **84**, 617, (1998).
54. P E Ramirez-Lopez, P.D. Lee, K.C. Mills, ISIJ Intl., **50** (3), 425, (2010).
55. M Kawamoto, T Mizukami, M Hanao, H Kikikuchi, T Watanabe, Ironmaking and Steelmaking**,** **29**,199, (2002).

56. K Watanabe, K Tsutsumi, M Suzuki, H Fujita, S Hatori, T Omoto, ISIJ Intl., **54**, 865, (2014).
57. T. Kitagawa, M. Ishiguro, Proc. 4th Japan-Germany Seminar, (ISIJ, Tokyo, 1980) p. 249.
58. T. Kajitani, K. Okazawa W. Yamada, H. Yamamura, ISIJ Intl., **46**, 250 and 1432 (2006).
59. IR Lee, JW Kim, J Choi, D Kwon, YK Shin, Proc. Conf. on Continuous casting in developing countries, Beijing, 1993, (SEAISI, Singapore, 1993) p. 814.
60. T. Mizukami, K Kawakami, T Kitagawa, M Suzuki, S Uchida, Y Komasu, Tetsu-to Hagane **72**, 1862 (1986).
61. M Suzuki, H Mizukami, T Kitagawa, K Kawakami, S Uchida, Y Komatsu, ISIJ Intl., **31**, 254, (1991).
62. M. Wolf, "*Effects of mould oscillation*" presented Discussion Group on Continuous casting of mould fluxes, Inst. of Metals, London (1984).
63. M. Wolf, Proc. Conf. Continuous casting of steel in developing countries, Beijing, China (1994) p. 69.
64. K. Tsutsumi, T. Nagasaka, M. Hino, ISIJ Intl., **39**, 1150, (1999).
65. T. Emi, H Nakato, K Suzuki, Y Iida, Proc. NOH- BOS Conf. (1978) p. 350.
66. H. Nakato, I. Muchi, Tetsu-to- Hagane, **66**, 33, (1980).
67. H. Nakato T Sakuraya, T Nozaki, T Emi, H Nikoshawa, *Mould fluxes for continuous casting and bottom pour teeming* (ISS, Warrendale, PA, 1987) p. 23.
68. K. Noguchi, K. Sawamura, Proc. 4th Intl. Conf. Cont. Casting, Brussels (1988) (CRM/ VDEh) p. 65.
69. Y. Kobayashi, S. Maruhashi," Effects of operational on oscillation mark of continuously cast, stainless steel slabs" Proc. 4th Japan-CSSR Seminar, Ostrava. (1983) p. 249.
70. S. Shimizu, Y. Imada et al., Proc. 6th Intl. Iron and Steel Congress (1990) p. 487.
71. M. Suzuki, S Miyahara, T Kitagawa, S Uchida, K Okimoto, Tetsu-to Hagane, **78**, 113, (1992).
72. M Ikeda, K Asano, T Nakano, M Fuji, S Mizoguchi, H Mizumi, Trans. ISIJ, **21,** B 511, (1981).
73. T. Mallaband, Metallugica. UK, private communication cited in AB Fox thesis [1].
74. T. Mukongo, C Pistorius, A Garbers-Craig, Ironmaking Steelmaking, **31**,135, (2004).
75. Q. Wang, Y. Lu, S. He, K.C. Mills, Z.S. Li, Ironmaking and Steelmaking, **38**, 297, (2011).
76. H Lei, Y Zhao, DQ Geng, ISIJ Intl., **54**, 1629, (2014).
77. T Kishi, H Takeuchi, M Yamamiya, H Tsuboi, T Nakano, T Ando, Nippon Steel Tech. Report, **34**, 11, (1987).
78. M Hanao, Y Tsukaguchi, M Kawamoto, Proc. 4th Intl. Congress Science and Technol., 2008, Gifu, Japan (ISI J, Tokyo, 2008), p. 694.
79. R. Koldwein, Unpublished Corus Internal Rept (2007) cited in KC Mills, J Kromhout, A Hamoen, R Boom: Proc. Admet Conf, Dnipropetrovsk, 2007(Natl. Metall. Acad. Ukr., Dnipropetrovsk, 2007) vol 2 p 174.
80. M Washio, M Sugizawa, S Moriwaki, K Kariyaa, S Idogawa, S Takeuchi, Revue de Metallurgie, CIT, **90** (April), 507, (1993).
81. D W van der Plas, C Platvoet, B Diesesme, JP Radot, JM Galpin, Proc. 2nd Europ. Conf. Continuous casting, Dusseldorf, 1994, Metec Congress'94 (VDEh, Dusseldorf, 1994) p. 109.
82. MY Ha, SG Lee, SH Seong, J. Mater. Processing Technol. **133**. 322, (2003).
83. G Bocher, U Hoffman, P Muller, Proc. 2nd Europ. Conf. Continuous casting, Dusseldorf, 1994, Metec Congress'94 (VDEh, Dusseldorf, 1994)p. 102.
84. J Kromhout, RS Schimmel, Proc. 8th Europ. Conf. Continuous casting, Graz, Austria, 2014 (Austrian Met Mater. Soc., Vienna, 2014).
85. M Tani, T Toh, K Umetsu, K Tanaka, M Zeze, K Tsunenari, K Hayashi, S Fukunaga, Nippon Steel Technical Report, **104**, 62, (2013).
86. JK Park, JW Cho, KH Moon, SH Lee, KH Kim, HS Jeong, Proc. 7th Intl. Conf. Clean Steel, Balatonfured, Hungary, 2007, (Hung. Min. Metall. Soc, Budapest, 2007) p. 264.
87. M. Wolf, Proc. 2nd Europ. Conf. Continuous Casting, METEC Congress'94 held Dusseldorf, 1994 VDEh, Dusseldorf, 1994), vol 1, p 78.

88. H Uchiyama, Proc. AISI Technical Committee on Strand casting, 1995 p.
89. J Sardemann, H Screwe, Stahl u Eisen, **111**, (11), 39, (1991).
90. H Yamamura, T Kajitani, J Nakashima, M Yamasaki, S Mineta, Nippon Steel Technical Report, **104,** 54, (2013).
91. K Tsutsumi, J Ohtake, M Hino, ISIJ Intl., **40**, 601, (2000).
92. H Steinruch, C Rudischer, W Schneider, Non-linear Analysis, Theory, Methods and Applications, **30** (8), 4915, (1997) see also BHM 141 (1996)(9) 399.
93. H Steinruch, C Rudischer, W Schneider Proc. Conf. Modelling of Casting Welding and Advanced Solidification processes VIII (MCWASP)(Minerals, Metals and Materials Soc. 1998).
94. PP Sahoo, S Basu, ISIJ Intl., **46**, 219, (2006).
95. M Wolf, Trans ISIJ, **22**, B204, (1982).
96. TJ Billany AS Normanton, KC Mills, P Grieveson, Ironmaking and Steelmaking **18**, 403, (1991).
97. K. Sorimachi, Proc. 5th Intl. Conf. Molten slags, fluxes and salts, Sydney,1997, (ISS, Warrendale, PA, 1997) p. 781.
98. PV Riboud, Y Roux, *Fundamental study of the behaviour of casting powders*. Report EUR 9560,1985 (Eur. Comm. Sci and Tech. Publ., Luxembourg, 1985).
99. R.D.'Haeyer, *Influence of chemical composition of continuous casting powders* Report EUR 10326 EN (1987) (Eur. Comm. Sci and Tech. Publ., Luxembourg, 1987).
100. Y Nakamori et al, Nippon Steel Tech. Report, **34**, 53, (1987).
101. B Mairy, D Ramelot, M Dutrieux, Proc, Technol. Conf., Measurement and Control Instrumentation in the Iron and steel Industry, Detroit, 1985 (ISS, Warrendale, 1985) p. 101.
102. G. Saucedo et al, Proc.74th Steelmaking Conf. (1991) (ISS, Warendale, PA, 1991) p 79.
103. D. Bowen: Proc. Seminar on Mould powders for continuous casting, held British Steel Teesside Laboratories, Sept (1989) Paper 8.
104. L. Hering, HP Heller, HW Fenske., Stahl u Eisen, **17,** 61, (1992).
105. S. Ogibayashi et al, Proc.78th Steelmaking Conf., Nashville, TN, 1995, (ISS, Warrendale, PA,1995) p. 451.
106. H Mizukami, M Komatsu, T Kitagawa, K Kawakami, Trans ISIJ, **24**, B 181, (1984).
107. H Mizukami, K Kawakami, S Miyahara, M Suzuki, T Kitagawa, O Terada, Trans. ISIJ, **25**, B 300, (1985).
108. H Mizukami, A Ozeki, A Kurabayashi, N Hsebe, S Uchida, T Kitagawa, Trans ISIJ, **25**, B 301, (1985).
109. T Sohlgren, private communication, Sweden, 2015.

Chapter 3
Heat Transfer in the Mould and Shell Solidification

Abstract The condition of the shell is paramount in continuous casting. The heat transfer from the shell is important because it determines how thick the shell is and, consequently, how strong the shell is. Four aspects of the heat transfer are considered here, namely, (i) horizontal heat transfer, (ii) shell solidification, (iii) vertical heat transfer and (iv) the variability in heat transfer. The *horizontal heat transfer* occurs across the slag film separating the shell from the mould. Heat is transferred by two mechanisms, *lattice conduction* and *radiation conduction*. The latter is usually controlled by (i) manipulation of the amounts of glassy (f_{gl}) and crystalline phases (f_{crys}) in the slag film and (ii) to a lesser extent, by incorporating transition metal oxides into the mould powder to absorb the IR radiation. The key properties of the slag film are (i) the *thickness of solid slag film* (which is dependent on the solidification or break temperature) and (ii) the *fraction of crystalline phase* (f_{crys}) formed in the film (which (a) reflects the IR radiation and (b) creates an interfacial resistance, ($R_{Cu/sl}$) during crystallisation). In practice, all of these factors increase with increasing basicity (*C/S*) of the mould slag. Other factors like the effect of casting conditions (e.g. casting speed, metal flow pattern) on the heat flux (q_{hor}) are discussed. The factors affecting *shell thickness* (d_{shell}) are discussed below. There are two regimes controlling shell growth (i) a period of slower growth for the initial period $t \leq 0.05$ s and (ii) a subsequent period of linear growth of d_{shell} which exhibits a linear relation with $t^{0.5}$. *Vertical heat transfer* is controlled through (i) the *depth of the bed,* (ii) *use of exothermic mould powders,* (iii) to a less amount, by the *particle size of the powder* and (iv) by *use of electromagnetic braking (EMBr)* in the mould which reduces the efficiency of vertical heat transfer. Gaseous convection is shown to be a major contributor to the vertical heat flux (q_{vert}) and the permeability of the powder bed is a key factor affecting q_{vert}. Local *variations in heat flux* are an issue because an uneven shell can lead to longitudinal cracking. The various causes of local variations in shell thickness are discussed.

The original version of this chapter was revised: For detailed information please see correction. The correction to this chapter is available at https://doi.org/10.1007/978-3-319-53616-3_12

Within this book minor changes to some of the figure permissions and the readability of the chapters have been made and several figures have been replaced with new versions.

K.C. Mills and C.-Å. Däcker, *The Casting Powders Book*,
DOI 10.1007/978-3-319-53616-3_3

Symbols, Abbreviations and units

a	Thermal diffusivity ($m^2\ s^{-1}$)
d_{bed}	Depth of bed (m)
d_l	Thickness of liquid slag film (m)
d_{menis}	Distance from meniscus (m)
d_s	Thickness of solid slag film (m)
d_{shell}	Thickness of steel shell (mm)
k	Thermal conductivity ($W\ m^{-1}\ K^{-1}$)
n	Refractive index
q	Heat flux density ($W\ m^{-2} = J\ s^{-1}\ m^{-2}$)
q_{hor}	Horizontal heat flux ($W\ m^{-2}$)
q_{vert}	Vertical heat flux ($W\ m^{-2}$)
q_R	Heat flux by radiation conduction
q_{total}	Integral (total) heat flux (J)
R	Thermal resistance ($m^2\ K\ W^{-1}$)
$R_{Cu/sl}$	Interfacial thermal resistance
T	Temperature (°C)
T_{br}	Break temperature (°C)
t	Time (s)
t_{mould}	Dwell (residence) time (s)
V_c	Casting speed ($m\ min^{-1}$)
α^*	Absorption coefficient (m^{-1})
η	Viscosity (d Pas)
HC	High-carbon steel
Hi Al	High-Al steel
IR	Infrared
LC	Low-carbon steel
MC	Medium-carbon steel
SEN	Submerged entry nozzle

Subscripts, superscripts

Cu/sl	Interface between Cu and slag
film	Slag film
l	Liquid
s	Solid

3.1 Introduction

The various aspects of heat transfer and solidification are discussed in this chapter. The control of the heat extraction from the newly formed shell is very important in the continuous casting process. Its importance stems from the effect it has on the shell. It is essential when casting peritectic, steel grades to reduce the horizontal heat flux to a satisfactory level to create a thin, uniform shell, in order to avoid longitudinal cracking. However, to avoid sticker breakouts in HC steels, where the shell is weak, it is necessary to increase the horizontal heat flux to create a thicker, stronger shell. Thus, control of the heat extraction from the shell is essential in obtaining good surface quality and in minimising process problems. It will be shown below that this is principally achieved by controlling the nature of the slag film formed between the steel shell and the mould.

The continual flow of molten steel into the water-cooled mould provides the energy required for the process. For solidification to occur, heat must be lost. It can be seen from the meniscus region of Fig. 3.1 that the solidified shell near the meniscus has a curved shape, thus heat losses follow a radial path [1, 2]. However, for simplicity and convenience, heat losses are usually divided into two contributions, namely, (i) *horizontal heat flux* (through the solid and liquid slag films to the mould) and (ii) the *vertical heat flux* (through the slag pool and powder bed to the atmosphere). The horizontal and vertical heat fluxes are related to one another because they have a common source of heat (i.e. the flow of molten steel).

The importance of the horizontal heat transfer is due to the fact that it determines the thickness of the solidified shell formed (especially in the meniscus region). The

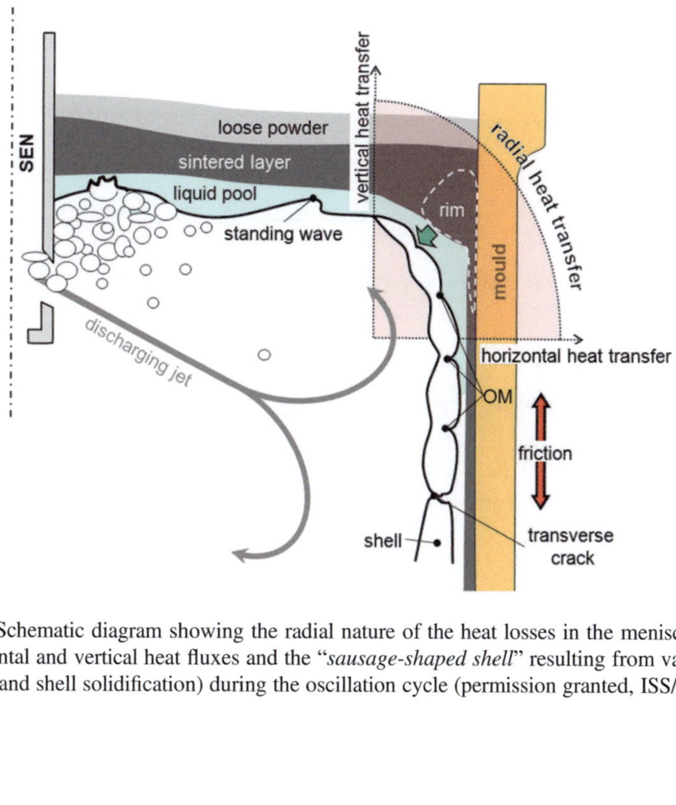

Fig. 3.1 Schematic diagram showing the radial nature of the heat losses in the meniscus region, the horizontal and vertical heat fluxes and the "*sausage-shaped shell*" resulting from variations in heat flux (and shell solidification) during the oscillation cycle (permission granted, ISS/AIST, [2])

thickness of the shell (d_{shell}) can be calculated approximately by Eq. 3.1 where d is the distance below the meniscus and K is a solidification constant; thus d_{shell} decreases with increasing casting speed (V_c) [3]:

$$d_{shell} = K(d/V_c)^{0.5}. \quad (3.1)$$

3.1.1 Heat Flux

The magnitude of heat transfer is usually expressed in terms of the heat flux (q). For *steady-state* conditions (where the temperature gradient (dT/dx) is constant), the heat flux between two points (in the x direction) is usually given by Fourier's first law (Eq. 3.2) where k is the thermal conductivity and A is the area:

$$q = -kA(dT/dx). \quad (3.2)$$

Heat flow is often considered in terms of electrical analogues and the heat flux can be considered to be the equivalent of current (and (dT/dx) as potential).

For *transient conditions* [where the temperature gradient is changing with time (t)], the rate of heat extracted (or supplied) $\left(\delta^2 T/\delta x^2\right)$ is usually given by Fourier's second law shown in Eqs. 3.3–3.5 for 1-, 2- and 3-dimensions, respectively, where a is the thermal diffusivity.

1-dim:

$$(\delta^2 T/\delta x^2) = (1/a)(dT/dt) \quad (3.3)$$

2-dim:

$$(\delta^2 T/\delta x^2) + (\delta^2 T/\delta y^2) = (1/a)(dT/dt) \quad (3.4)$$

3-dim:

$$(\delta^2 T/\delta x^2) + (\delta^2 T/\delta y^2) + (\delta^2 T/\delta z^2) = (1/a)(dT/dt). \quad (3.5)$$

The total amount of heat supplied is usually calculated from Eq. 3.6 where m, C_p, V, ρ and $(T_2 - T_1)$ are mass, heat capacity, volume, density and temperature rise, respectively:

$$q_{total} = m \cdot C_p(T_2 - T_1) = \rho V C_p(T_2 - T_1). \quad (3.6)$$

3.2 Horizontal Heat Flux

The horizontal heat flux is generally regarded as being more important than the vertical heat flux since it determines the thickness of the steel shell formed in the meniscus region. There are two issues here (which affect the surface quality of the steel product) which are linked but are treated separately here, namely,

(i) The *magnitude* of the horizontal heat flux (described in Sects. 3.2 and 3.3) and
(ii) The *variability* of the horizontal heat flux (described in Sect. 3.4).

3.2.1 Heat Transfer Mechanisms Involved in Horizontal Heat Transfer

The horizontal heat transfer refers to the transport of heat from the shell to the water-cooled mould. Initially, oil was used to prevent the shell from sticking; the oil evaporated to form a gaseous film which formed a thermal barrier to heat flow [4]. The replacement of oil by mould powders was found to improve the control of heat extraction from the shell. This control is provided by the slag film formed between the shell and the mould. The slag film consists of a mixture of crystalline and glassy phases and the control of the horizontal heat flux is obtained from manipulation of (i) the ratio of glass and crystalline phases and (ii) the thickness of the slag film.

The mechanism of heat transfer between the shell and the mould is complex since it involves at least three heat transfer mechanisms, namely (i) *lattice conduction* through the slag film, (ii) *radiation conduction* and (iii) *convection* in the liquid layer of the slag film.

The solid slag film formed in the first moments of casting is predominantly glassy because of the high cooling rates involved when it solidifies in the mould/strand gap. Heat transfer can occur by two mechanisms in solid glasses, i.e. (i) normal, *phonon or lattice conduction* and (ii) *radiation conduction.* The latter mechanism was first detected when measured thermal conductivities of glasses were observed to increase with increasing sample thickness until a critical point was reached [5] (Fig. 3.2a). The sample is denoted as "*optically thick*" at this critical point, beyond which the measured thermal conductivity remained constant (Fig. 3.2a) [5].

Consider a glass sample consisting of a number of layers. Radiated heat falls on the surface layer of the sample and this causes the temperature of layer 1 to increase. This layer now is at a higher temperature than layer-2 and so layer-1 emits radiation to layer-2; the radiant energy is absorbed and thus the temperature of layer-2 rises (Fig. 3.2b). The temperature of layer-2 is now higher than that for layer-3 so layer-2 radiates to layer-3 and so on. Thus, radiation conduction is a process of absorption and re-emission (Fig. 3.2b).

A sample is usually considered to be *optically thick* when $\alpha^*d > 3$ where α^* is the absorption coefficient and d is the thickness of the sample. For *optically thick*

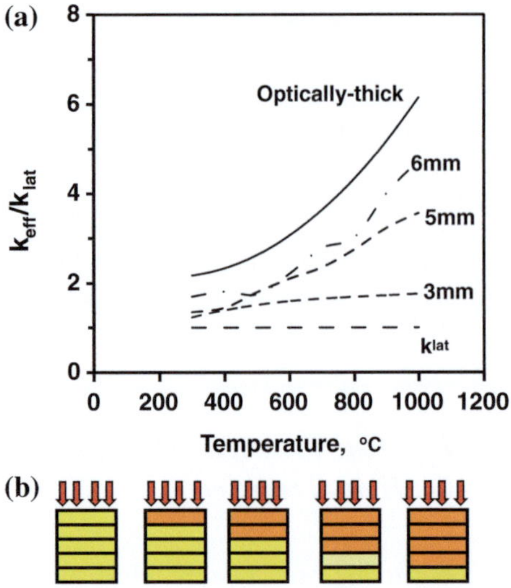

Fig. 3.2 Schematic drawings showing **a** thermal conductivity ratio(k_{eff}/k_{lat}) as a function of temperature showing both the exponential effect of temperature and sample thickness on magnitude of radiation conduction; k_{lat}, lattice conductivity is for thin sample, drawn from data due to [5] and **b** schematic drawing showing the mechanism of radiation conduction)

samples ($\alpha^*d > 3$) the radiation conductivity (k_R) can be calculated from Eq. 3.7 where σ = Stefan–Boltzmann constant, and n = refractive index which usually has a value of ca. 1.58 for mould slags [6, 7].

$$k_R = 16\,\sigma n^2 T^3/3\alpha^* \tag{3.7}$$

Since k_R is a function of T^3, k_R tends to increase sharply with increasing temperature (Fig. 3.2a). However, the values of k_R are reduced by

(i) The presence of transition metal oxides (e.g. FeO, NiO, CrO_x) which absorb radiation [6, 8] and
(ii) Crystalline phases (e.g. *cuspidine*) which scatter or reflect the IR radiation (Fig. 3.3) at the interface of the slag film with the liquid [8].

Usually, the heat transfer through the slag film is controlled by the reflection mechanism but the transition metal oxides have been used to control the heat flux in the casting of round billets [10]. However, additions of FeO, MnO, etc. tend to

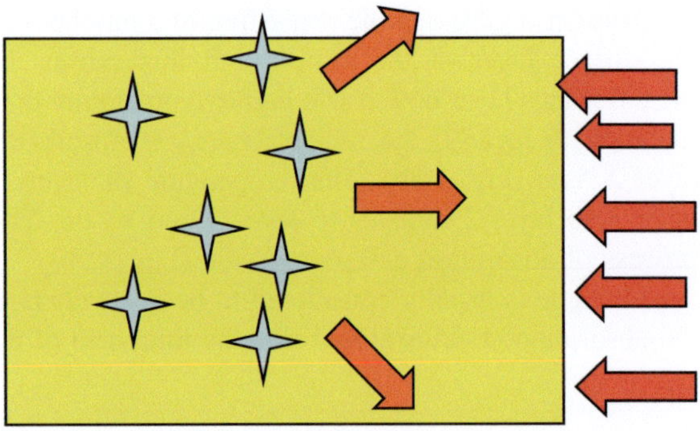

Fig. 3.3 Schematic diagram showing the scattering of IR radiation by crystallites

increase the amount of glassy phase and thereby reduce the reflection of IR radiation [11–13]. The scattering and reflection have been reported to increase with increasing crystal size [12].

If the sample is glassy, it is possible to calculate the magnitude of k_R from Eq. 3.7, providing the value of the absorption coefficient of the slag is known and the sample is optically thick. However, it is much more difficult to calculate the effect of crystalline phases on k_R. Some workers have replaced the absorption coefficient (α^*) in Eq. 3.7 by the extinction coefficient ($E = \alpha^* + s$), where s is the scattering coefficient. However, the process of radiation scattering is different to that of radiation absorption. A parameter known as the *albedo (or reflection coefficient)* is needed for the calculation and there are no values for the *albedo* of casting slags [7]. It has been shown that for slag films containing crystalline phases that radiation is increasingly reflected (i.e. k_R decreases) with (i) increasing crystalline fraction, (ii) increasing grain size of the crystals, (iii) increasing film thickness and (iv) decreasing transition metal oxide (e.g. FeO_x) content (this is a consequence of their tendency to increase f_{glass}) [11, 12, 14, 16]. Susa et al. [12] calculated that the heat flux arising from radiation conduction, $q_R = 0.29$ M W m^{-2} for a 1-mm-thick slag film with mean crystal grain size of 2.5 μm. This compares with a heat flux via conduction, q_c = of 1.45 M W m^{-2}, i.e. 20% of the heat was transferred by radiation conduction. Similar values for the relative contribution of q_R (= q_R/q_{total}) have been obtained by other workers (see [14, 16, 17]) for crystalline slag films, e.g. <10% (5–19%) [15] with (q_R/q_{tot}) values decreasing with increasing fraction of crystalline phase (f_{crys}) in the slag film. Hanao et al. [17] cite results for (q_R) which indicate they constitute ca. 20% of q_{tot} for crystalline slag films but ca. 60% of that for purely glassy films. Reported values of (q_R/q_{tot}) for the liquid film are much larger, e.g. 37–50% [14] but the liquid layer is probably optically thin.

The thermal conductivity (k) of slags in the crystalline state is about twice that of the glassy phase (i.e. $k_{crys} = 2\, k_{gl}$) [18–20]. Despite this, the most common method for reducing the horizontal heat flux is to increase the crystallinity of the slag film (to reduce q_R); this underlines the importance of crystallinity (in the slag film) in reducing the magnitude of the radiation conductivity.

3.2.2 *Interfacial Thermal Resistance* ($\mathbf{R}_{Cu/Sl}$)

An interfacial thermal resistance ($R_{Cu/sl}$) is created by the crystallisation of the glassy solid in the slag film (formed by the infiltration of molten slag between the shell and the copper mould). Some workers have suggested that this interfacial resistance arises from the contraction of the shell on solidification. This is not correct since any gap formed by shrinkage of the shell will automatically be filled with molten slag, except in exceptional circumstances (e.g. overcooled corners and where there is insufficient, liquid slag flow, e.g. in the lower mould) [21, 22]. Crystalline phases have higher densities than those of the glass and thus crystallisation results in shrinkage. This shrinkage causes (i) the creation of pores near the crystallites (Figs. 3.4b and 3.5) and (ii) some *wrinkling* of the slag surface (often referred to as "*surface roughness*") resulting in a "*gas gap*" at the interface between the slag and the copper mould (Fig. 3.4b). The magnitude of the interfacial resistance ($R_{Cu/sl}$) is related to the width of this gas gap ("*surface roughness*") as can be seen from the trend shown in Fig. 3.6. Since the interfacial resistance $R_{Cu/sl}$ is

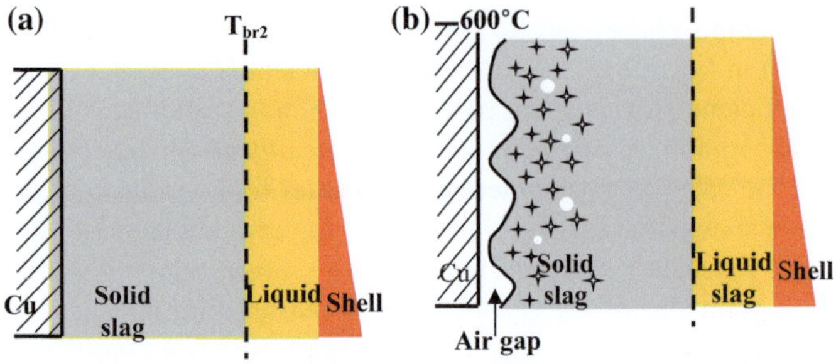

Fig. 3.4 Schematic diagrams showing **a** glassy slag film **b** partially crystallised slag film showing pores and the formation of an "*air gap*" or "*surface roughness*"

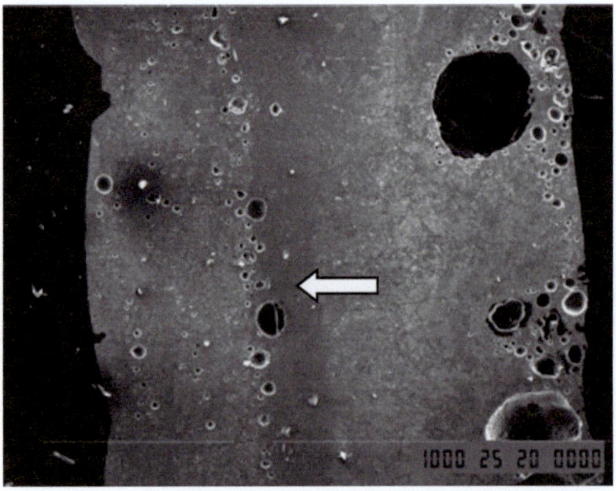

Fig. 3.5 Photograph of pores (shown by the *arrow*) formed around crystalline phase in a slag film taken from the mould and the surface roughness on the mould (*left*) interface (courtesy of Carboox)

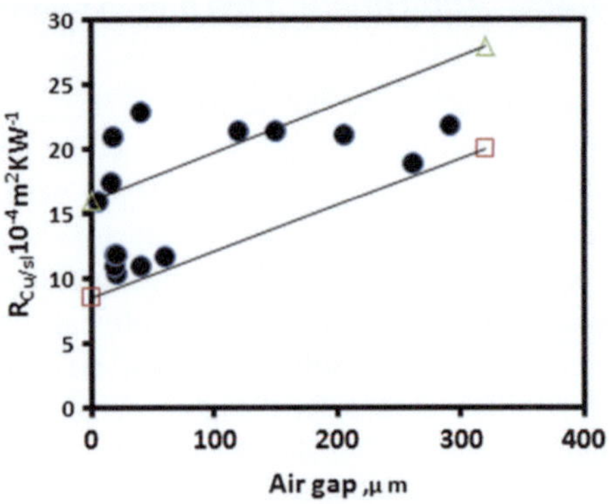

Fig. 3.6 The interfacial resistance ($R_{Cu/sl}$) as a function of the thickness of the air gap; ● = measurements; −−Δ, □ = trendlines [27]; (permission granted, ISIJ, [27], re-drawn)

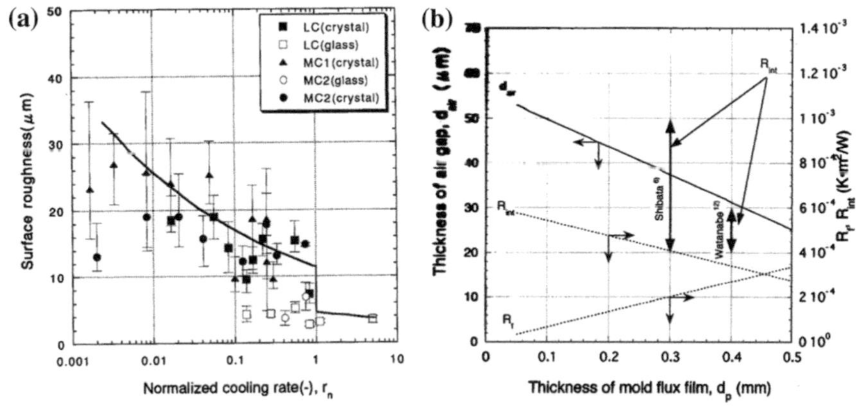

Fig. 3.7 Schematic diagrams showing **a** surface roughness as a function of the normalised cooling rate (= actual cooling rate/critical cooling rate); symbols: *open* = glass; *closed* = crystalline; ■ = flux for LC steels (*C/S* ≈ 1) ▲, ● = fluxes for MC steels (C/S ≈ 1.45) and **b** gas gap width (d_{air}) and the thermal resistances of the mould slag (R_f) and interface ($R_{int} = R_{Cu/sl}$) as a function of slag film thickness [27]; Note: Slag films taken from the mould are usually about 2 mm thick (permission granted, ISIJ, [27])

related to the shrinkage, $R_{Cu/sl}$ increases with increasing proportion of crystalline phase (f_{crys}) and increasing slag film thickness (d_s) [16, 23–26]. Thus, d_s increases with increasing break temperature (T_{br} or T_{sol}). The interfacial resistance ($R_{Cu/sl}$) also tends to increase with increasing T_{br} (or T_{sol}) [3, 23, 27–29].

Tsutsumi et al. [27] used an infrared image furnace to obtain cooling rates similar to those encountered in the mould. They also used a confocal microscope to measure the surface roughness of the quenched sample. It was reported [27] that the surface roughness increased as:

(i) The fraction of crystalline phase increased (Fig. 3.7),
(ii) The cooling rate decreased (Fig. 3.7) and;
(iii) The basicity (C/S) of the mould flux increased (Fig. 3.7).

Tsutsumi et al. [27] also calculated the width of gas gap (d_{air}) from reported values of $R_{Cu/sl}$ and reported the surprising result that d_{air} decreased as the thickness of solid slag film increased ($d_{air} \downarrow$ as $d_s \uparrow$) which is contradictory to the findings of Cho et al. [23] and others [30–33]. However, the reported values of d_{air} were found to be similar to those recorded for the surface roughness (ca. 30 μm).

There has been some discussion on the relative importance of the two mechanisms responsible for reducing heat flux in the slag film, *i.e.* interfacial resistance ($R_{Cu/sl}$) and the reflection by crystals in the slag film [3, 17, 28, 34–36]. However, Hanao and Kawamoto [3, 17] carried out elegant, pilot plant measurements of the heat flux and extracted the slag film from the mould. A value for the interfacial resistance was obtained by analysis of these heat flux measurements. It was found that the interfacial resistance value was smaller than values reported by other workers using simulation experiments (Table 3.1). It was concluded that the ferrostatic pressure on the shell effectively reduced the thickness of the "*gas gap*" and

Table 3.1 Values of interfacial resistance, $R_{Cu/sl}$ (10^{-3} m^2 K W^{-1}) reported by various investigators [3, 17, 27, 31, 33, 35, 37–40]; values in parenthesis = d_s (mm). *indicates liquid phase value

Reference	Watanabe [31]	Inoue [33]	Yamauchi [37]	Kawamoto [38]	Nakato [39]	Cho [23, 28]	Stone [40]	Yamamura [35]	Hanao [3, 17]
$R_{Cu/sl}$	0.4–0.6	0.8	0.7	0.7	0.5	0.6–1	1.5	0.2*–0.7	0.4
d_s	(0.2)	(0.25)	(0.3)	(0.1)	(0.1)	(0.7)	(1.0)	(1.0)	(1.0)

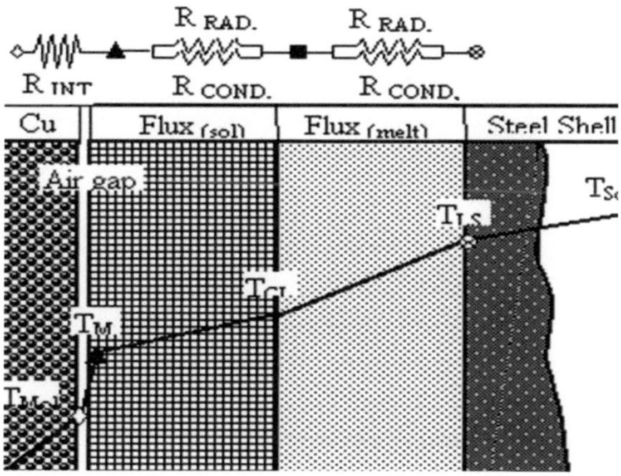

Fig. 3.8 Schematic representation of the thermal resistance of the various layers between the shell and the mould (permission granted, ISIJ, [14])

thereby reduced the $R_{Cu/sl}$ value. Pilot plant trials reflect continuous casting conditions more closely than simulation experiments, so preference is given to the pilot plant data.

Susa et al. [11–13] considered that much of the reduction in horizontal heat flux came from the reflection and scattering of the IR radiation back towards the shell. They reported that the reflection increased as the (i) fraction of crystalline phase increased and (ii) grain size of the crystallites increased (i.e. $q_{hor} \downarrow$ as $D_{crys} \uparrow$).

It is possible to represent the thermal resistance (R_{total}) between shell and mould as a series of resistances (i.e. using Ohm's Law) where R is the thermal resistance and $R_{Cu/sl}$ is the interfacial resistance of the "*gas gap*", d is the thickness of the slag film layer, and subscripts Cu, sl, gl and crys represent the mould, slag, glass and crystalline layers, respectively (Fig. 3.8) [14, 41]. The terms $(d/k)_{shell}$ and the contact resistance at the shell (R_{shell}) are negligible and $(d/k)_{liq}$ is small because d is small (ca. 0.1 mm) and the effective thermal conductivity is high:

$$R_{total} = (d/k)_{shell} + (d/k)_{liq} + (d/k)_{gl} + (d/k)_{cryst} + R_{Cu/sl} + R_{shell}. \tag{3.8}$$

Radiation conduction can be taken into account using a parallel circuit in the Ohm's law circuit diagram [14, 15, 41].

In practice, the glass and crystalline phases are partially intermixed. Consequently, the terms $(d/k)_{gl}$ and $(d/k)_{cryst}$ can be replaced by $((1 - f_{crys}) \cdot d_{solid}/k_{gl})$ and $(f_{crys} \cdot d_{solid}/k_{crys})$, respectively, where f_{crys} = fraction of crystalline phase present in the slag film. Thus Eq. 3.8 can be rewritten

$$R_{total} = (d/k)_{liq} + (1 - f_{crys}) \cdot (d_{solid}/k_{gl}) + (f_{crys} \cdot d_{solid}/k_{crys}) + R_{Cu/sl}. \tag{3.9}$$

In an analysis of heat transfer measurements during castings on pilot plant, R_{total} was considered to be composed of two terms $(R_{total} = R_{Cu/sl} + R_{film})$, where

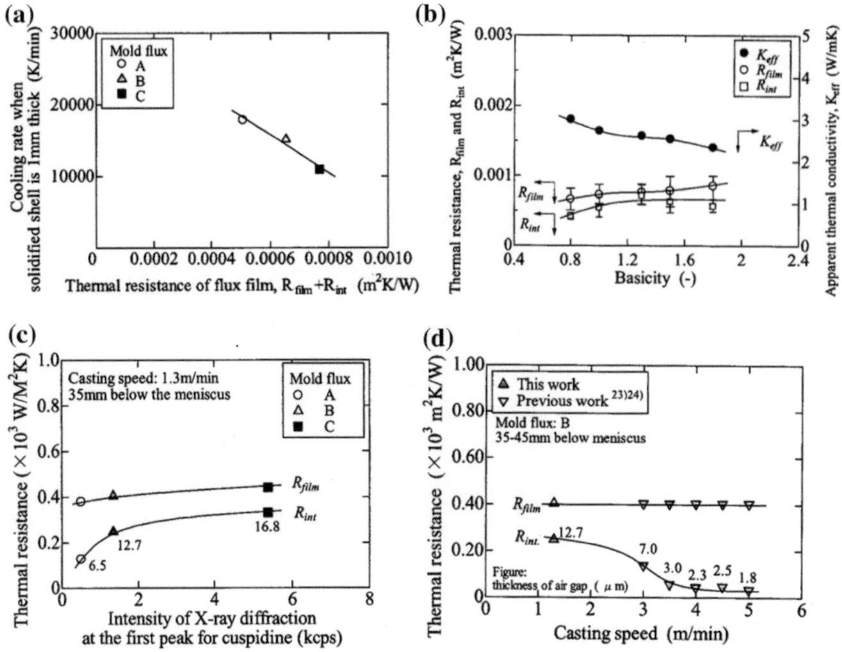

Fig. 3.9 Thermal resistance R_{total} as a function of **a** cooling rate, **b** R_{film}, $R_{\text{Cu/sl}}$ and k_{eff} as functions of basicity (*C/S*), **c** R_{film} and $R_{\text{Cu/sl}}$ as functions of f_{crys} (represented by the intensity of the *cuspidine* diffraction peak), **d** R_{film} and $R_{\text{Cu/sl}}$ as functions of casting speed [17, 42]. *Note* values cited in (**c, d**) refer to thickness of air gap (µm) (permission granted, ISIJ, [17, 42])

$(R_{\text{film}} = (d_s/k) + R_R)$ and R_R is the thermal resistance associated with the scattering and reflection of IR radiation [17]. In this study [17, 42] it was found that

(i) R_{total} increased as the cooling rate increased (Fig. 3.9a);
(ii) R_{film} and $R_{\text{Cu/sl}}$ both increased with increasing basicity (C/S) (Fig. 3.9b); R_{film} was found to increase only slightly with increasing f_{crys} whereas $R_{\text{Cu/sl}}$ increased with increasing f_{crys} (Fig. 3.9c);
(iii) R_{film} was found to be essentially independent of casting speed whereas $R_{\text{Cu/sl}}$ decreased with increasing casting speed (Fig. 3.9d); and
(iv) R_{film} was greater in magnitude than $R_{\text{Cu/sl}}$ (Fig. 3.9d).

In summary, the interfacial resistance can be controlled by (i) the amount of crystalline phase formed (f_{crys}) and (ii) the slag film thickness (d_s which is dependent upon T_{br}). In conventional (F-containing) mould slags these parameters are determined largely by the basicity (C/S) and the F content of the mould slag (Fig. 3.8b, c). Thus, increasing basicity (and hence f_{crys}) results in both increased scattering of IR radiation and increasing $R_{\text{Cu/sl}}$.

3.2.3 *Factors Affecting the Horizontal Heat Flux*

There are a significant number of factors affecting the horizontal heat flux. These are described below in order of grouping rather than in order of importance.

3.2.3.1 Mould Position

The magnitude of the heat flux varies with position, in both the horizontal and vertical directions. The cooling is greater in the corners than in the centre of the mould; a typical horizontal profile of heat flux obtained during high-speed casting is shown in Fig. 3.10a [30]. The heat flux density is not uniform around the mould; q_{hor} is greatest in the regions near the narrow faces and decreases gradually to the region around the centre of the wide face. The increased shrinkage in the corner regions results in an increase in slag film thickness (d_s) and, subsequently, causes a decrease in q_{hor} at the narrow faces. The mould flux ratio [43], defined in Eq. 3.10, is used to measure the uniformity of the heat flux in the mould:

$$\text{Heat Flux Ratio} = \left(q_{hor}^{narrow}\right)/\text{average}\left(q_{hor}^{wide}\right). \tag{3.10}$$

The taper must compensate for the shrinkage, which, in turn, is related to the heat flux (see Sect. 3.2.3.5 (iii) below). Thus, the required taper changes according to the mould slag used. An increase in taper from 1.1 to 1.4% was observed to increase the heat flux ratio from 0.6 to 0.8 [44].

In the vertical direction, the heat flux is at a maximum at the point where the metal flow impacts with the shell; this occurs around 45 mm below the meniscus [42]. Heat fluxes and mould temperatures decrease as the distance from the meniscus (d_{menis}) increases; typical results are shown in Fig. 3.10b [42].

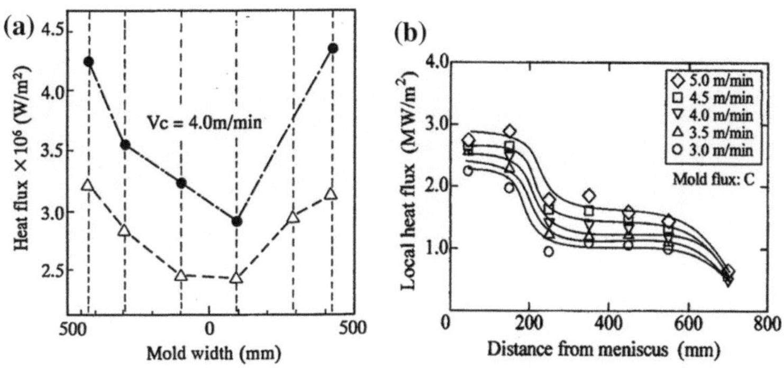

Fig. 3.10 Local heat flux in the mould as **a** in the horizontal direction (wide face) ● = MCsteel with cracks; Δ = LCAK steel, and **b** in the vertical direction, for different casting speeds (permission granted, (**a**) JSPS and M Kawamoto [30] (**b**) ISIJ, [42])

3.2.3.2 Shell Conditions

Oscillation marks are formed in the early stages of shell solidification [2]. The oscillation marks are filled, or partially filled, with slag and this reduces the heat flux in the deeper parts of the mark. Similar effects are to be expected when transverse depressions are formed. Thomas et al. [45] studied the effects of both oscillation marks and transverse depressions on the horizontal heat flux.

3.2.3.3 Casting Conditions

(i) ***Casting speed*** (V_c)

The effect of casting speed on the horizontal heat flux has been studied by several investigators [30, 46–51]. The effect of casting speed (V_c) can be quite confusing since an increase in V_c results in

(i) An *increase* in the *rate of heat flow* (q_{hor}, with units of W m^{-2} = J m^{-2} s^{-1}, sometimes referred to as heat flux density) as can be seen from Fig. 3.11a [46] and
(ii) A *decrease* in *total heat flow* ($q_{total} = A \cdot q_{hor} \cdot L_{mould}/V_c$ with units of J) shown in Fig. 3.11b [47]; this is a consequence of the shorter residence time of the steel in the mould and results in a decrease in q_{total} [43].

The effect of increased casting speed is best viewed through its effect on the thickness of the shell $\left(d_{shell} = K(d_{menis}/V_c)^{0.5}\right)$, where K is a constant and d_{menis} is the distance to the meniscus. Thus, an increase in casting speed results in a thinner shell ($d_{shell} \downarrow$ as $V_c \uparrow$), i.e. since ($d_{shell} \downarrow$ as $q_{total} \downarrow$) it follows that ($q_{total} \downarrow$ as $V_c \uparrow$) and thus the *total heat flux* is the key factor affecting shell thickness. Since shell thickness is affected by residence time in the mould (t_{mould}) it follows that the length

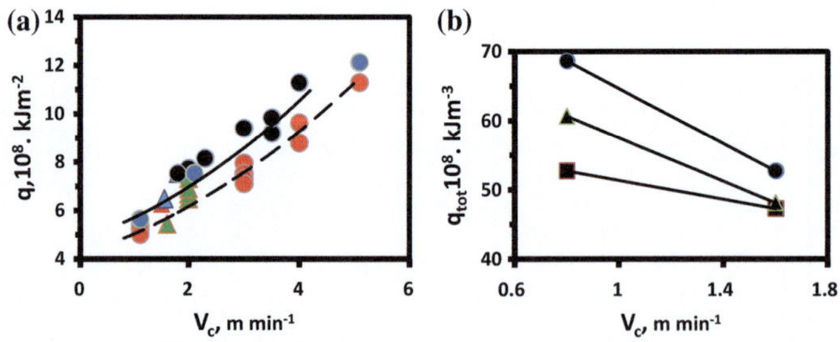

Fig. 3.11 The effect of casting speed (V_c) on **a** the heat flux density (q) for casting MC steels (▲, ▲, ▲ and *dashed line*) and LC steels (•, •, •, *solid line*) (reproduced with permission from ISIJ [46]) and **b** for total or integral heat fluxes for three mould powders (reproduced with permission from ISS/AIST [47])

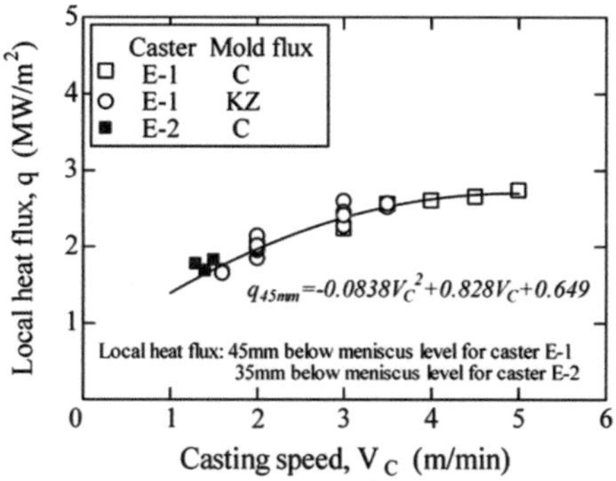

Fig. 3.12 Local heat flux at a point 45 mm from the meniscus as a function of casting speed (permission granted, ISIJ, [42])

of the mould should also be taken into account (see Sect. 3.2.3.5 (iv) below); longer moulds are required for high-speed casting [52].

The heat flux density at a point 45 mm below the meniscus (i.e. maximum value of q_{hor} where the metal flow impacts the shell) can be calculated by Eq. 3.11 (see Fig. 3.12) for casting speeds between 1 and 5 m min^{-1} [42]:

$$q_{45\,mm}\left(\mathrm{M\,W\,m^{-2}}\right) = 0.649 + 0.828\,V_c - 8.38x10^{-2}V_c^2. \qquad (3.11)$$

(ii) ***Superheat***

Superheat tends to increase the heat flux since it increases the thermal gradient between molten steel and mould [50, 51]. However, superheat will also tend to delay solidification to a lower position in the mould.

A statistical analysis of plant data [48] established that increasing superheat caused a significant increase in horizontal heat flux for both the narrow and wide faces of the mould.

(iii) ***Oscillation characteristics (stroke length, frequency)***

A statistical analysis of plant heat flux measurements showed that the horizontal heat flux increased with increasing stroke length for both the narrow and wide faces of the mould [48]. A possible explanation is given below in Sect. 3.4.1. The prediction of a mathematical model [1, 2] indicates that as the mould starts descending a downward, "Arctic" slag flow is produced which causes increases in both heat flux and shell solidification. Thus an increased stroke length will cause both an increased period where this flow is in operation and bring the slag flow closer to the shell tip; both of these factors will increase the horizontal heat flux.

Preliminary measurements obtained with a mould simulator [53] indicated that

(i) Increase in frequency (2.14 to 2.43 Hz) resulted in a increase in mould temperature ($f \uparrow \rightarrow t_n \downarrow \rightarrow T_{mould} \uparrow$) and
(ii) Increase in stroke (3.4 to 6.8 mm) resulted in a decrease in mould temperature ($s \uparrow \rightarrow T_{mould} \downarrow$).

These results are consistent with mould temperature increasing as the negative strip time (t_n) decreases ($T_{mould} \uparrow$ as $t_n \downarrow$). Non-sinusoidal oscillation was found to give a 10% decrease in q_{hor} [54].

3.2.3.4 Slag Characteristics

(i) ***Thickness of solid slag film*** (d_s)

It can be seen from Eq. 3.9 that the thermal resistance of the slag film increases as the thickness of the solid layer (d_s) increases. The principal factor affecting d_s is the break (or solidification) temperature (T_{br}); it can be seen from Figs. 3.13 and 3.14b that an increase in T_{br} results in an increase in d_s. The effect of break temperature on the heat flux density can be clearly seen in Fig. 3.14a ($q \downarrow$ as $T_{br} \uparrow$); an increase in T_{br} of 150 °C has been found to result in a 10–20% decrease in q_{hor} [17, 32, 55]. It has been reported that the slag film thickness (d_s) remains constant for all mould positions (d_{menis}) below d_{menis} = (35–45) mm (i.e. d_{menis} > 35–45 mm) [34].

(ii) ***Interfacial Resistance***

The interfacial resistance ($R_{Cu/sl}$) has been discussed above in Sect. 3.2.2. In summary, the magnitude of $R_{Cu/sl}$ increases with (i) increasing f_{crys} [14, 16], (ii) increasing solid film thickness, d_s [14, 16] and (iii) increasing gas gap thickness (d_{gas}) (Fig. 3.6). There is still some dispute about (i) the magnitude of $R_{Cu/sl}$ values obtained in simulation experiments and in pilot plant trials and (ii) the relative magnitudes of $R_{Cu/sl}$ and R_{film}. The pilot plant data are preferred here because they simulate the conditions in the mould more closely than those in simulation experiments; thus (i) a value of $R_{Cu/sl} = 0.4 \times 10^{-3}$ m^2 K W^{-1} [3, 17] and (ii) $R_{Cu/sl}$

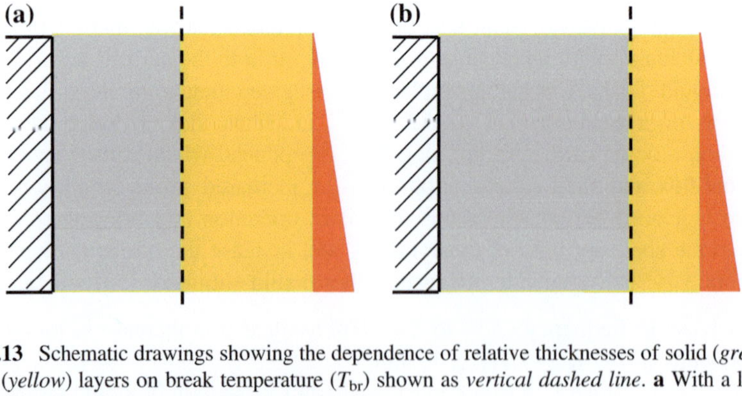

Fig. 3.13 Schematic drawings showing the dependence of relative thicknesses of solid (*grey*) and liquid (*yellow*) layers on break temperature (T_{br}) shown as *vertical dashed line*. **a** With a low T_{br} and **b** high T_{br} with larger *solid layer* thickness

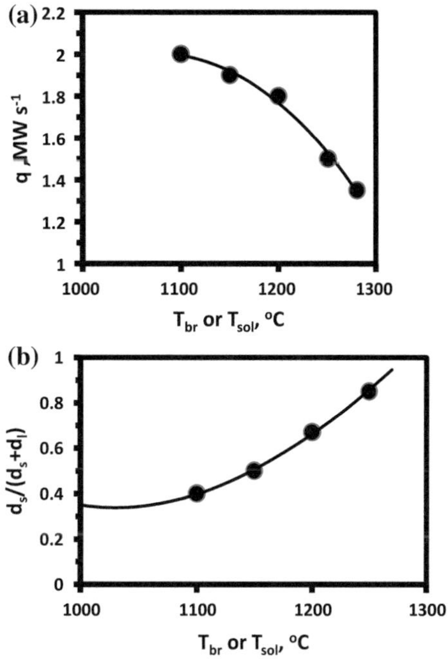

Fig. 3.14 The effect of solidification (or break) temperature on **a** heat flux density and **b** $f_{solid} = d_s/(d_s + d_l)$ in slag (Re-drawn from [17], permission granted, ISIJ)

< R_{film} (where $R_{film} = (d_s/k) + R_R$) are preferred. The ratio of the two resistances ($R_{film}/R_{Cu/sl}$) was found to increase with increasing casting speeds as a result of the decrease in $R_{Cu/sl}$ with increasing casting speed (Fig. 3.9d) [17].

Cho et al. [23] reported that the interfacial resistance was slightly lower when the copper mould was coated with Ni–Fe coating.

(iii) ***Crystallisation and porosity of slag film***

The principal way to control the horizontal heat flux is via the control of the amount of crystallinity in the slag film. This is usually done by selecting the appropriate chemical composition (or (%CaO)/(%SiO_2) or (C/S) ratio) of the casting powder. *Cuspidine* ($3CaO{\bullet}2SiO_2{\bullet}CaF_2$ or C_3S_2Fl) is the principal crystalline phase formed. Other phases are formed; the phase formed depends upon the composition. A phase diagram has been reported to identify these phases for powders with a C/S ratio of ca. 1 [56]. The basicity of casting powders varies between 0.6 and 1.3, for instance, (C/S) ratios of around 1.3; (0.7–0.9) and 1.0, respectively, are used when casting, peritectic (MC), sticker-sensitive (HC) and other steels (LC, ULC) [57]. Environmental concerns have led to a drive to reduce F levels in casting powders; this has led to a search for a crystalline phase to replace *cuspidine* for powders used to cast slabs. One possible candidate is *perovskite*, $CaTiO_3$ [58, 59] but it has been reported that TiO_2 reacts in the mould to form Ti (CN) which results in sticker breakouts [60].

Crystallisation of the glassy phase of a slag film results in

(i) Reflection of the IR radiation by the crystals with a resultant decrease in q_R;
(ii) The formation of porosity and an interfacial thermal resistance, $R_{Cu/sl}$, (caused by the resultant shrinkage) which both cause a reduction in k_R and q_R; and
(iii) Substantial increases in the lattice thermal conductivity (k_{lat}).

The overall effect of crystallisation is a significant decrease in the horizontal heat flux.

Further studies [11–13] showed the following:

(i) Crystals tended to increase the reflectivity and decrease the transmissivity, whereas FeO additions tended to increase the absorption of IR radiation and decrease the reflectivity; FeO additions promote glass formation, which enhances q_R and thereby opposes the reduction in q_R resulting from reflection but simultaneously reduce q_R through increased IR absorption [11];
(ii) For FeO-free fluxes, the reflectivity of IR radiation at higher wavelengths tended to increase with increasing *cuspidine* grain size (i.e. $k_R \downarrow$ as $d_{grain} \uparrow$) but the effect was less marked at for slags containing FeO [12]; and
(iii) The IR radiation tends to be reflected at the liquid/ crystal interface [9].

Crystallisation is not the only cause of porosity. Casting powder with a high moisture content (or water leaks) can lead to hydrogen pores in the slag film. Despite the fact that $k_{H_2} > k_{air}$, the increased porosity leads to a decrease in heat flux (Fig. 3.15) and to sticker breakouts [61]. Blowholes containing CO (g) can be formed from reaction of C and FeO in the slag.

3.2.3.5 Slag Viscosity

Hanao et al. [17] have reported a slight increase in slag film thickness as the mould slag viscosity was increased (Fig. 3.15); this results in a decrease in q_{hor}. Further evidence for this proposition comes from a statistical analysis of plant data which indicated that the heat flux decreased with increasing slag viscosity ($q\downarrow$ as $\eta\uparrow$) [48] and from other sources [1, 50]. It was suggested that the thickness of the solid layer increases with increasing viscosity which would result in a decreased heat flux. Another possible explanation for this opposition is that increased viscosity would

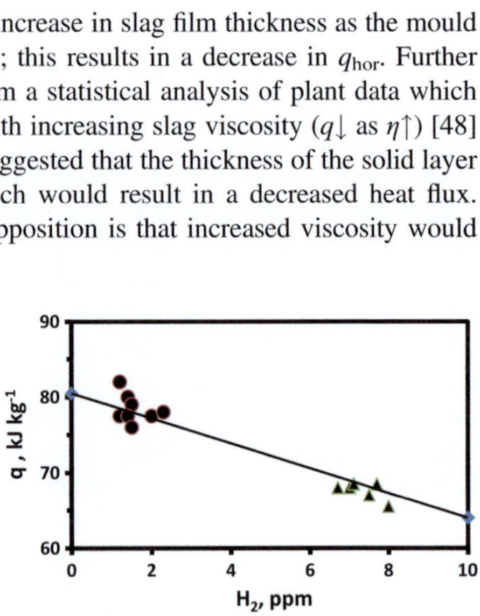

Fig. 3.15 Mould heat removal, q, as a function of hydrogen in the slag film (a measure of the porosity) for melts which were degassed (●) and where no degassing was carried out (▲) [61]; V_c = 1 m min^{-1} (permission granted, EDP Sciences, [61])

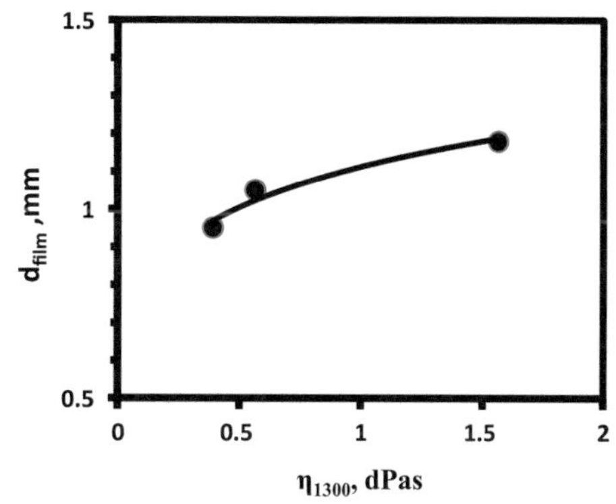

Fig. 3.16 Slag film thickness as a function of slag viscosity at 1300 °C for three mould slags (permission granted, ISIJ, [17])

oppose the cold downward slag flow in the pool when the mould is descending (see Sect. 3.4.1) and would hence reduce the heat flux (Fig. 3.16).

3.2.3.6 Shell Condition

The solidification of the shell occurs during the period when the mould is descending (i.e. during negative strip time) but there is little solidification when the mould is ascending. Consequently, the variation in solidification results in a shell with variable thickness. In MC peritectic grades, the shell is wrinkled and this too leads to variability of the heat flux. Furthermore, oscillation marks will tend to increase the variability of the heat flux [45].

3.2.3.7 Mould Conditions

(i) ***Position in the mould***

The heat flux varies in both the horizontal and vertical directions (see Sect. 3.2.3.1 and Fig. 3.10a). The heat flux is higher in the corners than in the mid-face; this results in a thicker film in the narrow faces *cf.*, the broad face and lower value of q; this affects the heat flux ratio (defined in Eq. 3.10). In the vertical direction the maximum mould temperature occurs at ca. 45 mm below the meniscus and decreases with increasing distance from the meniscus (Fig. 3.10b).

(ii) ***Water flow rate***

One practice used to obtain "*mild cooling*" (in order to reduce longitudinal cracking) is to reduce the water cooling rates to reduce the horizontal heat flux [50, 62, 63]. However, Lu et al. [62] found that reduction of water cooling rates made only a small reduction in shell thickness (Fig. 3.17).

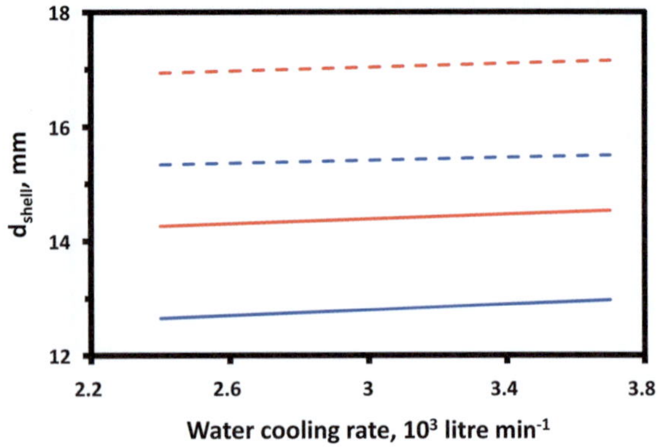

Fig. 3.17 The effect of water cooling flow rate on the thickness of the shell for two different superheats (20 °C = *blue* and 60 °C = *red*) during casting of stainless steel 304 (- - -) and CS 1026(——) [62]; $V_c = 0.9$ m min^{-1} (permission granted, ISS/AIST [62])

(iii) ***Mould Taper***

When the steel solidifies, the shell tends to shrink away from the mould. Consequently, the narrow faces are tapered, so that they can provide support to the shell. If the taper is insufficient, a depression is usually formed 50–100 mm from the edge of the slab which gives rise to cracks. When the taper is insufficient to compensate for shrinkage, a "*gas gap*" is formed, providing there is insufficient liquid slag to flow into the gap; this is a particular problem in both the corners (where overcooled corners can lead to transverse cracks [64]) and in the lower mould. It is important that the taper should match the shrinkage (see Fig. 11.48), which, in turn, is related to the heat flux Tapers are dependent on the steel being cast (e.g. Taper $_{MC}$ < Taper $_{HC}$ due to higher q_{hor} [65]). Mathematical models have been developed to calculate the appropriate taper for the steel and casting conditions [64, 65].

A statistical analysis of plant measurements, relating cracking and taper, indicated that increased taper produced a thicker and more uniform shell and, consequently, resulted in fewer cracks [48].

(iv) ***Mould length***

The mould length (L_{mould}) is an important factor when minimising the incidents of sticker breakout. Mould lengths vary from 0.7 to 1.5 in m [52]. The residence (or dwell) time (t_{mould}) determines the shell thickness at the mould exit and is calculated by ($t_{mould} = (L_{mould} - 0.10)/V_c$) where ($L_{mould} - 0.10$) m is the *active* mould length. The heat flux density (q) decreases with increasing t_{mould} and the relation (q_{total} (M W m^{-2}) $= 7.3/t_{mould}^{0.5}$ where t_{mould} is in seconds) was obtained [52]; however, the integral or total heat flux increases with increasing viscosity. It was found that the heat flux q was ca. 30% higher for high-speed casting ($q_{total} = 9.5/t_{mould}^{0.5}$) than for conventional slab casting [52]. This was attributed to a thinner slag film thickness (d_s) resulting from lower slag infiltration during high-speed casting [52]. High-speed casting requires a longer mould.

(v) ***Low conductivity coatings and refractories***

Coating the mould [27, 66–68] with high-melting metal (such as nickel, chromium or molybdenum) tends to reduce the magnitude of the horizontal heat flux because the conductivities (k_{25}) of these metals (= 83, 90, 139 W m^{-1} K^{-1}, respectively) are much lower than that of copper (= 402 W m^{-1} K^{-1}). The thermal resistance of the coating is given by (d/k) where d is the thickness of the coating and k is the conductivity. Thus the thermal resistance of a Ni coating is 5X that of copper. Coating the mould with these metals also improves mould life [68]; coatings of WC were found to be particularly effective in improving mould life but were expensive [69]. However, it has been reported that coating the mould with Fe/Ni caused a slight decrease in the interfacial resistance ($R_{Cu/sl}$) [23].

(vi) ***Meniscus free casting***

The term "*Meniscus free casting*" is applied to the practice of lining the mould at the meniscus level with a refractory material (usually boron nitride) [61, 70, 71]. Meniscus free casting reduces the horizontal heat transfer but it can only be used for billets and blooms because the refractory is not available in pieces long enough for use with slabs. The reduction of horizontal heat flux in Meniscus free casting causes an increase in metal temperature at the meniscus level. This causes solidification to occur lower down the mould (away from the turbulent meniscus region) and has been reported to give the following benefits:

(i) Shallow oscillation marks and
(ii) A reduction in inclusions trapped by the shell at meniscus level.

(vii) ***Grooved moulds***

In the quest to produce a thin, uniform shell, to minimise any unevenness in the shell, (especially when casting MC peritectic steels) several workers have used machined grooves in the meniscus region of the copper mould to reduce the heat flux in this region [72–79]. The aim is to (i) thicken the slag film and (ii) produce gas gaps between the slag film and the mould surface which reduce the horizontal heat transfer and decrease the unevenness in shell thickness (Fig. 3.18). Grooved moulds give a ca. 10% decrease in heat flux [73, 80]. Sugitani et al. [73] have discussed the best designs for grooved moulds.

Cho et al. [74] studied the effect of grooves on the slag film using vertical grooves in the central portion of the mould for distance 0 to 150 mm below the meniscus. The rapidly cooled slag in the groove tended to form a glassy phase. They found that

(i) Both the total film and glass film thicknesses decreased exponentially with increasing distance from the meniscus (Fig. 3.19);
(ii) The slag infiltration into the groove tended to be semi-circular and thus did not fill the whole groove;

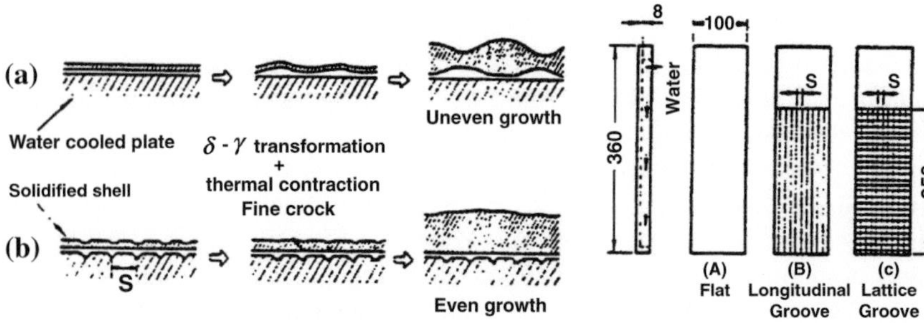

Fig. 3.18 Schematic diagrams showing **a** growth of a shell formed during the quench on a flat plate and that formed on a lattice grooved plate and **b** various plates used in the quench and the arrangements of the grooves *(D* = depth of groove, *W* = width of groove (0.5 mm) (0.3, 0.5 and 1.0 mm) and *s* = pitch of grooves (3, 5, 10, 15, 30 mm) [75] (permission granted, ISIJ, [75])

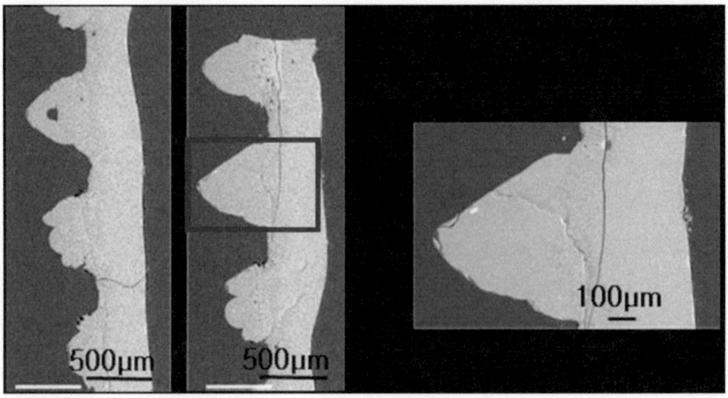

Fig. 3.19 Photographs of the slag film showing thicknesses of the total slag film and the glassy film as functions of distance from the meniscus (permission granted, ISS/AIST, [74])

(iii) The calculated infiltration ratio (d_{slag}/d_{groove}) increased as (a) the surface tension of the slag decreased and (b) groove width increased and
(iv) The calculated infiltration ratio increased from ca. 80 to 95% for positions 50–150 mm below the meniscus.

3.2.4 *Measurement and Calculation of Heat Fluxes*

3.2.4.1 Plant Trials

The total heat flux (q_{total}) in the mould can be determined by monitoring the temperatures of the cooling water at the entrance and exit of the mould (T_{in} and T_{out}, respectively), the volume flow rate (d*V*/d*T*) of the cooling water and using Eq. 3.12 where C_p and ρ_T are the heat capacity and the density of the water at the mean temperature (= 0.5 (T_{out} + T_{in})) and *t* is the duration of the flow [3, 17, 43, 49, 81]:

$$q_{total} = (dV/dT)t \cdot \rho_T C_p (T_{out} - T_{in}). \tag{3.12}$$

Mould thermal monitoring (MTM) is widely used for detection of potential sticker breakouts [3, 17, 82–84]. An array of thermocouples provides heat flux measurements at different locations in the mould. Thus, local values of heat flux are available with these instrumented moulds. Hanao et al. [3, 17] carried out a thorough investigation of the horizontal heat flux using a pilot caster with an instrumented mould in which the slag film was removed and characterised. The MTM technique is widely used and the heat flux of mould powders has been used to compare heat fluxes for CS- and CA-based fluxes for casting high Al (Trip) steels ($q_{CS} > q_{CA}$) [84].

3.2.4.2 Simulation Experiments

A number of simulation experiments have been carried out to gain an understanding of the various factors affecting the horizontal heat transfer. The experiments have tended to become more sophisticated with time. The thickness of the slag specimen is important because of its effect on (i) thermal resistance of the slag sample (d_s/k) and (ii) optical thickness ($\alpha^* d_s > 3$). Consequently, it is important that d_s for the simulation experiment should be similar to that in the slag film; there are some indications that slag films may be optically thin [85]. Some of the simulation experiments described below involve both molten steel and slag, whereas others involve only slag [58, 81]. It has been proposed that the low values of interfacial resistance values recorded in pilot plant trials [3, 17] are due to the ferrostatic pressure exerted on the shell; this will be absent in experiments which exclude molten iron.

(i) ***"Cold finger" experiments*** [35, 86–89]

In these experiments a water-cooled, copper mould is lowered, mechanically, through a slag pool into a bath of molten iron. The passage of the cold finger through the slag layer results in the formation of a frozen slag film around the finger (Fig. 3.20a). The slag film was found to be of similar thickness to that formed in the continuous casting mould [86–88]. Heat fluxes were determined by measuring the water temperatures at the entrance and exit of the finger and the water flow rate. The instrument has been modified to allow the effect of oscillation to be identified [88]. The tests which involve only molten slag have been used on various types of slags including those used for casting HiAl steels [59, 81, 90] and F-free slags containing TiO_2 [91].

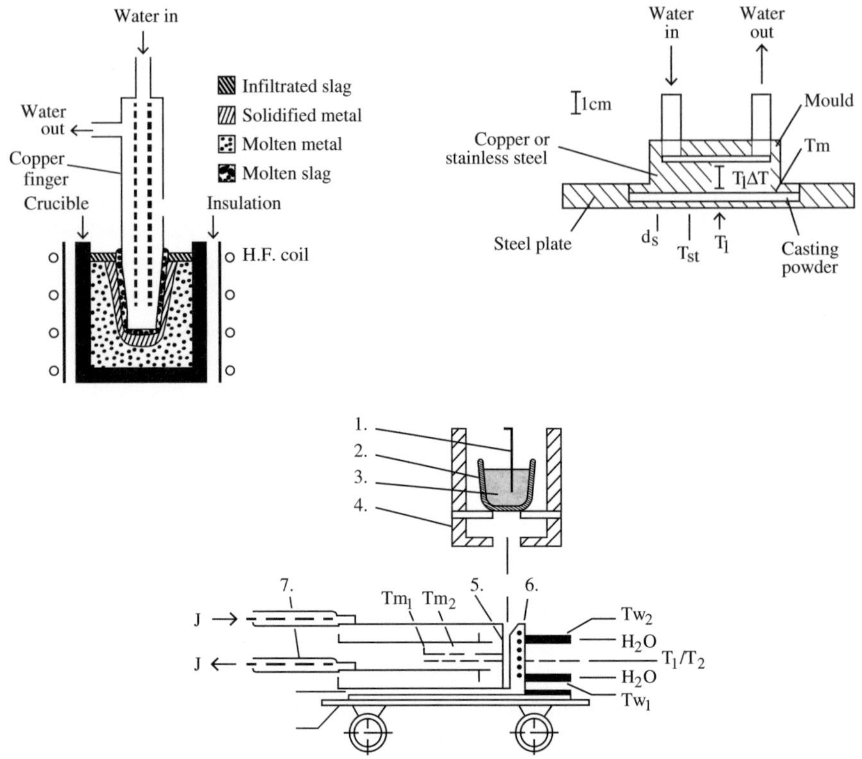

Fig. 3.20 Schematic drawings of **a** cold finger [86], **b** parallel plate [92] and **c** vertical plate experiments [94] J = water in and out (permission granted, **a** IOM/Taylor & Francis [86] **b** and **c** Verlag Stahleisen GmbH [92] and [94], resp.)

(ii) ***Parallel plate experiments*** [17, 92, 93]

The thickness of the slag film can be accurately controlled in this experiment. A steel plate (representing the shell) is machined to form a well (of known thickness) to contain the mould slag when it is introduced. It is then heated to melt the slag (Fig. 3.20b). A water-cooled copper plate (i.e. the mould) is then lowered onto the surface of the molten slag. The heat flux is determined from the temperature rise of the water and the water flow rate. An equation relating the thermal conductivity of the slag film to the composition was obtained with this technique (Eq. 3.13) where %CaO* = Σ % (CaO + MgO + MnO + Na_2O).

$$q_{1200\,^\circ\mathrm{C}}\left(\mathrm{Wm^{-1}K^{-1}}\right) = 2.03\text{–}0.459(\{\%\mathrm{CaO}^*/\%\mathrm{SiO_2}\} + \%\mathrm{B_2O_3}) - 0.1695(\%\mathrm{FeO}) - 0.034(\%\mathrm{Al_2O_3}). \quad (3.13)$$

In these experiments it was determined that horizontal heat fluxes were considerably greater for glassy slags than those for crystalline or partially crystalline slags [92].

The steel plate has been replaced by both an AlN plate and by a tungsten plate [17] and the mould by stainless steel [93] in similar tests.

(iii) ***Vertical plate experiments*** [40, 94]

In these experiments the slag film thickness is controlled and sited vertically by pouring molten slag into a vertical rectangular gap (Fig. 3.20c). The plate is heated on one side and water-cooled on the other as shown in Fig. 3.20c [94]. Another variation of this technique was developed by Stone [40].

(iv) ***Bottom-filling simulators*** [95–97]

These bottom-filling methods have been used mostly in studies of the formation of oscillation marks [95] and ripple marks [97] and have not been used for heat flux measurements.

(v) ***Mould simulators*** [53, 98–101]

Mould simulators [98, 53, 99–101] are essentially small-scale models of a continuous caster. The mould consists of two water-cooled, copper plates which contain a number of thermocouples embedded in the copper to measure local heat fluxes. A cover plate (connected to the extractor system) separates the two copper plates. The mould can be oscillated, and the assembly is lowered into a bath of molten steel (at the target temperature) which is covered with mould powder (Fig. 3.21a). The cover plate is then lowered and thus the copper plate is exposed to the molten steel and a steel shell is formed (Fig. 3.21b). The cover plate and shell are then lowered a further (60–100 mm) which results in further growth of the shell (Fig. 3.21c). At the end of the experiment the assembly is raised and removed from the steel and the shell is allowed to cool. The simulator is fitted with a device to

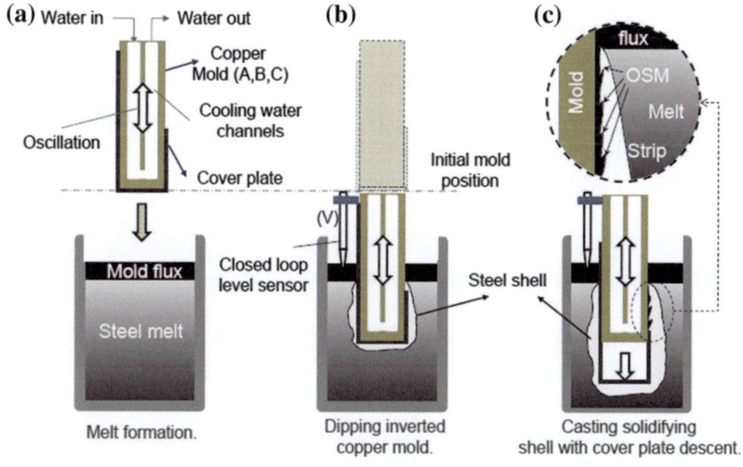

Fig. 3.21 Schematic diagram showing the various stages in the casting process using a mould simulator: **a** mould is lowered into steel, **b** cover plate in place (no shell solidification) and **c** cover plate is lowered and shell solidification occurs. (Reproduced with permission from Springer [53])

maintain a constant mould level. Mould simulators have been used to study (i) heat flux variations during the oscillation cycle [99, 100] and (ii) the effect of casting speed and oscillation characteristics on the heat flux [53].

3.2.4.3 Physical Property Measurements

Measurements of the physical properties of mould slags proved very useful in understanding the heat transfer mechanisms involved in the heat transfer between the shell and mould. These property measurements are covered in detail in Chap. 9. The key properties affecting the horizontal heat transfer are as follows:

(i) The *thermal conductivity/diffusivity* of the slag and the slag film [7, 8, 14, 15, 19, 20, 102–107];
(ii) The *optical properties* of the slag (particularly the absorption coefficient, reflectivity and the refractive index) since they affect the magnitude of both the radiation conduction [6–8, 14] and the optical thickness; the absorption coefficients (α^*) increase with increasing % transition metal oxide, e.g. FeO, [6, 13, 104, 108–110];
(iii) The *fraction of crystalline phase* (f_{crys}) in the slag film since this affects the magnitude of (a) the thermal conductivity ($k_{crys} \approx 2k_{glass}$), (b) the radiation conductivity (k_R, with $k_R \uparrow$ as $f_{crys} \downarrow$) and (c) the porosity and surface roughness (or interfacial resistance,' $R_{Cu/sl}$) of the slag film which tend to decrease the horizontal heat flux; and
(iv) The *solidification or break temperature* (T_{br}) since this affects the thickness of the solid slag film (d_s) [111].

These data have been used to gain both an understanding of the mechanisms of heat transfer occurring in the slag film and to model their effect on the heat transfer [11, 12, 112]. Values of $k_{eff} = 2.3$ to 3 W m^{-1} K^{-1} were obtained from an analysis of heat flux data recorded in pilot plant trials [17] on three mould slags with basicities of 0.8, 1.7 and 1.8. The results of this study indicate that k_R contributes ca. 20% to k_{eff} for highly crystallised slags but ca. 60% in glassy slags.

3.2.4.4 Mathematical Modelling of Heat Transfer in the Mould

Mathematical modelling of the heat transfer in the mould has evolved over time [1, 2, 16, 32, 113–120]. The more recent models couple shell solidification and heat transfer for the fluid flow in the steel and slag phases. Furthermore, no assumptions are made [1, 2] concerning (i) the formation of a slag film and its thickness and (ii) the shape of the meniscus. It can be seen from Fig. 3.22 that the predicted heat fluxes are in excellent agreement with those obtained in plant trials [1, 120].

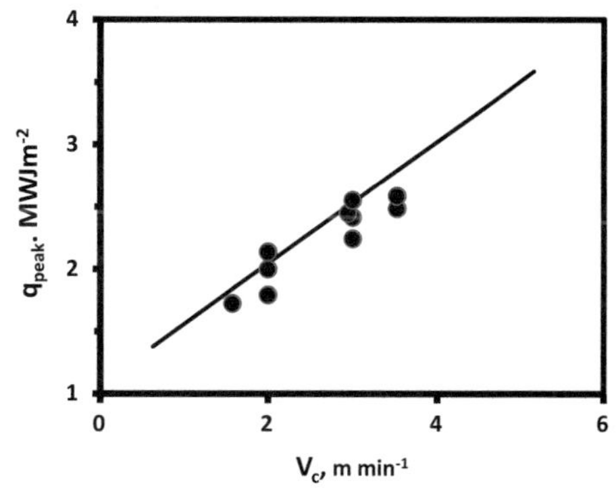

Fig. 3.22 The peak heat flux (i.e. at 45 mm below meniscus) as a function of casting speed; ● = plant measurements [121, 125] and *solid line* = predicted values [1, 2, 120] (permission granted, ISIJ [121] re-drawn)

3.3 Shell Solidification and Growth

Slag films, mould fluxes are important because they affect the nature of the shell; the condition of the shell is all-important in continuous casting and slag film performance should be regarded in terms of how it affects the shell. The heat flux is important because it determines the thickness (and hence the strength) of the solidified shell. Shell solidification occurs principally in the period when the mould is descending and grows little during the rest of the cycle. Thus the shell does not have uniform thickness but tends to have a "*sausage-shaped*" profile (Fig. 3.1). Unevenness in the shell can be harmful in certain steel grades, particularly those prone to longitudinal cracking.

Two methods are commonly used to measure shell thickness (d_{shell}). In the first method the steel is doped with FeS alloy at a specific time and the solidified steel product at that time is identified and retrieved. The sample is then cut length-wise and a sulphur print of the section is prepared. The width of the white section near the surface is taken to be the thickness of the solidified shell and is accurately measured [42].

In the second method, the growth rate of the shell (δ {d_{shell}/dt}) is calculated from the cooling rate (dT/dt) at specific locations. The cooling rate is determined by measuring either the arm spacings of the primary or the secondary dendrites [17, 34, 42]. Samples are taken from regions close to the centreline of the slab. The samples are taken parallel to the surface for measurements of primary arm spacings [44]. In contrast, samples are taken perpendicular to the casting direction where measurements were made of the secondary arm spacings [42]. The samples were polished and etched and the secondary arm spacings were measured on photographs of the optical micrographs [42]. The cooling rate ($\delta T/dt$) is related to the temperature gradient ($\delta T/dx$) and the rate of solidification (δ {d_{shell}/dt}) as shown in Eq. 3.14 where x is distance. The cooling rate at different distances can be determined in this way:

Fig. 3.23 Thickness of the solidified shell (45mm below meniscus); o, as a function of square root of solidification time; o, □ = 4 and 5 m min^{-1}, respectively; *solid line* above $t^{0.5} > 0.06$ is d_{shell} value calculated by $30t^{0.5} - 1.76$ (permission granted, ISIJ, [42])

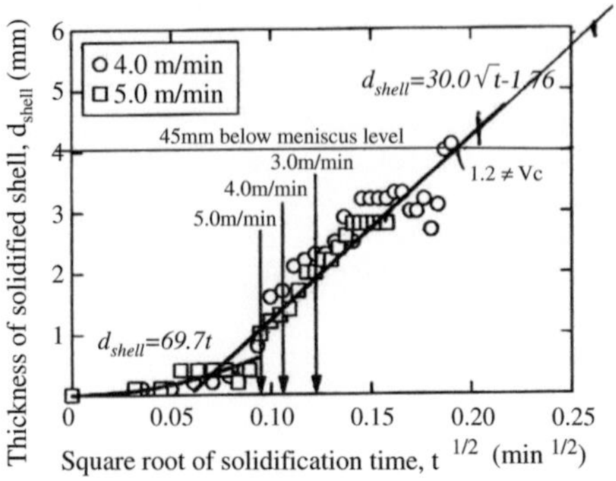

$$(\delta T/\mathrm{d}t) = (\delta T/\mathrm{d}x)(\delta\, d_{\text{shell}}/\mathrm{d}t). \tag{3.14}$$

$$q = -k(\delta T/\mathrm{d}x) \tag{3.15}$$

$$(\delta T/\mathrm{d}t) = (-q/k)(\delta d_{\text{shell}}/\mathrm{d}t). \tag{3.16}$$

Figure 3.23 a shows the results reported by Hanao et al. [42]. The solidification rate is given by the gradient of the plot ($\delta\ d_{\text{shell}}/\delta\sqrt{t}$).

An alternative method to calculate the cooling rate is by the use of the relation between it and the secondary dendrite arm spacing (d_{SDA}) reported by Suzuki [122, 123]:

$$d_{\text{SDA}} = 710(\delta T/\delta t)^{-0.39}. \tag{3.17}$$

It is customary to express the thickness of the shell in the form of Eq. 3.18 (see [124]), where K is the solidification coefficient and a' is a constant (usually with a negative value):

$$d_{\text{shell}} = K\sqrt{t} + a'. \tag{3.18}$$

Relationships relevant to shell thicknesses have been reported by various investigators and the results are summarised in Table 3.2. Most of the results show that the initial growth rate of the shell is slow and then at a certain point (around 0.07 min$^{-0.5}$) the growth rate suddenly increases as shown in Fig. 3.23 [42]. Values for ($d_{\text{shell}}/\mathrm{d}t^{0.5}$) = 69.7$t$ for $0 < t \leq 0.005$ and ($d_{\text{shell}}/\mathrm{d}t^{0.5}$) = $30\sqrt{t} - 1.76$) for $t > 0.0055$. Most of the results of the studies reported in Table 3.2 show similar behaviour [17, 42, 122, 125–130]. One possible reason for the slow initial growth is that the initial shell has a curved profile (see Fig. 3.1) and the early shell is further from the mould wall than later but this separation distance will decrease as the shell moves downward and reaches the gap. If this is correct, the delay time would be expected to

Table 3.2 Reported relationships for the thickness of the shell; *denotes mean values; A, B, C = 3 different mould powders with (*C/S*) = 0.8, 1.8.1.7, respectively

Reference	Steel	Casting speed (m min^{-1})	Dimensions w × t (m)	Super heat (°C)	d_{shell} (mm)	K (m min$^{-0.5}$) a′ (mm)
Nagaoka [126]	0.15C	0.5–0.65	1.6 × 0.20		$=19.1\sqrt{t} - 2.1$	
Suzuki [122]	HC			45–55	$=k\sqrt{t} - a'$	k = 34.7 to 38 a′ = 2.7 to 3.2
Okano [127]	MC	0.65	0.11 × 0.11		$=18.4\sqrt{t} - 1.5$	
Mori [128]			1.6 × 0.23		$=23.9\sqrt{t} - 0.86$	
Fujii [129]	LC	0.65	2.1 × 0.25	20	$=28.3\sqrt{t} - 8.6$	
Narita [130]	LC	0.65	1.6 × 0.23		$=27.5\sqrt{t} - 2.8$	
M Suzuki [125]	LC	2.0–2.5	1.55 × 0.22	25–30	$=23\sqrt{t} - 2.72$	
Hanao [17, 42]	MC MC	1.1–1.6	0.9* × 0.1* 1.9* × 0.25* 0.8 × 0.1	12–36 80–89 80–89	$=30\sqrt{t} - 1.76$ A: $= 24.8\sqrt{t} - 1.47$ B: $= 20.6\sqrt{t} - 1.14$ C: $= 18.7\sqrt{t} - 0.99$	

decrease with increasing superheat since solidification will be delayed to a point further down the mould. The shell thickness increases with increasing time (Eq 3.18) in the mould and thus increases as the distance from the meniscus (d_{menis}) increases. An empirical relation, Eq. 3.19 was derived from an analysis of plant data [52] for the thickness of the shell at the mould exit (d_{shell}^{exit} in mm) where K' is a constant with a value of 7.3 for conventional slab casting and 9.5 for high-speed casting:

$$d_{shell^{exit}} = K' \, t_{mould}^{0.5}/3.6. \tag{3.19}$$

The effect of increased casting speed on shell thickness can be clearly seen in Fig. 3.10 [42] where increased casting speed results in a thinner shell at $t^{0.5}$ = 0.07 (min) $^{-0.5-}$or t = ca. 5 × 10^{-4} min because of the shorter residence time (t_{mould}).

The nature of the mould slag was found to affect the dendritic structure of the solidified shell with a more crystalline slag film giving an increase in primary arm spacing [34].

3.4 Variability in Heat Flux

Variations in heat flux can cause longitudinal cracking, longitudinal corner cracking and star cracking. These variations in heat flux cover those arising from thermal gradients (d*T*/d*x*) and from variations over time (d*T*/d*t*). Some variations are inevitable (e.g. those occurring through an oscillation cycle) whereas other sources of variability can be minimised or partially controlled. The various sources of variability in heat flux in the mould are discussed below. Heat flux variations result in (i) variations in shell thickness and longitudinal cracking in the upper mould [131, 132] and (ii) star cracking, in the lower mould [133].

3.4.1 Variations in Heat Flux (q_{Hor}) During the Oscillation Cycle

The heat flux varies through the oscillation cycle [99, 100]. The predictions of a recent mathematical model [1, 2] are shown in Fig. 3.24. The oscillation characteristics are usually defined in terms of negative (t_n) and positive strip time (t_p) shown as a solid line in Fig. 3.24b but can also be defined in terms of position in the mould (shown as a dotted line in Fig. 3.24b). It can be seen from Fig. 3.24c that the heat flux (shown in red) is at its lowest value when the mould is at its highest position (in late t_p). The heat flux continues to increase as the mould descends (throughout t_n) and reaches its highest value when the mould is at its lowest position (in early t_p). The heat flux continues to decrease as the mould ascends.

Shell solidification follows the heat flux since a high heat flux results in a thick shell. Thus the variation of heat flux through the oscillation cycle results in a shell of varying thickness which is responsible for the corrugated or "*sausage-shaped*" appearance of the shell.

These variations in heat flux during the oscillation cycle result from changes in the direction of the flow in the slag pool. It can be seen from Fig. 3.25a that when the mould is ascending the slag flow is outward (towards the narrow face) and upward; such a flow is warm (tropical) because it moves over the surface of the molten steel. In contrast when the mould is descending the direction of flow in the slag pool is downward (Fig. 3.25b) [1, 2]. This flow will be cold (Arctic) because it originates in the colder parts of the slag pool. This downward, arctic flow is responsible for increasing heat flux and shell solidification and these will achieve a maximum value when the mould is at its lowest position (in early t_p) shown in Fig. 3.25c.

3.4.2 Thermal Gradient Variations Arising from Metal Flow and Other Causes

The variations in metal flow are initiated in the SEN. The first metal flowing through the SEN strikes the baseplate and rebounds; this results in the formation of a vortex in the sump of the SEN as shown in Fig. 3.26a [134]. This vortex affects the flow of metal through the ports of the SEN, sometimes favouring the flow through one port and at other times favouring the flow through the other port.

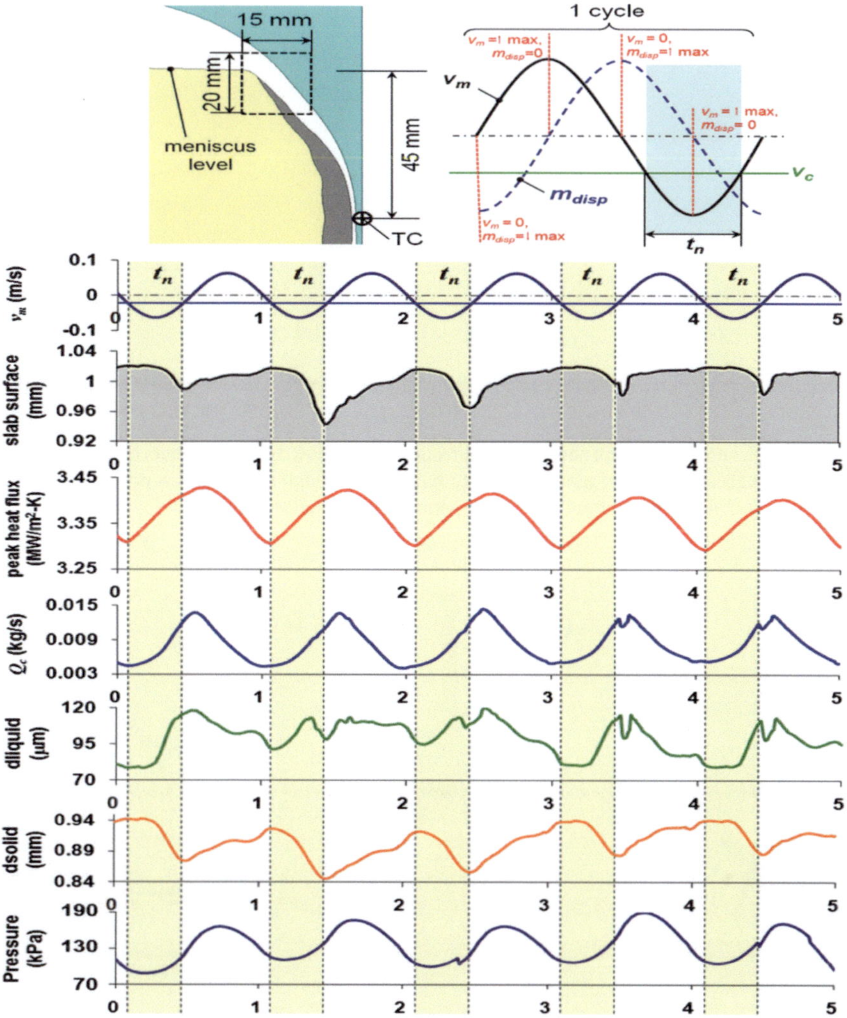

Fig. 3.24 Schematic drawings showing **a** the shell and slag rim, **b** the relation between mould velocity (*solid curve*) and position (*dotted curve* during an oscillation cycle) and **c** mathematical model predictions [1, 2] from the top of (i) profile of strand surface, (ii) heat flux, (iii) powder consumption in kg s^{-1}, (iv) liquid film thickness, d_l, (v) solid slag film thickness, d_s and (vi) pressure during five oscillation cycles (permission granted, ISIJ, [1])

Recently, the predictions of a mathematical model of the metal flow in the SEN indicated that the vortex creates a swirl in the metal which favours the flow towards the loose side [135] as shown in Fig. 3.26b. Thus the metal flow in the mould in the mould is continually fluctuating. Furthermore, the metal flow rebounds from the narrow face wall and this causes the formation of waves and in some cases, vortices in the meniscus region. These waves and vortices tend to move either side of the

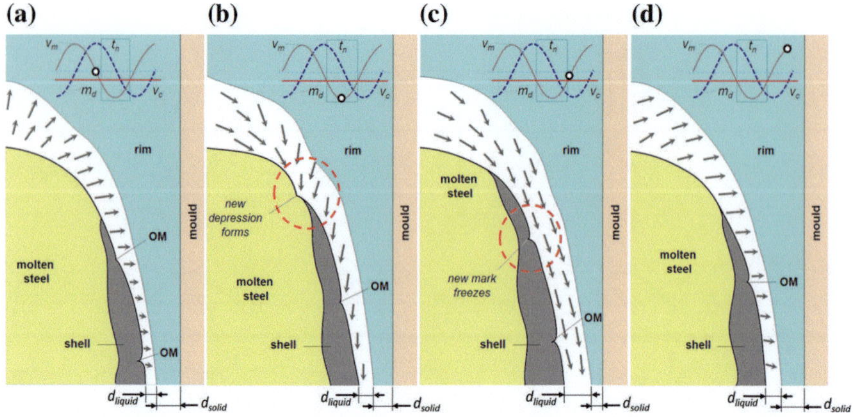

Fig. 3.25 Schematic diagram showing the direction of slag flow at different parts of the oscillation cycle **a** at highest position of mould in late t_p, **b** halfway through t_n, **c** at lowest position of mould in early t_p and **d** midway through t_p [1, 2] (permission granted, ISIJ, [1])

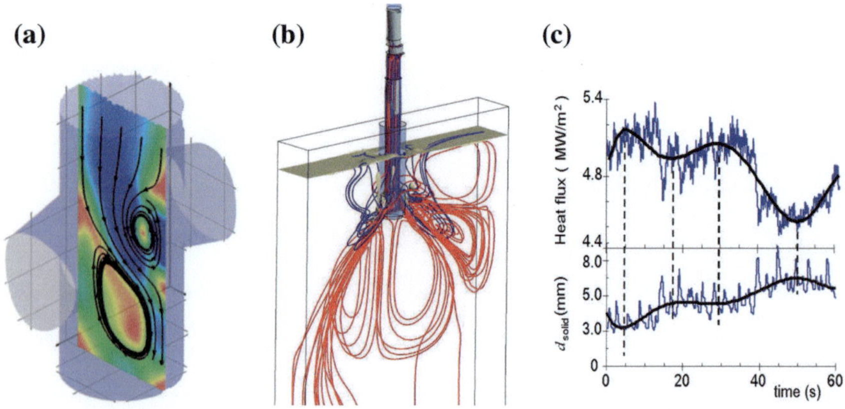

Fig. 3.26 Schematic drawings showing **a** the fluid flow in the SEN and the formation of vortex in the sump of the SEN [134] and model predictions showing, **b** the bias in metal flow even with EMBr [135] and **c** low-frequency variations of heat flux and solid slag film thickness over a number of oscillations [136] (permission granted, **a** ISIJ [134], **b** Verlag Stahleisen GmbH [135], **c** IOM/Taylor and Francis, [136])

SEN in response to which flow is the more dominant at the time. In fact, the metal flows and heat flows are continually fluctuating and interacting with other parameters so the situation is one of continual changes in response to the existing conditions. These transient fluctuations increase with increasing casting speed. In extreme cases, when metal flow variations combine with other variations (e.g. mould level variations) and all the effects are acting in collusion, they can lead to a calamitous event such as a breakout. This has been cited as an example of "*the butterfly effect*" [136]. Such variations through one oscillation are often referred to as high-frequency variations and exemplified by the variations in oscillation mark depth shown in Fig. 3.24c (i). However, model predictions indicate that there are also low-frequency variations going on in the mould which occur over a period of ca. 30 min; these can be seen in Fig. 3.26c [1].

Transient thermal fluctuations (i.e. d*T*/d*t*) can also occur in the lower mould when the liquid slag film has become too thin and the resulting frictional forces cause "*spalling*" and fracture of the solid slag film. This, in turn, results in variability in the horizontal heat flux (as shown in Fig. 3.27) and in star cracking [133, 137].

Thermal gradients (d*T*/d*x*) in the mould can occur locally in response to

(i) Overcooled corners in billets and at "*corners*" in both beam-casting moulds and football-shaped moulds used in thin-slab casting (see Chap. 11; Figs. 11.13, 11.14 and 11.15);
(ii) Conduction via high thermal conductivity SENs causing local "*cold spots*" in the mould [132] and
(iii) The impact of the metal flow on the shell resulting in both a very thin shell and a large thermal gradient in the corner (see Fig. 3.31) which can lead to longitudinal corner cracking [138].

3.4.3 *Mould Level Variations*

Model level variations can also cause variations in the heat flux. Level variations can arise through (i) poor meniscus level control or (ii) in response to bulging and

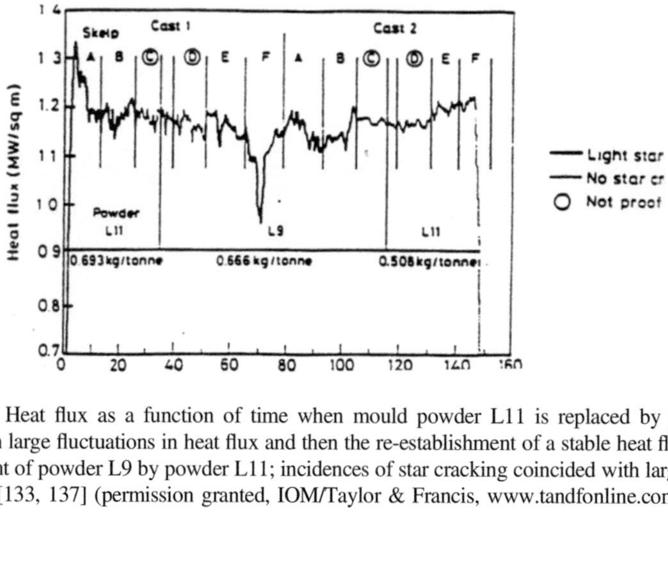

Fig. 3.27 Heat flux as a function of time when mould powder L11 is replaced by powder L9 resulting in large fluctuations in heat flux and then the re-establishment of a stable heat flux after the replacement of powder L9 by powder L11; incidences of star cracking coincided with large heat flux variations [133, 137] (permission granted, IOM/Taylor & Francis, www.tandfonline.com [133])

squeezing at the first rolls [139] when the shell at the mould exit is too thin (usually when casting MC steels). Large mould level variations can result in the capture of the slag rim which causes a rapid local decrease in horizontal heat flux as it passes down the mould (see Sect. 11.7.2).

3.4.4 Carbon Content of Steel

When molten steel solidifies δ-Fe (*ferrite*) is formed and this phase then transforms to *austenite* (γ–Fe) via the peritectic reaction:

$$\text{Liquid}\, Fe + \delta - Fe(\text{ferrite}) = \gamma{-}Fe(\text{austenite}). \tag{3.20}$$

The peritectic reaction (δ → γ) is particularly important in steels containing 0.08–0.17%C (although the range is affected by the presence of other elements). The packing in the FCC, *austenite* phase is tighter than that in the bcc, δ-*ferrite*, so the transformation is accompanied by a volume decrease of 0.4–0.6%. The shrinkage produces stresses which cause distortion in the shell (see Fig. 3.28) [74] which can result in longitudinal cracking [140]. These distortions cause unevenness in the shell (Fig. 3.28a) and this unevenness is maintained by the strength of the austenite phase.

Fig. 3.28 Photographs showing **a** wrinkled shell resulting from the peritectic reaction [141] and **b** unevenness of shell formed by (i) a steel with %C = 0.057 quenched onto copper, (ii) steel with %C = 0.122 and quenched onto copper, (iii) steel with %C = 0.133 and quenched onto SUS 304 stainless steel (slower cooling rate) and (iv) as in **c** but SUS304 heated to 200 °C [141] (permission granted, ISS/AIST, [141])

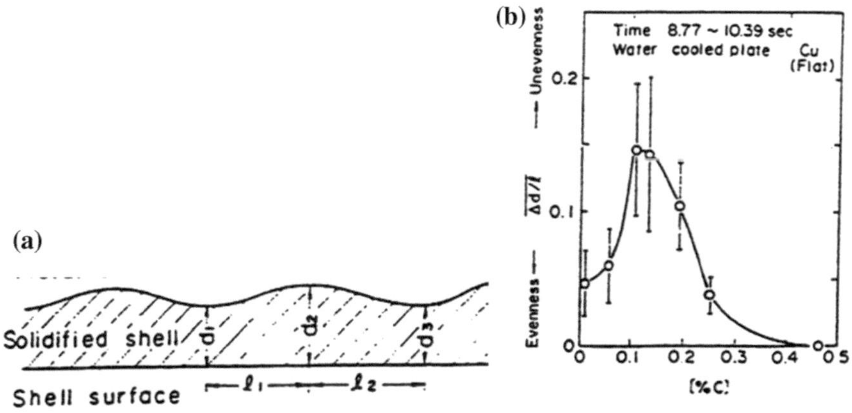

Fig. 3.29 **a** Schematic diagram showing the measurements made to derive the unevenness parameter $(\Delta d/l)_{mean}$ and **b** values for the parameter $(\Delta d/l)_{mean}$ as a function of the C-content of the steel [75] (*courtesy* of ISIJ)

Murikami et al. [75] carried out dip tests in which water-cooled plates were dipped into molten steels and the resulting shells were examined. The unevenness of the shells produced was quantified in terms of the average value of the parameter $(\Delta d/l)_{mean}$ (= Σ ($d_1 - d_2/l$) see Fig. 3.29a). It can be seen from Fig. 3.29b that the parameter $(\Delta d/l)_{mean}$ represents the unevenness of the shell and is much greater for steels in the peritectic range.

The unevenness in the shell also appeared to increase with increasing cooling rate. Hanao et al. [17, 34] used a different measure of unevenness, the Uneveness Index (= δ d_{shell}) where $\delta\ d_{shell}$ is the standard deviation of differences in d_{shell}. This index was found to (i) become significant when $d_{shell} \geq 1$ mm, (ii) was greater for glassy slags than for predominantly crystalline slags (since both q_{hor} and the cooling rate were higher) and (iii) increased gradually with increasing d_{shell} [17, 34].

This unevenness in the shell causes variability in the mould thermal monitoring (MTM) traces (Fig. 3.30) [82]; since longitudinal cracking is associated with thick,

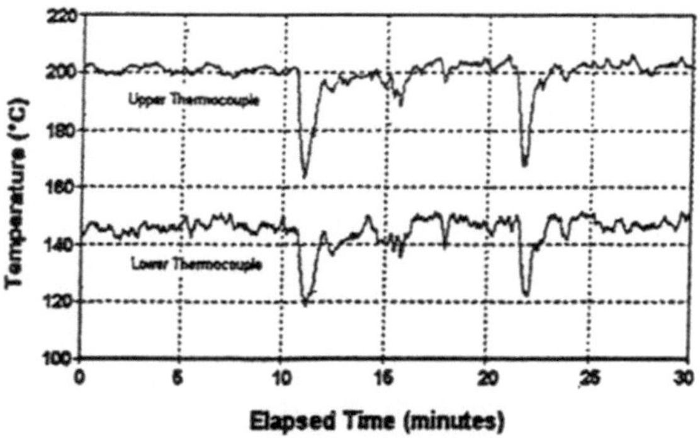

Fig. 3.30 Mould temperature as a function of elapsed time in mould thermal monitoring (MTM) traces (permission granted, IOM/Taylor & Francis, www.tandfonline.com [82])

(a)

(b)

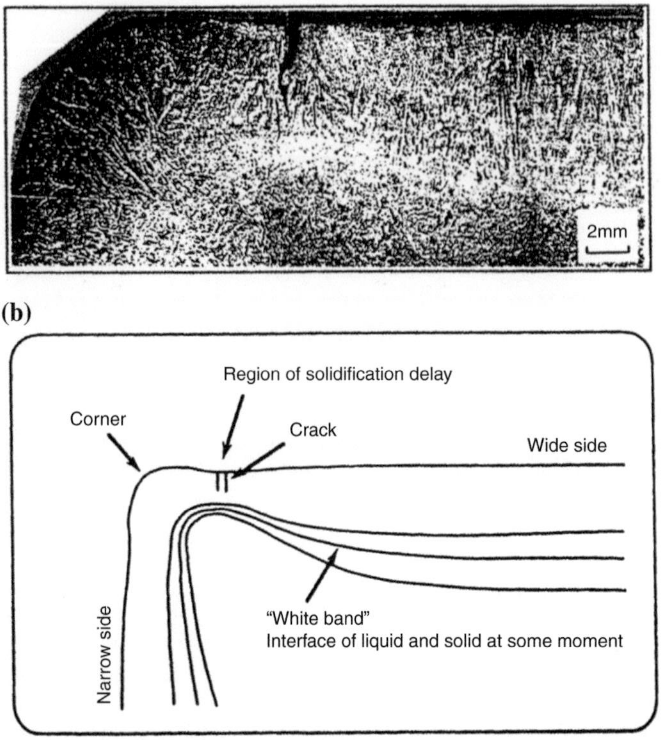

Fig. 3.31 **a** Photograph of sulphur print showing shell profile and the location of the longitudinal corner cracking, **b** schematic drawing showing the "*shell thinning*" caused by the metal flow [138] (permission granted, UNESID, [138])

uneven shells, it has been proposed that variability in MTM traces provides a way of monitoring for outbreaks of longitudinal cracking. It would appear that when shells (usually for MC steels) develop a certain thickness, the unevenness in the shell is sufficient to cause variability in the MTM trace and this signals the possibility of longitudinal cracking.

3.4.5 Thermal Gradients in the Mould

Some forms of cracking are related to variations in the local heat flux along the shell. These heat flux variations can arise from a variety of conditions:

(i) The presence of a high conductivity, SEN (containing carbon) can extract heat and if the SEN is close to the mould, it can create a cold spot in the mould, which, in turn, can cause large, local differences in shell thickness

Fig. 3.32 The effect of EMS stirring velocity on the incidence of longitudinal cracking (permission granted, Nippon Steel Sumitomo Metal Corp. NSSMC [142])

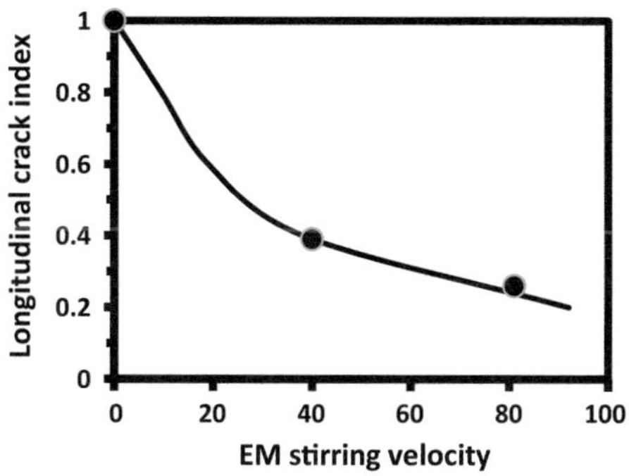

which create stresses and lead to longitudinal cracking (especially in MC steels, see Chap. 11) [80].

(ii) Excessive cooling in the corners (especially in billet casting) can result in variations in shell thickness which can lead to longitudinal corner cracking.

(iii) In slab casting, variations in the shell thickness can result from the flow of hot metal into the corner region which causes "*shell thinning*" in these corner regions and the resultant variations in shell thickness can lead to longitudinal corner cracking [138].

(iv) It has been reported that mould coatings (e.g. Ni) tend to reduce thermal gradients in the mould [68].

(v) EMS and EMBr help to reduce thermal gradients in the mould and thus help to minimise longitudinal cracking caused by steep differences in shell thickness. The beneficial effects of EMS in reducing local thermal gradients can be clearly seen in Fig. 3.32 where the incidence of longitudinal cracking is decreased with increasing stirring velocity [142].

3.4.6 Fracture of Slag Films

Some variability in the heat flux is caused by the fracture of the slag film [143, 144]: These fractures can occur in both the upper and lower parts of the mould; they are treated separately below.

Fractures can occur in the border region between slag rim and slag film [143, 144]. Following the fracture, the space between rim and film is filled by liquid slag which results in a glassy, repaired region [143, 144] of the slag film. These fractures are characterised by increases in heat flux (due to the high radiation conduction contribution through the new glassy slag film) followed by a gradual decrease in heat flux (due to the crystallisation of the slag film and the decrease in k_R) as shown in Fig. 3.33 [144, 145].

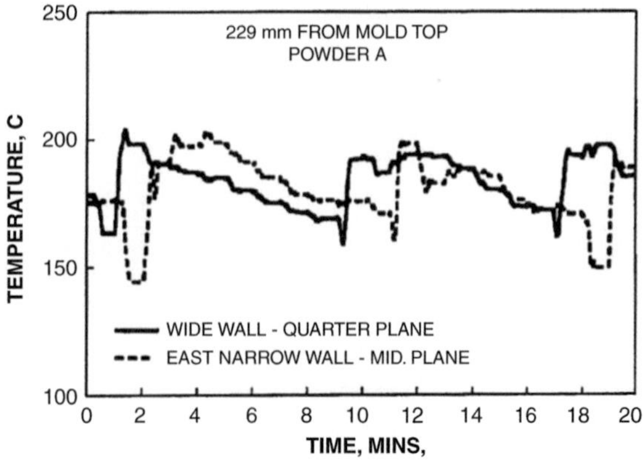

Fig. 3.33 Mould temperature as a function of time showing periodic increases in mould temperature caused by fracture of the slag film followed by a gradual decrease in temperature during the repair period [145] (permission granted, ISS/AIST [145])

Fracture in the lower mould occurs when enhanced heat extraction in the upper part of the mould results in the loss of liquid lubrication in the lower mould (i.e. liquid slag film is too thin). This results in increased friction and in "*spalling*" and fracture of the solid slag film [133, 137] which causes marked variations in heat flux (see Fig. 3.27). The repair of the slag film can take a significant time (e.g. 20 min) because of the poor supply of liquid slag. The variations in heat flux result in "*star cracking*" *and* "*spongy cracking*" [133, 137] (Chap. 11, Sect. 11.8).

3.5 Vertical Heat Flux

The upward transfer of heat from the steel meniscus through the liquid pool, sintered and powder layers is referred to as vertical heat transfer. The formation of the bed is an important step in the process. The powder is usually dispensed automatically and various types of dispensers are available; the strengths and weaknesses of various dispensers have been reviewed [146]. A device for monitoring the thickness of the bed has also been reported [147].

3.5.1 Heat Transfer Mechanisms Involved in Vertical Heat Transfer

Several heat transfer mechanisms are simultaneously operative in both the liquid and solid layers of the bed (shown in Fig. 3.34). In the liquid slag pool, convection, lattice conduction and radiation conduction occur, simultaneously. In the powder bed (consisting of sintered and powder layers) lattice conduction and gaseous conduction are the principal mechanisms involved.

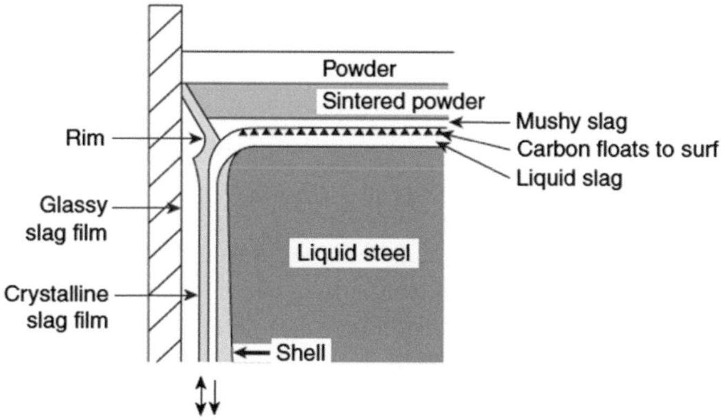

Fig. 3.34 Schematic drawing showing the various layers formed in the powder bed

3.5.2 *Factors Affecting Vertical Heat Transfer*

The vertical heat flux is linked to the horizontal heat flux since they share a common source of heat, i.e. the molten metal discharging from the SEN. The vertical heat flux is determined by several factors as follows:

(i) The source of heat, i.e. the flow rate of the molten steel, the metal flow pattern and the degree of superheat;
(ii) The efficiency of heat transfer; and
(iii) The thermal insulation provided by the bed.

These factors are themselves affected by other parameters, e.g. thermal insulation is affected by (a) the temperature gradients in the beds and (b) the average thermal conductivities of the sinter and powder beds.

The various factors affecting the vertical heat transfer are detailed below.

3.5.2.1 Steel Flow Rate and Superheat

It is self-evident that the vertical heat flux density (q_{vert} with units of W m^{-2}), like the horizontal heat transfer, will increase with increasing casting speed. It should be noted the *total* horizontal heat extracted (J m^{-2}) in the mould tends to decrease with increasing casting speed because of the shorter residence time [47].

The heat flux increases with increasing temperature gradient (dT/dx see Eq. 3.2) and superheat increases the temperature gradient. Thus increased superheat will result in an increase in heat flux density (q_{vert}).

The metal flow pattern is important too. The "*double roll*" pattern is preferred usually to the "*single roll*" pattern since the latter brings cold metal to the surface which can result in freezing of the metal (i.e. *skull* formation) [148]. The metal flow pattern is determined by (i) steel flow rate, (ii) SEN immersion depth, (iii) SEN port design and (iv) argon flow rate and these factors are interactive. Recent work has shown that small changes in Ar flow rate can cause changes in the metal flow pattern (see Sect. 3.5.2.7).

3.5.2.2 Efficiency of Heat Transfer

Heat transfer through turbulent flow is more efficient than that with laminar flow. Thus vertical heat transfer is expected to increase with casting speed increase since the metal flow will become more turbulent and this, in turn, will induce more turbulent flow in the molten slag pool. Convection in the slag pool makes a significant contribution to the vertical heat flux; convection in the slag pool is inversely dependent upon the slag viscosity. Some mathematical models assume convection of the pool is a multiple of the conductivity of the liquid slag ($k_{conv} = C\ k_{lat}$) and reported values for the constant C vary between 10 and 100 with an optimum value of 15 [149].

Electromagnetic braking (EMBr) is used to reduce both the metal flow velocities and the turbulence in the mould when using high-casting speeds. EMBr reduces (i) the turbulent flow at the interface and (ii) the penetration depth of the metal flow; the latter results in a 5–10 °C increase in the steel meniscus temperature. It has been calculated that EMBr causes a 30% decrease in vertical heat flux [150]. However, this will be partially offset by increased convection in the slag pool resulting from the lower slag viscosity arising from the increased meniscus temperature [151] (Fig. 3.35).

3.5.2.3 Thermal Insulation of Beds

If we consider the bed to consist of two layers (i.e. powder and sintered slag), the thicknesses of these two layers and the liquid slag pool (d_{pool}) have been studied by a number of investigators [152–158]. The vertical heat flux for the bed is given by

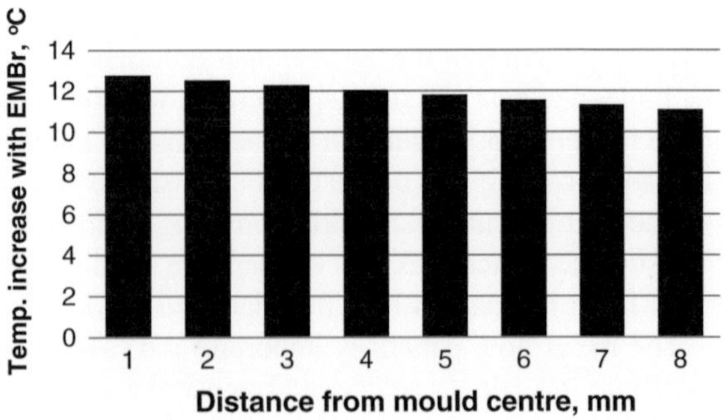

Fig. 3.35 Increase in meniscus temperature resulting from use of EMBr as a function of position in the mould (distance from centre); showing increase in meniscus temperature is ca. 10–12 °C (drawn using data due to [151])

$$q_{\text{vert,bed}} = k_{\text{bed}}\left(T_{\text{bed/pool}} - T_{\text{air/bed}}\right)/d_{\text{bed}}, \tag{3.21}$$

where k_{bed} = mean thermal conductivity for the two layers, $T_{bed/pool}$ = the temperature at the interface between the pool and the bed, $T_{air/bed}$ is the temperature at the air/bed interface (usually between 200 and 400 °C) [154]) and d_{bed} is the distance between the interfaces of the bed with the air and the molten pool. It is apparent that thermal insulation of the bed ($1/k_{bed}$) increases as the depth of the bed increases (see Fig. 4.60 [152–158]) as k_{bed} decreases and the temperature gradient decreases.

The same finding would be obtained if we applied the analysis to the heat flux to any specific layer. In practice, the thickness of these layers is determined by the heat flux entering and leaving each layer.

The vertical heat transfer has been studied using (i) simulation experiments [29, 159, 160] (see Sect. 9.8.4 and Fig. 9.52) and (ii) mathematical modelling [155–158]. In the latter, the temperature gradients in the various layers of the bed are affected by both the occurrence and the site of the various endothermic and exothermic reactions [152, 155, 157, 158, 161]. The exothermic, carbon combustion is affected by the size of the carbon particles (D_C) with larger particles taking longer to combust. The kinetics of carbon combustion has been studied by both experiments [157, 158, 162] and through mathematical models of the bed [155–158, 163] in which the permeability of the bed is taken into account [157, 158]. The rate of sintering of the oxide particles in the sinter layer has also been studied through experiments [152, 162, 164] and mathematical modelling [163, 164].

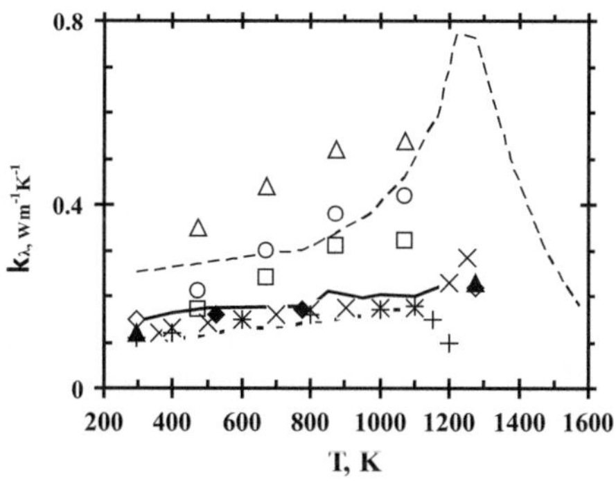

Fig. 3.36 Thermal conductivity of casting powders as a function of temperature (in *K*); *dotted* and *solid lines*, = Andersson values [167]; +, X; ♦ = k_{THW} due to Macho [166]; *dashed line* = Nagata values [101] □, Δ, ○ = values due to Taylor [102]; ▲, ◇ = Neumann [165] granules and powder, respectively (permission granted, Taylor & Francis, www.tandfonline.com [167])

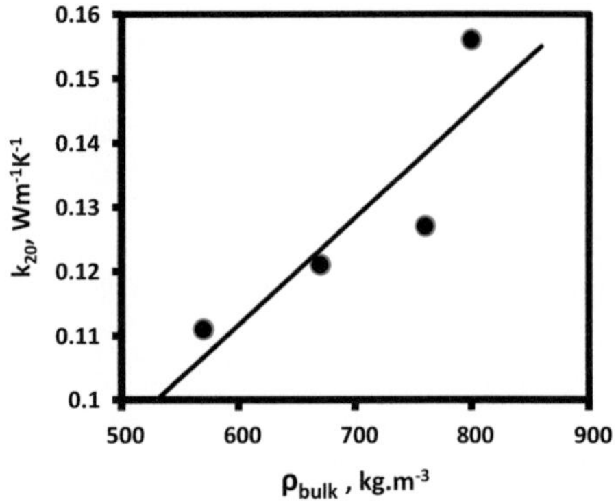

Fig. 3.37 Thermal conductivity of powders at 20 °C as a function of bulk density (permission granted, Taylor & Francis, www.tandfonline.com [167])

Thermal conductivity measurements have been carried out on casting powders by several workers [102, 103, 165–167]; reported values are shown in Fig. 3.36. With normal packing, the thermal conductivities of the powders (k_{powd}) at 25 °C have values around 0.12–0.15 W m^{-1} K^{-1} [167]. Values of k_{powd} increase with increasing temperature (dk dT = 9 × 10^{-5} W m^{-1} K^{-2}). Thus approximate values for k_{powd} for temperatures in the range (25–750 °C) are given by Eq. 3.22:

$$k_{powd\,T}\left(\mathrm{Wm}^{-1}\mathrm{K}^{-1}\right) = 0.13 + 9x\,10^{-5}(T - 25\,°\mathrm{C}). \quad (3.22)$$

Values of k_{powd} also increase with increasing powder density (ρ_{bulk}) due to the improved packing density and improved contact between particles (Fig. 3.37). The following equation was obtained by Andersson [167]:

$$k_{powd\,25}\left(\mathrm{Wm}^{-1}\mathrm{K}^{-1}\right) = 0.01 + 1.69x\,10^{-4}\rho_{bulk}. \quad (3.23)$$

The thermal conductivities of the sintered layer will, in general, be higher than those for the powder bed. The thermal conductivity of the sinter layer will decrease with increasing porosity (or decreasing density). The porosity in the sinter layer may decrease (or thermal conductivity increase) with increasing local temperature in the sinter bed. Thermal conductivity values for the sintered layer will lie between those for powders (ca. 0.15 W m^{-1} K^{-1}) and values reported for sintered slags with higher densities (1–2 W m^{-1} K^{-1}); in the absence of measured data a value of ca. 0.5 W m^{-1} K^{-1} is suggested for k_{sint}.

However, thermal insulation tests give thermal conductivity values for the bed in the range k_{bed} = 1.0 (for powders) to 1.3 W m^{-1} K^{-1} (for granules) [160] which are almost an order of magnitude higher than the values for powders given in Fig. 3.36. This discrepancy is probably due to the contribution from gaseous convection which would be expected to be large in tests where the powder is heated from below and where there is a large temperature gradient between the lower and upper surfaces (as there is in the thermal insulation tests). The question arises which value should be used to calculate the vertical heat flux. It is suggested that

(i) The higher k_{bed} value be used if gaseous convection is not being calculated separately and
(ii) The lower k_{bed} value be used if gaseous convection is being calculated separately.

3.5.2.4 Depth of Bed

It can be seen from Eq. 3.19 that the heat flux decreases (i.e. thermal insulation improved) as the depth of the powder bed increases (see Fig. 4.6) [153–156, 158]. Thus using a deeper powder bed is one way of decreasing the vertical heat flux and can be used to shorten the size of the solidified meniscus in order to reduce the entrapment of inclusions and bubbles and the depth of oscillation marks [165].

3.5.2.5 Powder/ Granule Size and Packing Density

The thermal insulation of the bed has been reported to increase as the mean powder/ granule size (D_{bed}) decreased [160]. This is due to the fact that smaller grained powders pack more densely and thereby reduce the permeability of the bed to gases passing through it. This, in turn, reduces the effective thermal conductivity of the sample. Thus thermal insulation is better for powder mould fluxes than for granulated fluxes since powder particles have smaller diameters.

3.5.2.6 Powders Containing Exothermic Agents

It can be seen from Eq. 3.19 that the heat flux increases with increasing temperature gradient (dT/dx). Consider two locations in the bed, with site 2 close to the interface with the liquid pool and site 1 higher up the bed; the temperature gradient is given by $\{(T_2 - T_1)/(d_2 - d_1)\}$ where d is the distance from the powder/air interface.

Exothermic powders usually contain silicon in the form of silicides (e.g. Fe/Si). At a certain temperature these silicides react with oxygen to form SiO_2. This reaction is very exothermic and the heat given out causes the local temperature to rise and thereby reduces the temperature gradient and hence, the vertical heat flux. The combustion of carbon is also exothermic and incidences of skull formation have been reported when free carbon contents have been lowered too much in the search for faster melting rates.

By the same token, the decomposition of carbonates and the melting of slag are both endothermic and thus these will cause decrease in local temperature increase in temperature gradient thereby increasing the vertical heat flux.

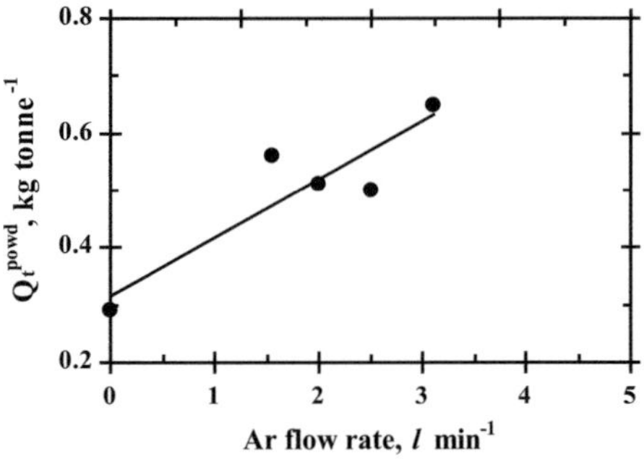

Fig. 3.38 The powder consumption, Q_t, as a function of Ar flow rate [170] (*courtesy* of A.B. Fox)

3.5.2.7 Argon Blowing

Argon blowing increases the vertical heat flux due to the promotion of gaseous convection. It is known that bigger bubbles emerge close to the SEN and smaller bubbles emerge near the mould walls (where they may be incorporated into the metal flow [168]). It would thus be expected that the vertical heat flux due to Ar flow would be greater near the SEN than at the edges of the mould. However, Ar bubbling also affects the metal flow in the mould. The Ar bubbles tend to "*cushion*" the metal flow. However, it has been reported that an increase in Ar flow rate from 4 to 5 L min^{-1} causes the metal flow pattern to change from a "*double roll*" to "*single roll*" pattern [135, 169]; this is accompanied by a decrease of the steel temperature (at the slag/steel interface) and has been reported to lead to skull formation on occasions.

Since gaseous convection contributes to the vertical heat flux, it is necessary to consider the permeability of the powder bed since this too affects the magnitude of the contribution. It is known that (i) the thermal insulation of a bed derived using powders is greater than that for granules and (ii) thermal insulation improves as the diameter (D) of the particle or granule decreases ($k_{\text{bed}} \downarrow$ as $D \downarrow$). This might help to explain the increased powder consumption with increased Ar flow rate (Fig. 3.38) [170]. Increased vertical heat flux (i.e. gaseous convection) leaving the molten slag pool would be expected to increase the melting rate of the powder in the bed and thus lead to higher powder consumption.

3.6 Summary

Control of the heat transfer from shell to mould is essential in order to minimise incidences of longitudinal cracking and sticker breakout. The heat transfer is usually divided into horizontal and vertical heat transfer with the horizontal heat flux being the more important. The factors affecting four aspects of heat transfer in the mould are considered here, namely (i) horizontal heat transfer, (ii) shell solidification, (iii) vertical heat transfer and (iv) the variability in heat transfer.

Although many factors affect horizontal heat transfer, control is usually exerted through manipulation of the slag film formed between the steel shell and the mould. The principal factors involved are (i) the *thickness of solid slag film* (which is dependent on the solidification or break temperature) and (ii) the *fraction of crystalline phase* (f_{cryst}) formed in the film (which reduces the radiation conduction and creates an interfacial resistance, $R_{Cu/sl}$). In practice, the heat flux is usually reduced by increasing the basicity of the mould powder.

The metal flow is also an important factor but usually the factors controlling metal flow (such as SEN immersion depth) are optimised to provide a satisfactory "*double roll*" pattern and so metal flow tends not to be an issue. However, SEN blockages and clogging can lead to unsatisfactory metal flow patterns and excessive Ar flow rates can lead to the formation of "*single roll*" patterns.

Vertical heat transfer can be controlled through (i) the *depth of the bed,* (ii) *use of exothermic mould powders,* (iii) to a less amount by the *particle size of the powder* and (iv) by the *use of EMBr* which reduces the efficiency of heat transfer by retarding the metal flow velocity. Gaseous convection is a contributor to the vertical heat flux, which, in turn, is affected by the Ar flow rate, the carbonate content of the mould powder and the permeability of the powder bed

References

1. PE Ramirez-Lopez, KC Mills, PD Lee, B Santillana, ISIJ Intl., **50,** 1797, (2010).
2. PE Ramirez-Lopez, KC Mills, PD Lee, B Santillana, Met. Mater. Trans. **43**, 109, (2011).
3. M Hanao, M Kawamoto, ISIJ Intl., **48**, 180, (2008).
4. H Jacobi, Arch. Eisenhüttenwesen, **47**, 441, (1976).
5. R. Gardon, Proc. 2nd Intl. Conf. Thermal Conductivity, Ottawa, 1962, p. 167 and J. Amer. Ceram. Soc. **44**, 305, (1961).
6. M Susa, K Nagata, KC Mills, Ironmaking and Steelmaking, **20**, 312, (1993).
7. S Ozawa, M Susa, T Goto, R Kojima, KC Mills, ISIJ Intl., **46**, 413, (2006).
8. M. Susa, KC Mills, MJ Richardson, R Taylor, D Stewart, Ironmaking and Steelmaking, **21**, 279, (1994).
9. M Hayashi, K Matsuo, K Nagata, H Nakada, Proc. 8th Intl. Conf. Molten Slags, fluxes and salts, Santiago, Chile, 2009, (GECAMIN, Santiago, Chile, 2009) p. 1091.
10. M. Kawamoto, Paper presented at Intl. Workshop on "*Thermophysical data for the development of mathematical models of solidification*", Gifu City, Japan, Oct. (1995) p. 10.
11. M Susa, A Kushimoto, H Toyoto, M Hayashi, R Endo, Y Kobayashi, ISIJ Intl., **49,** 1722, (2009).
12. M Susa, A Kushimoto, R Endo, Y Kobayashi, ISIJ Intl., **51**, 1587, (2011).
13. Y Kobayashi, R Maehashi, R Endo, M Susa, ISIJ Intl., **53**, 1725, (2013).
14. JW Cho, T Emi, H Shibata, M Suzuki, ISIJ Intl., **38**, 268, and 440 (1998).
15. H Shibata, K Kondo, M Suzuki, T Emi, ISIJ Intl., **36**, S 179, (1996).
16. MS Jenkins, Proc. 78th Steelmaking Conf. (1995) p 669.
17. M Hanao, M Kawamoto, A Yamanaka, ISIJ Intl., **52**, 1310, (2012).
18. K Nishioka, T Maeda, M Shimizu, ISIJ Intl., **46**, 427, (2006).
19. P. Andersson, C Eggertson, Ironmaking and Steelmaking, **42**, No 6, (2015), p. 456.
20. P Andersson, *Thermal conductivity of powders used in continuous casting of slabs and billets*. To be published.

21. R. Saraswat, A.B. Fox, K.C. Mills, P.D. Lee, B. Deo, Scand., J Met., **33**, 85, (2004).
22. IV Samarasekera, JK Brimacombe, Intl. Metals Review, **286**, (1978).
23. J. W. Cho, H. Shibata, T. Emi, M. Suzuki, ISIJ Int., **38** (5), 440 (1998),
24. K Watanabe *et al.* Proc.79th Steelmaking Conf.,1996, (ISS, Warrendale, PA, 1996) p. 265.
25. H Shibata, JW Cho, T Emi, M Suzuki, Proc. 5th Intl. Conf. Molten slags, fluxes and salts, Sydney, 1997, (ISI –AIME, Warrendale, PA, 1997) p. 771.
26. H Nakada, K Nagata, ISIJ Intl., **46**, 441, (2006).
27. K Tsutsumi, T Nagasaka, M Hino, ISIJ Intl., **39**, 1150, (1999).
28. JW Cho, T Emi, H Shibata, M Suzuki, ISIJ Intl., **38**, 834 (1998).
29. KH Spitzer, JF Holzhauser, FU Bruchner, B Siera, HJ Gretthe, K Schwerdtfeger, Stahl u Eisen, **108** (9), 71, (1988).
30. M Ichihashi, M Kawamoto, Y Tsukaguchi, N Nishida, T Kanazawa, S Hiraki, Jap. Soc. Promotion of Science (1995).
31. K Watanabe, M Suzuki, K Murakami, Tetsu-to-Hagane, **83**, 115, (1997).
32. A Yamauchi, K Sorimachi, T Yamauchi, Ironmaking and Steelmaking, **29**, 203, (2002).
33. T Inoue, K Noro, Y Akita, I Katano, Nippon Steel Technical Rept. 12, **86**, (1978) 86/96 see also Seitetsu-Kenkyu: **294**, (1973) 12412.
34. M Hanao, M Kawamoto, A Yamanaka, Proc. 8th Europ. Conf. Continuous Casting, Graz, 2014, (Austrian Metals Mater. Soc., 2014).
35. H Yamamura, T Kajitani, J Nakashima, M Yamasaki, S Mineta, Nippon Steel Technical Report, **104**, 54, (2013).
36. J Kromhout, RS Schimmel, Proc. 8th Europ. Conf on Continuous Casting, Graz, 2014, (Austrian Metals Soc., 2014).
37. A Yamauchi, K Sorimachi, T Sakuraya, T Fujii, ISIJ Intl., **33**, 140, (1993).
38. M Kawamoto, A Hakari, T Watanabe, CAMP-ISIJ, **10**, 890, (1997).
39. H Nakato, M Ozawa, K Kinoshita, Tetsu-to Hagane, **67**, 1200, (1981).
40. DT Stone, BG Thomas, Canad. Met. Quart., **38** (5), 363, (1999).
41. A Yamauchi, T Yamauchi, S Seetharaman, Proc. 6th Intl. Conf. Molten slags, fluxes and salts, Stockholm, 2000, paper 148.
42. M Hanao, M Kawamoto, ISIJ Intl., **49**, 365, (2009).
43. G Volk, K Wunnenberg, Klepzig Fachberichte, **80** (10), 491, (1972).
44. JA Kromhout, ER Dekker, M Kawamoto, R Boom, Ironmaking and Steelmaking, **40**, 206, (2013).
45. BG Thomas, B Ho, G Li, Proc. McLean Symp., Toronto, (1998) (ISS, Warrendale, PA, 1998) p 177.
46. K Nakajima, S Hiraki, T Kanazawa, T Murakami, CAMP-ISIJ, **5**, 1221, (1992).
47. R Bommaraju, Proc. 74th Steelmaking Conf. (1991) p. 131.
48. CA Dacker, T Sohlgren, Steel Research Intl., **81**, 899, (2010).
49. R Alberny, A Leclercq, D Amoury, M Lahousse, Revue de Metall., **73** (July/Aug), 545, (1976).
50. HP Narzt, R Martinelli, C Furst, G Xia, H Pressinger, Proc. 3rd Europ. Conf. Continuous Casting, Madrid, 1998 (UNESID, Madrid, 1998) p 747.
51. S. Ogibayashi, Proc. 85th Steelmaking Conf. (2002) (ISS/AIME Warrendale, PA, 2002) p 175.
52. MM Wolf, Iron and Steelmaker, **23** (2), 47, (1996).
53. EY Ko, J Choi, JY Park, I Sohn, Met. Mater. Intl. **20**, 141 and 1103 (2014).
54. M Suzuki, S Miyahara, T Kitagawa, S Uchida, K Okimoto, Tetsu-to- Hagane, **78**, 113, (1992).
55. WL Wang, AW Cramb, ISIJ Intl., **45**, 1864, (2005).
56. P Grieveson, S Bagha, N Machingawuta, K Liddell, KC Mills, Ironmaking and Steelmaking, **15**, 181, (1988).
57. KC Mills, AB Fox, ISI J Intl., **43**, 1479, (2003).
58. GH Wen, S Sridhar, P Tang, X Qi, TQ Liu, ISIJ Intl., **47**, 1117 (2007).
59. H Nakada, K Nagata, ISIJ Intl., **46**, 441, (2006).

60. Q Wang, YJ Lu, SP He, KC Mills, Z Li, Ironmaking and Steelmaking, **38**, 297, (2011).
61. JN Pontoire, JP Radot, V Delvaux, E Dehaussy, Revue Metall., CIT, **93**, 1237, (1996).
62. MJ Lu, JH Chen, DH Tsai, CH Huang, Proc. 78th Steelmaking Conf., 1995, (ISS, Warrendale, PA, 1995) p. 359.
63. J Nagai, M Ohnishi, T Yamamoto, K Hirayama, S Ohto, T Fujiyama, Tetsu-to- Hagane, **69**, S158, (1983).
64. CS Li, BG Thomas, Proc. ISS Tech Conf. (2003), Indianapolis, (ISS, Warrendale, PA, 2003) p. 685.
65. BG Thomas, C Ojeda, Proc. ISS Tech. Conf., Indianapolis, 2003, (ISS, Warrendale, PA, 2003) p. 295.
66. A Yamauchi, S Itoyama, Y Kishimoto, H Tozawa, K Sorimachi, ISIJ Intl., **42**, 1094, (2002).
67. T Watanabe, CAMP- ISIJ, **14**, 153, (2001).
68. A Cristallini, R Langella, A Polk, Proc. 3rd Europ. Conf. Continuous Casting, Madrid, Spain, 1998, (UNESID, Madrid, 1998) p. 589.
69. S Riaz, Private communication on "*Reducing horizontal heat flux and increasing mould life by coatings on mould*". Tata Steel Teesside Research Laboratories, Middlesborough (2008).
70. C Bertolletti, P Courbe, JM Jolivet, PP Naveau, A Oper, E Perrin, C Salaris, AL Spierings, E Wiesseldinger, Proc. 3rd Europ. Conf Continuous Casting, Madrid, 1998, (UNESID, Madrid, 1998) p. 65.
71. M M'Hamdi, J Lessoult, E Perrin, JM Jolivet, ISIJ Intl., **36**, S197, (1996).
72. K Nakai T Sakashita, M Hashio, M Kawasaki, K Nakajima, Y Sugitani, Tetsu-to-Hagane, **73** (3), 498, (1987).
73. Y Sugitani, M Nakamura, M Okuda, M Kawasaki, K Nakajima, Trans. ISIJ, **25**, B91, (1985).
74. JW Cho, HT Jeong, Met. Mater. Trans. B, **44B**, 146, (2013).
75. H Murukami, M Suzuki, T Kitagawa, S Miyahawa, Tetsu-to- Hagane, **78**, 105, (1992).
76. T Mochida, S Kohriyama, S Itoyama, T Yamashita, M Suzuki, Y Kishimoto, CAMP-ISIJ, **20**, 851, (2007).
77. K Ichikawa, A Morita, Y Kawabe, Shinagawa Tech. Report, **36**, 283, (1993).
78. K Watanabe, M Suzuki, K Murakumi, H Kondo, A Miyamoto, T Shiomi, Tetsu-to-Hagane, **83**, 115, (1997).
79. M Hanao, M Kawamoto, M Hara, T Murakami, H Kikuchi, K Hanazaki, Tetsu-to-Hagane, **88**, 23, (2002).
80. V Guyot, JF Martin, A Ruelle, A d'Anselme, JP Radot, M Bobadilla, JV Lavant, JN Pontoire, ISIJ Intl., **36**, S 227, (1996).
81. GH Wen, P Tang, B Yang, XB Zhu, ISIJ Intl., **52**, 1179, (2012).
82. AS Normanton, PN Hewitt, NS Hunter, B Harris, D Scoones, Proc. 4th Europ. Conf. Continuous Casting, Birmingham, 2002, (IOM, London, 2002) p. 561.
83. M Hanao, M Kawamoto, M Hara, T Murakami, H Kikuchi, A Yamanaka, Proc. 5th Europ. Conf. Cont, Casting, Nice, France (2005) see also Tetsu-to Hagane, **88** (1), 23, (2002).
84. CB Shi, MD Seo, JW Cho, SH Kim, Proc. 8th Europ. Conf. Continuous Casting, Graz, 2014, (Austrian Metals Soc. 2014).
85. KC Mills, ISIJ Intl., **56** (1), 1 and 14, (2016).
86. S Bagha, N Machingawuta, P Grieveson, Proc. 3rd Intl. Conf. Molten slags and fluxes, Glasgow, 1988, (Inst. of Metals, London, 1988) p. 235.
87. N Machingawuta, S Bagha, P Grieveson, Proc. 74th Steelmaking Conf.,1991, (ISS-AIME, Warrendale, PA, 1991) p. 163.
88. MS Jenkins, PhD thesis, *Heat transfer in the continuous casting mould*. Monash Univ., Australia, (1999).
89. D Bouchard, FC Hamel, JP Nadeau, S Bellemare, F Dreneau, DA Tremblay, D Simad, Met. Mater. Trans. B, **32B**, 111, (2001).
90. X Yu, GH Wen, P Tang, H Wang, Ironmaking and Steelmaking, **36**, 623, (2009).
91. X Qi, GH Wen, P Tang, J Non- Cryst. Solids, **354**, (52/54), 5444, (2008).

92. JF Holzhauzer, KH Spitzer, K Schwerdtfeger, Steel Research, **70**, 252, (1999).
93. A Yamauchi, K Sorimachi, T Sakuraya, Proc. 4th Intl. Conf. Molten slags and fluxes, Sendai, Japan, 1992 (ISIJ, Tokyo, 1992) p. 415.
94. R Scheel, W Korte, Metall. Plant and Technol., **1987** (6), 22, (1987).
95. H Tomono, H Ackermann, W Kurz, W Heinemann, *Solidification Technology in the foundry and Cast house* (Metals Soc., Warwick, 1983) Book 273; p. 524.
96. PJ Wray, Met. Trans B, **12B**, 167, (1981).
97. DK Stemple, EC Zuluta, MC Flemings, Met. Trans B, **13B**, 503, (1982).
98. I Saucedo, *Initial Solidification and strand surface quality of pertectic steels. Continuous Casting*. Vol. 9, (ISS-AIME Warrendale, PA, 1997) p. 131.
99. A. Badri, T.T. Natarajan, C.C. Snyder, K.D. Powers, F.J. Mannion, A.W. Cramb: Met. Mater. Trans. B, **36B**, 355, (2005).
100. A. Badri, BT Natarajan, CC Snyder, KD Powers, FJ Byrne, M Byrne, AW Cramb, Met. Mater. Trans B, **36B**, 373, (2005).
101. I Sohn, TT Natarajan, TJ Piccone, KD Powers, CC Snyder, Proc. 6th. Europ Conf. Continuous Casting, Riccone, Italy, 2008.
102. K Nagata, Proc. 2nd Intl. Symp. Metallurgical slags and fluxes, Lake Tahoe, NV, 1984, ed. HA Fine and DR Gaskell, (Met. Soc. AIME, Warrendale, PA. 1984) p. 875.
103. R Taylor, KC Mills, Ironmaking and Steelmaking, **15**, 187, (1988).
104. H Ohta, K Masuda, K Watanabe, K Nakajima, H Shibata, Y Waseda, Tetsu to Hagane, **80** (6), 33, (1994).
105. Y Waseda, K Masuda, K Watanabe, H Shibata, H Ohta, K Nakajima, High Temp. Materials & Processes, **13,** 267, (1994).
106. M Gonerup, M Hayashi, CA Dacker, S Seetharaman, Proc. 7th Intl. Conf. Molten slags, fluxes and salts, Capetown, 2004, (SAIMM, Johannesburg, 2004) p. 745.
107. BJ Monaghan, RF Brooks, Ironmaking and Steelmaking, **29**, 115, (2002).
108. J Diao, B Xie, J Xiao, C Ji, ISIJ Intl., **49**, 1710, (2009).
109. J Chen, J Xie, F He, E Wan, Z Xu, C Yang, G Liu, Ironmaking and Steelmaking, **42**, 128, (2015).
110. R Endo, Y Kono, Y Kobayashi, M Susa, S Mineta, H Yamamura, Tetsu- to Hagane, **100**, 571, (2014).
111. S Sridhar; KC Mills; ODC Afrange, HP Lorz, R Carli, Ironmaking and Steelmaking, **27**, 238, (2000).
112. H Nakada, M Susa, Y Seko, M Hayashi, K Nagata, ISIJ Intl., **48**, 446, (2008).
113. S Ogibayashi, Proc. 85th Steelmaking Conf. 2002, (ISS, Warrendale, PA, 2002) p. 175.
114. S Ogibayashi, CAMP-ISIJ, **18**, 126, (2003).
115. R Saraswat, DM Maijer, PD Lee, KC Mills, ISIJ Intl., **47**. 96, (2007).
116. BG Thomas, Proc. McLean Symp., Toronto, 1998, (ISS, Warrendale, PA) p. 177.
117. Y Meng, BG Thomas, Met. Mater. Trans., B, **34B**, 681 and 707, (2003).
118. BG Thomas, Proc. ISS Tech Conf., 2003, (ISS, Warrendale, PA, 2003) p. 295.
119. C Li, BG Thomas, Met. Mater. Trans. B, **35B**, 1151, (2004).
120. KC Mills, PE Ramirez- Lopez, PD Lee, High Temp. Mater, and Processes, **31**, 221, (2012).
121. M Hanao, K Kawamoto, ISIJ Intl. **48**, 180, (2008).
122. A Suzuki, T Suzuki, Y Nozaki, Tetsu-to Hagane. **55**, S 110, (1969) and **56**, S 427, (1970).
123. A Suzuki, T Suzuki, Y Nagaoka and Y Iwata: J Jap. Inst. Metals, **32**, 1301, (1968).
124. DR Poirier, EJ Poirier, *Heat transfer fundamentals for metal casting*. (TMS, Warrendale, PA, 1994).
125. M Suzuki, S Miyahara, T Kitagawa, S Uchida, K Okimoto, Tetsu-to- Hagane, **78**, 113, (1992).
126. N Nagaoka, K Iwamoto, G Usui, S Nemota, Y Ohkawa, Tetsu- to- Hagane, **55**, S 109, (1969).
127. S Okano, T Nishimura, H Ooi, T Chino, Tetsu- to- Hagane, **61**, 2982, (1975).
128. T Mori, K Ayata, M Fujimaki, T Soejima, M Kawahara, Tetsu- to- Hagane, **62**, S 132, (1976).

129. S Fujii, T Ohashi, Tetsu- to- Hagane, **64**, S 646, (1978).
130. K Narita, T Mori, K Ayata, J Miyazaki, M Fujimaki, T Shiomi, Tetsu- to- Hagane, **64**, S 659, (1978).
131. H Nakato, Y Habu, H Kitaoka, K Kinoshita, T Emi, Trans. ISIJ, **21**, B 393, (1981).
132. JA Moore, C Cimeno, S Diehl, RJ Phillips, D Piwinski, Proc. 79th Steelmaking Conf., 1996, (ISS, Warrendale, PA, 1996) p. 259.
133. TJ Billany AS Normanton, KC Mills, P Grieveson, Ironmaking and Steelmaking, **18**, 403, (1991).
134. C Real, R Miranda, G Vulchis, M Barron, L Hoyes, J Gonzalez ISIJ Intl., **46**, 1183, (2006).
135. E van Vliet, DW van der Plas; SP Carless, A A Kamperman, AE Westendorp, Proc. 7th Europ. Conf. Cont. Casting, Dusseldorf, (2011) Session 4.
136. PD Lee, PE Ramirez-Lopez, KC Mills, B Santillana, Ironmaking and Steelmaking, **39** (4), 244, (2012).
137. RJ O' Malley, J Neal, Proc. Conf. on new developments in Metall. Proc. Technol., Dusseldorf,1999, METEC Congress, p. 73.
138. H Tai, M Morashita, T Miyake, Proc. 3rd Europ. Conf. Continuous Casting, Madrid, 1998, (UNESID, Madrid,1998) p. 447.
139. T Matsumiya, ISIJ Intl., **46**, 1800, (2006).
140. SN Singh, KE Blazek, J Metals, **26** (10), 17, (1974).
141. M Suzuki, JW Cho, H Sato, H Shibata, T Emi, Proc 81st Steelmaking Conf., 1998, (ISS, Warrendale, PA., 1998) p. 165.
142. S Ogibayashi, K Yamaguchi, T Mukat, T Takahashi, Y Mimura, K Koyama. Y Nagano, T Nagano., Nippon Steel Technical Report, **34**, 1, (1987).
143. J Kromhout, PhD Thesis, "*Mould powders for the high speed continuous casting of steel*". Univ. of Delft, (2011) p. 165.
144. C-A Dacker, P Andersson, C Eggertson, *The evaluation of the mould slag film during continuous casting of steel*, Proc. ISIJ-VDEh-Jernkontoret Joint Symposium, Osaka (2013), p. 182.
145. MR Ozgu, B Kocatulum, Iron and Steelmaker, **21** (5), 77, (1994).
146. F Mantovani, S Spagnul, M Padovan, A Bianco, Proc 8th Europ. Conf Continuous casting, Graz, 2014 (Austrian Metals Soc., 2014).
147. T Lamp, M Tamminga, H Kochner, S Schiewe, D Kirsch, Proc. 8th Europ. Conf. Continuous Casting, Graz, 2014 (Austrian Metals Soc., 2014).
148. NS Hunter, JD Madill, PN Hewitt, AS Normanton, Proc. 3rd Europ. Conf. Continuous casting, Madrid, 1998 (UNESID, Madrid, 1998) p. 289.
149. R Saraswat, PhD Thesis, "*Modelling the effect of mould flux on steel shell formation during continuous casting*", Imperial College, London (2006).
150. R Koldewijn Unpublished Corus Internal Rept (2007) cited in KC Mills, J Kromhout, A Hamoen, R Boom, Proc. of Admet Conf., Dnipropetrovsk, Ukraine, 2007, (Natl. Metall. Acad. Ukr., Dnipropetrovsk,2007) vol. 2, p. 174.
151. H Take, H Osanai, J Hasanuma, T Yamamoto, H Bada, H Tozaka, Proc.. Conf. Quality Improvement Technol for ordinary steel in Iron and steelmaking process, Bangkok, 1994 Session 3 Paper 1.
152. K Schwerdtfeger, "*Giessen und Erstarren von Stahl III*". Final Report EUR 8569 (1981) Research contract to ECSC, 7210. CA/112 9 (Europ. Comm. Sci. & Tech. Publ., Luxembourg, 1981).
153. T Sakuraya, T Emi, T Imai, K Emoto, M Kodama, Tetsu-to- Hagane, **67**, 1220, (1981).
154. R Shah, JG Williams, G Hecko, Proc. ISS Tech. Conf., 2003, (ISS, Warrendale, PA, 2003) p. 555.
155. NN Viswanathan, S Sridhar, KC Mills, S Du, Scand. J Metall., **31**, 191, (2002).
156. RM McDavid, BG Thomas, Met. Mater. Trans B, **27B**, 672, (1996).
157. H Nakato, S Takeuchi, T Fujii, T Nozaki, N Washio, Proc. 74th Steelmaking Conf., Washington, DC, 1991 (ISS, Warrendale, PA, 1991) p. 639.

158. H Nakato, T Sakuraya, T Nozaki, T Emi, H Nikoshawa, Proc. 69th Steelmaking Conf., Washington, DC, 1986, (ISS, Warrendale, PA, 1986) p. 137.
159. KH Spitzer, K Schwerdtfeger: *Investigation of the isolating properties and melting behaviour of mould powders for continuous casting*. Tech Univ. Clausthal Report 7249 (Dec 1989).
160. S Diehl, JA Moore, RJ Phillips, Proc. 78th Steelmaking Conf., Nashville, TN,1995, (ISS, Warrendale, PA, 1986) p. 351.
161. Y Matsushita, T Takahashi, Kobe Research & Development, **43** (2), 123 (1993).
162. M Supradist, AW Cramb, K Schwerdtfeger, ISIJ Intl., **44**, 817, (2004).
163. MB Goldschmidt, JC Gonzalez, EN Dvorkin, Ironmaking and Steelmaking **20**, 379, (1993).
164. M Kawamoto, K Nakajima, CAMP-ISIJ, **5**, 1297, (1992).
165. F Neumann, J Neal, MA Pedroza, AH Castiliejos E., FA Acosta: Proc.79th. Steelmaking Conf., 1996, (ISS, Warrendale, PA, 1996) p. 249.
166. JJ Macho, G Hecko, B Golinmowski, M Frazee: *Development of a new generation of no free carbon continuous casting fluxes*. Preprints 33rd McMaster Symp. On Iron and Steelmaking, Hamilton, Ont., Canada, (2005) (McMaster Univ. Hamilton, CN, 2005) p. 131.
167. P Andersson, Ironmaking and Steelmaking **42**, 6 (2015), p. 465.
168. K Pericleous, private communication, University of Greenwich, UK, (2009).
169. PE Ramirez- Lopez, PN Jalali, J Bjorkvall, U Sjostrom, C Nilsson, Proc. 8th Europ. Conf Continuous Casting, Graz, Austria, 2014 (ASMET, Vienna, 2014).
170. T Mallaband, Unpublished results from Metallurgica, UK, cited in A Fox: PhD Thesis, Dept. of Materials, Imperial College, London (2003).

Chapter 4
How to Manipulate Slag Behaviour in the Mould

Abstract The mould slag plays a key part in the continuous casting process. The mould slag carries out a series of tasks and it is essential that the slags perform each of these tasks efficiently. This chapter looks at these tasks and analyses the key factors affecting performance and suggests ways in which the slag can be manipulated to perform efficiently. The following processes are examined: (1) Thermal insulation of the powder bed and vertical heat transfer (2) Melting rate of the mould powder (3) Formation of a slag pool (4) Control of powder consumption and lubrication of the shell (5) Control of Horizontal heat transfer and the effect of the thickness of the solid slag film (6) Control of crystalline and glass phases in the slag film and (7) Control of both horizontal and vertical heat transfer in order to delay solidification of the shell. The proposed actions are summarised in a table.

Symbols, Abbreviations and Units

D	diameter (m)
d	depth, thickness (m)
f_{crys}	fraction crystalline
k	Thermal conductivity ($Wm^{-1}K^{-1}$)
Q_{MR}	Melting Rate (kg min^{-1} or kgs^{-1})
Q_s	Powder consumption (kg_{slag} m^{-2})
q	Heat flux density (Wm^{-2})
R_{th}	Thermal resistance (m^2KW^{-1})
$R*$	(Surface area/volume) of mould (m^{-1})
T	Temperature (°C)
t	thickness of mould (m) or time (s)
V_c	Casting speed (m min^{-1})
w	width of mould (m)
AC	Alternating current

The original version of this chapter was revised: For detailed information please see correction. The correction to this chapter is available at https://doi.org/10.1007/978-3-319-53616-3_12

Within this book minor changes to some of the figure permissions and the readability of the chapters have been made and several figures have been replaced with new versions.

K.C. Mills and C.-Å. Däcker, *The Casting Powders Book*,
DOI 10.1007/978-3-319-53616-3_4

DC	Direct current
EMBr	Electro-magnetic braking
EMC	Electro-magnetic casting
SEN	Submerged entry nozzle

Subscripts and Superscripts

gran	granule
hor	horizontal
lat	lattice
pool	slag pool
powd	powder (layer or pulverised)
req	required
sint	sintered layer
slag	slag *cf.* powder
vert	vertical

Chemical Formula for Slags and Minerals

A	Al_2O_3
B	B_2O_3
C	CaO
F	FeO
F*l*	CaF_2
K	K_2O
L	Li_2O
M	MgO
Mn	MnO
N	Na_2O
S	SiO_2
T	TiO_2

4.1 Introduction

There are many factors affecting the performance of the mould powder in the continuous casting mould. Furthermore, these factors tend to be interactive. When certain defects occur in continuous casting it is necessary to take remedial action (e.g. if the steel product has a large inclusion content, the problem can be partially solved by reducing the length of the steel meniscus and this can be achieved by reducing the vertical heat flux). The necessary, remedial actions, such as the reduction of vertical heat flux, are discussed in this chapter and the ways in which the mould powder can be manipulated to achieve this and other targets, are described. Some of these topics have been discussed in other chapters; the aim here has been to place all remedial actions together in one location. The possible remedial actions are summarised in a single table at the end of the chapter to provide the reader with quick access to the different possible remedial actions which are available. In this approach, only the immediate effects of the remedial action are described, the effects of collateral interactions are largely ignored.

The various topics are arranged in terms of their location in the mould, starting at the top of the mould and working downwards.

4.2 Vertical Heat Flux and Thermal Insulation of Bed

These two topics are intimately linked since the thermal insulation supplied by the powder bed is used to control the vertical heat flux. The resistance to heat flow is that provided by the slag pool, the sinter layer and the powder bed (Fig. 4.1).

4.2.1 Vertical Heat Flux

There are several mechanisms involved in the vertical heat transfer. In the liquid slag pool, heat transfer occurs by (i) lattice conduction (ii) thermal convection and (iii) radiation conduction; the latter two mechanisms dominate in the liquid pool. In the sinter and powder layers, the principal mechanisms are thermal conduction and gaseous convection.

The thermal insulation provided by the mould powder can be considered as that provided by the three layers (powder, sinter and liquid) of thickness, *d*. The vertical heat flux density (q_{vert}) from the surface of the steel to the powder/air interface (Fig. 4.1) can be calculated by Eq. 4.1 (assuming 1-dim heat transfer) where

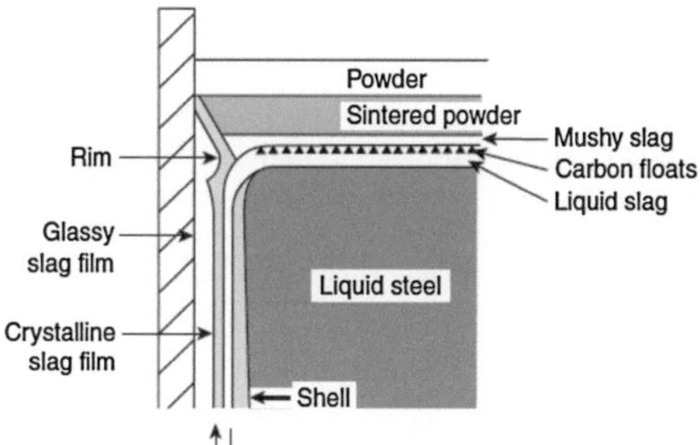

Fig. 4.1 Schematic drawing showing the various slag layers formed in the powder bed and the slag film; Note (i) glassy phase and liquid shown in *white* and crystalline phase as grey (ii) high cooling rates promote glass formation and so glass forms initially on the mould side of slag film but *cuspidine* is also precipitated at high temperature on mould side and, subsequently, crystal growth occurs in the high temperature regions of the slag film, over time

d_{bed} = (d_{pool} + d_{sint} + d_{powd}) and k_{bed} is the mean value for the three layer system and A = area = width × thickness of the mould.

$$q_{vert} = -k_{bed}A\left(T_{steel/slag} - T_{powd/air}\right)/d_{bed} \tag{4.1}$$

The thermal resistance, R_{th} = (d_{layer}/k_{layer}) of the three layers in the bed (Eq. 4.2) can be obtained by assuming a series resistance for Ohm's Law (in a similar way to that applied to the slag film [1–3]) and where k for the various layers refers to the effective thermal conductivity (i.e. containing contributions from conduction, convection and sometimes, radiation). The thermal resistance for each layer increases with increasing thickness (d) and decreasing effective thermal conductivity, k.

$$R_{th} = (d/k)_{pool} + (d/k)_{sint} + (d/k)_{powd} \tag{4.2}$$

The depth of the slag pool and the thicknesses of the sintered and powder layers are affected by the differences between the heat entering and the heat leaving that layer (Eq. 4.3). The depth of the slag pool decreases with decreasing q^{in}_{vert}, as witnessed by the lower pool depths associated with the application of EMBr. The depth of any layer increases with increasing, Δq_{vert}; thus, the depth of the layer increases with increasing q^{in}_{vert} and decreasing q^{out}_{vert}

$$\Delta q_{vert} = q^{in}_{vert} - q^{out}_{vert}. \tag{4.3}$$

In the case of the sintered and powder layer, porosity will have conflicting effects on the vertical heat flux since increased porosity enhances gaseous convection and decreases lattice thermal conduction.

4.2.1.1 Importance of Vertical Heat Flux

Control of the vertical heat flux is essential in order to ensure that:

(i) The steel surface does not solidify (due to an excessive q_{vert}).
(ii) The slag pool has sufficient depth to (a) cover the surface waves formed in the steel meniscus (b) provide good slag consumption and (c) avoid C-pick-up by the shell [4–6].
(iii) The solid shell/meniscus is not excessively long since a long meniscus can lead to (a) inclusion and bubble entrapment (b) deep oscillation marks (and transverse cracks) [7].

4.2.1.2 Control of Vertical Heat Flux

The vertical heat flux is largely controlled through the thermal insulation of the bed. It can be seen from Eq. 4.2 that thermal insulation is largely determined by the thickness and effective thermal conductivity of the bed.

The effective thermal conductivity of the liquid pool is determined by the contribution from convection (some models assume $k_{eff} = N\ k_{lat}$ where N is given values between 10 and 100 but usually optimises at *ca.* 15 [8]). The convective contribution will decrease with increasing viscosity of the slag pool and radiation contributions will decrease with additions of transition metal oxides [9, 10].

Values of k_{eff} for the solid layers contain contributions from the lattice conduction and gaseous convection; the latter is affected by the permeability of the layer and so is affected by the packing density (or bulk density, $\rho_{bulk,}$ the powder size (D_{powd})) and the thermal gradient across the layer. The results of thermal insulation tests [11] suggest that gaseous convection is the dominant mechanism in the solid layers (see Sects. 4.2.3 and 9.4.5.8); thus, q_{vert} decreases as both permeability and D_{powd} decrease and as ρ_{bulk} increases.

Exothermic reactions cause local increases in bed temperature, which reduce the temperature gradient and thereby reduce q_{vert} (i.e. improve the thermal insulation) and increase the depth of the molten slag pool. Endothermic reactions, such as carbonate decomposition, will increase the heat losses by (a) decreasing the local temperature gradient and thereby increasing the thermal gradient and (b) creating more porosity and (c) by adding to the gaseous convection.

The vertical heat flux is also affected by the following casting conditions:

Steel throughput and Casting speed

Increases in steel throughput are manifested as casting speed increases which result in an increase in the heat flux density (q_{vert}). Since both the enthalpy demand (ΔH, defined in Eq. 4.4) and cross-sectional area of the mould remain constant, the ratio (q_{vert}. $A/\ \Delta H$) must increase and so q_{vert} must increase, in response to an increase in casting speed.

Steel Temperature

It can be seen from Eq. 4.1 that the heat flux is dependent upon the temperature gradient ($T_{steel/slag} - T_{powd/air}$). Thus, a decrease in steel temperature will reduce this gradient and hence, the heat flux. Steel liquidus temperatures vary from (*ca.* 1510 °C for LC steels) to 1440 °C (for *austenitic* stainless steels); values of q_{vert} also increase with increasing superheat. Good thermal insulation is essential when casting steels with low T_{liq} or when casting at low speeds to ensure that that the steel meniscus does not freeze.

Metal Flow

The vertical heat flux is affected by both the flow pattern and the turbulence developed in the metal flow. Freezing of the steel meniscus is more common with a *single roll* pattern than with a *double roll*. The values of q_{vert} increase with increasing turbulence in the metal flow, since heat transfer for turbulent flow is more efficient than that for laminar flow.

Electromagnetic Braking (EMBr)
Electromagnetic braking is used in thin slab- and high-speed casting to suppress turbulent flow in the molten metal; it has been calculated to cause a 30% reduction in q_{vert} in thin slab-casting [12] due to the fact that laminar flow is a less-efficient form of heat transfer than turbulent flow. Thus, the application of EMBr would be expected to reduce both q_{vert} and the melting rate.

Ar Flow Rate
The Ar flow contributes to the vertical heat flux by gaseous convection; it increases the vertical heat flux leaving the metal but also increases the heat flux leaving the pool (or that leaving the sinter layer) due to increased gaseous conduction. Argon flow rate has also been reported to alter flow patterns in the metal phase when it exceeds a critical value (*ca* 5 L min^{-1}) [13, 14].

4.2.1.3 Ways of Controlling the Vertical Heat Flux

The vertical heat flux leaving the metal (and entering the slag pool) can be decreased by:

- Decreasing the casting speed and superheat.
- Decreasing the Ar flow velocity.
- Using EMBr to reduce turbulence in the metal flow.

The vertical heat flux leaving the slag pool can also be reduced by improving the thermal insulation of the bed (see Sect. 4.2.2.2) by:

- Increasing the depth of the bed.
- Using granules of smaller diameters to increase packing density and reduce gaseous permeability.
- Adding exothermic agents to the powder to reduce the thermal gradients for q^{out} and by minimising the carbonate content of the powder.

4.2.2 Thermal Insulation of the Bed

It can be seen from Eq. 4.2 that the thermal resistance (or thermal insulation) of each layer increases as the ratio (*d/k*) increases. It follows that thermal insulation will increase if the depths of the powder and sintered layers are increased and the values of the thermal conductivity are low. The only actions open for the liquid pool are to reduce the convection by using a high-viscosity slag or to reduce the radiation by adding FeO or MnO to the slag.

In order to heat the mould powder from room temperature to the steel temperature, the endothermic energy must be sufficient to provide for the following:

- To heat the slag components from 25 °C to the steel temperature (ΔH_{endo}).
- The enthalpy of fusion of the slag formed (ΔH^{fus}).
- The heat of vaporisation of the moisture (ΔH^{vap}).
- The enthalpy of sublimation for the carbonates (ΔH^{sub}).
- To heat any O_2 and N_2 present in the bed from 25 °C to the temperature of reaction (ΔH_{gas}).

However, the following exothermic reactions (ΔH_{exo}) reduce the energy budget:

- The oxidation of the free carbon to form CO(g) and CO_2(g).
- The oxidation of any exothermic agents (e.g. Ca, Si, Fe) present.

Thus, the energy demand (ΔH) is given by:

$$\Delta H = \Delta H_{endo} + \Delta H^{fus} + \Delta H^{vap} + \Delta H^{sub} + \Delta H_{gas} - \Delta H_{exo}. \tag{4.4}$$

Once the powder bed has achieved steady state, the vertical heat flux (q_{vert}) must meet this energy demand (denoted ΔH) and thus, $\Delta H = q_{vert} \cdot A$ where A = cross-sectional area of the mould. The ratio ($q_{vert} \cdot A/\Delta H$) has units of kg s^{-1}, i.e. identical to that of the melting rate (Q_{MR}).

The thicknesses of the various layers in the powder bed in Fig. 4.1 represent the balance between the required energy and the vertical heat flux at steady state.

Ar blowing will tend to increase the gaseous conduction and thereby, increase the heat flow into the sinter layer (q^{in}_{vert}) but it will also increase the heat loss from the sinter to the powder layer (q^{out}_{vert}).

4.2.2.1 Importance of Thermal Insulation

See that for vertical heat transfer–Sect. 4.2.1.1.

4.2.2.2 The Effects of Powder Characteristics on Thermal Insulation

In addition to the casting conditions, the vertical heat flux is also affected by the powder bed characteristics. Thus, the heat flux density leaving the molten slag pool is reduced by increasing the thermal insulation of the sinter and powder layers. The thermal insulation is improved (i.e. q_{vert} decreases) by the actions summarised above in Sect. 4.2.1.3, namely, (i): increased bed depth [15] (ii) using smaller particles with lower permeability (iii) using powders containing exothermic agents (iv) minimising the carbonate content since their decomposition is endothermic and (v) reduce convection and radiation in the pool by increasing the viscosity or the FeO, MnO contents, respectively.

4.2.3 *Measurements of Thermal Insulation of Powders*

Measurements methods tend to fall into two classes, namely:

(i) Measurements of the thermal conductivity of powders using the transient hot wire (THW) [7, 16–20] or the laser pulse methods [21] (See Sect. 9.4.5.4).
(ii) Hot plate test (HPT) measurements [11, 22–24], where the temperatures are measured at different positions in the powder bed lying between the hot plate and powder surface (See Sect. 9.8.4).

The values of the effective thermal conductivity (k_{eff}) obtained in the two experiments show large differences [18] (e.g. k_{eff} ($Wm^{-1}\ K^{-1}$) values of 0.1–0.2 (in THW studies) *cf.* 1–1.3 (for HPT) [11]. These differences are due to the gaseous convection contribution which is large in the HPT studies and much smaller in THW and LP (Laser Pulse) studies where the temperature gradient across the specimen is small (in contrast to HPT where it is *ca.* 1000 °C. The HPT method provides a closer simulation of k_{eff} for the bed but the THW studies give a more accurate value for k_{lat} of the powder (See Sect. 9.4.5.8).

Increased bulk density of the sample results in:

(i) An increase in $k_{powd} = (0.010 + 1.69 \times 10^{-4}\ \rho_{bulk})$ [18].
(ii) A decrease in gaseous convection; the gaseous conduction appears to dominate. Bulk density increases with decreasing powder size (D_{powd}).

4.2.4 *Ways of Improving the Thermal Insulation of the Bed*

The various ways of improving thermal insulation are identical to those given for reducing vertical heat flux in Sect. 4.2.1.3.

4.3 Melting Rate of the Powder (Q_{MR})

The melting rate is an important factor since:

(i) The melting rate should match the required powder consumption (Q_s^{req}) to ensure that good lubrication is supplied to the newly formed shell ($Q_s^{req} \leq Q_{MR}$).
(ii) The melting rate also affects the slag pool depth.

The melting rate (Q_{MR}) is dependent upon the following factors:

- Properties of the mould powder.
- Vertical heat flux, which, in turn, is affected by a variety of factors including, casting speed, steel temperature, electromagnetic braking and Argon flow rates.

A number of tests have been developed to provide a value, or a ranking, of the melting rate for various mould powders. These are listed below:

(i) ***Combustion boat tests*** in which 1.5 g of the mould powder is placed in a combustion boat with the flat-end cut off to provide easy viewing of the sample. The sample is placed in a muffle furnace at a known temperature; the time taken for the sample to melt is inversely related to the melting rate.

(ii) ***Molten slag drip tests*** [25–29] using the apparatus shown in Fig. 4.2a, where the graphite block is heated to 1500 °C. The slag melts and drips through the graphite block and then is collected and weighed at different time intervals. The melting rate results derived may be affected by the fluidity ($1/\eta$) of the slag.

(iii) ***Vitrification rate tests*** [30] in which 25 g of the mould powder is placed in a graphite crucible (30 mm × 40 mm) and heated with a uni-directional heat source for 7 min and the amount of powder melted determined. The measured rate is denoted the *vitrification rate* [30].

(iv) ***Sumitomo tests*** [31] are carried out using 1 tonne of steel (Fig. 4.2b); the steel surface is divided by refractory boards to allow different powders to be tested at the same time. A known weight of powder is added and the amount of slag formed in a certain time is determined using dip tests with steel and Cu wires.

(v) ***Pradhan test*** [29] is a small scale, Sumitimo test, where a mould powder is held within a refractory ring on the surface of the molten steel and the refractory ring is oscillated to allow the effect of mould oscillation to be explored.

(vi) ***Däcker test*** [32] in which a constant depth (20 mm) of powder is maintained in a graphite crucible, the latter contains machined slots in the base to allow the liquid slag to drip off. The crucible base is maintained at 1500 °C and a regulated flow of air is directed on to the powder surface to aid the

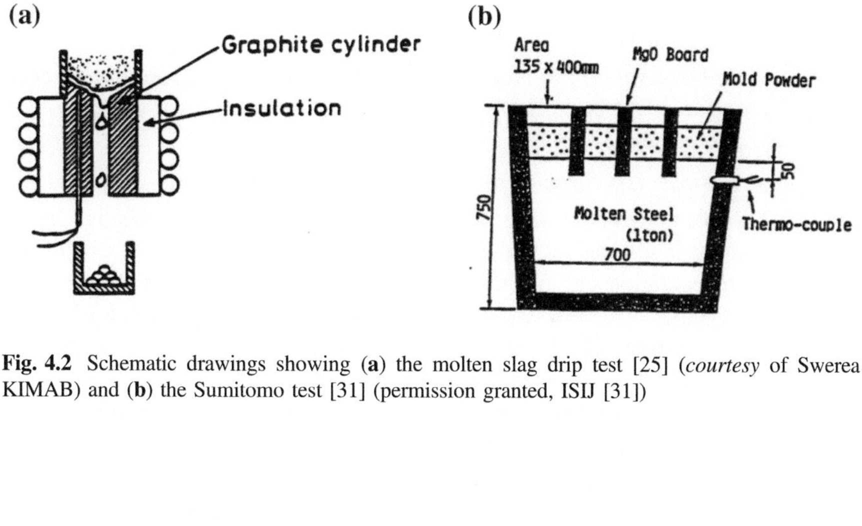

Fig. 4.2 Schematic drawings showing (**a**) the molten slag drip test [25] (*courtesy* of Swerea KIMAB) and (**b**) the Sumitomo test [31] (permission granted, ISIJ [31])

combustion of carbon. A siphon system is used to maintain the height of the slag at 10 mm and the mould powder stock is weighed continuously which allows the consumption to be measured as a function of time.

(vii) ***Kromhout method*** [33] contains two tests. In the *softening test,* a pressed cylinder of mould powder is put in a furnace at a known temperature and an alumina rod is placed on the top surface of the sample; the time taken for a 20% displacement of the rod is measured. In the *melting test*, the mould powder is held in a steel crucible with a base made of Al foil. The crucible is placed on liquid steel at a known, controlled temperature. The time for the sample to melt was observed using a camera mounted above the crucible. The two tests were used in combination [33].

4.3.1 The Effect of Mould Powder Properties on Melting Rate

The effect of the powder properties on the melting rate has been studied by a number of investigators [34–45].

Several properties of the powder affect the melting rate, namely:

(i) The *free carbon content* of the flux (C_{free}%) with Q_{MR} increasing as %C_{free} decreases [35].

(ii) The *mean diameter* of the carbon particles (D_C) with $Q_{MR,}$ to a less extent, increasing as D_C increases [35].

(iii) The *bulk density* (ρ_{bulk}) with Q_{MR} increasing as ρ_{bulk} decreases which is probably due to the increasing gaseous conduction resulting from an increasingly permeable powder bed. (Sect. 4.2.1.2) [35].

(iv) The *carbonate content* (%CO_3^{2-}) with Q_{MR} increasing as (%CO_3^{2-}) content increases which again may be associated with increased gaseous conduction and the formation of a more permeable powder bed [35].

(v) *The liquidus temperature* has been reported to have a slight effect on the melting rate with Q_{MR} increasing as T_{liq} decreases [29].

(vi) The *contact angle* (θ, of slag on carbon) and *energy of adhesion* have been reported to affect the melting rate, with Q_{MR} increasing as θ decreases and E_{adhes} increases [37].

Carbon particles are non-wetting to molten slag globules and hinder the agglomeration of the globules (Fig. 4.3). However, at high temperatures the carbon particles oxidise and gradually disappear. Smaller carbon particles, obviously, provide more separation per unit mass of carbon. However, smaller particles also have a much larger (surface area/ mass) ratio and consequently oxidise more rapidly. The fact that melting rate decreases with decreasing size, indicates the separation provided by carbon particles is more important than the faster oxidation rate. Physical and mathematical modelling of the kinetics of carbon oxidation has

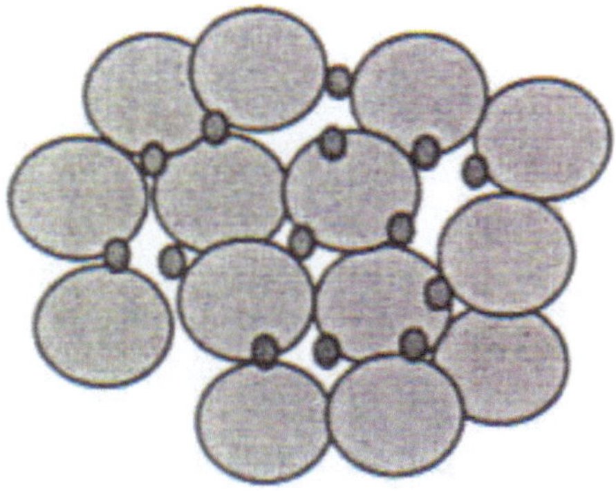

Fig. 4.3 Schematic drawing showing non-wetting carbon particles (●) preventing the agglomeration of slag globules (●)

been carried out [44]. The average size of the globules is a linear function of the melting rate [28].

4.3.2 The Effect of Casting Conditions on Melting Rate

The principal factor affecting the melting rate is the vertical heat flux (q_{vert}); so any variable which increases the heat coming into the pool (q_{vert}^{in}) or reduces the heat leaving (q_{vert}^{out}) will increase the melting rate. The q_{vert} is increased by:

(i) Increasing casting speed (increases (q_{vert}^{in}).
(ii) Turbulent heat transfer (*cf.* laminar heat transfer) increases q_{vert}^{in} (EMBr reduces q_{vert}^{in}).
(iii) Improved thermal insulation of powder bed (e.g. a deeper bed or more dense packing reduces q_{vert}^{out}).

The effect of Ar flow is more uncertain since it increases both q_{vert}^{in} and q_{vert}^{out} but it probably enhances the melting rate overall.

4.3.3 Ways of Increasing Melting Rate

The principal ways of increasing the melting rate are:

- To decrease the %C_{free} in the mould powder but care must be taken since C is an exothermic agent and large decreases in %C_{free} may lead to freezing of the steel meniscus.
- To use carbon particles with a larger mean diameter ($D_C\uparrow$).
- To decrease the bulk density of the mould powder; this can be achieved by increasing the particle size of the granules ($D_{gran}\uparrow$).

- To increase the carbonate content of the powder.
- Increase the vertical heat flux (e.g. Increase casting speed but this will cause other collateral changes).

4.4 Depth of Molten Slag Pool

4.4.1 Molten Slag Pool

The molten slag pool acts as a reservoir of molten slag which is, subsequently, pumped into the channel between shell and mould during the period of oscillation cycle when the mould is descending [46].

It should be noted that the slag pool does not have a uniform depth (d_{pool}); reported values of d_{pool} indicate that [47, 48]: d_{pool} (SEN) > d_{pool} (midway) > d_{pool} (large face) > d_{pool} (narrow face) (Fig. 4.4). This has been attributed to the flow separation, which occurs *ca.* 200 mm from the narrow face and results in a cold spot [48]. The application of EMBr has been found to produce a more uniform slag

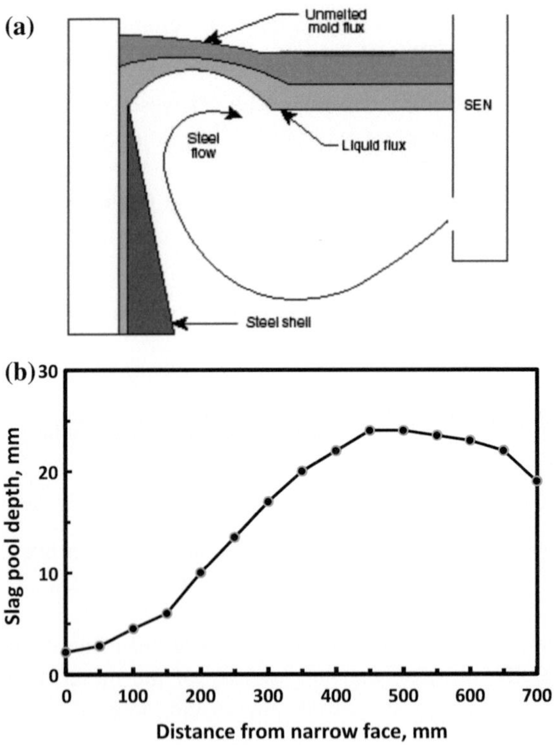

Fig. 4.4 Schematic drawing showing the depth of the molten slag pool as a function of position in the mould (**a**) General view of liquid slag layer (shown in *light grey*) and (**b**) experimental measurements (●) of slag pool depth and predicted values (▬) versus distance from narrow face ((permission granted, (**a**) ISS/AIST [11]) (**b**) ISS/AIST [48])

layer [49] presumably, by reducing the size of the surface waves on the steel meniscus.

The pool depth, at steady state, represents the balance between (i) the *heat* coming in and leaving the liquid zone and (ii) the *fluid flow* in and out of the pool (i.e. the melting rate and slag consumption, respectively); thus an increase in melting rate or a decrease in Q_s will result in a deeper pool.

Several methods have been proposed for measuring the depth of the slag pool (d_{pool}). The following methods have been used.

4.4.1.1 Dip Tests

These usually contain two or three wires (of steel, Cu and Al which are poked through the bed into the steel for a few seconds and then removed (Fig. 4.5a). The melting point of Cu (1085 °C) is close to the T_{liq} of most mould slags, so the distance between the Cu and steel wires represents the depth of the slag pool. When Al wire is used its height represents the 660 °C temperature contour.

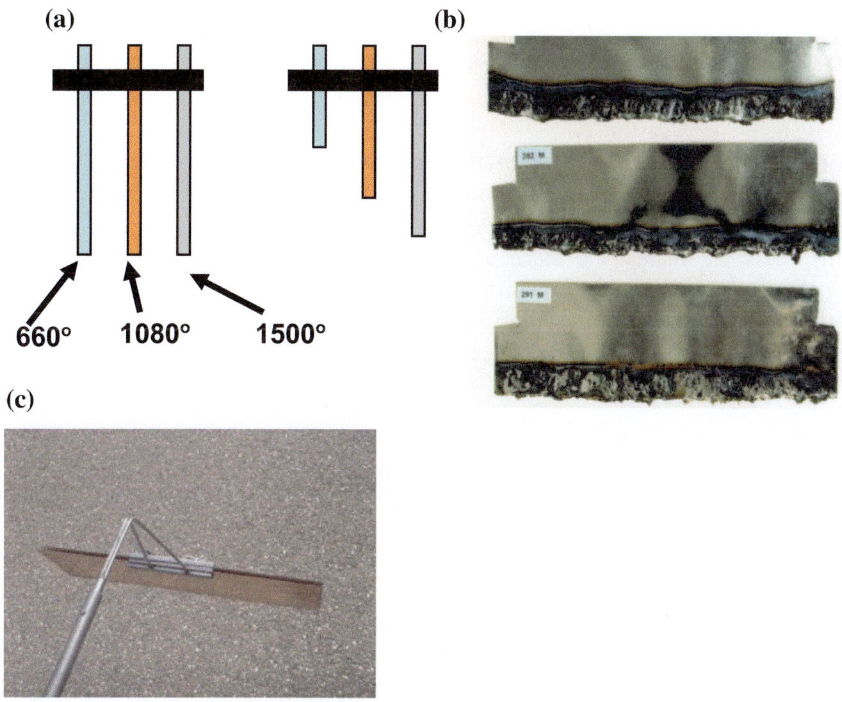

Fig. 4.5 (**a**) Schematic diagram and photographs showing (**a**) before and after the dip test (**b**) plate-dip test and (**c**) photographs of plates removed from the mould, showing slag attached to the plate; note that the steel surface is not flat (*courtesy* of F Shahbazian, Swerea KIMAB [119])

4.4.1.2 Plate-Dip Tests

In these tests a thin steel plate (*ca.* 1 mm thick) is pushed through the powder bed and held for *ca.* 1 s and then removed. The slag sticks to the steel plate (see Fig. 4.5b) and its depth can be measured (Fig. 4.5b).

4.4.1.3 Nakamori Test

A device has been reported that uses two eddy current sensors, the low frequency sensor reveals the position of the steel surface and the high frequency sensor locates the upper surface of the slag pool [50]. More recently, the slag pool thickness has been determined using two sensors, an EM sensor to detect the steel level and a sensor to detect radioactive signals for the slag level [51].

4.4.2 *Importance of Depth of Molten Slag Pool*

Manipulation of slag pool depth is needed to:

(i) Ensure good powder consumption since it is known that a shallow slag pool leads to poor powder consumption (a value of $d_{pool} = \geq 10$ mm is often cited as being necessary but >20 mm is cited as being necessary for high-speed casting).
(ii) Minimise carbon pick-up by the shell and steel surface in order to separate the layer of carbon particles floating at the top of the pool from the solidifying shell.
(iii) Dissolve inclusions, since a large slag volume is needed when casting (a) steels containing Ti which form Ti(CN) or *Perovskite* ($CaO \cdot TiO_2$) which has low saturated solubility in slag and (b) High-Al steels where large amounts of Al_2O_3 are produced.

4.4.3 *Factors Affecting Slag Pool Depth*

The principal factors affecting the depth of the molten pool (d_{pool}) are the magnitudes of the heat flux and the fluid flow entering and leaving the liquid pool. The heat transfer involves both the thermal insulation properties of the bed and the casting conditions. The fluid flow involves the melting rate ($Q_{MR,}$ coming in) and the powder consumption (flowing out). Obviously, the depth of the slag pool will tend to increase as the liquidus temperature of the mould slag deceases [48].

Consider the molten slag pool layer:

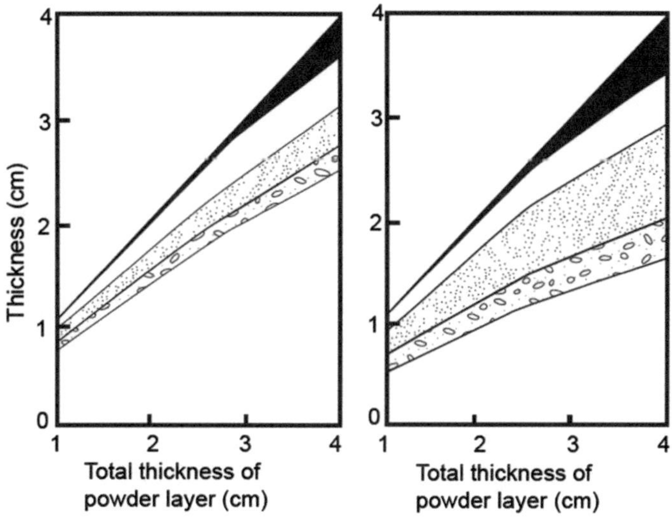

Fig. 4.6 Schematic drawings showing the effect of powder depth on the thicknesses of various layers in the powder bed for two mould powders; *black* Powder; *white* Sintered; *dotted layer* molten; *larger dots* indicate mushy zone (permission granted, ISIJ [15]; re-drawn Swerea KIMAB)

(i) heat is supplied to the pool by the metal (q^{in}_{vert})
(ii) heat is lost principally through the sinter layer and powder bed (q^{out}_{vert}).

Obviously, an increase in q^{in}_{vert} will result in a higher melting rate (see Sect. 4.2.2.2) which will result in an increase in the slag pool depth (d_{pool}). The application of EMBr has been calculated to cause a 30% decrease in q^{in}_{vert}, [12] which, thus, results in a decrease in pool depth (d_{pool}).

An improvement in thermal insulation will cause a decrease in q^{out}_{vert} and this too will result in an increase in d_{pool}. This can be clearly seen in Fig. 4.6 [18] where the improved thermal insulation, resulting from a deeper powder bed [52,48], causes an increase in d_{pool}. In addition, it can be seen from Fig. 4.6 that improving the thermal insulation of the bed also increases the thicknesses of the sinter and powder layers. A similar argument can be applied to the sinter and powder layers where q^{in}_{vert} refers to the layer below and q^{out}_{vert} to the layer above.

Several different types of mathematical models have been developed to predict the depth of the molten pool [37, 48, 52, 53]. One model considers the bed to be a packed bed and solves the heat and mass transfer of the gaseous and condensed phases [52] to calculate the temperature contours in the bed; the pool depth can be calculated from the these contours. A coupled heat and fluid flow model has been used to determine the depth of the pool in other studies [48, 53]. A mathematical model for the slag pool depth formed in thin slab-casting has been reported [54].

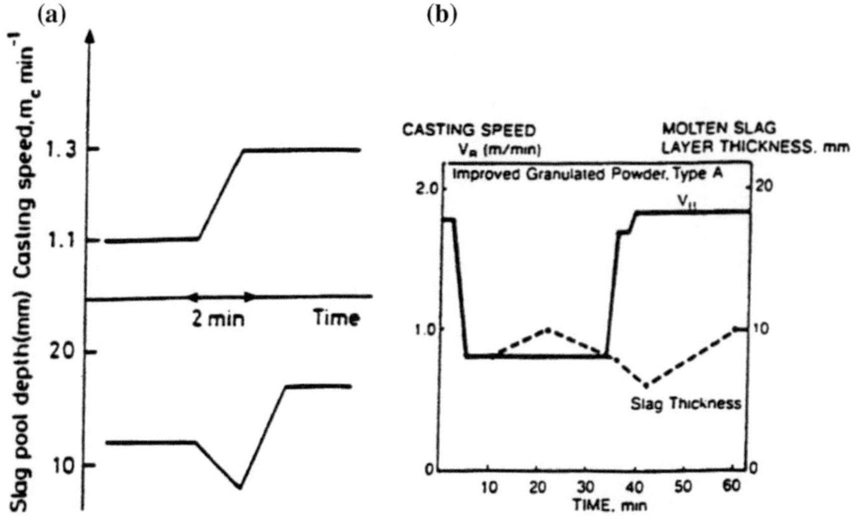

Fig. 4.7 Time-transients for the depth of slag pool and the casting speed, showing the effects of (**a**) an increase in V_c and (**b**) a decrease in casting speed (*solid line*) and d_{pool} = *dashed line* (permission granted, (**a**) Europ. Sci. Tech. Publ. [47] and (**b**) ISS/AIST [55])

4.4.4 *The Effect of Casting Speed and Oscillation Characteristics*

When the casting speed is increased abruptly, it causes increases in heat flux density, i.e. for both q_{vert} and q_{hor}. An increased q_{vert} needs time to work its way through to the sinter and powder layers; ultimately, an increased melting rate results in an increase in d_{pool}. This process takes time and there is a time lag (of 2–5 min (see Fig. 4.7) before it responds [47, 55]. The same applies to q_{hor} and the changes in slag film thickness. When steady state has been reached an increase in casting speed results in a deeper slag pool [38, 47, 56].

It can be seen from Fig. 4.7 that as the casting speed increases the slag pool depth initially decreases in response to the higher demand for slag infiltration [47, 55]. However, after a while the increased q_{hor}, results, sequentially, in a thinner solid slag layer (d_s), thicker liquid layer (d_l) and a higher melting rate. Thus, d_{pool}, after its initial fall, starts to increase and reaches an enhanced value when steady-state conditions are attained. Similarly, a decrease in casting speed results in an initial increase in d_{pool}, which is followed by a decrease in d_{pool} when the decreased q_{vert} starts to take effect.

It has been found that oscillation caused a *ca*. 20% increase in slag pool depth recorded in simulation tests [29].

It has been reported that the pool depth increases with increasing % negative strip [53, 57].

4.4.5 *The Effect of Thermal Insulation of Bed on Pool Depth*

As mentioned above, the depth of any layer (pool, sinter or powder) will increase if q_{vert}^{in} increases or q_{vert}^{out} decreases; improved thermal insulation affects the latter. The following actions lead to improved thermal insulation:

- Using a deeper powder bed.
- Replacing granules with powder or powders with a smaller diameter ($D_{powd}\downarrow$) which reduces the gaseous conduction.
- Adding exothermic agents to the powder to reduce the temperature gradient in the sinter layer.
- Using, a slag with a high T_{liq} and viscosity since these will tend to reduce the contribution from convection to the vertical heat transfer.

4.4.6 *Ways of Increasing the Melting Rate*

An increased melting rate will increase the depth of pool; this can be achieved by:

- Reducing the free carbon ($\%C_{free}$) but it should be noted that C combustion is exothermic and a reduction of *C* would lead to an increase in the temperature gradient and q_{vert}^{out}.
- Increasing the size of the C particles ($D_C\uparrow$).
- Decreasing the bulk density (by increasing the particle size of the granules, $D_{gran}\uparrow$).
- Increase the carbonate content of the powder.
- Increase the vertical heat flux by increasing the casting speed, the superheat or the Ar flow rate (providing it remains below 5 L min^{-1}.

4.5 Powder Consumption (*Q*) and Liquid Film Thickness (d_l)

Molten slag from the slag pool infiltrates into the shell/mould channel during the downward movement of the mould and the slag rim. The molten slag entering the channel, subsequently, partially solidifies to give a slag film with a solid layer (*ca.* 2 mm thick) and a liquid layer (*ca.* 0.1 mm thick).The liquid slag film moves downward during the descent of the mould and is the major contributor to the powder consumption, although some downward movement of the solid slag layer may also occur [58].

Most slag films consist of a crystalline and a glass phase. It should be noted that glass phases form a super-cooled liquid (*scl*) at temperatures above T_g, the glass transition temperature. These *scl*'s will provide some lubrication, despite their high viscosities, and some glassy, high-viscosity slags are used in billet casting.

A description of the methods used to determine powder consumption are given in Chap. 2 (Sects. 2.1.2 and 2.5.1, respectively).

4.5.1 Reasons for Controlling Powder Consumption

Powder consumption provides a measure of the lubrication supplied by the molten slag to the newly formed shell. It can be seen from Eq. 4.5 that the liquid frictional force applied to the shell (F) decreases with increasing liquid slag film thickness (d_l) and decreasing viscosity (η), where A is the area of shell and V_c and V_m are the casting speed and mould velocity, respectively.

$$F = A(V_m - V_c)\eta/d_l \tag{4.5}$$

Inadequate powder consumption can lead to longitudinal and transverse cracking and to star cracking in the lower half of the mould.

It should also be noted that not all of the mould powder forms slag and it is the slag which provides the lubrication; thus it is necessary to differentiate between slag and powder. There are several powder consumption terms in common use which are defined below in Eqs. (4.6)–(4.9). The term, Q_t^{powd} is the quantity usually measured on plant, Q_s^{slag} is the best measure of the lubrication supplied to the shell and this can be related to the melting rate Q_{MR}^{slag} via Eq. 4.9. It should be noted that the liquid friction becomes dependent on $\eta^{1.5}$ when empirical relations (e.g. $Q_s = 0.55/\eta^{0.5} V_c$) are coupled with Eqs. 4.5 and 4.8.

$$Q_t^{slag} = f * Q_t^{powd}, \tag{4.6}$$

where

$$\begin{aligned} f^* &= (100 - \%C_{free} - \%H_2O) - \{(44/12)(\%C_{total} - \%C_{free})\}/100 \\ &= (100 - \%LOI)/100 \end{aligned} \tag{4.7}$$

$$Q_s^{slag}(\text{kg m}^{-2}) = f * 7.6\, Q_t^{powd}/R^* = \rho\, d_l \approx 2550 \cdot d_l. \tag{4.8}$$

$$Q_{MR}^{slag}(\text{kg s}^{-1}) = 2(w + t) Q_s\, V_c/60 \tag{4.9}$$

The *required powder consumption* is that which is necessary to ensure good lubrication of the shell. Wolf [59, 60] introduced the concept of *minimum powder consumption* and recommended values of 0.25 [59] and 0.4 [60] kg m^{-2}, respectively, for casting round billets and heavy plates. There are no reliable values for Q_s^{min} but one possible rule would be $Q_s^{min} = 0.5\ Q_s^{req}$.

4.5.2 *Factors Affecting Powder Consumption*

There are a number of factors affecting both the required powder consumption for the casting conditions and the actual powder consumption achieved in practice; in practice, powder consumption can be constrained by factors like the free carbon content of the powder.

4.5.2.1 Mould Dimensions

The slag consumption per tonne of steel is approximately constant at $Q_t^{slag} = 0.48$ kg tonne steel^{-1} [61]. However, the *required powder consumption per* m^2 *of mould* (Q_s in kg (flux) m^{-2}) increases with increasing surface area of mould (or decreasing surface area/volume, denoted as R^* [62] as shown in Fig. 4.8 and Eqs. 4.10 and 4.11. It is important that the free carbon content of the casting

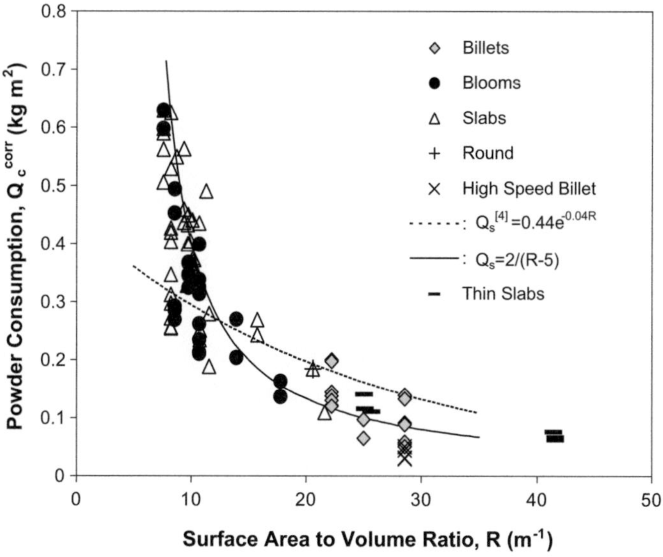

Fig. 4.8 Powder consumption, $Q_{s\ req}^{slag}$ as a function of R^*, the ratio of (surface area/volume) of the mould, diamond billets; ● = blooms; Δ = slabs; ▬ = thin slabs; + = rounds; + =high-speed billets; *solid line* Eq. 4.10 [62] *dotted line* [7] (permission granted, UNESID [62])

powder should be at a level where it allows the required, powder consumption to be attained, i.e. $Q_{MR}^{slag} \geq Q_{MR}^{slag\ req}$.

$$Q_{s\,req}^{slag} = 2/(R^{*} - 5) \tag{4.10}$$

$$R^{*} = 2(w + t)/wt \tag{4.11}$$

4.5.2.2 Casting Speed (V_c) and Viscosity (η)

The powder consumption, Q_s, is affected by several casting variables (as can be seen from Eqs. 4.12 to 4.16). However, the most influential terms are the casting speed and the viscosity. Several empirical rules have been proposed for the required powder consumption involving these two parameters, [59–64] such as those shown in Eqs. 4.12 and 4.13.

$$Q_{s\,req}^{slag} = 0.55/\eta^{0.5} \cdot V_c \tag{4.12}$$

$$Q_{s\,req}^{powd} = 0.6/\eta \cdot V_c \tag{4.13}$$

The required viscosity (η_{req}^{slag}) tends to decrease as the casting speed is increased (rough values *of* η_{req}^{slag} can be calculated from the following rules: $\eta_{req}^{slag} = 2/V_c$ or $= (5/V_c^2)$).

As mentioned above, the combination of Eqs. 4.5 and 4.8 give Eq. 4.14, showing that the frictional forces acting on the shell are proportional to $\eta^{1.5}$.

$$F = A\eta(V_m - V_c)\rho_l\eta^{0.5}V_c/0.55 \tag{4.14}$$

The viscosity of non-Newtonian slags decrease with increasing shear rate. Thus, they provide a high viscosity in the low-shear rate (meniscus) region (and thereby reduce slag entrapment) and a lower viscosity in the high-shear rate, region (at the mouth of the shell/mould channel) which reduces friction acting on the shell.

4.5.2.3 Solidification (or Break) Temperature (T_{br})

If the solidification (or break, T_{br}) temperature (T_{sol}) of the solid slag film is increased, the thickness of the solid slag film (d_s) is increased and the thickness of the liquid layer (d_l) is reduced ($d_s\uparrow$ and $d_l\downarrow$ as T_{br} (or T_{sol})$\uparrow$). Since $Q_s = \rho_l d_l$ and ρ_l is reasonably constant (at 2550–2600 kg m^{-3}) it would be expected that Q_s is inversely related to T_{br}. Equation 4.15 was derived from physical modelling studies and shows an inverse relationship [64] between Q_s and T_{sol}. Equation 4.16 was obtained from a statistical analysis of plant data and also indicates a similar inverse relation between Q_s and T_{br}.

$$Q_s = \frac{k\beta}{T_{sol}^{1.6}\eta^{0.5}} \frac{s^{0.4}}{V_c} \cos^{-1}\left[\frac{1000V_c}{2\pi fs}\right], \tag{4.15}$$

$$Q_s = \left(1/\eta V_c^{0.46}\right).\left(1/f^{0.49}\right)\left(1/s^{1.37}\right)\left(1/T_{br}^{3.48}\right)\exp^{28.81}. \tag{4.16}$$

4.5.2.4 Oscillation Characteristics

There is no agreement on the effect of stroke length (s) on the powder consumption but a statistical analysis of plant data indicated that Q_s decreases with increasing stroke length. Mathematical models of powder consumption based on Navier–Stokes indicate that Q_s increases with increasing oscillation frequency (f). However, most plant data show the reverse trend, i.e. Q_s decreases as frequency increases (e.g. Eq. 4.16). It has been suggested that "*tide-changes*" of the flow in the slag pool are followed by periods of "*confused flow*" where little slag infiltration occurs and the models fail to account for this [46]; an increase in frequency will increase the number of these periods resulting in less time for slag infiltration.

There is no agreement on the effects of negative and positive strip times (t_n, t_p). A recent mathematical model of slag infiltration indicated that it occurred in the period when the mould was descending [46]; thus a higher t_n value might thus be expected to increase Q_s.

Powder consumption increases of 10% have been reported when using non-sinusoidal oscillation [65–68].

4.5.2.5 Slag Pool Depth and Vertical Heat Flux

It might be expected that powder consumption, Q_s, will increase with increasing slag pool depth (d_{pool}) as slag is pushed into the shell/mould channel during the descent of the slag rim. However, there is no direct evidence for such a relationship. It is known, however, that values of d_{pool} and Q_s both tend to be low in thin slab- casting; this provides some circumstantial evidence for the view that Q_s increases with increasing d_{pool}.

The slag pool depth tends to increase if $\Delta q_{vert} = (q_{vert}^{in} - q_{vert}^{out})$ increases and thus increasing vertical heat flux might be expected to increase the powder consumption.

The effect of vertical heat transfer on powder consumption is not clear. Increased q_{vert} will cause increases in melting rate and d_{pool}. However, increased powder consumption has been reported to increase when EMBr is applied which has been attributed to the 10 °C increase in meniscus temperature and its effect on slag viscosity but the reduced q_{vert} will also result in a shallow d_{pool}. If the reduced viscosity is the cause of enhanced Q_s then superheat would be expected to produce the same effect. This problem needs resolving.

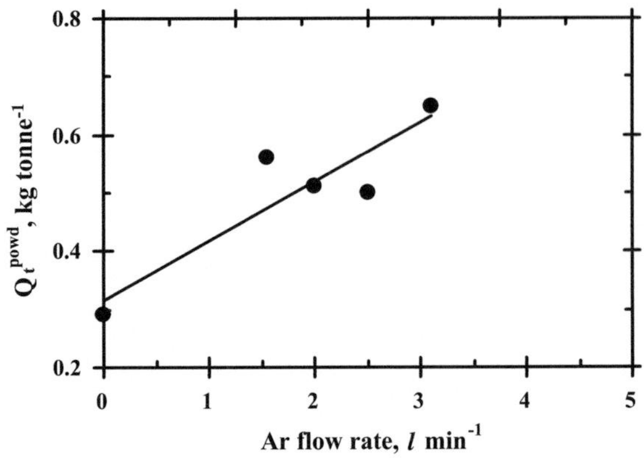

Fig. 4.9 Powder consumption, Q_t^{powd} as a function of Ar flow rate (*courtesy* of A.B. Fox [69])

4.5.2.6 Argon Flow Rates

It has been reported that powder consumption increases with increasing Ar flow rate (Fig. 4.9) [69]. This is probably associated with the increase of gaseous conduction in the molten metal phase which will enhance the vertical heat flux. This will increase the heat entering the slag pool (q_{vert}^{in}). However, the Ar flow rate will also increase the heat leaving the molten slag pool (q_{vert}^{out}) to the sinter layer. If $\Delta q_{vert} = (q_{vert}^{in} - q_{vert}^{out})$ increases with increasing Ar flow rate, then the higher heat flux will result sequentially in a higher melting rate, a deeper pool and increased powder consumption. However, recent work has shown that the Ar flow interacts with the metal flow and can alter the metal flow pattern when the Ar flow rate exceeds 5 L min^{-1} [13] which could lead to a higher q_{vert}.

4.5.2.7 Electromagnetic Casting (EMC)

It has been reported that the application of EMC results in a 20% increase in powder consumption which has been attributed to an increase the temperature of *ca.* 10 °C at the steel/slag interface and change in meniscus shape [70]. One possible explanation of the enhanced powder consumption is that the enhanced meniscus temperature reduces the slag viscosity slightly.

4.5.2.8 Blockage to Slag Infiltration

It has been observed that powder consumption is frequently lower than predicted when casting steel grades containing Ti (ULC and stainless steels). This is thought to be due to the formation of TiN or Ti(CN) or *Perovskite* ($CaTiO_3$). TiN has a low solubility in the slag pool and $CaTiO_3$ has a high melting point, so a large number of solid particles are present in the slag pool [71, 72]. These particles agglomerate through turbulent collisions. If these agglomerates collect in the mouth of slag/mould channel they will restrict the slag flow in the channel (see Fig. 2.16). It has also been suggested that Ti(CN) can be formed in the mould [73] through the reaction of N_2 in F-free slags, where CaF_2 has been replaced with 6% TiO_2 [73, 74]. A similar mechanism probably applies when casting TRIP steels with >1% Al, where large amounts of Al_2O_3 are formed which only partially dissolve and, subsequently, impede the flow of molten slag [75] into the shell/mould channel.

In these cases, the formation of a deep slag pool will be beneficial since it will:

- Dissolve more TiN or Al_2O_3 resulting in fewer agglomerates to impede the flow of molten slag into the mould/shell channel.
- Seal off the metal from gases in the sinter layer more efficiently.

4.5.3 Ways of Controlling the Powder Consumption

The powder consumption can be increased by the following:

- Decreasing slag viscosity (η).
- Decreasing the casting speed (V_c).
- To a less extent, by decreasing frequency (f) and stroke length (s).
- Decreasing the Solidification (or Break) temperature (T_{br}).
- Increased slag pool depth (d_{pool}) when casting Ti- steels and TRIP steels to dissolve Ti(CN) or Al_2O_3, respectively, which tend to block slag infiltration.
- Increasing the Argon flow rate (providing it does not exceed 5 L min^{-1}).
- Use of non-sinusoidal oscillation.
- Use of EMBr and EMC.

4.6 Solid Slag Film and Horizontal Heat Flux

The slag film formed between the shell and the mould is probably initially glassy but crystallises over time. The mixture of glassy and crystalline phases is essential to control of both the horizontal heat flux and the provision of some lubrication to the shell. The horizontal heat flux between shell and mould involves two heat transfer mechanisms, namely, *phonon (or lattice) conduction* and *radiation conduction* (see Sect. 3.2). The lattice conduction can be treated by assuming the various layers in the slag film (liquid, crystalline and glass) are viewed as thermal resistances in series. In addition, there is also an interfacial resistance ($R_{Cu/sl}$) created at the copper/slag interface as a result of the shrinkage caused by the transformation of glass to crystalline phase; this too is treated as a series resistance [1–3]. The radiation conduction has been accounted for by assuming it to be a parallel resistance [3]. The thermal resistance of the slag film is calculated by Eq. 4.17.

$$R_{th} = (d/k)_{liq} + (d/k)_{glass} + (d/k)_{cryst} + R_{Cu/sl} \quad (4.17)$$

The glass and crystalline phases are intermixed in the slag film with highest concentrations of crystals located on the shell (high temperature) side and the glass on the mould side [76] because high temperatures favour crystal growth.

4.6.1 *Reasons for Control of Slag Film Thickness and Horizontal Heat Flux*

It is important to control the horizontal heat transfer across the slag film to ensure that the newly formed shell has the optimum thickness for the type of steel being cast and the casting conditions. Thus for the casting of the following steel grades:

(i) In MC (Medium Carbon), peritectic steels, a thin, even shell is needed to avoid longitudinal and other surface cracking and this is obtained with a low horizontal heat flux density (q_{hor}).

(ii) In HC (High Carbon) steels the shell is relatively weak and a thick shell is required to provide mechanical strength; failure to do so can result in sticker breakouts and hot tears.

(iii) In steels where the primarily solidification results in the formation of *ferrite* (e.g. ULC steels (Ultra Low carbon)) the shell is soft and a strong cooling is needed to create a strong shell to minimise the risk of sticker breakouts.

The horizontal heat flux is also involved in delaying solidification to reduce both the inclusion content and depths of oscillation marks (Sect. 4.8).

4.6.2 *Factors Affecting of Slag Film Thickness and Horizontal Heat Flux*

The horizontal heat flux between the shell and the water-cooled copper mould is affected by the following factors.

4.6.2.1 Thickness of the Solid Slag Film (d_s)

It can be seen from Eq. 4.17 that the thermal resistance of the slag film increases as the ratio (*d/k*) increases for both the glass and crystalline phases. Since the thickness of the solid slag (d_s) is composed of both glass and crystalline phases, q_{hor} will decrease as (k/d_s) decreases ($q_{hor}\downarrow$ *as* $d_s\uparrow$ *as* $k\downarrow$). The slag film thickness is usually between 1 and 3 mm [58, 76–79] but a value of d_s = 0.3 mm has been reported for thin slabs [80]. The thickness, d_s, can be increased by increasing the break or solidification temperature (T_{br}). In practice, this is usually achieved by increasing the basicity (C/S) of the powder; frequently, powders with a C/S ratio in the range, 1.1–1.35, are used to cast steels prone to longitudinal and other forms of surface cracking.

4.6.2.2 The Fraction of Crystalline Phase (f_{crys})

The presence of crystalline phases in the slag film has two effects on the horizontal heat flux:

(i) Crystals in the slag film scatter the IR radiation and thus reduce k_R.
(ii) Crystallisation of a glassy slag film increases the interfacial resistance ($R_{Cu/sl}$) and creates some porosity in the slag film ($k_{lat}\downarrow$) due to the shrinkage (since $\rho_{crys} > \rho_{glass}$); the interfacial resistance and the associated surface roughness of the Cu/slag interface increase with increasing f_{crys} [81].

Although, $k_{cryst} \approx 2\ k_{glass}$, the value of q_{hor} is significantly lower for slag films containing crystals; this demonstrates the importance of reducing k_R. It has been estimated that the k_R contribution to the total heat flux for partially crystalline mould fluxes is 10–20% [10]. The fraction crystallinity, f_{crys}, is usually increased by increasing both basicity (C/S) and the CaF_2 content but large pick-ups of Al_2O_3 when casting high Al steels are also reported to increase f_{crys} [82].

4.6.2.3 Porosity of the Slag Film

Some porosity of the slag film results from the crystallisation of an initially glassy slag film [83]; porosity reduces both the thermal conductivity and q_{hor} by lowering the lattice conductivity and by the scattering of IR radiation by the pores. However, gas pores can also be formed in the slag film when the mould powder has high moisture content or there is a water-leak in the process. Sticker breakouts (accompanied by reduced q_{hor}) have been reported to occur when hydrogen contents in the slag film are high [84–86].

4.6.2.4 Viscosity of the Mould Slag

There is inconsistency in the reported effects of slag viscosity on the thickness of the slag film (d_s), and hence the horizontal heat flux (i.e. q_{hor}) has been reported to increase with increasing viscosity [87] and also with decreasing viscosity [88, 89].

4.6.2.5 Casting Speed and Superheat (ΔT)

An increase in casting speed results in a higher *rate* of heat flux ($Js^{-1}\ m^{-2}$) but the *total* heat loss (Jm^{-2}) is lower because of the shorter residence time. The effect of V_c is best viewed through its effect on the shell, i.e. an increase in V_c produces a thinner shell ($d_{shell}^{exit} = K'\ t_{mould}^{0.5}/3.6$) where K' = constant with values of 7.3 for slab casting and 9.5 for high-speed casting [90]. Thus a high V_c will result in a lower value for t_{mould} and lead to a thinner shell. The heat flux also increases with increasing superheat and leads to thinner shell due to the delay in solidification [91].

4.6.2.6 Non-sinusoidal Oscillation

The use of non-sinusoidal oscillation has been reported to result in a 10% decrease in q_{hor} [92].

4.6.2.7 Taper and Uniformity of Mould Temperature

The heat flux density is not uniform around the mould; q_{hor} is greatest in the regions near the narrow faces and decreases gradually to the region around the centre of the wide face (see Fig. 3.10a). The heat transfer is highest near the corners of the mould. Thus, the shrinkage of the shell is greater near the narrow faces than in centre of the wide face. The increased shrinkage results in increased slag film thickness (d_s) and, subsequently, results in a decrease in q_{hor} at the narrow faces. The mould flux ratio, defined in Eq. 4.18 is used to measure the uniformity of the heat flux in the mould.

$$\text{Heat Flux Ratio} = \left(q_{hor}^{narrow}\right)/\text{average}\left(q_{hor}^{wide}\right) \tag{4.18}$$

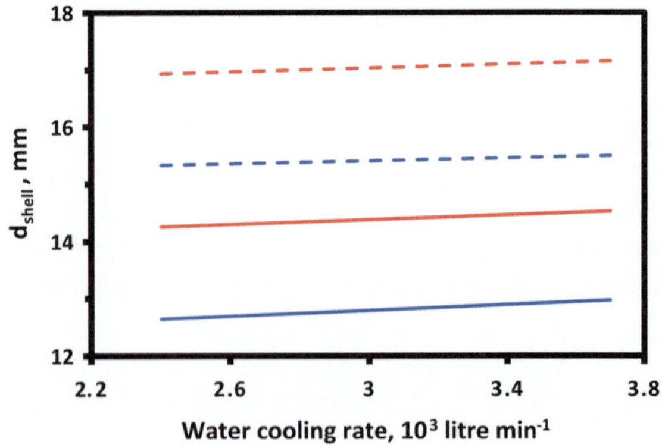

Fig. 4.10 The effect of water cooling flow rate on the thickness of the shell for two different superheats (20 °C = *blue* and 60 °C = *red*) during casting of stainless steel 304 (- - -) and CS 1026(——) [94]; V_c = 0.9 m min^{-1}. (permission granted, ISS/AIST [94] re-drawn)

The taper must compensate for the shrinkage, which in turn is related to the heat flux and will change according to the mould slag used. An increase in taper from 1.1 to 1.4% was observed to increase the heat flux ratio from 0.6 to 0.8 [93].

4.6.2.8 The Mould Water-flow Rate

The water-flow rate has a small effect on both q_{hor} and the shell thickness, as can be seen from Fig. 4.10 [94]. The flow rate must also be high enough to avoid the formation of steam bubbles which are detrimental to the cooling and can cause excessive temperatures in the copper mould.

4.6.2.9 Coating Moulds, Grooved Moulds

Copper has a very high-thermal conductivity, so the horizontal heat flux can be reduced by coating the mould with a metal or material with lower conductivity (e.g. Nickel) [95]. In Meniscus-free casting (MFC) a refractory is placed at the top of the mould to reduce q_{hor} and delay solidification to a site further down the mould [96, 97].

Grooved moulds have also been used to reduce q_{hor} in the meniscus region [98–103].

4.6.3 Measurement of Horizontal Heat Flux

A number of techniques have been used to study the horizontal heat flux across the slag film, these are described in Sect. 3.2.4.

In summary; the horizontal heat flux can be reduced by:

- Increasing the slag film thickness (by increasing the T_{br}).
- Increasing the fraction of crystalline phase (f_{crys}) by increasing the C/S ratio and %CaF_2.
- Decreasing the water-flow rate (small effect).
- Reducing the casting speed.
- Applying coatings to the mould and machining grooves in the mould.

4.7 Crystallinity in Slag Film

The slag film formed in the shell/mould channel usually contains a mixture of glass and crystalline phases. The crystal phase aids the control of the heat transfer and the glass phase aids the lubrication of the shell. Glass forms a super-cooled liquid (*scl*) at temperatures above T_g (*ca.* 630 °C) and the *scl*, despite its high viscosity, provides some measure of lubrication to the shell. This is a particular issue when casting high Al steels where the large Al_2O_3 pick-up results in a fully crystalline slag which provides little lubrication. Slag films with a high f_{crys} tend to be fragile and break along a line of pores (created during the crystallisation process) but glassy slag films are less prone to fracture [104].

The first slag film formed in the shell/mould channel is probably glassy because of the high cooling rates involved. However, the slag film crystallises over time until f_{crys} in the slag film reaches a steady state.

In conventional, (F–containing) slags, *cuspidine* ($3CaO{\cdot}2SiO_2{\cdot}CaF_2$) is precipitated during primary solidification and other phases form during secondary solidification [83]. The *cuspidine* is precipitated at high temperature (on the mould side of the slag film) and secondary precipitation occurs in hotter regions (on shell side) of the slag film [76] since high temperatures promote crystal growth.

Crystallisation involves two mechanisms, namely, *nucleation* and *growth.* Temperature has different effects on these two processes; Nucleation is promoted by high undercooling (lower temperatures) whereas growth is enhanced by high temperatures. Thus, crystals formed at high temperatures (just below T_{liq}) tend to be few but large (due to strong growth) whereas many finer crystals are formed at lower temperature due to the high undercooling and slow growth. Some oxide particles (e.g. TiO_2 and ZrO_2) have been reported to aid nucleation [105].

Agitation has been found to affect primary solidification (presumably, due to the removal of local undercooling) but has little effect on secondary solidification [106].

The application of an electrical potential to a slag has been reported to change f_{crys} in the sample [107, 108]; it was reported that when using DC, increasing voltage resulted in increases in both f_{crys} and D_{crys}, whereas with AC it caused increasing f_{crys} and decreasing D_{crys} [107].

4.7.1 Importance of Crystallinity to the Casting Process

The horizontal heat flux from the shell to mould is largely controlled by (i) the thickness of the slag film (d_s) (ii) the fraction crystalline phase (f_{crys}) and the size of the crystallites (D_{crys}) in the slag film and (iii) the interfacial resistance ($R_{Cu/sl}$) and porosity formed by the shrinkage of slag during crystallisation. The thickness of the solid slag is a linear function of the break (T_{br}, or solidification) temperature and the horizontal heat transfer can be controlled through T_{br}. The principal issues are (i) controlling f_{crys} and (ii) replacing *cuspidine* ($C_3S_2F_1$) with a suitable alternative, crystalline phase. Increasing basicity (C/S) causes both d_s and f_{crys} to increase. However, the replacement of *cuspidine* has proved difficult. Mould slags have been divided into three categories in Sect. 4.7.2, and crystallisation is considered individually, for each type of slag.

4.7.2 *Factors Affecting* f_{crys}

The *required* values of the powder consumption (Q_s^{req}), viscosity (η^{req}) and the break temperature (T_{br}^{req}) are determined by the mould dimensions, the casting speed and the steel grade being cast; these *required* values apply to all three types of slag, discussed below. The mould slag must possess these required properties plus an appropriate value of f_{crys}.

4.7.2.1 Conventional (F-Containing) Mould Slags

Cuspidine ($3CaO{\cdot}2SiO_2{\cdot}CaF_2$ or $C_3S_2F_1$) is precipitated during primary crystallisation in these slags. Other phases are crystallised later during secondary crystallisation [109]. The compositions of most mould slags lie outside the *cuspidine* phase field and moving the slag composition into this field results in a reduction in the horizontal heat flux [110].

Both f_{crys} and the crystal size (D_{crys}) increase with increasing basicity (C/S) for the slag film. However, additions of other oxides promote glass formation (which tends to increase the horizontal heat flux). Consequently, the various slag constituents are divided into *cuspidine promoters* and *glass promoters*:

(i) *Cuspidine promoters* (i) CaO (ii) SiO_2 (when <CaO) and (iii) CaF_2 and (iv) large amounts of Al_2O_3 [111].
(ii) *Glass Promoters*: (i) B_2O_3 (ii) Na_2O, (iii) Li_2O (iv) FeO [112] and (v) >4% MnO and >7% MgO [113].

Although increasing basicity (C/S) is the main tool in controlling both f_{crys} and q_{hor}, care must be taken since a high (C/S) can lead to an increased T_{liq} which can result in the formation of large slag rims. High (C/S) slags have the advantage of possessing low oxygen potential.

The glass promoters also tend to slow down the crystallisation process (see Sect. 9.3.4) but Na_2O is thought to reduce the incubation time (i.e. accelerate crystallisation [114]). Additions of >5% TiO_2 have been reported to result in the formation of $CaO{\cdot}TiO_2$ which, subsequently, suppresses *cuspidine* precipitation [115].

Melilite slags have a high basicity (C/S) but have been reported to maintain some glassy phase, which proved useful (by providing some lubrication) when casting high Al steels where the large amounts of Al_2O_3 formed result in a highly crystalline slag film [111].

4.7.2.2 F-Free (FF) Mould Slags

The replacement of *cuspidine* has proved difficult. The phases, *perovskite* (CT) [74] CST [114] and NC_2S_3 and *melilite* [111] have been proposed as replacements for *cuspidine*, in order to reduce q_{hor}. The problem with *perovskite* is that it has a high melting point and in the casting of Ti- stabilised steels, the basicity must be kept below (C/S) = 1 to ensure that CT is not formed. *Melilite* slags have been used for casting round billets (see Sect. 5.2.9) and high Al steels (see Sect. 6.2.12). The formation of a glassy phase is an advantage in both cases. Additions of Na_2O tend to decrease incubation times for crystallisation [116, 117].

4.7.2.3 Calcium Aluminate (CA) Mould Slags

These slags have been developed for the casting of high Al steels to minimise the formation of Al_2O_3. There is a eutectic in the CaO–Al_2O_3 system at *ca.* 50% CaO (close to $C_{12}A_7$) and <10% SiO_2 and fluxes are added to the eutectic to reduce the melting temperature. These CA-type slags have proved to be effective in casting high Al steels but still need some development to address the following problems:

(i) The fluxes are restricted to CaF_2, Na_2O, Li_2O and K_2O because Al in the steel reacts with other potential fluxes (e.g. B_2O_3) to form Al_2O_3.
(ii) The slag films are crystalline, fragile and provide little lubrication to the shell; thus it is necessary to find a way of introducing some glassy phase.

The thickness of the slag film can be kept to a minimum by ensuring (C/A) ratio lies between 0.9 and 2 [118]. The incubation time for crystallisation was found to be at a maximum using a slag with (C/A) = 1.5 [118] and had the weakest crystallisation tendency.

An equation has been reported for calculating the value of f_{crys} in the slag film from the chemical composition (in terms of the modified (NBO/T) of the mould powder (Eq. 4.19) [113].The modified (NBO/T)* is defined in Eq. 4.20, where the contents in the curly bracket { }, are included in the numerator if MgO is <7% and MnO is <4% and in the denominator when they exceed these values:

$$\%\,\mathrm{crys} = 100 f_{crys} = -284 + 141.1(\mathrm{NBO/T}) \tag{4.19}$$

$$\begin{aligned}\mathrm{Modified(NBO/T)}^* = {} & 2(X_{CaO} + X_{FeO} + X_{Na_2O} + X_{K_2O} + X_{CaF_2} + \{X_{MgO} + X_{MnO}\} \\ & - X_{Al_2O_3})/\left(X_{SiO_2} + 2X_{Al_2O_3} + \{X_{MgO} + X_{MnO}\}\right)\end{aligned} \tag{4.20}$$

According to Eq. 4.19, the slag film will be totally glassy (f_{crys} = 0) when (NBO/T)* $\leq$ 2 and totally crystalline (f_{crys} = 1) when (NBO/T)* $\geq$ 2.75.

4.7.3 *Ways of Increasing Crystallinity in Slag Film*

In conventional, F-containing powders the usual method of increasing the fraction of crystalline phase, f_{crys}, is to increase the {%CaO/%SiO_2} ratio from about 1.0 for casting LC grades to 1.3 (up to 1.6) for casting MC, peritectic steel grades.

4.8 Delaying Solidification and Shortening the Length of Shell

There is considerable turbulence in the meniscus region which results in inclusion and bubble entrapment. Delaying the solidification of the shell, to a position lower in the mould, provides significant benefits since it results in a shorter shell (like that shown in Fig. 4.11). The shorter shell/meniscus has the following advantages:

- It results in fewer inclusions being trapped in the surface of the steel product.
- It results in shallow oscillation marks (since delayed solidification results in an increase in (slag rim-shell tip distance) and consequently, results in shallow oscillation marks.

Delaying solidification involves both vertical and horizontal heat fluxes.

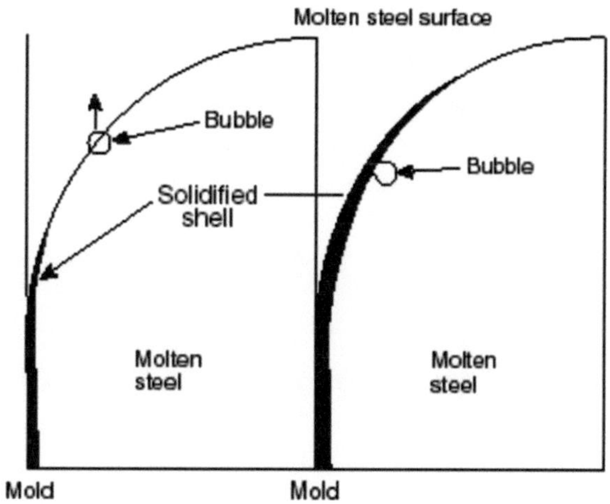

Fig. 4.11 Schematic figure showing that a shorter meniscus shell leads to less slag entrapment (permission granted, ISS/AIST [7])

4.8.1 *Factors Affecting Shell Length*

The following methods can be used to delay solidification and to shorten the length of the meniscus tip.

4.8.1.1 Reducing the Vertical Heat Transfer

This can be achieved by improving the thermal insulation of the bed (Sects. 4.2.1.3 and 4.2.4) by:

(i) Using a deeper powder bed.
(ii) Using smaller granules and lower Ar flow rates to reduce gaseous conduction.
(iii) Using exothermic agents in the powder to reduce the thermal gradient in bed.
(iv) Using EMBr to reduce turbulent flow which also decreases q_{vert}.

4.8.1.2 Reducing the Horizontal Heat Transfer

The horizontal heat flux can be reduced by using:

- A thick, crystalline slag (i.e. with a high T_{br} and f_{crys}) which also creates a thick slag rim.
- Meniscus-free casting (where a refractory lining is added to the meniscus region of the mould [96, 97]—see Sect. 3.2.3.3)
- Electromagnetic casting (EMC) where the "*pinch effect*" causes a reduction in q_{hor}.

4.8.1.3 Superheat

Solidification can also be delayed by increasing the superheat.

4.8.2 *Ways of Controlling the Length of Meniscus/Shell*

Solidification can be delayed by:

(i) Reducing vertical heat transfer.
(ii) Reducing horizontal heat transfer in the meniscus region.
(iii) Increasing superheat.

Table 4.1 Summary of the remedial actions that can be undertaken to bring about certain tasks or objectives

Task	Variable	Remedial action	Repercussions, Comments
Vertical heat flux and thermal insulation of the powder bed	$q_{vert}\downarrow$ Insulation $\uparrow$	(a) *From metal* by (i) $\downarrow V_c$ and $\downarrow$ superheat (ii) $\downarrow$ Ar flow rate (iii) use EMBr $\rightarrow \downarrow$ turbulence and q_{vert} (b) *From slag pool* by improving thermal insulation (i) Increase depth of bed (ii) Use smaller granules to $\downarrow$ gaseous convection (iii) Incorporate exothermic agents into powder (iv) Reduce CO_3^{2-} content	Gaseous convection probably dominant mechanism in bed
Melting rate (MR)	$Q_{MR}\uparrow$	(i) Decrease $\%C_{free}$ (ii) Increase C particle size ($D_C\uparrow$) (iii) Decrease bulk density ($\rho_{bulk}\downarrow$) (iv) Increase carbonate content (v) Increase q_{vert} into slag pool ($q_{vert}^{in}\uparrow$) See vertical heat flux above-note EMBr $\rightarrow q_{vert}\downarrow$	(a) Also decreases exothermic heat—may cause *T* gradient in powder bed to increase $\rightarrow$ steel meniscus may solidify
Depth of slag pool	$d_{pool}\uparrow$	increasing $\Delta q_{vert} = (q_{vert}^{in} - q_{vert}^{out})$ by (i) Increasing q_{vert}^{in} by (a) $V_c\uparrow$ (b) Superheat $\uparrow$ (ii) Improving bed thermal insulation by (a) $\uparrow d_{bed}$ (b) $D_{gran}\downarrow$ (c) Using exothermic agents	(a) $V_c\uparrow \rightarrow$ turbulence $\uparrow$; $\rightarrow t_n\downarrow$; $\rightarrow Q_s\downarrow$; (b) d_{pool} tends to be low for thin slabs
Powder consumption and liquid slag film	$Q_s\uparrow$	(i) Decrease viscosity ($\eta\downarrow$) (ii) Decrease V_c ($V_c\downarrow$) (iii) $f\downarrow$ $s\downarrow$ and $T_{br}\downarrow$ (iv) Increase Ar flow rate (v) Use Non-sinusoidal osc'n (vi) Use EMBr, EMC (vii) Large amounts of Ti (CN) + $Al_2O_3 \rightarrow$ form blockages $\rightarrow$ use $\uparrow d_{pool}$ to dissolve more solid	(a) ($\eta\downarrow$) Slag entrapment danger (b) Keep flow rate <5 L min^{-1} (c) Q_s low for thin slabs
Horizontal heat flux and solid slag film	$q_{hor}\downarrow$	By: (i) $T_{br}\uparrow$ (*via.* C/S$\uparrow$) (ii) $f_{crys}\uparrow$ (*via.* C/S$\uparrow$ + $CaF_2\%$ $\uparrow$) (iii) Use Non-sinusoidal oscillation (iv) Coating mould and MFC (v) Use EMC (vi) Use of grooved mould	(a) For Thin slabs of MC steel- danger of bulging

(continued)

Table 4.1 (continued)

Task	Variable	Remedial action	Repercussions, Comments
Crystallisation of slag film	$f_{crys}\uparrow$ $D_{crys}\uparrow$	(a) *Cuspidine*- (i) use high (C/S) and CaF_2 (ii) Maximise *cuspidine* formation;% $Na_2O\uparrow \rightarrow T_{incubn}\downarrow$ i.e. accelerates crystallisation (iii) Large amounts of $Al_2O_3 \rightarrow \uparrow f_{cryst}$ (b) *FF- fluxes*- TiO_2 suppresses *Cuspidine* (c) CA *fluxes*- have high f_{crys} (e) D_{crys} high at high T D_{crys} fine at lower T	(a) Shrinkage occur during crystallisation → Porosity↑ (b) Important slag film keeps some glass phase for lubrication Glass promoted by SiO_2, B_2O_3; >4%MnO, FeO, >75MgO (c) Slag films with high f_{crys}—fragile and prone to cracking
Delayed solidification and shortening shell tip	$q_{vert}\downarrow$ $q_{hor}\downarrow$	(i) Reducing q_{vert} by (a) $d_{bed}\uparrow$ (b) $D_{gran}\downarrow$ (c) Use exothermic agents (d) EMBr (ii) Reducing q_{hor} by (a) $T_{br}\uparrow$ (b) $f_{crys}\uparrow$ (c) Ar flow↓ (d) using MFC and EMC (iii) Increasing superheat	leads to $N_{incl}\downarrow$ and $d_{OM}\downarrow$

Note in right-hand column (i) etc. refers to (i) in column to left; *MFC* Meniscus-free casting; ↑ increase; ↓ decrease

4.9 Summary

The remedial actions which can be taken to achieve certain tasks or objectives are summarised in Table 4.1.

References

1. JW Cho T Emi, H Shibata, M Suzuki, ISI J Intl. **38**, 440, (1998).
2. JW Cho H Shibata, T Emi, M Suzuki, ISI J Intl. **38**, 268 (1998).
3. A Yamauchi, T Emi, S Seetharaman, ISIJ Intl., **42**, 1084, (2002).
4. C Lefevre, JP Radot, JN Pontoire, Y Roux, Revue de Metall. CIT, **94**, 489 (1997).
5. Y Tanizawa, CAMP-ISIJ, **3**, 257, (1991).
6. K Yamaguchi, CAMP-ISIJ, **4** , 181 (1990).
7. F Neumann, J Neal, MA Pedroza, AH Castillejos, FA Acosta, Proc. 79th Steelmaking Conf. (ISS ,Warrendale, PA, 1996), p. 249.
8. R Saraswat, PhD Thesis,“ *Modelling the effect of mould flux on steel shell formation during continuous casting*”, Imperial College, London , 2006.
9. M. Susa, K Nagata, KC Mills , Ironmaking and Steelmaking, **20**, 372, (1993).
10. M. Susa, KC Mills, MJ Richardson, R Taylor, D Stewart, Ironmaking and Steelmaking, **21**, 279, (1994).

11. S Diehl, JA Moore, RJ Phillips, Proc. 78th Steelmaking Conf. (ISS , Warrendale, PA, 1995) p. 351.
12. R Koldewijn Unpublished Corus Internal Rept (2007) cited in KC Mills, J Kromhout, A Hamoen and R Boom: Proc. Admet Conf held Dnipetrovsk, Ukraine, 2007.
13. PE Ramirez- Lopez, PN Jalali, J Bjorkvall, U Sjostrom, C Nilsson, ISIJ Intl. **54**, 342. (2014) and Proc. 8th Europ. Conf Continuous Casting , Graz, Austria, 2014 (ASMET, Vienna, 2014).
14. E van Vliet, DW van der Plas, SP Carless, A A Kamperman, AE Westendorp: Proc. 7th Europ Conf. Cont. Casting, Dusseldorf, 2011 (VDEh, Dusseldorf, 2011) Session 4.
15. T Sakuraya, T Emi. T Imai, K Emoto, M Kodama, Tetsu-to- Hagane, **67**,1220, (1981).
16. K Nagata, KS Goto, Proc. 2nd Intl. Conf. Metallurgical slags and fluxes, Lake Tahoe, NV, (1984) ed. by HA Fine and DR Gaskell (TMS-AIME, Warrendale, PA, 1984) p. 875.
17. KS Goto, HW Gudenau, K Nagata, KH Lindner, Stahl u Eisen, **105**, 1387, (1985).
18. SP Andersson, Ironmaking Steelmaking, **42**, 465, (2015).
19. P Andersson, to be published in Ironmaking Steelmaking.
20. JJ Macho, G Hecko, B Golinmowski, M Frazee, Preprints 33rd McMaster Symp. on Iron and Steelmaking,. McMaster Univ., Hamilton, CN, June (2005).
21. R Taylor, KC Mills: Ironmaking and Steelmaking, **15**, 187, (1988).
22. KH Spitzer , K Schwerdtfeger, *Investigation of the isolation properties and melting behaviour of mould powders used in the continuous casting of steel.* Report 7249,Tech. Univ. Clausthal, DE, (1989).
23. KH Spitzer, K Schwerdtfeger, Stahl u Eisen, 108, 441, (1988).
24. M Supradist , AW Cramb , K Schwerdtfeger, ISIJ Intl., **44** , 817, (2004).
25. P Hasselström, "*On the formation of a lubrication film from mould powders used in continuous casting.*" Swedish Inst. Metals Research Report IM-1467, (1980).
26. H Lidefeldt , P Hasselström, Proc. Continuous Casting; 4th Intl. Iron and Steel Congress, (The Metals Society, London , 1982) p. 10.
27. H Litterscheid, "*Untersuchung des Verhalten beim Strangeisen, Giesen und erstarren von stahl III*" Report EUR 8569, Contract 7210- CA/112 (EC Sci. Tech. Publ., Luxembourg,1984) p. 34.
28. JW Kim, SK Kim, YD Lee, Proc. 4th Europ. Conf. on Continuous Casting, Birmingham, UK, 2002, (IOM, London, 2002) p. 371.
29. N Pradhan, M Ghosh, DS Basu, S Mazumdar, ISIJ Intl., **39**, 804, (1999).
30. K Koyama, K Nagano, Y Nagano, T Nakano. Nippon Steel Technical Report, **34**, 41, (1987).
31. M Kawamoto, K Nakajima, T Kanazawa, K Nakai, ISIJ Intl., **34**, 593, (1994).
32. CÅ Däcker, C Eggertsson, J Lonnqvist, VIII Intl. Conf. Molten slags, fluxes & salts, Santiago, Chile, Jan., 2009 (GECAMIN, Santiago, 2009) p. 1111.
33. JA Kromhout, DW Plas, Ironmaking and Steelmaking, **29**, 303, (2002).
34. T Emi, H Nakato, K Suzuki, Y Iida, T Ueda, Tetsu- to Hagane, **60** (7), 981, (1974) (Henry Brutcher Translation, HB9357).
35. M Kawamoto, K Nakajima, T Kanazawa, K Nakai, ISIJ Intl., **34**, 593, (1994). see also Proc. 75th Steelmaking Conf., Toronto, (1992) (ISS, Warrendale, PA, 1992) p. 389.
36. T Kishi, Y Nagano, T Nagano, Trans. ISIJ, **25**, B 302, (1985).
37. T Nakano, M Fujii, K Nakano, T Matsuyama, Nippon Steel Tech Rept., **34**, 21, (1987).
38. J Sardemann, H Screwe, Stahl u Eisen, **111** (11), 39, (1991).
39. R Sato, Bull. Jap. Inst. Metals, **12** (6), 391, (1973).
40. D Singh, P Bhardwaj, Y Dong, A McLean, M Hasegawa, M Iwase, Proc. 8th Intl. Conf. Molten Slags, fluxes and salts, Santiago, Chile, 2009 (GECAMIN, Santiago, Chile, 2009) p. 1073.
41. EF Wang, YD Yang, CL Feng, ID Sommerville, A McLean, J Iron Steel Res. Intl. **13** (2), 22, (2006).
42. QC Wei, YQ Ding, KD Peng, J Chongqing Univ., **4**, 110, (1995).

43. B Xie, J Wu, Y Gan, Proc. 74th Steelmaking Conf., Washington, DC, 1991 (ISS, Warrendale, PA, 1991) p. 647.
44. K Schwerdtfeger, A Jablonka, Steel Research Intl., **64**, 77, (1993).
45. K Schwerdtfeger, J Sardemann, HJ Grethe, Stahl u Eisen, **114**, 57, (1994).
46. PE Ramirez-Lopez, KC Mills, PD Lee, B Santillana, Met. Mater. Trans. B, **66**, 109 (2011).
47. P.V. Riboud ,Y. Roux: *Fundamental study of the behaviour of casting powders* Report 9560, (1985) (Europ. Comm. Sci. Tech. Publ., Luxembourg, 1985).
48. RM McDavid , BG Thomas, Metall. Trans. B, **27B**, 672, (1996).
49. M Washio, M Sugizawa, S Moriwaki, K Kariyaa, S Idogawa, S Takeuchi, Revue de Metallurgie, CIT, **90** (April), 507, (1993).
50. Y Nakamori *et al*, Nippon Steel Tech Report, **34**, 53, (1987).
51. F Mantovani, S Spagnul, M Padovan, A Bianco, Proc. 8th Europ. Conf. Continuous casting , Graz, Austria, (2014) (Austrian Soc. Metals, 2014).
52. N Viswanathan, AB Fox , KC Mills, Proc. Mills Symp., London, (2002) ed. RE Aune and S Sridhar, (IOM, London 2002) p. 513.
53. B Zhao, SP Vanka, BG Thomas, Intl. J Heat and Fluid Flow, **26**, 105, (2005).
54. E Maclas, AH Castillejos, FA Acosta, GM Herrera, F Neumann, Ironmaking and Steelmaking, **29**, 347, (2002).
55. R Bommaraju , Proc. 74th Steelmaking Conf., Washington, DC, 1991, (ISS, Warrendale, PA, 1991) p. 95.
56. JA Moore, C Camino, S Diehl, RJ Phillips , D Piwinsky, Proc. 79th Steelmaking Conf., Pittsburgh, PA , 1996, (ISS, Warrendale, PA, 1996) p. 259.
57. T Nakano, M Fuji, K Nakano, S Mizoguchi, T Yamamoto, K Asano, Trans. ISI J, **21**, B 307, (1981).
58. C-Å Däcker, A Salwén, P Andersson, C Eggertsson, Proc. 7th Europ. Continuous Casting Conf., Düsseldorf, (2011) Session 12, pp. 1–8.
59. M Wolf, AIME Electrical Furnace Proc., 40, 335, (1982).
60. M Wolf, Proc. 2nd Europ. Conf. on Continuous casting, Dusseldorf, 1994, p. 335.
61. AB Fox, PhD Thesis, *Mould fluxes, their properties and performance.* Imperial College, London (2003).
62. S. Sridhar, K.C. Mills, V. Ludlow, S.T. Mallaband, Proc. 3rd Europ. Conf. Continuous Casting, Madrid, (UNESID, Madrid, 1998) p. 807.
63. S Ogibayashi, K Yamaguchi, T Mukai, T Takahashi, Y Mimura, K Koyama. Y Nagano, T Nagano, Nippon Steel Technical Report, **34**, 1, (1987).
64. K. Tsutsumi, H Murakami, S Nishioka, M Tada, M Nakada, M Komatsu, Tetsu- to Hagane, 84 , 617, (1998).
65. H Maeda, T Hirose *et al*, CAMP-ISIJ, **6**, 280, (1993).
66. T Mizukami, K Kawakami, T Kitagawa, M Suzuki, S Uchida, Y Komasu, Trans ISIJ, **26**, B 164, (1986).
67. R D'Haeyer, *Influence of powder consumption on physical properties of continuous casting powder*, Report EUR 10326 (Europ. Comm.Sci. Tech. Publ, Luxembourg, 1987).
68. M Suzuki, S Miyahara, T Kitagawa, S Uchida, K Okimoto, Tetsu-to Hagane, **78**, 113, (1992).
69. T Mallaband: Unpublished results from (Metallurgica, UK) cited in A Fox: PhD Thesis, Imperial College, London. (2003).
70. M Tani, T Toh, K Umetsu, K Tanaka, M Zeze, K Tsunenari, K Hayashi, S Fukunaga, Nippon Steel Technical Report, **104**, 62, (2013).
71. T. Mukongo, MEng Thesis , "*Effect of titanium pick-up on mould flux viscosity in continuous casting of titanium-stabilised stainless steel.*", Faculty of Eng., Univ. Pretoria, South Africa, (2003).
72. RC Nunnington, N Sutcliffe, Proc. 59th Electric Furn. Conf., Pheonix, AZ, Nov (2001).
73. Q Wang, YJ Lu, SP He, KC Mills, Z Li, Ironmaking and Steelmaking, **38**, 297, (2011).
74. GH Wen, S Sridhar, P Tang, X Qi, TQ Liu, ISIJ Intl., **47**, 1117, (2007).
75. JW Cho, KE Blazek, MJ Frazee, HB Yin, JE Park , SW Moon, ISIJ Intl., **53**, 62, (2013).

76. M Hanao, M Kawamoto, ISIJ Intl., **48**, 180, (2008).
77. MV Fonseca, ODC Afrange, A Lavinas, AA Ramos, CA Valadares, Proc. 5th Intl. Conf. Molten Slags, fluxes and salts , Sydney, Australia,1997 (ISS, Warrendale, PA, 1997) p. 851.
78. ODC Afrange, MVA Fonseca, H Polivanov, « *Differential thermal analysis applied in design, quality control and establishment of prefused and mixed mould powders operational parameters*. Intl. Congress Metall. Technol. Materials, 1994, p 623.
79. MCC Bezerra, ODC Afrange, AC Valadares, Proc. Mills Symp., London, (2002) ed. RE Aune and S Sridhar, (IOM, London, 2002) p. 293.
80. J Kromhout, RS Schimmel, Proc. 8th Europ. Conf on Continuous casting, Graz, Austria, June, 2014 (Austrian Soc. Metals, 2014).
81. K Tsutsumi, R Nakamura, J Ohtake, T Nagasaka, M Hino, ASIA Steel International Conference, 2000 (ASIA Steel' 2000), (Chinese Society for Metals, Beijing, 2000) Vol. C, (2000), p. 366.
82. HB Yin, K Blazek, J Macino, J Cotrell, Proc. 8th Europ. Conf. Continuous casting, Graz, Austria, 2014, (Austrian Soc. Metals, 2014).
83. M Hanao M Kawamoto, M Hara, T Murakami, H Kikuchi, K Hanazaki, Tetsu-to-Hagane, **88**, 23, (2002).
84. JN Pontoire, JP Radot,V Delvaux, E Dehaussy, Revue de Metallurgie, CIT. **93**, 1237, (1996).
85. Y Uemiya, T Mizoguchi, T Kajitani, Proc. 9th Intl. Conf. Molten Slags, fluxes and salts (Molten 12) Beijing, 2012 (Chinese Soc. Metals, 2012).
86. H Yamamura, T Kajitani, J Nakashima, M Yamasaki, S Mineta, Nippon Steel Technical Report, **104**, 54, (2013).
87. HP Narzt, R Martinelli, C Furst, G Xia, H Pressinger, Proc 3rd Europ. Conf. Continuous casting, Madrid, Spain, 1998, (UNESID, Madrid, 1998) p. 747.
88. L Hering, HP Heller, HW Fenske. Stahl u Eisen **17**, 61 (1992).
89. P E Ramirez-Lopez, P.D. Lee, K.C. Mills, ISIJ Intl., **50** (3), 425, (2010).
90. M Wolf, Iron and Steelmaker, **23**, 47, (1995).
91. M Hanao, M Kawamoto, A Yamanaka, ISIJ Intl., **52**, 1310, (2012).
92. M Suzuki, S Miyahara, T Kitagawa, S Uchida, K Okimoto, Tetsu-to- Hagane, **78**, 113, (1992).
93. JA Kromhout, ER Dekker, M Kawamoto, R Boom, Ironmaking and Steelmaking, **40**, 206, (2013).
94. MJ Lu, JH Chen, DH Tsai, CH Huang, Proc. 78th Steelmaking Conf. (ISS, Warrendale, PA., 1995) p.359.
95. A Cristallini, R Langella, A Polk, Proc. 3rd Europ. Conf. Continuous casting, Madrid, Spain 1998, (UNESID, Madrid, 1998) p 581.
96. C Bertoletti, P Courbe, JM Jolivet, PP Naveau, A Oper, E Perrin, C Salaris, AL Spierings , E Weisseldinger, Proc. 3rd Europ. Conf. Continuous Casting, Madrid, 1998 (UNESID, Madrid, 1998) p. 65.
97. P Courbe, JM Jolivet, G Lesoult, E Perrin, P Nilles, Steel Research, **70** (8/9), 338, (1999).
98. K Nakai T Sakashita, M Hashio, M Kawasaki, K Nakajima, Y Sugitani, Tetsu-to- Hagane, **73** (3), 498, (1987).
99. Y Sugitani, M Nakamura, M Okuda, M Kawasaki,K Nakajima, Trans. ISIJ, **25**, B 91, (1985).
100. JW Cho, HT Jeong, Met. Mater. Trans. B, **44B**, 146, (2013).
101. H Murukami, M Suzuki, T Kitagawa, S Miyahawa, Tetsu-to- Hagane, **78**, 105, (1992).
102. T Mochida, S Kohriyama, S Itoyama, T Yamashita, M Suzuki, Y Kishimoto, CAMP-ISIJ, **20**, 851, (2007).
103. K Ichikawa, A Morita, Y Kawabe, Shinagawa Tech. Report, **36**, 283, (1993).
104. J Kromhout, PhD thesis *Mould powders for high speed continuous casting of steel*. Univ of Delft, NL (2011) Chapter 5.
105. P Rocabois, JN Pontoire, J Lehmann, H Gaye, J Non- Cryst. Solids, **283**, 98, (2001).

106. Y Harada, K Kusada, S Sukenaga, H Yamamura, Y Ueshima, T Mizoguchi, N Saito, K Nakashima, ISIJ Intl., **54**, 2071, (2014).
107. Y Wang, Yu Wang, L Fan , B Xie, XP Liang, X Zeng, ZW Wei, Proc, 9th Intl. Conf. Molten slags, fluxes and salts , Beijing, China (2012) ((Chinese Metals Soc., 2012).
108. S Riaz, Ironmaking and Steelmaking, **39**, 409, (2012).
109. P Grieveson, S Bagha, N Machingawuta, K Liddell, KC Mills, Ironmaking Steelmaking, **15**, 181, (1988).
110. M Hanao, M Kawamoto, T Watanabe, ISIJ Intl., **44**, 827, (2004).
111. M Hanao, Y Tsukaguchi, M Kawamoto, Proc. ICS 2008, Gifu, Japan (2008).
112. M Susa, A Kushimoto, H Toyota, M Hayashi, R Endo ,Y Kobayashi, ISIJ Intl., **49**, 1722, (2009).
113. Z Li, R Thackray, KC Mills, Proc. 7th Intl. Conf. Metallurgical slags, fluxes and salts , Capetown, 2004 (SAIMM, Johannesburg, 2004) p. 813.
114. H Nakada, K Nagata, ISIJ Intl., **46**, 441, (2006).
115. Z Wang, QF Shu, XM Hou, KC Chou, Ironmaking and Steelmaking, **39**, 210, (2012).
116. JL Klug, R Hagermann, NC Heck, JFC Viela , PR Scheller, ICS Conf. 2012 , Dresden, DE, 2012.
117. JL Klug, R Hagermann, NC Heck, JFC Viela, PR Scheller, Proc. 9th Intl. Conf. Molten slags, fluxes and salts, Beijing, China (2012) (Chinese Metals Soc.).
118. XJ Fu, H Wen, P Tang, Q Liu, ZY Zhou, Ironmaking and Steelmaking, **41**, 342, (2014).
119. F. Shahbazian, private communication concerning plate dip tests, SWEREA, KIMAB, Stockholm, Oct. (2016).

Chapter 5
Effect of Casting Variables on Mould Flux Performance

Abstract Continuous casting is a complex process affected by a number of factors and it is necessary to optimise these factors (e.g. shell lubrication and heat extraction) to minimise product defects and process problems. The mould powder is expected to alleviate these problems. However, the casting conditions have a pronounced effect on factors like slag infiltration and heat transfer and this requires careful selection of mould slag properties to provide optimum casting conditions. The effect of individual, casting variables (e.g., casting speed) on the casting process is discussed in this chapter and the following casting variables are examined and discussed: (i) mould characteristics (e.g., dimensions, taper, coatings) (ii) oscillation characteristics (iii) casting speed (iv) metal flow characteristics (v) metal-level variations (vi) fluctuations (in metal flow, mould level, etc. (vii) the application of electromagnetic devices (viii) steel grade being cast (ix) water flow rate. The effects of these individual factors are discussed and how the mould powder can be used to optimise the casting process.

Symbols, Abbreviations and Units

d_s	Thickness of solid slag layer (m)
d_s	Thickness of liquid slag layer (m)
f	Frequency (cpm or Hz)
f_{crys}	Fraction crystalline phase
Q_s	Powder consumption (kg m^{-2})
q	Heat flux density (Wm^{-2})
$R*$	Mould (surface area/volume) (m^{-1})
s	Stroke length (m)
T	Temperature (°C)
$T_{br} = T_{sol}$	Break, Solidification temperature
t	Time (s) or thickness of mould (m)

The original version of this chapter was revised: For detailed information please see correction. The correction to this chapter is available at https://doi.org/10.1007/978-3-319-53616-3_12

Within this book minor changes to some of the figure permissions and the readability of the chapters have been made and several figures have been replaced with new versions.

K.C. Mills and C.-Å. Däcker, *The Casting Powders Book*,
DOI 10.1007/978-3-319-53616-3_5

t_n	Negative strip time (s)
V_C	Casting speed (m min^{-1} or ms^{-1})
$\boldsymbol{V}_m$	Velocity of mould (ms^{-1})
w	Width (m)
η	Slag viscosity (dPas)
EMBr	Electromagnetic braking
EMC	Electromagnetic casting
EMS	Electromagnetic stirring
HC	High-carbon steel
LMF	Level magnetic field
MC	Medium carbon steel

5.1 Introduction

There are many factors affecting the control of the continuous casting process. Frequently, the mould flux is asked to alleviate the problems arising from the casting conditions (e.g. those due to metal flow turbulence). Consequently, it is essential to have knowledge of how the various casting conditions affect (i) the heat transfer, powder consumption, metal flow etc. (ii) the mechanisms responsible for process control problems and defect formation and (iii) to identify the limitations of any proposed solution involving mould flux use. It should be noted that many of the factors itemised below are interactive and steps taken to solve one problem may have an adverse effect in triggering other problems or defect mechanisms. In this chapter, the effects of individual casting variables on the casting process are examined and how mould fluxes can be used to alleviate these problems or defects.

5.2 Mould Characteristics

5.2.1 Mould Dimensions

It is customary to classify the cast steel products according to their mould geometries, e.g. into (i) *slabs* (ii) *blooms* (iii) *billets* and (iv) *thin slabs*. The liquid friction forces acting on a specific face of the shell tend to increase with increasing distance from the corner (Fig. 5.1) [1, 2]. Consequently, the amount of lubrication required (i.e. required powder consumption, Q_s^{req}) increases with increasing width (w) of the wide face. Since the required lubrication is usually expressed in terms of the required powder consumption Q_s^{req} the hierarchy (for Q_s^{req}) is slabs > blooms > billets.

Although the distance from the corner has been used to represent the effect of mould geometry on the required powder consumption, Q_s^{req}, [2] it is customary to use the parameter $R*$, (defined in Eq. 5.1) [3]; $R*$ represents the ratio (surface area/volume) of the mould. Typical values for $R*$ values for slabs are, slabs *ca.* 10, blooms (10–18), billets (22–30), thin slabs *ca.* 40. Values of $R*$ for round billets can be calculated using $R* = 4/D$ where D is the diameter of the billet.

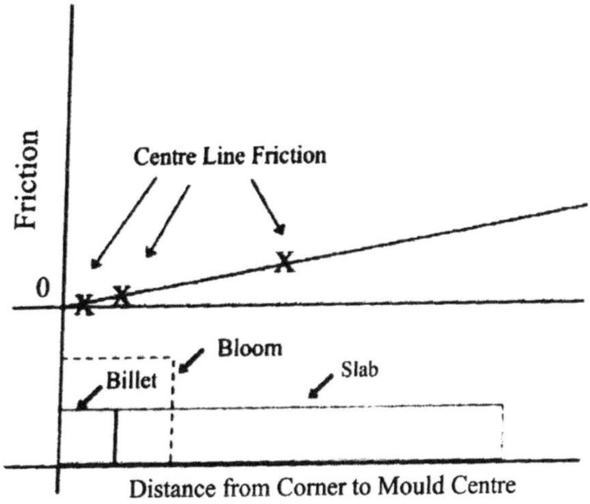

Fig. 5.1 Diagram showing the centre-line friction increasing with increasing distance from the corner; *x* friction at centre line for 3 mould geometries [1, 2]; *solid line* billet; *dotted line* bloom and *solid line* slab (courtesy of A.B. Fox [2])

$$R^{*}(m^{-1}) = 2(w + t)/w \cdot t \tag{5.1}$$

Since slabs have a large surface area (*cf.* those for blooms and billets), it follows that a higher powder consumption [in kg m^2 (of mould)] is required to lubricate slabs than for other mould dimensions. The relationship between Q_s and R^* shown in Fig. 5.2 can be described by the following equation [4, 5].

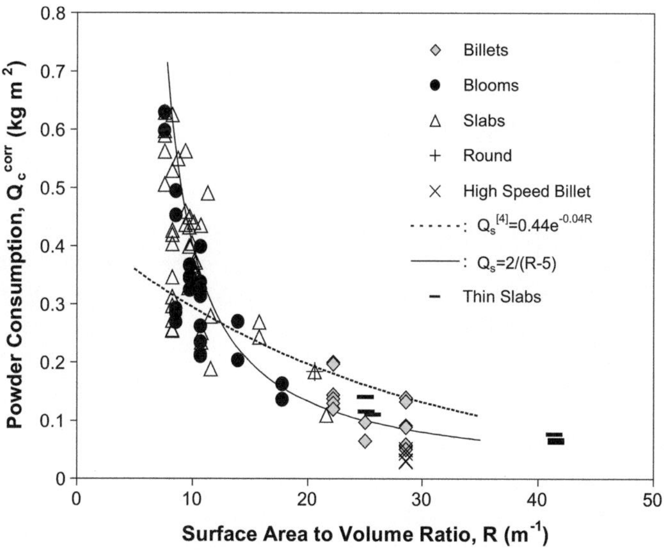

Fig. 5.2 Required powder consumption, Q_s^{req}, as a function of the parameter, R^* for different mould geometries; ◇ = billets; o = blooms; Δ = slabs; ▬ = thin slabs; + = Rounds; x = high speed billets; *dotted line* ($Q_s = 0.44 \exp^{-0.44R^*}$) [3]; *solid line* ($Q_s = 2/(R - 5)$ [4, 5]; (permission granted, UNESID [5])

$$Q_{\mathrm{s\ req}}^{\mathrm{slag}} = 2/(R^{*} - 5) \tag{5.2}$$

Thus, the required powder consumption is principally dependent upon the mould dimensions.

5.2.2 *Mould Length* ($\mathbf{L}_{mould}$)

It is important that (i) the shell is lubricated by liquid slag throughout the mould and (ii) the steel shell at the mould exit is sufficiently thick to ensure that no bulging occurs (which leads to mould level variations) [6]. The mould length affects both of these issues. Wolf [7] has discussed the importance of mould length in the continuous casting process; the required L_{mould} tends to increase with increasing casting speed.

5.2.3 *Mould Taper* ($\mathbf{L}_{mould}$)

Mould taper is important since an insufficient mould taper for the steel being cast leads to (i) hot spots at off-corner sites and (ii) to sub-mould bulging [6, 8]. In contrast, an excessive taper leads to overcooled corners and transverse cracking [8] (see Sect. 11.1.6 and Fig. 11.48). The ideal taper for the steel is dependent upon the steel composition, the casting speed and the mould length; a parabolic taper is recommended for the meniscus region [8].

5.2.4 *Mould Coatings*

Mould coatings (such as Mo, Cr or Ni) are usually applied to the upper part of the mould in order to extend the life of a mould. However, these coatings also help to reduce the horizontal heat flux since the conductivity of these metals is significantly lower than that of copper; coatings are therefore beneficial in reducing longitudinal cracking in MC steel grades (see Sect. 3.2.3.7(v)).

5.3 Casting Speed (V_c)

The casting speed (V_c) has a marked effect on the casting process. Increased productivity is attained by using higher casting speeds. However, an increase in casting speed results, simultaneously, in the following changes:

- A decrease in powder consumption, i.e. less lubrication for the shell.
- An increase in the *rate* of heat transfer (q_{hor}, in Wm^{-2} or Js^{-1} m^{-2}).
- A shorter residence time in the mould (resulting in decrease in the *total* heat flux) and which leads to a thinner shell (despite the increase in q_{hor}) and which could result in improved lubrication in the lower mould.
- Increased metal flow turbulence and higher metal flow velocities (which could result in slag entrapment and/or thinning of shell in the corner leading to longitudinal corner cracking).
- A decrease in negative strip time (t_n) which will reduce the depth of the oscillation marks formed.

It is apparent that casting speed changes affect mould performance in a variety of ways and these are discussed in more detail below.

5.3.1 Effect of Casting Speed on Powder Consumption

The effect of casting speed on powder consumption can be seen from the empirical rules reported for the required powder consumption, Q^{slag}_{sreq} [9, 10], e.g. Eq. 5.3 (or the alternative relation, $Q^{powd}_{s} = 0.60/\eta\ V_c$ [9–11]). Thus, the *required* powder consumption decreases with increasing casting speed and slag viscosity.

$$Q^{slag}_{s\,req} = 0.55/\eta^{0.5} \cdot V_c \tag{5.3}$$

Thus the slag viscosity must be selected to satisfy both the mould dimensions and casting speed used, e.g. by using Eqs. 5.2 and 5.3.

5.3.2 Effect of Casting Speed on Heat Transfer

Increasing the casting speed results in an increased *heat flux density* (q), but leads to a decrease in the *total heat flux* (q_{total}) because of the shorter residence time in the mould ($q \uparrow$ and $q_{total} \downarrow$ as $V_c \uparrow$). Increased casting speed results in a thinner shell ($d_{shell} \downarrow$ as $V_c \uparrow$).

The horizontal heat flux is principally controlled by the nature of the solid slag film formed between the shell and the mould; the heat flux decreases as (i) the solid film thickness (d_s) increases [which, in turn, increases as T_{sol} (or T_{br}) increases (i.e. $d_s\uparrow$ as $T_{br}\uparrow$)] and (ii) the fraction of the slag film forming crystalline phases (f_{crys}) increases.

However, the response of the slag film to a change in heat flux is not immediate since it takes time for the slag film thickness to adjust to these changes (Fig. 5.3)

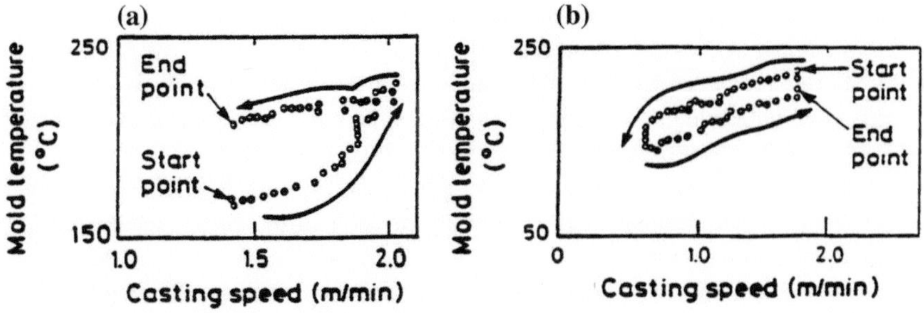

Fig. 5.3 Mould temperatures (related to heat flux) as functions of casting speed for a cycle (**a**) where casting speed, V_c, is increased then decreased and (**b**) where V_c is decreased then increased (permission granted, Nippon Steel Sumitomo Metal (NSSM) Corp. [11])

[11] and in some cases this has led to sticker breakouts [11]. This is particularly so in the lower mould where fracture of the slag film takes 20–30 min to repair since the rate of liquid slag supply is low in the lower mould [12].

5.3.3 *Effect of Casting Speed on Metal Flow Turbulence*

Increased casting speed results in increased metal flow turbulence which, in turn, produces:

- Waves on the surface of the steel with heights of up to 30 mm in slabs and 50 mm in thin slabs; these waves(result from the impact of the metal flow on the mould wall) [13].
- Vortices (shown in Fig. 5.4a) [13].
- Significant fluctuations in the metal flow.

Slag entrapment tends to become an ever-increasing problem as casting speeds are increased. Electromagnetic devices (e.g, EMBr, see Sect. 5.8) have proved useful in moderating metal flow velocities. High metal flow velocities can also lead to thinning of the shell in the corner regions (Fig. 5.4b) leading to longitudinal corner cracking [14].

5.3.4 *Effect of Casting Speed on Negative Strip Time*

The negative strip time (t_n) is calculated using Eq. 5.4 where f = frequency and s is the stroke length [15].

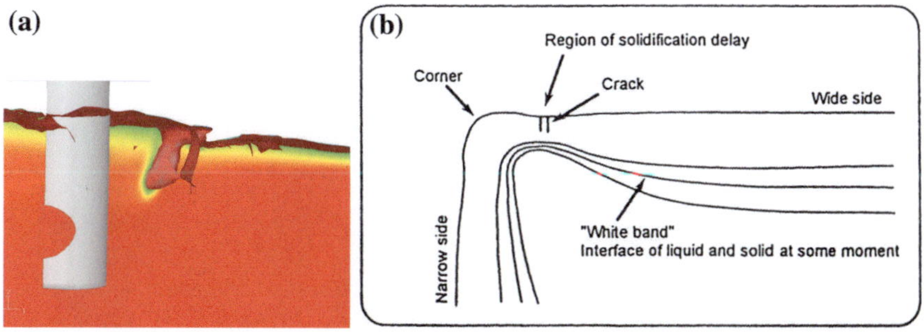

Fig. 5.4 Schematic diagrams showing (**a**) Karman vortex formed in steel and (**b**) thinning of the shell in corner regions caused by excessive metal flow (permission granted (**a**) ISIJ [13] and (**b**) UNESID [14])

$$t_n = \left(\frac{60}{\pi f}\right) \cdot \arccos\left(\frac{v_c}{\pi s f}\right) \tag{5.4}$$

The equation contains two terms, with the first term being more important and the second term acting so as to reduce the first term. Consequently, increases in the second (arc cos) term result in a decrease in t_n. Thus, an increase in casting speed results in a decrease in t_n. The depth of the oscillation marks (d_{OM}) increases with increasing t_n [16]. Transverse cracking tends to increase with increasing d_{OM}.

5.4 Oscillation Characteristics

The mould is oscillated to prevent the steel shell from sticking to the mould. The oscillation characteristics affect:

- The powder consumption (i.e. the lubrication supplied to the shell).
- The heat transfer.
- The depth of the oscillation marks (d_{OM}).

The oscillation characteristics are usually defined in terms of the positive (t_p) and negative strip times (t_n) [15].The negative strip time is the fraction of the oscillation cycle where the mould is descending faster than the mould (i.e. $V_m > V_c$) and is calculated by Eq. 5.4. The positive strip time constitutes the remainder of the oscillation cycle (Eq. 5.5).

$$t_{cycle} = t_n + t_p \tag{5.5}$$

The predictions of a mathematical model [13, 16, 17] have shown that the slag flow in the region of the mouth of the shell/mould gap is affected by the movement of the mould. When the mould is ascending the slag flow is hot ("*tropical*") and

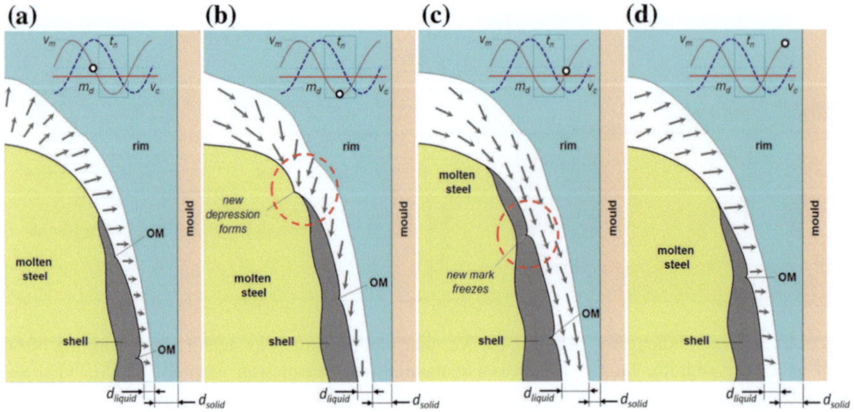

Fig. 5.5 Schematic diagram showing the direction of slag flow at different parts of the oscillation cycle (**a**) at highest position of mould in late t_p (**b**) halfway through t_n (**c**) at lowest position of mould in early t_p and (**d**) midway through t_p [13, 16, 17]; (Permission granted, ISIJ [13])

radially outward as shown in Fig. 5.5a. However, when the mould and the slag rim reach their highest positions (in late t_p) there is a *tide change* and the slag flow becomes downward. Since this flow originates from the cooler parts of slag pool, the resulting flow is cold or "*arctic*" (Fig. 5.5b, c). This arctic flow persists until the mould reaches its lowest point in early t_p. Tide changes (which occur at the highest and lowest positions of the mould) are not instantaneous and there is a *period of confused flow* before the new flow is established.

It has been reported that the distance between the slag rim and the liquid steel adjacent to the shell tip ($d_{rim/tip}$) is an important factor affecting both the depth of oscillation marks (since d_{OM} increases as $d_{rim/tip}$ decreases) and in delaying solidification of the shell [13].

The oscillation characteristics affect the powder consumption, the heat flux and depth of the oscillation marks (d_{OM}). The results of a parametric study using a mathematical model [16–18] are summarised in Table 5.1.

5.4.1 *Effect of Oscillation Characteristics on Powder Consumption*

Mathematical models based on Navier–Stokes calculations consistently predict that powder consumption (Q_s) increases with increasing frequency (f) [19–22]. However, plant observations indicate the reverse relationship (i.e. $Q_s\uparrow$ as $f\downarrow$) [23]. This discrepancy between predicted and actual dependence of Q_s on frequency has been explained in terms of the "*periods of confused flow*" following the tide change [16]. It has been proposed that slag infiltration is at a low level during these periods (following a tide change). Furthermore, increasing the frequency effectively

Table 5.1 Predicted cause and effect relationships during casting

Parameter		$\overrightarrow{V_c}$	$\overrightarrow{\Delta T}$	$\overrightarrow{t_n}$	$\overrightarrow{t_p}$	$\overrightarrow{f}$	$\overrightarrow{s}$	γ_{msl}	$\overrightarrow{\eta}$	$\overrightarrow{T_{br}}$
$\downarrow d_{OM}$	=	↑	↑	↓	↑	↑	↓	↑	↓	↓
$\uparrow Q_s$	=	↓	↑	↑	↑	↓	↑	↑	↓	↓
$\downarrow q_{peak}$	=	↓	↓	↑	↓	↑	↓	↓	↑	↑

↑ = increase; ↓ = decrease in values, e.g. $d_{OM}\downarrow$ as $V_c\uparrow$ [16–18]; ΔT = superheat; γ_{msl} = Interfacial tension between liquid metal and liquid slag; η = slag viscosity

increases the number of these periods of low slag infiltration per second. Consequently, increased frequency will lead to increase in the time where slag infiltration is low and so will cause a decrease in powder consumption. The failure to take into account the increased importance of these *periods of confused flow* (accompanied by low slag flow rates) with increasing frequency may well be the reason why slag infiltration models [19–22] give erroneous predictions of the effect of frequency on powder consumption.

The results of the parametric study given in Table 5.1 are in agreement with plant observations [23] with the exception of the stroke where a statistical analysis of plant observations indicated that Q_s increases as s decreases. This discrepancy could possibly be due the fact the stroke affects Q_s through its effect on t_n, so that an increase in stroke would lead to an increase in t_n which causes Q_s to increase (i.e. $s \uparrow\rightarrow t_n \uparrow\rightarrow Q_s \uparrow$).

The use of non-sinusoidal oscillation has been reported to bring about an increase in the liquid slag layer (i.e. Q_s) [24, 25].

5.4.2 *Effect of Oscillation Characteristics on Heat Flux*

Shell solidification occurs in the period of the oscillation cycle when the mould is descending [16, 17]. The heat flux increases throughout the negative strip period and peaks in early moments of positive strip (t_p^{early}) when the mould and slag rim attain their lowest positions. Thus, a long stroke may be expected to reveal that the heat flux, q_{hor}, increases with increasing stroke length; a statistical analysis of plant data has confirmed such a relationship (i.e. $q_{hor}\uparrow$ as $s\uparrow$) [26].

The results of a parametric study using a mathematical model are also in agreement with this finding ($q_{hor}\downarrow$ as $s\downarrow$) [17, 18]. The results of this study also indicated that ($q_{hor}\downarrow$ as $f\uparrow$) and ($q_{hor}\downarrow$ as $t_n\uparrow$).

The use of non-sinusoidal oscillation has been reported to bring about a decrease in both the heat flux and the thickness of the solidified shell [25].

It has been reported that variability in q_{hor} tends to increase with increasing depth of oscillation mark [27].

5.4.3 *Effect of Oscillation Characteristics on Oscillation Mark Depth (D_{OM})*

On the basis of the mechanism proposed by Ramirez et al. [16] the depth of the oscillation mark would be expected to increase as:

(i) The stroke length increases (i.e. $d_{OM}\uparrow$ as $s\uparrow$).
(ii) The distance between rim and the shell tip ($d_{rim/tip}$) decreases (i.e. $d_{OM}\uparrow$ as $d_{rim/tip}\downarrow$).

The depth of oscillation mark (d_{OM}) has been reported to increase with increasing negative strip time (i.e. $d_{OM}\uparrow$ as $t_n\uparrow$) [16] and that the effect of other oscillation variables on d_{OM} can all be predicted from their effect on t_n (i.e. $d_{OM}\uparrow$ as $t_n\uparrow$ or as $V_c\downarrow$, $s\uparrow f\downarrow$). This is consistent with the trends reported in Table 5.1 obtained in a parametric study using a mathematical model [16, 18].

5.5 Mould-Level Control

Good mould-level control is essential to the process, since poor mould-level control leads to a variety of problems. These include:

(i) decreases in mould temperature which lead to the formation of large slag rims [28, 29].
(ii) increased incidences of sticker breakouts [30] which are probably associated with the fracture of large slag rims.
(iii) the formation of longitudinal and transverse depressions which are also associated with the fracture and capture of the slag rim [31–33].
(iv) Increased concentrations of inclusions and bubbles (pinholes) in the steel [34, 35] and slag entrapment [36].
(v) Changes in the depth of oscillation marks [37, 38].
(vi) Increases in longitudinal cracking [29, 39].
(vii) The formation of bleeds [40].

There are two types of mould-level controllers in common use (i) eddy current and (ii) radioactive controllers. It is reputed that the former provide tighter control of the mould level. Mould-level fluctuations tend to be higher at the onset of casting especially when the removal of the dummy bar results in a sharp drop in the metal volume. Mould-level fluctuations can lead to variations in heat transfer (and to a corrugated shell with large differences in thickness). This, in turn, results in longitudinal cracking and the formation of depressions. The predictions of a mathematical model [37] indicate that a decrease in steel-level results in a thinner shell with a smaller pitch between oscillation marks.

Mould-level variations can also result from "*bulging*" of the strand which occurs when the shell (at the mould exit) is too thin to withstand the pressure exerted by

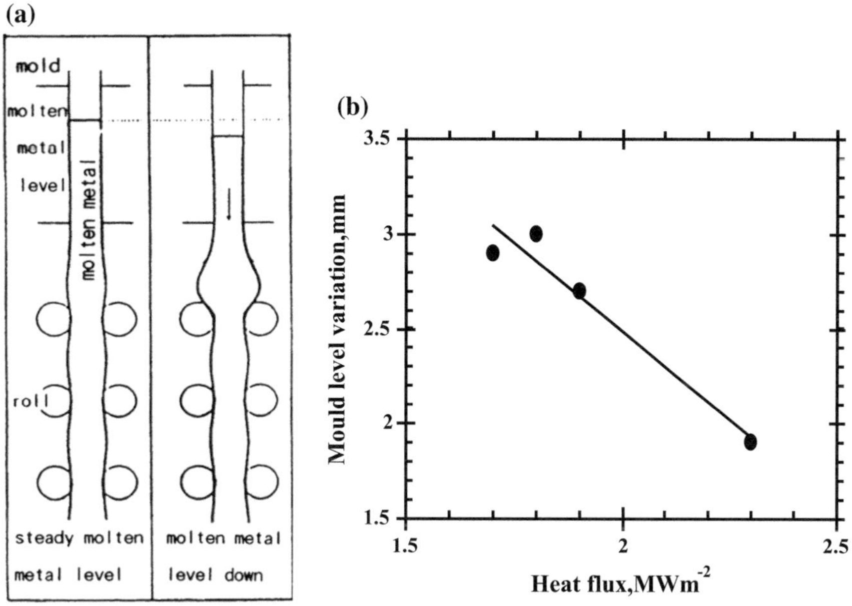

Fig. 5.6 **a** Schematic diagrams showing (**a**) mechanism for mould-level variations [6] when the shell leaving the mould is too thin and (**b**) mould-level variations as a function of the horizontal heat flux [41]; (permission granted (**a**) ISIJ [6] (**b**) drawn from data in [41])

the rolls [6]. It is considered good practice when casting, *peritectic*, medium-carbon steel grades to reduce the horizontal heat transfer in order to minimise longitudinal cracking. However, this low-heat flux practice can result in a very thin shell at the mould exit; this, in turn, leads to bulging and mould-level fluctuations (Fig. 5.6a) [6]. This mechanism is supported by the plant data for mould-level variations and horizontal heat flux obtained on a thin-slab caster using different mould fluxes [41]. It can be seen from Fig. 5.6b that a high heat flux is needed to minimise the effect of mould-level fluctuations arising from bulging of the shell [41]. In such cases, best practice requires a compromise. Some work has been carried out on developing a mould powder which produces a slag film which minimises the heat flux in the meniscus region but encourages higher heat fluxes further down the mould to provide a thicker shell at the mould exit [41].

Mould-level variations contain various frequency waves (f_1 = < 0.1 Hz; f_2 = 0.1–1 Hz; and possibly f_3 > 1 Hz) [42]. The long period waves are related to the surface velocity of the molten steel [36]; the f_1 frequency is linked to bulging and f_2 is possibly tied into surface flow and is affected by the mould dimensions [42]. The mould-level fluctuations can be minimised through control of SEN design and other casting variables [36]. These fluctuations can also be reduced using electromagnetic devices (e.g. EMBr [43–46] and EMS [35, 47]). Control of the upper poles of the EMBr provides the reduction in mould-level variation whereas the lower poles reduce the metal flow velocity [35, 42, 43] and the penetration into the mould which results in a reduction of slag entrapment and pinholes in the cast steel.

5.6 Metal Flow

The metal flow is very important in the control of the continuous casting process. The drive for improved productivity has led to the use of high casting speeds. It has been found that there are several metal flow patterns established in the mould. It is generally accepted that the pattern exhibiting a "*double roll*" provides the best conditions for continuous casting (Fig. 5.7a). The *Single-roll* flow pattern (Fig. 5.7b) has a tendency to lead to skull formation (i.e. a frozen steel meniscus) and does not provide sufficient time for the flotation of inclusions and gas bubbles to reach the molten slag pool. Asymmetric flows are very undesirable and are caused principally by clogging of the SEN (Fig. 5.7c) [48, 49].

High casting speeds lead to both turbulence and to considerable fluctuations in the metal flow velocities; these are discussed in Sect. 5.4. These fluctuations start in the SEN (Figs. 5.8, 5.9a and 5.10a). The impact of the metal flow on the SEN base plate results in the formation of vortices (see, for instance, Figs. 5.9a and 5.10a) and these vortices interfere, intermittently, with the flows passing through the SEN ports. Thus, the metal flow leaving the SEN is continually fluctuating. However, further turbulence is caused by the impact of the metal flow on the mould wall. This results in the formation of surface waves and vortices (Fig. 5.4a); the height of the surface wave (20–30 mm for slabs and up to 50 mm for thin slabs) increasing with increasing metal velocity.

The metal flow usually impacts against the shell around 40–50 mm below the metal surface and the hot metal flow can cause some thinning of the shell at this point (Fig. 5.4b) [14]. The principal factors affecting metal flows are:

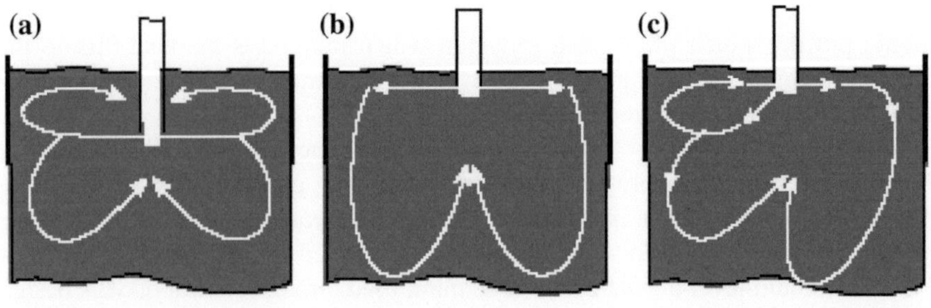

Fig. 5.7 Schematic diagrams showing various metal flow patterns (**a**) *double roll* (promoted by low Ar flow rate, deep SEN immersion and high V_c) (**a**) *single-roll* (promoted by high Ar flow rate, shallow SEN immersion and low V_c) and (**c**) asymmetric flow (promoted by SEN clogging) [48]; (permission granted, UNESID [48])

- The casting speed (turbulence increases as V_c increases).
- The immersion depth of the SEN (turbulence increasing at lower immersion depths) but it is customary to vary the immersion depth to spread the wear on the SEN.
- The SEN port-design, [14, 50].
- Argon flow rate (V_{Ar}) [51, 52] since the Ar flow tends to "*cushion*" the metal flow and the flow pattern has been found to transform from *double roll* to *single-roll* when $V_{Ar} > 5\ \mathrm{L\ min^{-1}}$ [51, 52].

Electromagnetic braking (EMBr) can be used to reduce metal flow turbulence. However, it is necessary to optimise the settings of the EMBR. It is possible to optimise the settings for the conditions using a reliable, validated mathematical model. Figure 5.8 is made up of frames taken from videos of the model predictions [53]. These frames show the flow patterns formed with (a) no EMBR (b) EMBR setting 91 mm (c) EMBR setting 121 mm [53]. The flow systems shown in Fig. 5.8a indicates the pattern obtained with no EMBR, the surface flow is a very slow, *single roll* which means there is a possibility of the metal surface freezing (i.e. *skull formation*). Figure 5.8b shows that the application of EMBR across the nozzle ports does create a *double-roll* flow, but in this case, the flow is very unsteady with occasional lapses into "*asymmetric flow*", which is undesirable [53]. Figure 5.8c shows a much more satisfactory and stable "*double-roll*" flow with the EMBR positioned lower, below the nozzle ports [53]. Thus, mathematical models can aid the setting of EMBr to provide a suitable metal flow.

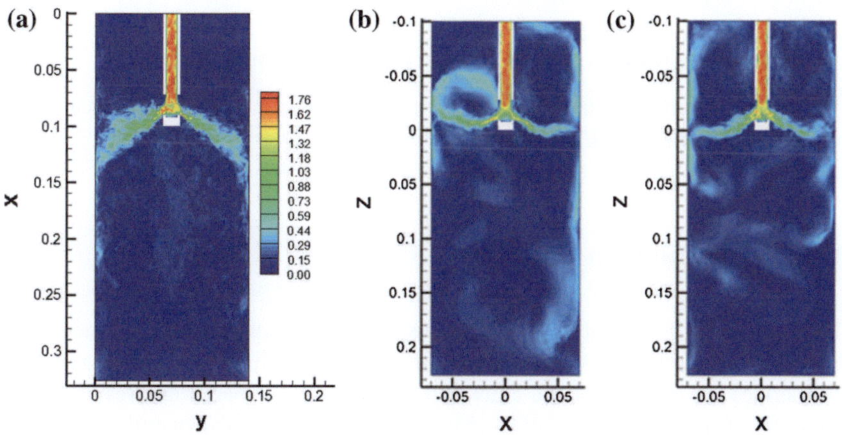

Fig. 5.8 Frames from video of model predictions for (**a**) no EMBR (**b**) EMBR setting 92 mm (across nozzle) (**c**) EMBR setting 121 mm (below nozzle) [53]; (permission granted, ISS/AIST [53])

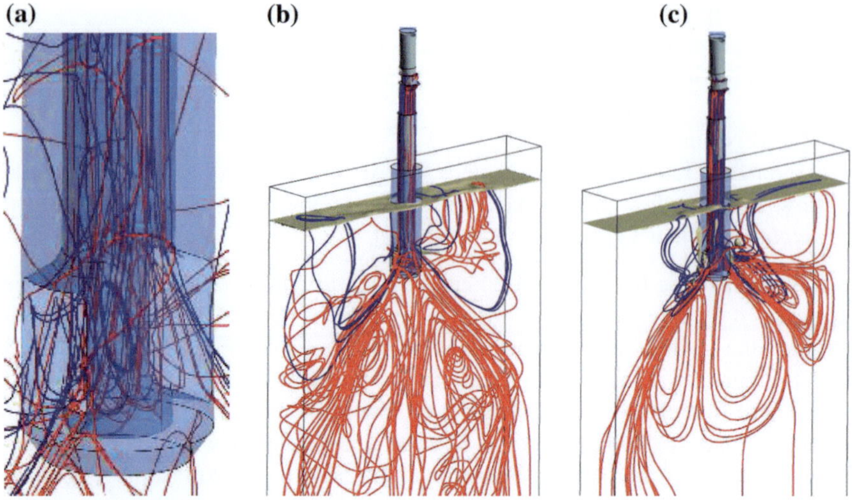

Fig. 5.9 Predicted metal flow patterns showing (**a**) the vortex formed in the sump of the SEN (**b**) *single-roll* pattern formed with no FC (=no EMBR) (**c**) *double roll* pattern formed with FC and the flow bias towards the loose, wide side of the mould [52]; (permission granted, VDEh, Verlag Stahleisen [52])

Mathematical modelling has also proved useful in optimising the effects of EMBr on the metal flow pattern in other studies (Fig. 5.9). It was shown that:

- EMBr improved the metal flow pattern (Fig. 5.9b, c) by transforming a single roll into a double-roll flow [52].
- The swirl in the sump of the SEN (Fig. 5.9a) was found to bias the flow, preferentially, towards the loose, wide face of the mould [52].

It has been reported that swirling flow SEN can reduce the turbulence of the metal flow leaving the SEN [54].

5.7 Fluctuations in Processes

The metal flow in the mould is in a constant state of fluctuation. As mentioned above, these fluctuations originate in the SEN, where the first metal flowing into the SEN strikes the bottom of the SEN and the subsequent backwash results in the formation of a vortex (Fig. 5.10a) [55, 56]. This vortex, subsequently, interferes, intermittently, with the metal flow through each port. This interference causes both fluctuations and asymmetry in the flows leaving the SEN [57, 58]. The flow in the mould is continually reacting to the current situation at any one time. It has been

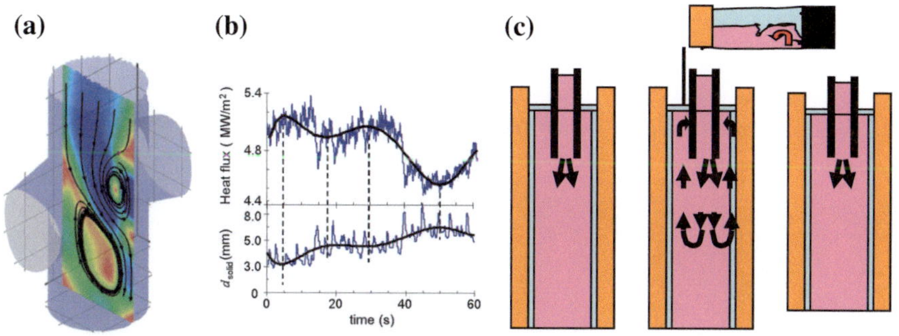

Fig. 5.10 Schematic drawings showing (**a**) the fluid flow in the SEN and the formation of a vortex in the sump of the SEN [55] (**b**) frequencies of oscillations [57] and (**c**) formation of periodic backflows in the mould during billet-casting resulting in periodic bouts of slag entrapment [58]; (permission granted, ISIJ (**a**) [55] (**b**) [17])

suggested that a catastrophic event (such as a breakout) can occur when all the factors are aligned together (i.e. "*the butterfly effect*" [59]).These fluctuations in metal flow velocity are also responsible for the movements of both surface waves and the vortices (like that shown in Fig. 5.4a).

In billet casting, the metal flows freely from an open SEN but it has been shown that fluctuations and vortex-formation also occur, intermittently, in this system and cause periodic "*backflows*" as shown in Fig. 5.10c which results in periodic bouts of slag entrapment [58].

When the metal flow hits the narrow face of the mould, the impact causes a wave to be formed on the surface of the liquid metal (a small wave can be seen in Fig. 5.11a); the position and the height of the waves fluctuate continually [60]. Vortices can also be formed as a result of fluctuating, asymmetric metal flows, an example can be seen in Fig. 5.4a and these vortices are continually changing position [60]. Vortex-formation and turbulent metal flow are major causes of the slag entrapment [61, 62]. Furthermore, turbulent flow plays an important part in (i) SEN erosion and (ii) Carbon pick-up (especially when casting low-carbon steel grades).

In the mould, the flows in the metal are in a continual state of fluctuation and these flows, at any instant, are responding to the flows in the previous time period. The effect of these fluctuations can be clearly seen in the varying depths of the oscillation marks shown near the top of Fig. 5.11b [13, 16]. Although, the aero-dynamic drag forces from the flow in the metal affect the flow in the slag phase, the slag flow is dominated by the movement of the mould (and slag rim) as shown in Fig. 5.5. When the mould is descending the slag flow is warm (*tropical*) and outward and upward and when the mould is descending the flow is cold (*arctic*) and downward.

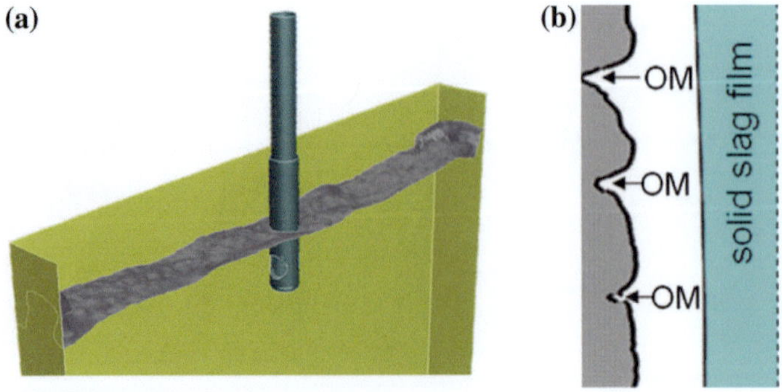

Fig. 5.11 Schematic diagrams showing (**a**) formation of waves on the surface of the steel and (**b**) predicted oscillation mark depth [13] (permission granted, ISIJ [13])

5.8 Application of Electromagnetic Devices

There are several electromagnetic devices used in continuous casting to improve process control and product quality and these have been described in a number of publications [35, 43–45, 62, 64–71].

5.8.1 Electromagnetic Stirring (EMS)

Electromagnetic stirring (shown in Fig. 5.12) is carried out with AC, usually at a frequency of 10–50 Hz, with the electromagnetic effect penetrating through the volume. A travelling magnetic field (i.e. the steel) induces an electric current in the molten steel. The interaction between this current and the travelling magnetic field

Fig. 5.12 Schematic diagram showing EMS applied to mould (permission granted, Nippon Steel, Sumitomo Metal (NSSM) Corp. [64])

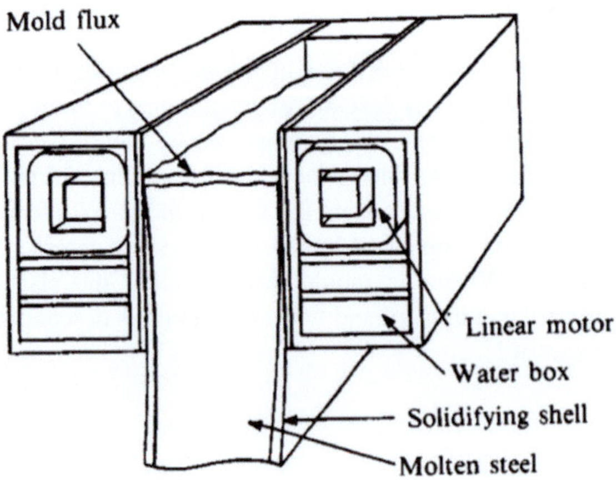

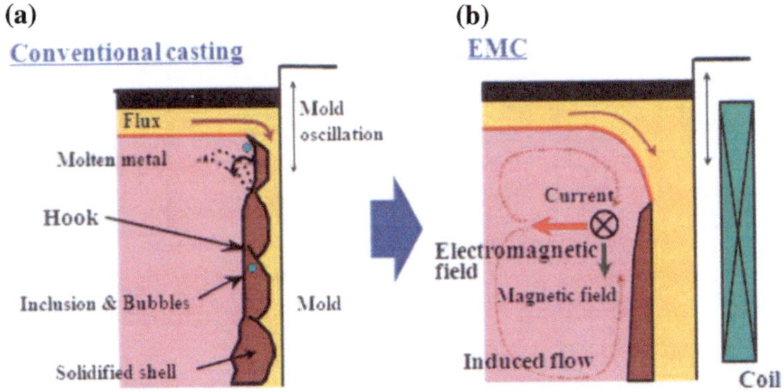

Fig. 5.13 Schematic diagrams showing the principle of EMC **a** conventional casting without EMC **b** EMC [67] (permission granted, Nippon Steel, Sumitomo Metal (NSSM) Corp. [67])

produces a Lorentz force which drives the stirring of the steel [6, 63–65]. The loss of magnetic flux density can be minimised by using (i) a frequency of $\leq$10 Hz and (ii) using a low-conductivity, stainless steel plate attached to the copper plate [62, 64]. EMS has proved effective in providing (i) a reduction in CO blowholes [62, 64] (ii) a fivefold reduction in inclusions and gas bubbles [6, 62, 64] and (iii) a threefold reduction in d_{OM} [64] (iv) homogenisation of the melt composition and temperature (leading to decrease in longitudinal cracking [64]) and (iv) refining of the microstructure (Fig. 5.13).

EMS of liquid metal causes Saffman forces to be exerted on the inclusions, the magnitude of the force increasing with increasing inclusion size [6]. Thus EMS aids the removal of non-metallic inclusions and minimises slag entrapment. Inclusion separation will occur if the Saffman velocity exceeds the velocity of the solidification front. EMS is successful in removing inclusions (and bubbles) with particle diameters >100 μm [6].

5.8.2 Level Magnetic Field (LMF)

This technique is also used in the removal of inclusions and bubbles. It involves applying a static, electromagnetic field (in the thickness direction) over the mould width. The magnetic field induces a current in the conducting fluid (steel) which interacts with the magnetic field so as to oppose the movement. Initially, LMF was used to reduce the velocity of the metal flowing out of the SEN. However, when LMF is applied lower down the mould it was found to reduce the penetration of the metal flow into the mould (see Fig. 5.14a). This reduced penetration with LMF promotes the flotation of inclusions and bubbles to the liquid slag pool where they are absorbed [6, 35, 66]. LMF is used in EMBr [6, 62–64]. The reduction in the

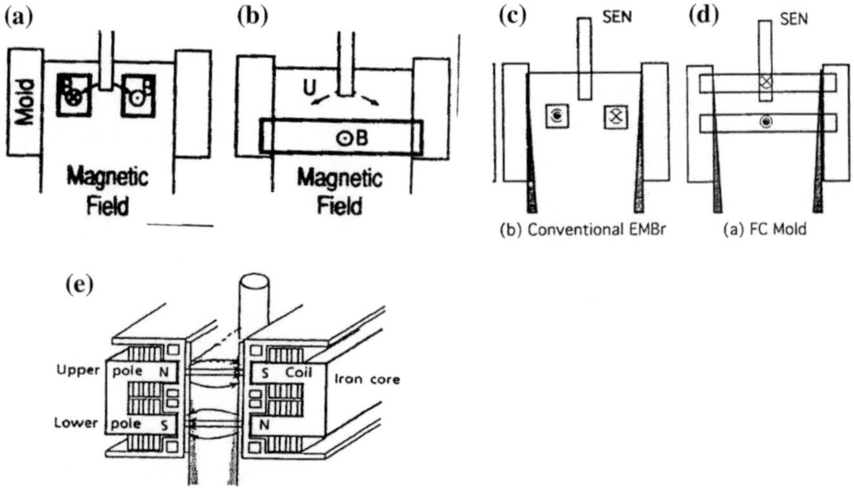

Fig. 5.14 Schematic diagrams showing arrangements for (**a**) Original EMBr [63] (**b**) Level magnetic field [63] (**c**) Conventional EMBr [44] (**d**) FC control [44] (**e**) Mould fitted with Flow control [44]; (permission granted, (**a**, **b**) JOM [63] (**c**, **d**, **e**) SEAISI [44])

penetration of the metal stream is also responsible for the *ca.* 10 °C increase in steel meniscus temperature [43, 45, 69, 71] which results in an increase in powder consumption.

LMF technology has been used to suppress mixing of two steels in the mould in the casting of, say a LC steel with a coating of stainless steel [62–64]. The two steels are held in separate tundishes and are fed through two SEN's and LMF is then used to separate the two layers of different steels in the mould.

5.8.3 Electromagnetic Casting (EMC)

EMC consists of an external, solenoid coil wrapped around a core at meniscus level, to which a high-frequency AC is applied (Fig. 5.13) [67]. This induces both a magnetic field and an induced current. The interaction of the electromagnetic field and the induced current produces a "*pinch force*" which pushes the shell away from the mould and thus increases the distance between shell and mould (i.e. it opposes the ferro-static pressure).The increased gap results in, sequentially, a thicker slag film and a reduced, horizontal heat flux which tends to delay the solidification of the steel. This delayed solidification takes the shell tip away from the turbulent interface which results in fewer entrapped inclusions and shallower oscillation marks (due to the increase in $d_{rim/tip}$).

However, the steel flow, induced by the electromagnetic field in EMC, is so rapid that it can cause the steel meniscus to become unstable and irregular [67]. In order to combat this problem, *pulsative* EMC was introduced [67].This entails passing an AC current through the solenoid, intermittently, at a frequency, of say, 10–30 Hz [67]. This allows the metal velocity (induced by the electromagnetic field) to be controlled. The induced electromagnetic field interferes with the eddy current sensors, so the mould levels have to be sampled in the periods when no current is flowing in the solenoid coil [67].

EMC has been largely used in billet casting but *pulsative* EMC has been successfully developed for slab casting [67]. It provides the following benefits (i) a fourfold decrease in defects [67] (ii) a threefold decrease in surface-roughness (i.e. d_{OM}) [67] (iii) a tenfold decrease in surface inclusions [67] (iv) a 20% increase in powder consumption [67] and (vi) a reduction in longitudinal cracking [6, 67].

5.8.4 Electromagnetic Braking (EMBr)

Electromagnetic braking uses DC and promotes resistance to the movement of the metal flow. When an electromagnetic field is applied to a moving, electrically conducting fluid, an electrical current is induced. The components of the induced current, which cross the magnetic field at right angles, exert a braking effect on the steel flow. EMBr was initially used to decelerate the steel stream discharging from the SEN (Fig. 5.14a). Although this was effective in reducing the metal flow turbulence and slag and gas bubble entrapment, problems related to the stability of the braking effect (and the associated metallurgical benefits) were encountered [62]. Consequently, Level Magnetic field (LMF) technology (Fig. 5.14b) was developed to achieve these targets. Further work resulted in the development of flow control EMBr (FC) [43, 62, 64, 68] and other designs (e.g. 3-pole arrangement [70]).

Flow control (FC) is shown in Figs. 5.14d, e. The upper magnetic field (pole) in Fig. 11.14e acts to: (a) decrease the metal velocity at the meniscus (thereby minimising slag entrapment in the steel below) and (b) stabilise the meniscus (thereby reducing mould-level variations, slag entrapment by the shell, and transverse cracking [43, 67].The lower magnetic field (pole) acts to decrease downward penetration of the metal flow which (c) increases the meniscus temperature by *ca.* 10 °C [43, 45, 69, 71] (thereby increasing powder consumption) and (d) promotes the flotation of inclusions and gas bubbles [43, 67]. It should be noted that EMBr results in a significant decrease in vertical heat transfer which can have effects on the depth of the molten pool and powder consumption [72].

EMBr has been reported to provide the following benefits (i) lower inclusion and pinhole concentrations in the cast steel [35, 43, 44, 65, 66, 68] (ii) reduced levels of longitudinal cracking [43, 44, 65] but also results in (iii) higher levels of SEN erosion [44].

5.9 Steel Grade

The behaviour of the mould slag during continuous casting differs according to the steel grade being cast. For the most part, the mould slag is tailored to provide the appropriate treatment for the transformation of *δ-ferrite* into *austenite* in various steel grades. However, recently new powders have been designed and developed to cast high-Al steels in order to minimise the amount of Al_2O_3 formed during casting.

5.9.1 Peritectic Steels

Inspection of the Fe–C phase diagram (given in Fig. 5.15) shows that for steels with compositions in the range (*C%* = 0.06–0.17) solidification of liquid steel results in the formation of a δ-Fe shell, which subsequently, undergoes a *peritectic* phase transition to γ-Fe (*austenite*) as shown in Eq. 5.6. *Austenite* shells are stronger but less ductile than *ferrite* shells.

$$\mathrm{Fe(liq)} + \delta - \mathrm{Fe} \rightarrow \gamma - \mathrm{Fe} \tag{5.6}$$

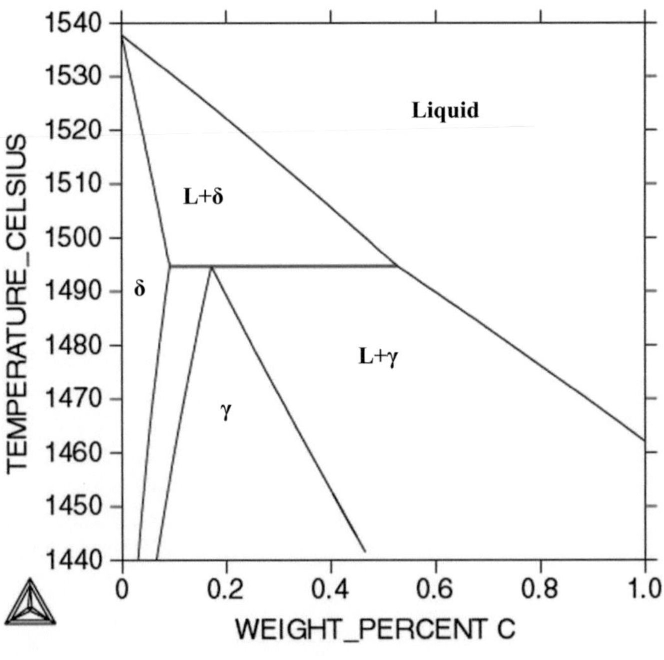

Fig. 5.15 The Fe–C phase diagram for the peritectic range (0.08 < wt% C < 0.55) calculated using ThermoCalc [75] (drawn Swerea KIMAB)

The transition ($L + \delta \rightarrow \gamma$) is accompanied by a volume change of (0.4–0.6%) due to the tighter packing in the *austenite* (*cf. ferrite*) and there is also a 4% mismatch between the thermal shrinkage coefficients of δ-Fe and γ-Fe [73]. *Austenite* is stronger and has better creep resistance than *ferrite* because of the better-packing of atoms in the FCC structure. However, *ferrite* (bcc-structure) is more ductile and exhibits less-micro-segregation than *austenite* [74].

The mismatch in the thermal shrinkage coefficients of δ-Fe and γ-Fe in peritectic, MC steels results in stresses. These stresses are much larger in thicker regions of the shell than in the thin regions and result in a hoop stress around the mould. This hoop stress is relieved by longitudinal cracking. The solidification process occurs during the period of the oscillation cycle when the mould is descending (in t_n and early t_p); little solidification occurs in the remaining portion of the cycle when the mould is ascending. This results in a corrugated "*sausage-like*" shell and the resulting variations in shell thickness lead to stresses. The usual strategy used to minimise longitudinal cracking, is to produce a shell which is both as thin and as uniform as possible. This is usually achieved by reducing the heat flux and this, in turn, is derived by using a powder which creates a thick and crystalline slag film [10]. The slag film thickness increases with increasing solidification temperature (or T_{br}).

In contrast, longitudinal cracking is much less prevalent in steels which do not undergo the *peritectic* reaction (shown in Eq. 5.6). Consequently, it is necessary to differentiate between those steels undergoing the peritectic reaction and those that do not.

Wolf [76, 77] introduced the term, "*ferrite potential*" as a measure of the amount of *peritectic* reaction occurring in the steel. Although, the ferrite potential is affected by the carbon content, it is also affected by other alloying elements; some of these elements stabilise the *ferrite* (Cr, W, Mo, Al and Si) whilst others stabilise the *austenite* (Ni, Mn, Co, N and Cu). For *low-alloy steels,* the ferrite potential (FP) can be calculated by Eq. 5.7 where CP is the carbon potential which is defined in Eq. 5.8.

Low alloy:

$$\mathrm{FP} = 2.5(0.5 - \mathrm{CP}) \tag{5.7}$$

$$\mathrm{CP} = (\mathrm{wt\%C}) + 0.04(\%\mathrm{Mn}) + 0.1(\%\mathrm{Ni}) + 0.7(\%\mathrm{N}) - 0.14(\%\mathrm{Si}) - 0.04(\%\mathrm{Cr}) - 0.1(\%\mathrm{Mo}) - 0.24(\%\mathrm{Ti}) \tag{5.8}$$

For *stainless steels* the FP is given by Eq. 5.9 where "Ni "and "Cr" are defined in Eqs. 5.10 and 5.11, respectively.

Stainless steel:

$$\mathrm{FP} = 5.26\left(0.74 - {''}\mathrm{Ni}{''}/{''}\mathrm{Cr}{''}\right) \tag{5.9}$$

$${''}\mathrm{Ni}{''} = (\mathrm{wt\,\%Ni}) + 0.31(\%\mathrm{Mn}) + 22(\%\mathrm{C}) + 17.5(\%\mathrm{N}) \tag{5.10}$$

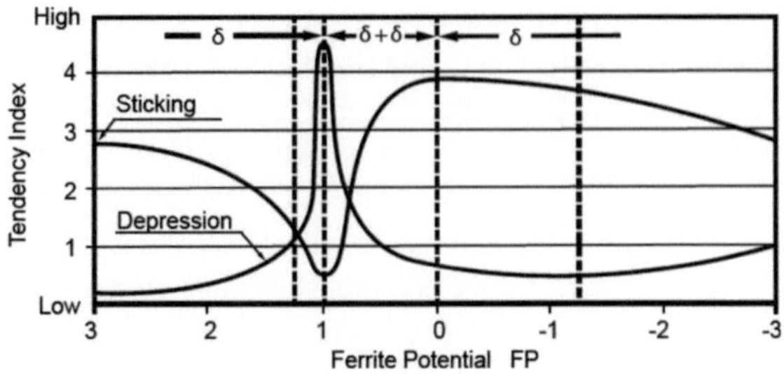

Fig. 5.16 The tendencies of steels with regard to cracking (denoted Depression) and sticker breakouts as functions of the ferrite potential [76, 77] (permission granted, ISS/AIST [76] and re-drawn by Swerea KIMAB)

$$''Cr'' = (wt\%Cr) + 1.5(\%Si) + 1.65(\%Mo) + 2(\%Nb)3(\%Ti) \quad (5.11)$$

The *peritectic* reaction is at a maximum at FP = 1.0 and the range, FP = (0.8–1.05) corresponds to the crack-sensitive range (Fig. 5.16).

The properties of various steels, as characterised by their ferrite potentials, are given in Table 5.2.

In addition to the Wolf routine, there are several other methods available for determining the amount of *peritectic* reaction occurring in specific steels. These various methods used to calculate the amounts of *peritectic* reaction have been reviewed by Santillana [78]. These methods tend to follow the Wolf routine but use different coefficients to calculate the carbon potential; these coefficients are given in Table 5.3 [78]. However, thermodynamic software has also been used to identify those steels where the *peritectic* reaction is likely to occur [78–80].

It was mentioned above, that *peritectic* (MC) steels are prone to longitudinal cracking and are often denoted as "*crack-sensitive*". For these MC steels, stresses are minimised by keeping the shell as thin and as uniform as possible. This is achieved by reducing the horizontal heat flux (using a flux which gives a thick (i.e. using a slag with a high T_{br}) and crystalline slag film).

In contrast, the shells formed by HC steels tend to be weak and are prone to "*sticker breakouts*". The strategy used here is to create a thick, strong shell by increasing the horizontal heat flux through the creation of a thin (i.e. low T_{br}) glassy, slag film.

A plot of viscosity versus T_{br} for mould fluxes (Fig. 5.17) is effectively a plot of (inverse lubrication) versus (inverse heat transfer) [4, 5, 10]. Figure 5.17 was derived from data for the fluxes used in a large number of casters to cast MC and HC steel grades [5]. It can be clearly seen that for a given viscosity the T_{br} values

Table 5.2 Characteristics of shells formed for different bands of Ferrite Potential (FP)

FP	Phase	Properties of shell
<0.8	*Austenite*	Strong and more ductile with increasing FP
0.8–1.05	Initially forms *ferrite*, then transforms to *austenite*	Irregular, strong shell
>1.05	*ferrite*	Weak, ductile shell

Table 5.3 Coefficients for selected equations to calculate Carbon potentials, after Santillana [78]

Ref.	Mn	Si	N	P	S	V	Ti	Cu	Cr	Ni	Mo	Others
Wolf [77]	0.01	−0.1			−0.7				−0.04	0.04	−0.1	
Trico	0.02	0.009	0.05	0.008	0.17	0.009	0.007	0.007	0.003	0.02	−0.007	Al_t0.05 Nb 0.04 Sn0.0006 B 1.32 Ca-0.24
BSSTC	0.043	−0.14	1.06	0.029	0.11	−0.13	−0.024	0.037	−0.083	0.1	−0.003	
SMS	0.14	−0.037		−0.04	0.222			0.003		0.023	−0.004	
Howe	0.04	−0.14	0.7				−0.24		−0.04	0.1	−0.1	

Al_t = total Al

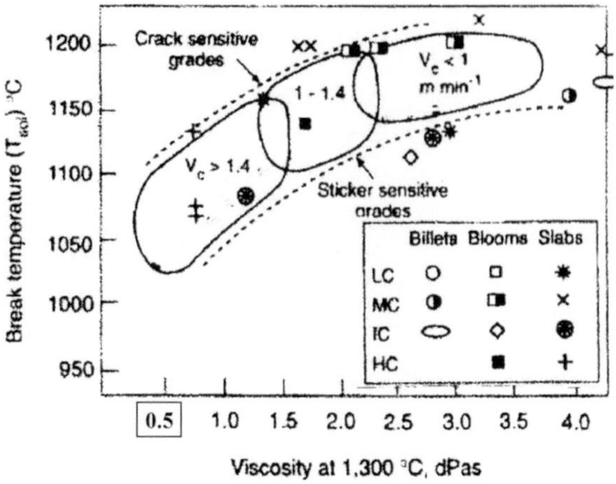

Fig. 5.17 The break temperature (T_{br}) as a function of slag viscosity for flux used in the casting of slabs, blooms and billets (permission granted, UNESID [5])

for crack-sensitive MC grades are 100 °C higher than those used to cast HC grades [4, 5]. This indicates that there has been a hidden logic underlying the development of mould fluxes. The T_{br} values of mould fluxes used to cast other steel grades tend to fall between these two curves. Thus T_{br} values for three categories of steels can be expressed as a function of viscosity (in dPas) shown in Eqs. 5.12 to 5.14.

Thus, it is possible to differentiate various categories of steels in terms of their carbon potentials or ferrite potentials [2, 76, 77].

Crack-sensitive:

$$\mathrm{CP} = 0.06 - 0.18 \quad T_{br}(^{\circ}\mathrm{C}) = 1157 + 60 \ln \eta \tag{5.12}$$

Sticker-sensitive:

$$\mathrm{CP} > 0.4 \quad T_{br}(^{\circ}\mathrm{C}) = 1051 + 76.4 \ln \eta \tag{5.13}$$

Others (intermediate):

$$\mathrm{CP}\, 0 - 0.06\, \mathrm{and}\, 0.18 - 0.4 \quad T_{br}(^{\circ}\mathrm{C}) = 1103 + 68.5 \ln \eta \tag{5.14}$$

5.9.2 *High-Al Steels*

High-Al steels combine good strength (from the *martensite* formed during plastic deformation of retained *austenite*) with high ductility (from the retained *austenite*). However, these steels have proved difficult to cast (giving rise to sticker breakouts, formation of depressions, erratic powder melting and to false sticker-breakout alarms). All of these casting problems are traceable to the large amount of Al_2O_3 formed via reaction 5.15 where the underline indicates in the steel.

$$2\underline{Al} + 3\underline{Si}O_{2\,slag} = 3Si + Al_2O_{3slag} \quad (5.15)$$

The alumina pick-up by the slag is typically 4% when casting normal (low-Al) steels and about 2% arises from the reaction shown in Eq. 5.15 and *ca.* 2% from steelmaking reactions [81]. However, for steels containing 1% Al, the Al_2O_3 pick-up can be greater than 30% due to the reaction of Al and SiO_2 in the slag. The reaction of Al is not confined to SiO_2 since similar reactions occur with most slag constituents (e.g. FeO, MnO, B_2O_3, etc.) lying above the Al/Al_2O_3 curve in the Ellingham diagram (see Fig. 6.21). Large amounts of Al_2O_3 in the slag:

- Cause an increase in both viscosity (thereby decreasing the lubrication supplied) and the liquidus temperature of the slag.
- Are slow to dissolve and leave solid particles in the slag which agglomerate and lead to large slag rims which fracture and, subsequently, cause depressions, sticker breakouts and false alarms on sticker-detector systems.
- Tend to produce highly crystalline slag films which provide little lubrication to the shell and furthermore, tend to fracture (where the fractured slag films/rims can lead to blockages, depressions, and sticker breakouts).

It has been reported that conventional fluxes can be used for casting steels with $\leq$0.6% Al [41]. However, for casting steels with $\geq$ 0.6 Al %, it is necessary to use some of the specialised powders which have been developed in recent years (see Sect. 6.3.12). The most promising powders appear to be those based on calcium aluminates with added fluxes (CaF_2, Na_2O) and with the SiO_2 content limited to *ca.* <10% to minimise formation of Al_2O_3 by the reaction shown in Eq. 5.15.

5.10 Water Flow Rate

In the past, the rate of water cooling has been reduced to produce mild cooling conditions in a quest to reduce longitudinal cracking. However, more recent studies have shown that the effect of reducing water flow rate tends to be small, as can be seen from the plant measurements [82] shown in Fig. 5.18.

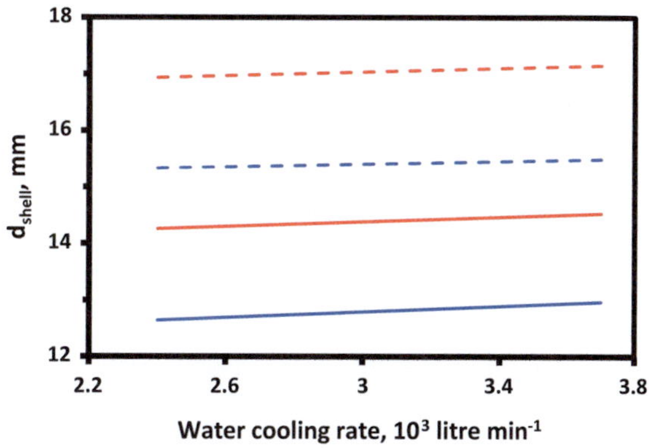

Fig. 5.18 The effect of water cooling flow rate on the thickness of the shell for two different superheats, (20 °C = *blue* and 60 °C = *red*) during casting of stainless steel 304 (- - -) and CS 1026(——); $V_c = 0.9$ m min^{-1}; (permission granted ISS/AIST [82]; re-drawn)

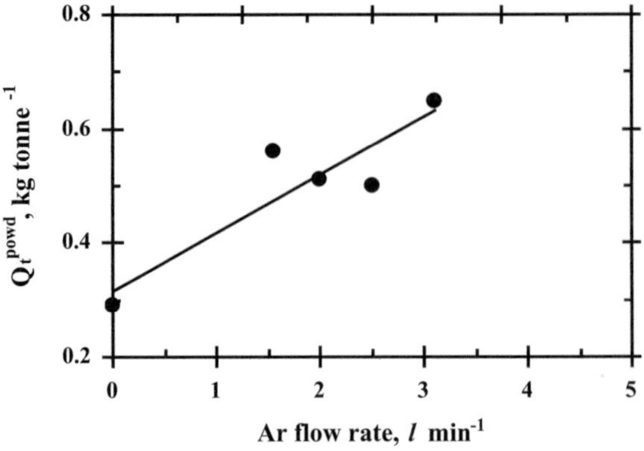

Fig. 5.19 Schematic diagram showing the effect of Argon flow rate on the powder consumption (Q_t) (courtesy of A.B. Fox [84])

5.11 Argon Flow Rate

Argon is introduced into the SEN to minimise clogging of the SEN and to aid the removal of inclusions from the steel. It has been reported [83] that the point of exit for the bubbles lies close to the SEN for large bubbles and moves further to the edge of the mould as the bubble size decreases. There is a possibility that small bubbles could be swept with the metal flow.

The argon flow rate has also been reported to affect the metal flow in the mould. Argon flow has been reported to have a "*cushioning effect*" on the metal flow and thereby retards the flow [83]. Recent studies [51, 52] have indicated that when Argon flow rates were increased from 4 to 5 litre min^{-1}, the metal flow changed from an *inward surface flow* (i.e. mould wall → SEN) to an *outward surface flow*, which is less desirable.

It has also been reported that powder consumption increased with increasing argon flow rate (Fig. 5.19) [84]. One possible explanation is that the argon bubbles enhance vertical heat transfer from the metal to the slag pool which results in a higher temperature for the slag pool and a reduced slag viscosity.

References

1. S Ogibayashi, T Mizoguchi, T Kajatani, Intl. Workshop on *Thermophys. Data for the Development of Mathematical models of solidification*, Gifu City, Japan, (1995).
2. AB Fox, PhD Thesis "*Mould fluxes- their properties and performance*" Imperial College, London, (2003).
3. F Neumann, J Neal, MA Pedroza, AH Castiliejos, E FA Acosta G, Proc.. 79th Steelmaking Conf., 1996, (ISS, Warrendale, PA, 1996) p. 249.
4. KC Mills, S Sridhar, AS Normanton, ST Mallaband, Proc. Brimacombe Conf., Vancouver, 2000, (Can. Inst. Min. Met. & Petrol., 2000), p. 781.
5. S Sridhar, KC Mills, V Ludlow, ST Mallaband, Proc. 3rd Europ. Conf. Continuous Casting, Madrid, 1998, (UNESID, Madrid,1998) p. 807.

6. T Matsumiya, ISIJ Intl., **46**, 1800, (2006).
7. MM Wolf, Iron and Steelmaker, **23**, (2), 47, (1996).
8. CS Li, BG Thomas, Proc. ISS Tech Conf., Indianapolis, 2003, (ISS, Warrendale, PA, 2003) p. 685, see also Met. Mater. Trans., **35B**, 1151, (2004).
9. M. Wolf, Proc. 2nd Europ. Conf. Continuous Casting, Dusseldorf, 1994, METEC Congress'94, (VDEh, Dusseldorf, 1994) vol. 1 p. 78,
10. KC Mills, AB Fox, ISIJ Intl., **43**, 1479, (2003).
11. S Ogibayashi, K Yamaguchi, T Mukai, T Takahashi, Y Mimura, K Koyama, Y Nagano, T Nagano, Nippon Steel Technical Report, **34**, 1, (1987).
12. RJ O'Malley, J Neal, Proc. Intl. Conf. on New Developments in Metallurgical Process Technol., Dusseldorf, METEC Congress (VDEh, Dusseldorf, 1999) p. 73.
13. PE Ramirez-Lopez, PD Lee, KC Mills, B Santillana, ISIJ Intl., **50**, 1797, (2010).
14. H Tai, M Morashita, T Miyake, Proc 3rd Conf. Continuous Casting, Madrid, 1998, (UNESID, Madrid, 1998) p. 447 see also T Miyake, K Nakayama, M Moroshita and H Tai, Kobelco Technol. Review, **21**, (April), 7, (1998).
15. ES Szerkeres, Iron and Steel Engineer, **73**, (7), 29, (1997).
16. PE Ramirez-Lopez, KC Mills, PD Lee, B Santillana, Met. Mater. Trans., **66**, 109, (2011).
17. PE Ramirez-Lopez, KC Mills, PD Lee, B Santillana, ISIJ Intl., **50**, 425, (2010).
18. P E. Ramirez-Lopez, P D. Lee, KC. Mills, C Puncreobutr, D Farrugia, B Santillana, U, Sjöström, Proc 7th Europ. Conf Continuous Casting, Dusseldorf, (2012) (VDEh, Dusseldorf).
19. Y Meng, BG Thomas, Proc. ISS Technol., Indianapolis, (2003) p. 589 see also Met. Mater. Trans., **34B**, 685, (2003).
20. K Schwerdtfeger, KH Tacke, Report EUR 9339 (Europ. Comm. Sci. Tech. Publ., Luxembourg, 1985).
21. S Ogibayashi, CAMP-ISIJ, **18**, 126, (2003) and Proc. 75th Steelmaking Conf. (2002) p. 175.
22. JM Hill, YH Wu, B Witwatanapataphee, J Eng. Math., **36**, 311, (1999).
23. R Saraswat, AB Fox, KC Mills, PD Lee, B Deo, Scand. J Metal., **33**, 85, (2004).
24. H Maeda, CAMP-ISIJ, 6, 280, (1993).
25. PE Ramirez- Lopez, J Bjorkvall, U Sjostrom, PN Jalali, PD Lee, KC Mills, B Jonsson, J Janis, M Petajajarvi, J Pirinen: Proc. 9th Intl Conf. on CFD on Minerals and Process Industries, CSIRO, Melbourne, 2012. paper available www.cfd.com.au/cfd_conf12.
26. CA Dacker, T Sohlgren, Steel Research Intl., **81**, 899, (2010).
27. BG Thomas, B Ho, G Li, Proc. McLean Symp., Toronto, 1998 (ISS, Warrendale, PA, 1998) p. 177.
28. JA Kromhout, C Liebske, S Meltzer, AA Kamperman, R Boom, Ironmaking and Steelmaking, **36**, 291, (2009).
29. J Kromhout, RS Schimmel, Proc 8th Europ. Conf Continuous casting, Graz, 2014 (Austrian Met. Mater. Soc., 2014).
30. Y Liu, XD Wang, M Yao, AB Zhang, H Ma, Z Wang, JC Ma, X Wang, GQ Shi, Ironmaking and Steelmaking, **41**, 748, (2014).
31. M Jenkins, Proc. 77th Steelmaking Conf., (1994), (ISS, Warrendale, PA, 1994) p. 337.
32. M S Jenkins, BG Thomas, Proc. 80th Steelmaking Conf. (1997) (ISS, Warrendale, PA, 1995) p. 285.
33. JW Kim, SK. Kim, D. S. Kim, Y. D. Lee, J. I. Eum, E. S. Lee, Proc. 78th Steelmaking Conf. (ISS, Warrendale, PA, 1995) p. 333.
34. S Chakraborty, W Hill, Proc. 77th Steelmaking Conf., 1994, (ISS, Warrendale, PA, 1994) p. 389.
35. K Miyazawa, Sci. & Technol. Advanced Materials, **2**, 59, (2001).
36. T Teshima, J Kubota, M Suzuki, K Ozawa, T Masoka, S Miyahara, Tetsu-to- Hagane, **79** (5), 576, (1993).
37. R Aigner, H Steinruck, EUROTHERM 82, Numerical Heat Transfer Gliwice-Cracow, Poland, 2005, ed. A Nowak, RA Bialecki: (Silesian Univ. Technol.) p. 1076.

38. G.G. Lee, H. Shin, S.-H. Kim, S.-K. Kim, W.-Y. Choi, B.G. Thomas, Ironmaking and Steelmaking, **36**, (1), 40, (2009).
39. V Guyot, JF Martin, A Ruelle, A d'Anselme, JP Radot, M Bobadilla, JV Lavant, JN Pontoire, ISIJ Intl., **36**, S 227, (1996).
40. H Takeuchi, H Mori, T Nishida, T Yanai, K Mukunashi, Trans. ISIJ, **19**, 274, (1979).
41. J Kromhout, PhD Thesis, "*Mould powders for high speed continuous casting of steel*" University of Delft, NL, (2011).
42. M Cervantes, H Gustavsson, Proc. 3rd Europ Conf. Continuous casting, Madrid, 1998, (UNESID, Madrid, 1998) p. 202.
43. M Washio, M Sugizawa, S Moriwaki, K Kariya, S Idogawa, S Takeuchi, Revue de Metallurgie, CIT, **90**, (April), 507, (1993).
44. H Take, H Osanai, J Hasunuma, T Yamamoto, H Bada, H Tozawa, Proc. Conf. "*Quality of ordinary steel in Iron- and steel-making*", Bangkok, (1994) Session 3, Paper 1.
45. D W van der Plas, C Platvoet, B Diesesme, JP Radot, JM Galpin, Proc. 2nd Europ. Conf. Continuous casting, Dusseldorf, 1994, Metec Congress'94 (VDEh, Dusseldorf, 1994) p. 109.
46. R Singh, BG Thomas, SP Vanka, Met. Mater. Trans. B, **44 B**, 1201, (2013).
47. M Morishita, CAMP- ISIJ, **20**, 867, (2007).
48. D Gotthelf, P Andrezjewski, E Julius, H Haubrich, Proc. 3rd Europ. Conf. Continuous Casting, Madrid, 1998, (UNESID, Madrid, 1998) vol. 2 p. 825.
49. BG Thomas, LF Zhang, ISIJ Intl., **41**, 1181, (2001).
50. QK Robinson, DM Gerstl, Proc. 3rd Europ. Conf. Continuous Casting, Madrid, 1998, (UNESID, Madrid, (1998) p. 1050.
51. PE Ramirez- Lopez, PN Jalali, J Bjorkvall, U Sjostrom, C Nilsson, ISIJ Intl. **54**, 342. (2014) and Proc. 8th Europ. Conf. Continuous Casting, Graz, Austria, 2014 (ASMET, 2014).
52. E van Vliet, DW van der Plas, SP Carless, A A Kamperman, AE Westendorp, Proc.. 7th Europ. Conf. Cont. Casting, Dusseldorf, (2011), Session 4.
53. R Choudhary, BG Thomas, SP Vanka, Met. Mater. Trans. B, **43B**, 532, (2012).
54. S Yokoya, S Tagaki, M Iguchi, K Marukawa, S Hata, Proc. 4th Europ. Continuous casting Conf., Birmingham, 2002 (IOM, London, 2002) vol 1, p. 149.
55. C Real, R Miranda, G Vulchis, M Barron, L Hoyes, J Gonzalez, ISIJ Intl. **46**, 1183, (2006).
56. A Ramos-Banderas, R Sánchez-Pérez, RD Morales, J Palafox-Ramos, L Demedices-García, M Díaz-Cruz, Met.. Mater. Trans. B, **35B**, 449, (2004).
57. PE Ramirez-Lopez, PD Lee, KC Mills, ISIJ Intl., **50**, 425, (2010).
58. E Torres-Alonso, RD Morales, S Garcia-Hernandez, A Najera- Bastida, A Sandoval- Ramos, Met. Mater. Trans. B, **39B**, 840, (2008).
59. PD Lee, PE Ramirez-Lopez, KC Mills, B Santillana, Ironmaking and Steelmaking, **39**, 244, (2012).
60. KC Mills, PE Ramirez-Lopez, PD Lee, High Temp. Materials & Processes, **31**, 221, (2012).
61. H Nakato *et al*, Continuous Casting, vol 6, (ISS, Warrendale, PA, 1992) p. 193.
62. S Feldbauer, I Jimbo, A Sharan, K Shimizu, W King, J Stepanek, J Harman, AW Cramb, Proc. 78th Steelmaking Conf., 1995, (ISS, Warrendale, PA, 1995) p. 655.
63. E Takeuchi, JOM, 1995 (May), 42, (1995).
64. E Takeuchi, H Harada, H Tanaka, T Ishii, T Toh, M Zeze, M Hojo, K Shigematsu, Nippon Steel Technical Report, **61**, 29, (1994).
65. E Favre, S Kunstreich, MC Nove, D Rotelec, W Courths, E Korte, Proc. 3rd Intl Conf. Continuous Casting, Madrid, 1998, (UNESID, Madrid, 1998) p. 595.
66. P Gardin, JM Galpin, MC Regnier, JP Radot, IEEE Trans. on Magnetics, **31**, 2088, (1995).
67. M Tani, T Toh, K Umetsu, K Tanaka, M Zeze, K Tsunenari, K Hayashi, S Fukunaga, Nippon Steel Technical Report, **104**, 62, (2013).
68. SG Kollberg, HR Hackl, PJ Hanley, Iron and Steel Engineer, **73**, (7), 24, (1996).
69. G Bocher, U Hoffman, P Muller, Proc. 2nd Europ. Conf. Continuous casting, Dusseldorf, 1994, Metec Congress'94 (VDEh, Dusseldorf, 1994) p. 103.
70. TWJ Peeters, R Koldwijn, JA Kromhout, AA Kampermann, Proc. 5th Conf Continuous Casting, Nice, 2005, (La Revue Metall., Paris, 2005) Vol 2, p. 515.

71. MY Ha, SG Lee, SH Seong, J. Mater. Processing Technol., **133**, 322, (2003).
72. R Koldewijn, Unpublished Corus Internal Rept. (2007) cited in KC Mills, J Kromhout, A Hamoen, R Boom, Proc. Admet Conf., Dnipetrovsk, Ukraine, May 2007, (Natl. Metall. Acad. Ukr., Dnipropetrovsk, 2007) vol 2, p. 174.
73. SN Singh, KE Blazek, J Metals, **26**, (10), 17, (1974).
74. E Schmidtmann, L Pluegel, Archiv. fur Eisenhuttenw., **51**, 49, (1980).
75. Thermocalc, www.thermocalc.com.
76. M Wolf, W Kurz, Met. Trans., **12**, 85, (1981).
77. M Wolf, Proc. 81st Steelmaking Conf.,1998, (ISS, Warrendale, PA, 1998) p. 53.
78. MB Santillana, PhD Thesis: *Thermo-mechanical properties and cracking during solidification of thin slab cast steel.* Univ. Delft, NL (2013) Chapter 3.
79. KE Blazek, O Lanzi, P Gano, D Kellog, Proc. AIST Tech. Conf., 2007, PR-351-141 – 2007.
80. AA Howe, PhD Thesis, *Micro-segregation in multi-component steels involving the peritectic reaction*, Univ. Sheffield, (1993).
81. G Skoczylas, Proc. 79th Steelmaking Conf., 1996, (ISS Warrendale, PA) p. 269.
82. MJ Lu, JH Chen, DH Tsai, CH Huang, Proc. 78th Steelmaking Conf., 1995,(ISS, Warrendale, PA) p. 359.
83. K Pericleous –private communication Univ. Greenwich, UK (2008).
84. T Mallaband (Metallurgica, UK) Unpublished results from cited by AB Fox, ref. 5–02.

Chapter 6
Different Types of Mould Powders

Abstract The mould powder must carry out a series of tasks. Arguably, the most important of these tasks is the formation of a slag film which provides the optimum level of lubrication and heat extraction from the shell. These properties are determined by the (i) mould dimensions (ii) the casting conditions and (iii) the steel grade being cast. Empirical rules have been developed to express the properties providing good casting performance; these include the *required* values for the powder consumption, slag viscosity and the break temperature but the fraction of crystal phase in the slag film (f_{crys}) should also be cited. The various types of mould powders are described, e.g. starter, exothermic and prefused powders. Most mould powders in use are denoted as *Conventional*; the properties of these powders would be expected to be (i) consistent with those derived from empirical rules and (ii) to contain fluorides to form *cuspidine* in the slag film. However, there are a number of specialist powders which have been developed to carry out specific tasks or to cover special steel grades or casting conditions. The specialist mould powders include the following: F-free powders; C-free powders, Non-Newtonian powders, and fluxes used to cast high–speed, thin slabs and round billets and powders used to cast steels with high Al and rare earth contents and stainless steels. Although the properties of all these fluxes are consistent with those predicted by the empirical rules, developments have been made to deal with special features for each type of powder.

Symbols, Abbreviations and Units

D	Diameter (m)
d	Thickness (m)
k	Thermal conductivity ($Wm^{-1}K^{-1}$)
Q_{MR}	Melting rate (kg s^{-1} or kg min^{-1})
Q_s	Powder consumption (kg m^2(mould))
q	Heat flux density (Wm^{-2})

The original version of this chapter was revised: For detailed information please see correction. The correction to this chapter is available at https://doi.org/10.1007/978-3-319-53616-3_12

Within this book minor changes to some of the figure permissions and the readability of the chapters have been made and several figures have been replaced with new versions.

K.C. Mills and C.-Å. Däcker, *The Casting Powders Book*,
DOI 10.1007/978-3-319-53616-3_6

t	Thickness of mould (m)
w	Width of mould (m)
γ	Surface tension (mNm^{-1})
η	Viscosity (dPas)
T	Temperature (°C)
T_{br}	Break temperature (°C)
T_{liq}	Liquidus temperature (°C)
T_{sol}	Solidification temperature (°C)
V_c	Casting speed ($m\ min^{-1}$)
C/S	$\%CaO/\%SiO_2$ = basicity
EMBR	Electromagnetic Braking
HC	High carbon (steel)
IR	Infrared
LC	Low carbon (steel)
MC	Medium carbon (steel)
SEN	Submerged entry nozzle
ULC	Ultra-low carbon (steel)

In chemical formulae for slags and minerals, the following abbreviations are used:

A	Al_2O_3
B	B_2O_3
C	CaO
F	FeO
F*l*	CaF_2
K	K_2O
L	Li_2O
M	MgO
Mn	MnO
N	Na_2O
S	SiO_2
T	TiO_2

6.1 Introduction

Mould powders were first used successfully in bottom-pouring castings in Belgium in 1958. These powders were based on *fly ash* (a waste product in power generation) and they were applied to continuous casting in 1963. It was found that the use of these powders reduced the heat losses significantly (*cf.* oil casting) and thus, the use of powders led to a significant decrease in the superheat of the steel.

In this chapter, we discuss the different types of mould powders which are available. These include conventional (F-containing) powders used for casting slabs, blooms and billets as well as specialist powders developed for casting specific steels or for very different casting conditions. The mould powders used to cast steels under normal conditions are denoted as "*Conventional*" in the following text; such powders usually contain CaF_2 to form the phase *cuspidine* in the slag film. Other powders have been developed to deal with specific casting conditions or for casting-specific steel grades; these will also be discussed below.

First, it is necessary to establish the functions and criteria which conventional fluxes must fulfil. Then, the properties and compositions will be examined to identify those providing a satisfactory performance of these tasks. When this has been successfully applied to conventional mould powders, a similar analysis can then be carried out for mould powders for the specialist casting.

6.1.1 Functions Carried Out by Mould Powder

The principal functions of the mould powder (discussed below with the aid of Fig. 6.1) are

(i) To prevent the steel meniscus from freezing (i.e. reducing vertical heat transfer).
(ii) To seal the liquid steel from the atmosphere to prevent oxidisation of the metal.
(iii) To lubricate the newly formed shell.
(iv) To control the horizontal heat transfer from the shell to the mould.
(v) To absorb inclusions from the steel and to minimise slag entrapment.

Although these are the primary functions, there are other secondary tasks which would be very welcome, if provided by the mould flux, namely

(vi) To minimise the pick-up of carbon from the casting powder.
(vii) To minimise SEN erosion rates.

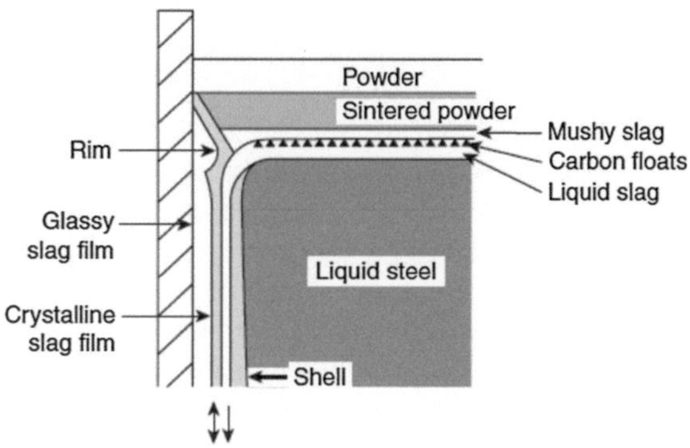

Fig. 6.1 Schematic drawing showing various layers formed in the powder bed and the slag film with liquid phases and glassy portion of slag film depicted in *white* and crystalline portion in *grey* (in practice, glassy and crystalline phases are mixed in the slag film) (Reproduced with permission from RFCS/ECSC [1])

6.1.2 Criteria Affecting Selection of Mould Powders

Mould powders are selected to provide the various functions listed above. When selecting a mould powder there are several criteria which must be addressed. Some compromises are needed in this selection process; for instance, a high-slag viscosity helps to reduce slag entrapment but will reduce the powder consumption (i.e. lubrication supplied to the shell). The various criteria affecting the selection process are summarised below (with particular emphasis on those criteria influenced by mould powder properties).

6.1.2.1 Thermal Insulation Provided by the Powder Bed

In order to avoid the freezing of the steel (i.e. to prevent *skull* formation) it is necessary to ensure that the bed provides the necessary thermal insulation. The thermal resistance (R_{th}) of the bed (containing, powder, sinter and liquid layers) is a measure of the thermal insulation and can be calculated from Eq. 6.1 where d is the depth; or thickness, and k is the thermal conductivity. The temperature at the mould powder surface ($T_{air/bed}$) usually lies between 200 and 400 °C and tends to increase with decreasing bed depth [2, 3].

$$R_{th} = (d/k)_{powder} + (d/k)_{sinter} + (d/k)_{pool} = (T_{steel/pool} - T_{air/bed})/q_{vert} \quad (6.1)$$

Vertical heat transfer from the surface of the steel is complex and involves a number of heat transfer mechanisms, e.g. (i) convection, radiation and conduction in the molten slag pool and (ii) conduction and gaseous convection in the powder bed and sinter. Consequently, there are a number of factors affecting the vertical heat flux and the thermal insulation, namely

(i) The *depth of the bed* (d_{bed}) (with $q_{vert} \downarrow$ as $d_{bed} \uparrow$).

(ii) The *particle size* of the mould powder (D) affects the packing of the bed and the bulk density (ρ_{bulk}) with ρ_{bulk} increasing as D decreases. Smaller particles have (a) *higher conductivity* (k_{powder} (Wm^{-1}K^{-1}) = 0.10 + 0.169 × 10$^{-3}\rho_{bulk}$ [4, 5]) and (b) *lower permeability* (P) to gaseous convection than bigger particles [6]. Thus, decreasing D results in an increase in q_{lat} and a decrease in q_{conv}, i.e. they have opposing effects. Since tests show that thermal insulation increases with decreasing particle size, gaseous convection must be the dominant mechanism in the powder bed.

(iii) The *reaction of O_2 with carbon* particles in the mould powder is exothermic. The heat given out causes increases in local temperatures and causes a decrease in the thermal gradient in the bed and in q_{vert}. There is *ca.* 3 × more energy given out in the formation of CO_2 *cf.* conversion to CO(g).

(iv) The introduction of *exothermic agents* (e.g. Ca/Si or Fe/Si) into the powder has the same effect as carbon, in that, heat generated by combustion, results

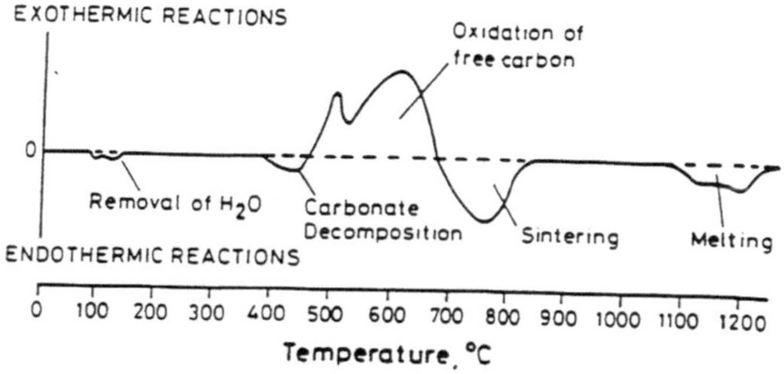

Fig. 6.2 DTA trace showing the various thermal events during heating of a mould powder (Reproduced with permission from RFCS/ECSC [8])

in a local increase in temperature which reduces the temperature gradient and hence, the vertical heat flux.

(v) Heat transfer in the *liquid pool* is dominated by convection and radiation (in some models, $k_{eff} = C\ k_{lat}$ is adopted with values of C varying between 10 and 100 and with a recommended value of 15 [7]. The vertical heat transfer can be reduced by increasing (i) the slag viscosity ($q_{conv} \downarrow$) or (ii) the FeO or MnO % in the slag ($q_R \downarrow$).

It should be noted that that there are a number of thermal events which occur at various levels in the powder bed (Fig. 6.2, [8]) namely

(i) The *endothermic* decomposition of carbonates in the range, 300–400 °C.
(ii) The *exothermic* oxidation of carbon particles to CO(g) and CO_2(g) between 500 and 700 °C and any exothermic agents present.
(iii) The *endothermic* sintering of particles at *ca.* 800 °C (this includes the enhanced C_p values for the supercooled liquid).
(iv) The *endothermic* melting of the powder at 1100–1200 °C.

Exothermic events will lead to a local heating resulting in decreases in both the local temperature gradient and hence q_{vert}. Endothermic events cause local cooling and increase in both the local temperature gradient and q_{vert}.

6.1.2.2 Formation of Molten Slag Pool

The molten slag pool must protect the steel from reactions with O_2, N_2 and CO_2(g). Thus, it is important that the pool should have sufficient depth to prevent the standing wave in the steel meniscus (resulting from turbulent metal flow) from poking into the powder bed (Fig. 6.3) which contains these gases.

The principal factors affecting the depth of the pool are (i) the vertical heat flux ($d_{pool} \uparrow$ as $q_{vert} \uparrow$) and (ii) the melting rate ($d_{pool} \uparrow$ as $Q_{MR} \uparrow$). The vertical heat flux, in turn, is affected by the depth of the powder bed, the particle size, and the presence of exothermic agents in the mould flux. It is also affected by factors like

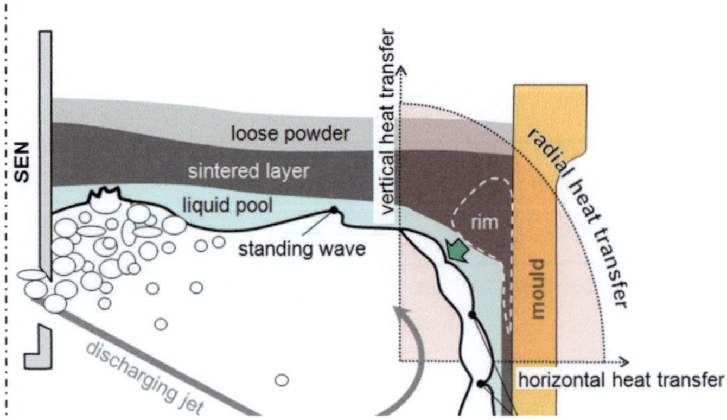

Fig. 6.3 Schematic drawing showing standing waves in the steel meniscus close to contact with the sintered layer (permission granted, ISS/AIST [9])

electromagnetic braking (EMBr) which affects the efficiency of vertical heat transfer. The principal properties of the mould powder affecting the melting rate are the C% and to a lesser extent, the size of the carbon particles ($Q_{\mathrm{MR}}\uparrow$ as $\%C\downarrow$ and $D_{\mathrm{C}}\uparrow$) the bulk density of the powder and the efficiency of heat transfer.

6.1.2.3 Slag Infiltration and Powder Consumption

The infiltration of liquid slag into the shell/mould gap provides the essential lubrication of the shell. The powder consumption, Q_s (in kg m^{-2} [of mould]) is equal to ($\rho.d_l$) where ρ is the density and d_l the thickness of the liquid layer of the slag film (see Chap. 2). Empirical rules have been developed to calculate a value for Q_s which provides a satisfactory level of lubrication [10, 11]. This required powder consumption (Q_s^{req}) is very sensitive to mould dimensions [width, (w) and thickness, (t)] but can be calculated from Eq. 6.2, where R^* is defined in Eq. 6.3 [12].

$$Q_s^{\mathrm{req}}\left(\mathrm{kgm}^{-2}\right) = 2/(R^* - 5) \tag{6.2}$$

$$R^* = 2(w + t)/wt \tag{6.3}$$

This calculated value of powder consumption can be related to the casting speed (V_c) and the slag viscosity at 1300 °C (η) through the empirical, Eq. 6.4 [12], thereby, providing the required slag viscosity for the given casting conditions (w, t and V_c).

$$Q_{\mathrm{s\,req}}^{\mathrm{slag}} = 0.55/\eta^{0.5} \cdot V_c \tag{6.4}$$

Low powder consumption can arise when casting steels containing Ti, or when using mould fluxes containing significant concentrations of TiO_2, because solid TiN (or TiCN) particles can be formed (which have a low solubility in the slag) and the solid particles block the infiltration of molten slag [13, 14] and increase the slag viscosity. This problem is discussed [12] below in Sects. 6.2.6 and 6.2.13 where the superscript ^{slag} refers to the liquid slag {and $Q^{slag} = f^* \, Q^{powd}$ where f^* is the fraction of powder forming liquid slag (see Chap. 2)}. Thus, it is possible to specify the slag viscosity needed for the casting conditions [12].

It is important that $Q^{slag}_{s\ req}$ should not be constrained by the rate of the melting rate, Q_{MR} which is linked to Q_s via Eq. 6.5.

$$Q^{slag}_{MR}(\text{kg s}^{-1}) = 2(w+t)Q_s V_c/60 \tag{6.5}$$

As seen above, the melting rate is principally controlled by (i) the % free carbon in the powder (ii) to a less extent, by the particle size of the carbon (D_C), i.e. ($Q_{MR} \downarrow$ as (% C_{free}) $\uparrow$ and $D_C \downarrow$) and (iii) the vertical heat flux ($Q_{MR} \downarrow$ as $q_{vert} \downarrow$). Consequently, it is important to ensure that the C_{free} % is low enough to ensure that the actual Q_{MR} is greater than the value of Q^{slag}_{MR} derived from Eq. 6.5.

6.1.2.4 Control of the Horizontal Heat Flux

Although the horizontal heat flux is affected by many factors, it is principally controlled by the slag film formed between the shell and the mould (Chap. 3).

The following properties are used to control the heat flux:

(i) the *solidification or break temperature* (T_{br}) of the slag which affects the thickness of the solid layer (d_s) of the slag film ($q_{hor} \downarrow$ as $d_s \uparrow$).
(ii) the *fraction of crystalline phase* (f_{crys}) formed in slag film since high values of f_{crys} result in increases in both the interfacial thermal resistance ($R_{Cu/sl} \uparrow$ as $f_{crys} \uparrow$) and reflection and scattering of radiation conduction ($q_R \downarrow$ as $f_{crys} \uparrow$).

In practice, q_R and q_{hor}, are reduced by increasing the basicity (*C/S*) of the powder since both the thickness of solid slag layer (d_s) and the fraction of crystalline phase of the slag film (f_{crys}) increase with increasing (*C/S*) ratio.

The appropriate level of horizontal heat flux (q_{hor}) to avoid longitudinal cracking varies according to the type of steel being cast [16, 17]; this is shown in Fig. 6.4 in which the lowest curve for the heat flux pertains to MC, *peritectic* steels and the higher curve for q_{hor} refers to casting LC steels which have a weak shell. The critical values of the required q_{hor} for HC steel grades lie at higher q_{hor} values.

6.1.2.5 Absorption of Inclusions and Minimising Slag Entrapment

Non-metallic inclusions reduce the mechanical strength of the steel. Thus, it is important to minimise the inclusion content of the steel. Inclusions arise from

Fig. 6.4 Index of longitudinal cracking as a function of heat flux (q_{hor}), curve at lower q_{hor} value and Δ refer to MC peritectic steels and curve for higher q_{hor} values and • are for a LC steel; (permission granted, ISS/AIST [16])

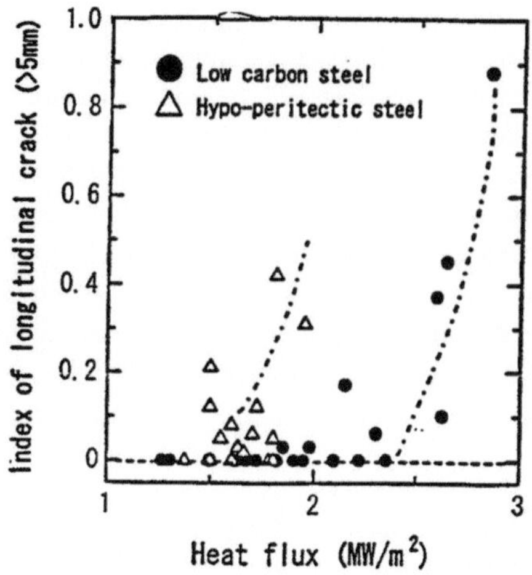

(i) oxidation reactions in steelmaking and casting (e.g. formation of Al_2O_3) [18] and (ii) entrapped slag. The objective here is to transfer as many of the inclusions as possible from the steel to the slag phase, where they can dissolve.

There principal strategy for reducing entrapment of inclusion particles (like Al_2O_3) is to minimise the length of the solidified meniscus (Fig. 6.5a, b). This reduces the probability of capture by the shell [19] and allows more time for the inclusion to migrate to the slag phase (Fig. 6.5). A short solidified meniscus can be achieved by reducing the vertical heat transfer by increasing the depth of the powder bed or by using exothermic agents in the powders. However, it has also been reported [20] that the number of slag inclusions trapped near the surface decreases with increasing basicity; this was attributed to the increase in interfacial tension (Fig. 6.5c) since $\gamma_{CaO} \approx 2\gamma_{SiO_2}$ where γ is the surface tension [21].

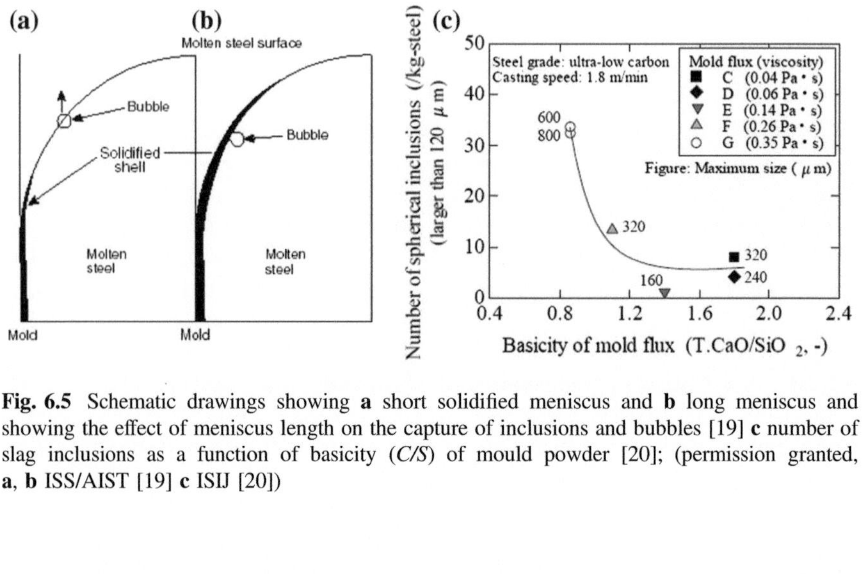

Fig. 6.5 Schematic drawings showing **a** short solidified meniscus and **b** long meniscus and showing the effect of meniscus length on the capture of inclusions and bubbles [19] **c** number of slag inclusions as a function of basicity (*C/S*) of mould powder [20]; (permission granted, **a**, **b** ISS/AIST [19] **c** ISIJ [20])

Some inclusions like TiN (or Ti(CN)) and ZrO_2 are not very soluble in the molten slag (*ca.* 0.5 and 2%, respectively) and consequently, solid particles tend to agglomerate through collisions in the slag pool. These agglomerates can block the entrance to the mould/shell channel and thereby reduce the powder consumption [13, 14]. In this case, the problem can be alleviated by increasing the volume of the molten slag pool (by increasing bed depth or using exothermic powders). This strategy is also used in casting TRIP steels where large amounts of Al_2O_3 are formed and must be encouraged to dissolve.

The amount of entrapped slag can be reduced by either increasing the slag viscosity (but this also decreases the powder consumption) or by increasing the interfacial tension. The latter can be achieved by reducing the concentrations of B_2O_3, K_2O, Na_2O, and SiO_2 which have low surface tensions [21].

6.1.2.6 Carbon Pick-up by the Shell

Carbon pick-up by the steel is a particular issue when casting LC and ULC steel grades. The principal sources of carbon pick-up are (i) the mould powder (ii) the carbon floating near the surface of the slag pool and (iii) the slag rim which contains glassy or amorphous carbon [22]. Carbon pick-up can be minimised by the following:

(i) Maintaining a deep slag pool.
(ii) Reducing the carbon content of the powder.
(iii) Replacing the carbon with nitrides [22–24].

6.1.2.7 Erosion of SEN

The principal causes of SEN erosion are [25, 26]

(i) High flow velocities in the metal and slag phases which impact on the SEN thus erosion rates decrease with increasing slag viscosity [25, 26].
(ii) The concentration difference ($C_{sat} - C_{slag}$) is the driving force for dissolution of the SEN refractory (e.g. ZrO_2).
(iii) High CaF_2 contents in the slag increase erosion rates (due to the attack of oxides stabilising ZrO_2); SEN erosion rates are lower in F-free slags [27] and are affected by the ZrO_2-stabiliser; erosion rates MgO > CaO > Y_2O_3 [26].
(iv) High O potentials for the slag increase erosion rates, so it is necessary to keep FeO, MnO in the slag to a minimum.

In some cases, where SEN erosion is an important issue, about 2% ZrO_2 can be added to the mould powder to bring the concentration, C_{ZrO_2}, close to C_{sat}; this minimises SEN dissolution but also promotes the nucleation of crystalline phases and increases T_{sol}.

6.1.2.8 Effect on Scale Formation

Certain mould slags adhere very closely to the surface of the strand and are not removed in the secondary cooling zone. Pockets of adhering slag become the sites where Fe_2SiO_4 is formed [28–38]; the Fe_2SiO_4 has a low melting temperature and readily penetrates the grain boundaries of the steel thereby providing a "*handhold*" for the scale to grip (see Sect. 11.11). Good adhesion between slag and steel is promoted by powder constituents with low surface tension (particularly, K_2O and Na_2O) which lower both the slag surface tension and interfacial tension [21]. Some mould powders, which fully meet the criteria for lubrication and heat transfer, have been rejected in thin-slab casting because of their effect on scale formation.

6.2 Selection of Mould Fluxes

Most mould flux compositions are selected to meet the criteria of

(i) Optimum powder consumption (i.e. lubrication of shell) for the given mould dimensions and casting conditions.
(ii) Optimum level of horizontal, heat flux extraction for the steel being cast.

It has been shown above, that the required powder consumption can be expressed as a function of the required viscosity at 1300 °C (η^{req}). Figure 6.6 shows the relation between η^{req} and the required break temperature (T_{br}^{req}) for the MC (crack-sensitive) steels and HC (sticker sensitive) steels. The *X*-axis of this figure represents the *reciprocal lubrication* and the *Y*-axis *the reciprocal of horizontal*

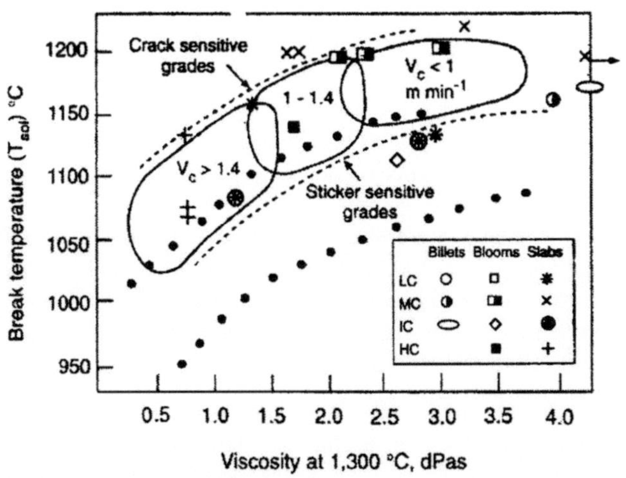

Fig. 6.6 Required break temperature as a function of required viscosity for mould slags used in casting slabs, blooms and billets; dotted curves: *upper* MC steels, *lower* HC steels; (permission granted, ISIJ [17])

heat transfer (i.e. thermal resistance) since the thickness of the solid layer increases with increasing T_{br}. The Eqs. 6.6–6.9 were obtained to calculate the optimum T_{br}^{req} for the mould slag from the η^{req} value.

The various steel grades can be classified in terms of the carbon potential (CP) which is defined in Eq. 6.6 [39] and then values T_{br}^{req} can be calculated for the appropriate steel grade (Eqs. 6.7–6.9).

$$CP = \%C + 10 - 2(2\%\text{Mn} + 4\%\text{Ni} - 10\text{Si}\%.) \quad (6.6)$$

$$\text{MC-steels CP}(0.06 - 0.18):\quad T_{br}^{req} = 1157 + 60\ \ln\ \eta \quad (6.7)$$

$$\text{HC steels CP}(>0.4):\quad T_{br}^{req} = 1051 + 76.4\ \ln\ \eta \quad (6.8)$$

$$\text{Other steels}:\quad T_{br}^{req} = 1103 + 68.5\ \ln\ \eta \quad (6.9)$$

As mentioned above, although most mould slag compositions are selected to give the appropriate values of T_{br}^{req} and η^{req}. However, sometimes a compromise is required, for instance, if slag entrapment is an issue, the slag viscosity may be increased to reduce the number of entrapped particles but this comes at the expense of a lower powder consumption.

The different types of powder available are discussed below. They include the powders used for regular casting and those powders used in the casting of specialist steel grades or used to combat certain casting problems or defects.

6.2.1 Conventional Mould Powders

As mentioned above, the mould powder composition is determined principally in terms of providing (i) the required mould slag viscosity for the given casting conditions and mould geometry and (ii) the optimum level of horizontal heat flux (or thermal resistance of the slag film) which can be represented by the break temperature of the slag. These conditions can be seen as minimising the chances of obtaining longitudinal cracks and sticker break-outs. Thus the composition of the powder is selected to give the required values of viscosity and T_{br}. The fraction of crystalline phase in the slag film should also be taken into account but the actions taken to increase f_{crys} are identical to those for increasing T_{br}.

The various mineral constituents can be classified into three groups, namely

(i) *Network formers* which includes SiO_2 and Al_2O_3
(ii) *Network breakers* which includes CaO, MgO, FeO, MnO and occasionally BaO
(iii) *Fluxes* to reduce T_{liq} and viscosity: Na_2O, K_2O, Li_2O, CaF_2 and B_2O_3.

The effects of individual constituents on the properties and the formation of glass or *cuspidine* phases (C_3S_2Fl) are summarised in Table 6.1.

Table 6.1 The effect of increases in individual slag constituents on their properties and structure

Oxide/fluoride	Role	Promotes	Effect on properties		
			η	T_{br}	f_{crys}
SiO_2	Network former	Glass formation—also C_3S_2Fl	↑	↓	↓
Al_2O_3	Network former	Glass formation—high Al_2O_3 → crystalline	↑	↓	small ↓ large ↑
CaO	Network breaker	C_3S_2Fl formation	↓	↑	↑
MgO	Network breaker	>7%—glass formation	↓	↓	↑?
Na_2O, K_2O, Li_2O	Flux & breaker	>4%—glass formation	↓	↓	↓
CaF_2	Flux	Cuspidine, C_3S_2Fl	↓	↓	↑
B_2O_3	Flux	Glass formation	↓	↓	↓
FeO, MnO	Network breaker	Glassy phases but also reduces k_R	↓	↓	↓
ZrO_2		Nucleates crystals	↑	↑	↑

Mould powders tend to have the following characteristics:

(i) The basicity, (denoted here as (*C/S*), i.e. (= %CaO*/%SiO_2) where CaO* is the total Ca content expressed as CaO) is often cited since it provides a measure of both the structure of the slag and the viscosity. Values of (*C/S*) for mould slags tend to fall within the range (0.6–1.3) but recently, experimental fluxes with *C/S* ratios as high as 1.7 have been used. The (*C/S*) ratio of the powder tends to be determined by the type of steel and casting speed, e.g. for (i) *C/S* = 0.6 for low-castings speeds and HC steels (ii) *C/S* = 0.9–1.0 for LC, ULC steels and casting speeds around 1–1.5 m min^{-1} and (iii) *C/S* = 1.3–1.7 for MC steels and high casting speeds.

(ii) The fluxes are added to reduce both the liquidus (T_{liq}) and the solidification (or "*break*", T_{br}) temperatures of the slag. However, they also reduce the viscosity of the liquid slag.

(iii) The FeO contents are usually low since they increase the oxygen potential of the slag but both FeO and MnO are used to reduce radiation conduction when it is not possible to use a crystalline slag film (e.g. when casting round billets) [40, 41]. The pick-up of FeO by the slag is typically around 1% but MnO pick-up can be appreciable when casting some Mn-containing steel grades but pick-up decreases with increasing basicity [41].

(iv) The B_2O_3 contents rarely exceed 5% because of the problems associated with B-pick-up by the steel.

(v) Other components, such as TiO_2, are usually present as impurities; TiO_2 might be expected to behave as a network former but, in practice, it reduces the viscosity, so TiO_2 is usually treated individually in the modelling of slag viscosity. TiO_2 has been used to form crystalline phases in fluorine-free mould fluxes (i.e. as a replacement for *cuspidine*). However, Ti tends to form high melting phases, like *perovskite*, ($CaTiO_3$) and it is necessary to keep (C/S) < 0.9 to ensure that $CaTiO_3$ is not formed [42]; Ti can also form Ti(CN) which tends to lead to sticker break-outs [15]. ZrO_2 is sometimes added (*ca.* 2–3%) to the slag to minimise SEN erosion rates but, both ZrO_2 and TiO_2 tend to increase the break temperature and consequently, may reduce the lubrication supplied to the shell.

(vi) In mould slags, Fluorine tends to bond exclusively with Ca^{2+} and Mg^{2+} ions and once these Ca–F bonds are formed, the CaF_2 has little effect on the structure or on structure-related properties, like viscosity. It is for this reason, the parameters, *NBO/T* and *Q* (measures of the degree of depolymerisation and polymerisation, respectively) are usually calculated by ignoring the CaF_2 in the slag.

(vii) Carbon has a fairly low solubility in liquid mould slags [43] and it is non-wetting to molten slag (Fig. 6.7). Thus, molten slag globules are unable to agglomerate until the carbon particles have been oxidised. Thus the melting rate decreases as (i) the % carbon in the mould flux increases and (ii) the carbon particle size decreases [44–46]. Carbon is usually added to regulate the melting rate (Q_{MR} in kg s^{-1} or kg min^{-1}). So it matches demand (which is determined by the mould dimensions (w, t) and casting speed (V_c) as can be seen from the combination of Eqs. 6.2, 6.3 and 6.5. Carbon was, fortuitously, present in the original fly ash. Powders for slab casting usually contain *ca.* 4% C_{free} but powders for billet casting can contain up to 25% C_{free}. The free carbon can exist in various forms (e.g. graphite, carbon black, coke-breeze, etc.). Different forms of carbon tend to

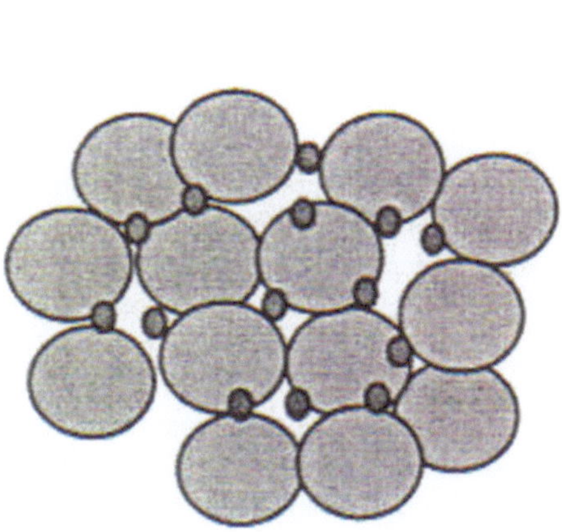

Fig. 6.7 Schematic drawing showing separation of molten slag globules (●) by graphite particles (•)

have different particle sizes and mould powders tend to contain a mixture of different carbon types (and different particle sizes). As mentioned above, the particle size of the carbon (D_C) is important with the melting rate decreasing as D_C decreases but smaller particles have a high (surface area/mass) ratio and will thus oxidise more rapidly than bigger particles. Since smaller carbon particles retard the melting rate, it is obvious that the enhanced separation is more significant than the rate of oxidation. The carbon also helps to maintain a reducing atmosphere in the mould.

(viii) Some mineral constituents are present as carbonates, these decompose around 400 °C to form oxides and $CO_2(g)$. The carbon contents are frequently cited as $\%C_{total}$ and $\%C_{free}$; the carbonate content $\%C_{CO_2}$ can be calculated from the equation

$$\%C_{CO_2} = \{44/12\}(\%C_{total} - \%C_{free}) \tag{6.10}$$

(ix) The moisture content (% H_2O) in the as-received powder is usually around 0.5%; it is important to ensure % $H_2O \leq 1\%$ since this can lead to hydrogen evolution in the slag film which can lead to sticker break-outs [47–49]. The hydration of calcium aluminate mould powders used in the casting of TRIP steels has been reported to cause *flaring* in the mould and it was found necessary to bake the powder at 250 °C prior to its use in order to eradicate the hydration [50].

(x) Zirconia, ZrO_2 additions (2–3%) are used (a) to nucleate crystals and increase T_{br} and (b) to minimise SEN erosion.

6.2.1.1 Mineralogical Constituents

The first mould powders to be produced were based on *fly ash*, to which fluxes, such as, Na_2O and B_2O_3 were added. These powders contained some carbon which, fortuitously, regulated the melting rate. In due course, more efficient power plants were introduced and these produced fly ash with lower carbon contents. This meant that fly ashes from different sources had to be carefully blended and it became more difficult to ensure consistent powder composition and quality.

This stimulated the development of *Synthetic* mould powders in which fly ash was replaced by silica, lime and bauxite and fluxing agents. These synthetic powders were made of mixtures of minerals and it is possible to achieve similar powder compositions from different mixtures of minerals. It should be noted that the slag pool and the slag film should be unaffected by the nature of the minerals used providing the composition is the same. For identical compositions produced with different minerals, the only differences expected would occur during the heating in the powder bed but once the minerals have been melted they should all behave in an identical manner. However, there are some general rules which are used in the manufacture of synthetic mould powders, namely, the powder should contain

(i) Mineral constituents with similar melting points (e.g. lithium *feldspars*, sodium *feldspars*, *wollastonite*) to provide uniform melting.
(ii) The minimum number of minerals to achieve the target composition in order to keep the recipe simple and to simplify quality assurance.
(iii) A minimum of mineral constituents which constitute a potential health hazard (e.g. silica).

Fly-ash powders are still available and are relatively cheap and are used extensively in ingot casting. However, in continuous casting, fly-ash powders have been largely replaced by ***synthetic fluxes*** where the chemical compositions are much better controlled.

6.2.1.2 Crystalline Phases Formed in Slag Film

When the molten slag solidifies in the shell/mould gap the first crystalline phase formed in slags containing CaF_2 is *cuspidine*, $3CaO{\cdot}2SiO_2$. CaF_2 (denoted here as C_3S_2Fl). The phase diagram containing the *cuspidine* phase has been reported [51]. *Cuspidine* is the first phase to solidify and it is formed on the mould side, but growth of large crystals takes place at high temperature and thus large crystals tend to be found on the shell side of the slag film [52, 53]. It can be seen from Fig. 6.8 that the slag film exhibits dendritic form on the shell side but much finer cubic crystals are formed on the mould side [53].

It can be seen from Fig. 6.9 that most mould slag compositions tend to lie outside the main crystallisation field for *cuspidine*. Thus, the powder composition

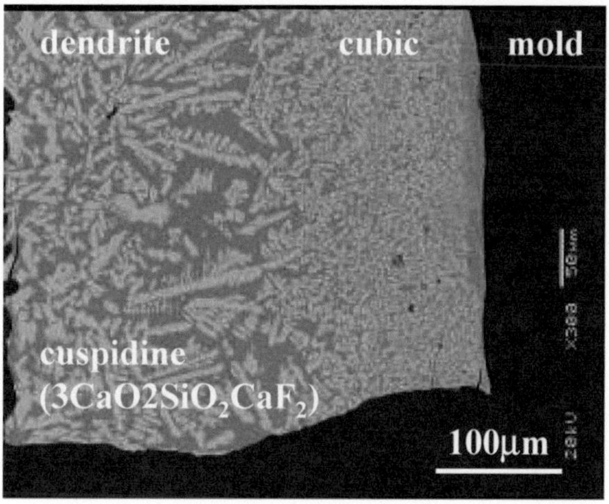

Fig. 6.8 Photograph of a section of slag film showing dendritic growth of crystals on shell, high-temperature side and fine cubic crystals on the mould, low-temperature, side (permission granted, ISIJ [53])

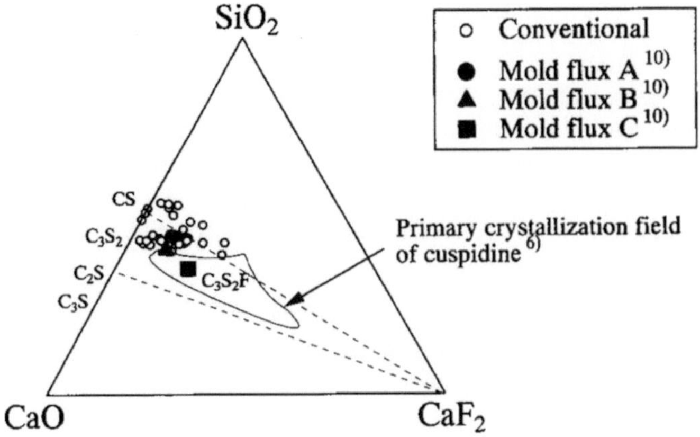

Fig. 6.9 Ternary diagram showing *cuspidine* phase field; it can be seen that most conventional mould slag compositions (o) lie outside the *cuspidine* phase field (*solid line* surrounding ▬); *dotted lines C/S* values of 1 and 2 [20]; (permission granted, ISIJ [20])

was moved into the *cuspidine* field [20]; the powder was used in high-speed casting ($V_c \leq 5$ m min^{-1}) and the slag was found to give lower horizontal heat fluxes (and less longitudinal cracking) than the original powder [20].

The resultant slag film contained *cuspidine* dendrites through the entire section, in contrast to the original powder which showed a glassy phase in addition to *cuspidine* (Fig. 6.10) [20].

Hayashi et al. [54] studied the effect of Na_2O on the formation of *cuspidine* (C_3S_2Fl) crystals using FMAS-NMR, DTA and XRD and found that (i) F^- ions tend to bond with Ca^{2+} ions (ii) the maximum amount of *cuspidine*, after annealing, was obtained with a sample containing 4% Na_2O and this was attributed to the effect of Na_2O on nucleation and growth rates (via a lower viscosity) (iii) T_g and T_{crys} were decreased with increasing % Na_2O.

Other mineralogical phases are formed during secondary crystallisation; a phase diagram (Fig. 6.11) of the various phases formed has been reported for mould slags with a basicity of *ca.* 1 [55].

6.2.1.3 Pick-up of Oxides by Mould Slag

The mould slag picks up various oxides from steelmaking reactions and reactions between the steel and slag phases. Mould powders usually contain about 5% Al_2O_3 but there is typically, a 4% pick-up of alumina by the slag during the cast; the

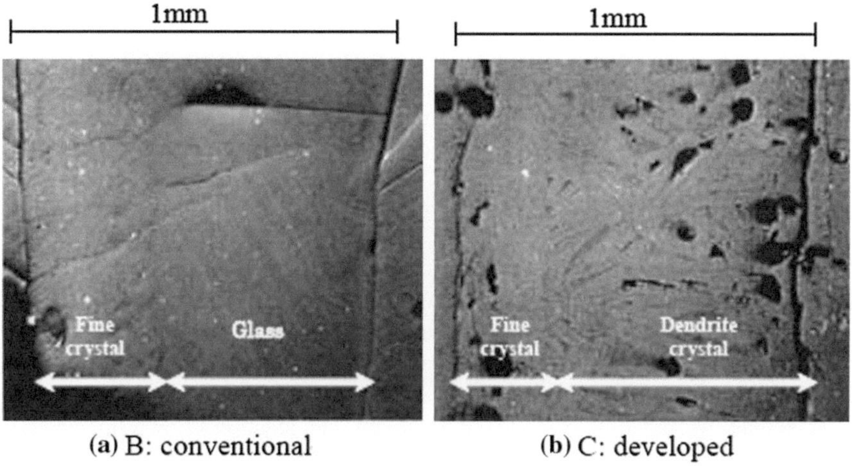

Fig. 6.10 Photographs of sections of slag films derived in high-speed casting with **a** conventional powder (outside *cuspidine* field) showing a large glassy layer and **b** the developed powder (inside *cuspidine* phase field) showing large crystalline layer; *Note* mould on *left* and shell on *right side*; (permission granted, ISIJ [20])

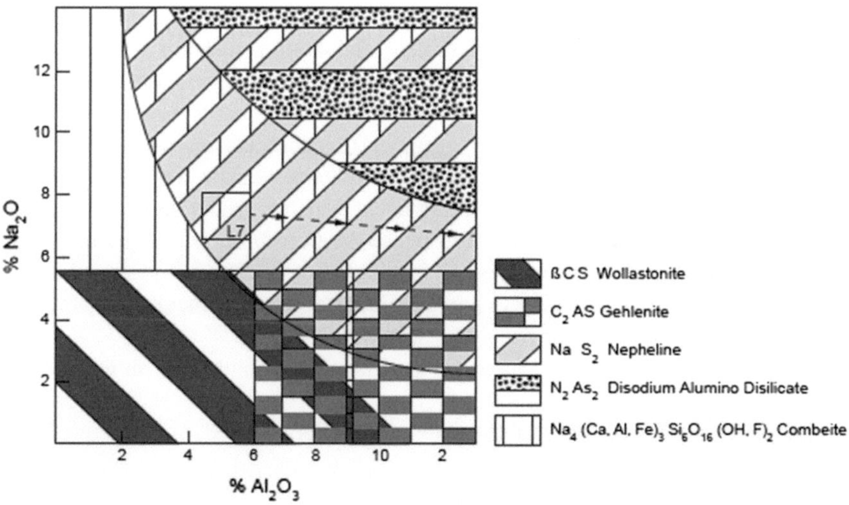

Fig. 6.11 The mineralogical phases formed in slags after *cuspidine* has been formed for mould powders with %CaO/%SiO_2 ratios of *ca.* 1.0 (permission granted, IOM/Taylor and Francis, www.tandfonline.com [55] re-drawn, Swerea KIMAB)

pick-up of inclusions arises, almost equally, [18] from (i) the steelmaking processes and (ii) from the reaction between Al in the steel (denoted by an underline in Eq. 6.11) and the slag pool. Slag viscosity increases with increasing Al_2O_3 content.

$$2SiO_2(slag) + 6\underline{Al} = 2Al_2O_3(slag) + 4\underline{Si}. \quad (6.11)$$

Aluminium reacts in a similar manner with both MnO and FeO in the slag.

There is also pick-up of FeO and MnO by the slag. The pick-up of Mn when casting Mn-steel grades occurs by the reaction shown in Eq. 6.12.

$$FeO\ (slag) + \underline{Mn} = MnO\ (slag) + \underline{Fe} \quad (6.12)$$

Thus, the composition of the slag pool and the liquid (which, subsequently, forms the slag film) is slightly different from that determined from the mould powder after allowing for the loss of carbon and carbonates. It is the slag pool composition which should be used (i.e. including pick-up of Al_2O_3, MnO and FeO) when calculating the properties of the slag film.

It should be noted that Al_2O_3 pick-up by the reaction shown in Eq. 6.11 becomes a major problem when casting TRIP and TWIP steels containing *ca.* 1% Al; this is discussed below in Sect. 6.2.12.

6.2.1.4 Nature of the Powder

Mould fluxes are usually supplied either in the form of ***powders*** or ***granules***. ***Powders*** are produced by grinding, mixing and homogenisation of the various minerals. ***Granules*** are usually prepared by mixing the flux ingredients into a slurry which is then either extruded or spray-dried. In the latter process the slag is sprayed as droplets (*ca.* 1 mm) which form granules or spheroids on drying. Extruded powders are produced by a similar process but are extruded instead of being sprayed. Expanding granules contain an expanding agent which expands on heating to produce irregular-shaped granules which decrease the flowability of the granules [6].

Flowability is a particularly important feature in ingot-casting powders where the powder must flow easily in order to prevent areas developing where there is no slag cover (i.e. no "*open eyes*" formed—see Sect. 7.4).

The advantages and disadvantages of powders and granules are summarised in Table 6.2.

The performances of the various forms of powders (mixed, sintered, prefused) have been compared [56, 57].

6.2.2 *Pre-melted Fluxes*

These are available but tend to be more expensive than granulated powders (or spherodised) fluxes because of the higher energy costs involved in melting the

Table 6.2 Advantages and disadvantages of powders and granules

Property	Powders		Granules	
	Advantages	Disadvantages	Advantages	Disadvantages
Cost	Cheaper			More expensive
Thermal insulation	Better	More dust,	More suitable for automated feeding	Not so good
Homogeneity		Worse-smaller particles gravitate to bottom of bag	Better	
Automatic feeding		Worse	Better	
Health and Safety		Dusts → Hazard	Less hazardous	
Flowability in mould	Lower			High—can expose metal in standing wave

mixture of oxides, and fluorides. However, pre-melted powders do reduce the rate of fluorine emissions on the steel plant since the chemical activity of CaF_2 in the fused powder is lower than that of CaF_2 particles present in the conventional powder.

6.2.3 *Starter Powders*

Starter powders, as their name suggests, are used at the beginning of the cast. Starter powders must melt rapidly and provide a pool of molten slag in order to (i) protect the surface of the steel from oxidation and (ii) encourage slag infiltration into the shell/mould gap. They have the following characteristics:

(i) They contain large concentrations of low-melting materials such as Na_2O, fluorides and borates.
(ii) The carbon contents are low (<1%) to ensure a high melting rate and minimise carburisation of the steel shell.
(iii) They sometimes contain exothermic agents to provide additional heat.
(iv) They have a tendency to form slag rims since low-melting materials (like Na_2O) tend to act like glue and incorporate solid materials into the rim; consequently, starter powders should not be used for longer than necessary.

The first slag film is formed from the slag produced by the starter powder but this will be modified in time by (i) any downward movement of the slag film which will encourage new slag film formation (ii) any fracture and movement of slag film which will encourage slag infiltration from the slag pool and (iii) gradual reaction with the liquid slag film.

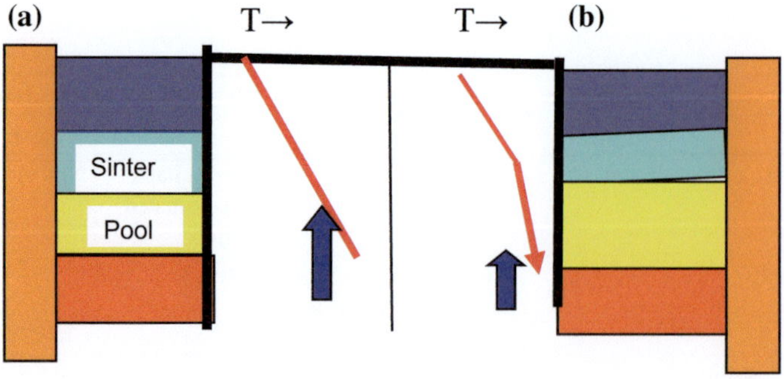

Fig. 6.12 Schematic diagram showing **a** powder bed and temperature gradient for **a** conventional powder and **b** powder containing exothermic agents—note how exothermic agents reduce both the temperature gradient and the vertical heat flux and increases the pool depth

6.2.4 *Exothermic Fluxes*

These contain exothermic agents such as Ca/Si or Fe/Si which give out heat when reacting with oxygen (Eqs. 6.13 and 6.14):

$$\mathrm{Si} + \mathrm{O_2(g)} = \mathrm{SiO_2} \quad \Delta H_{25^\circ\mathrm{C}} = -911\,\mathrm{kJ\,mol^{-1}} \tag{6.13}$$

$$\mathrm{Ca} + 0.5\,\mathrm{O_2(g)} = \mathrm{CaO} \quad \Delta H_{25^\circ\mathrm{C}} = -635\,\mathrm{kJ\,mol^{-1}} \tag{6.14}$$

This heat increases the local temperature in the bed and thereby reduces the thermal gradient ($\Delta T = T_{pool} - T_{local}$), and subsequently decreases the total heat flux in the mould ($q_{vert} = w.t.k_{bed}\ \Delta T$) where k_{bed} is the effective thermal conductivity of the bed and w and t are the width and thickness of the mould, respectively (Fig. 6.12).

Powders containing exothermic agents [23, 58, 59] are used to

(i) Increase the depth of the molten pool (e.g. to reduce C-pick-up by the shell and to improve both the absorption of inclusions and powder consumption).
(ii) Reduce the length of the solidified meniscus by reducing the vertical heat flux and thereby allowing the inclusions and gas bubbles to be transported to the slag layer (e.g. Fig. 6.5a, b) [58, 59].

It should also be remembered that the free carbon is also an exothermic agent and that conversion to CO_2(g) provides nearly 3× more heat than conversion to CO (g) (see Eqs. 6.15 and 6.16). In contrast, the decomposition of carbonates is endothermic and will lower the local temperature which will stimulate a higher heat flux in that locality of the bed at *ca.* 500 °C.

$$C + 0.5\,O_2 = CO(g): \quad \Delta H_{298} = -110\ \text{kJ mol}^{-1} \tag{6.15}$$

$$C + O_2 = CO_2(g): \quad \Delta H_{298} = -393\ \text{kJ mol}^{-1} \tag{6.16}$$

Liquid slag Feeding

The possibility of pouring liquid mould slag onto the surface of the steel has been explored [60–62]. Such a practice might be expected to provide the following benefits:

(i) Provision of extra heat to help maintain a high meniscus temperature.
(ii) Quicker establishment of steady state conditions.
(iii) The possibility of higher powder consumption and lower friction on the shell.
(iv) The possibility of creating a deep pool to absorb inclusions (especially when casting high-Al steels which produce large amounts of Al_2O_3).

A number of plant trials were carried out to cast blooms (0.38 × 0.225 m) at V_c = 0.65–0.7 m min^{-1} [60]. The following procedures were used in casting blooms:

(i) Enough mould powder was melted in a mobile furnace to produce 3 litres of liquid.
(ii) 1.5 litre of the liquid mould slag was poured onto the steel surface (this is equivalent to d_{pool} = *ca* 2 mm) and the casting started.
(iii) The remaining 1.5 litre of liquid was subsequently poured into the mould.
(iv) The regular mould powder was then added at the appropriate time.

Several successful casts were carried out and it was reported [60] that

(i) Friction on the shell was low using this practice (good liquid lubrication).
(ii) The surface quality of the casts was not very good with a large number of "*cavities*"; this was attributed to the freezing of mould slag to form "*lumps*" which subsequently infiltrated into the shell/mould channel and left an impression on the steel surface (Fig. 6.13).
(iii) Faint (very shallow) oscillation marks were observed but normal oscillation marks were obtained when the switch to regular mould powder was made; the faint oscillation marks suggests the slag pool depth was very low.

Liquid slag feeding casting of high-Al steels has also been reported [61, 62] and was found to provide good powder consumption. Unfortunately, we were unable to obtain these publications. However, more development of liquid slag feeding seems inevitable, if these high-Al steels, with low-melting temperatures, are ever to be continuously cast on a commercial scale.

Fig. 6.13 Photograph showing "*cavities*" in steel surface (*marked*) when using liquid mould slags (permission granted, Swerea MEFOS [60])

6.2.5 *Fluoride-Free Powders*

Mould powders consist of a mixture of oxides and fluorides (referred to as an oxy-fluoride system). Conventional powders tend to contain $\geq 7\%$ F which is usually added in the form of CaF_2. Oxy-fluoride systems are basically unstable since they react together to form gaseous fluorides (e.g. Eqs. 6.17–6.19) [63–65]. It has been calculated that the partial pressure of NaF(g) exceeds 1 atm at temperatures above 530 °C [64, 65]. The gaseous fluorides will also react with any moisture to form HF(g). Mass-spectrographic studies showed that safety limits for HF(g) (0.1 mg m^{-3}) soluble fluorides (e.g. NaF, 0.2 mg m^{-3}) and insoluble fluorides (e.g. SiF_4, 0.5 mg m^{-3}) were breached on plant [63].

$$Na_2O + CaF_2 = CaO + 2NaF(g) \tag{6.17}$$

$$SiO_2 + 2CaF_2 = 2CaO + 2SiF_4(g) \tag{6.18}$$

$$Al_2O_3 + 3CaF_2 = 3CaO + 2AlF_3(g) \tag{6.19}$$

$$H_2O + CaF_2 = CaO + 2HF(g) \tag{6.20}$$

The HF emissions cause corrosion of plant equipment [63, 66] and constitute a Health and Safety hazard to plant personnel [63]. Furthermore, the secondary cooling water leaches fluoride out of the spent slag film (at the mould exit or slag adhering to the steel strand) causing the cooling water to gradually increase in acidity [67–69]. Consequently, the cooling water must be treated with alkaline agents to reduce the acidity of the secondary cooling water [67].

Environmental worries about fluoride emissions resulted in a search for successful mould powders containing zero, or reduced, levels of CaF_2. The challenge has been to develop a F-free powder which could match the performance of the extant, F-containing powder in current use with regard to the following properties:

(i) To provide the same level of powder consumption (i.e. primarily determined by the required values of viscosity and T_{br}).
(ii) To provide the same level of horizontal heat transfer (involving T_{br}, $R_{Cu/sl}$ and f_{crys}).
(iii) To replace *cuspidine* with a suitable crystalline phase in order to meet the required thermal resistance (this involves both $R_{Cu/sl}$ and f_{crys} where $R_{Cu/sl}$ depends upon the density difference between the glass and the crystal phases $[\rho_{cryst} - \rho_{gl}]$).

It is the third criterion which is the most difficult to meet.

Two different approaches have been adopted in the development of F-free powders, namely

(i) To replace CaF_2 with other fluxes (e.g. B_2O_3, Li_2O) and then modify the composition so that the properties of the candidate powder match the values of the original powder for viscosity, T_{br} and heat flux (q_{hor}); one problem is that fluxes like B_2O_3 tend to promote the glass phase resulting in an increase in q_{hor} [27, 70, 71].
(ii) To identify a suitable crystalline phase (e.g. *perovskite* [72, 73]) to replace *cuspidine*, in order to reflect the IR radiation and create an interfacial thermal resistance, $R_{Cu/sl}$ (which is related to $[\rho_{crys}-\rho_{gl}]$) to give a similar level of q_{hor} to that obtained with a slag film containing *cuspidine*.

6.2.5.1 Replacement Fluxes Giving Similar Property Values

Several groups have used the approach where CaF_2 is replaced by other fluxes to provide similar values of viscosity and T_{br} to that of the F-containing slag, e.g. using B_2O_3, [27, 70, 71, 75, 76]; Li_2O [66, 77–79]; Na_2O [69, 80]; BaO [79, 81, 82]; and TiO_2 [72, 73, 76]. The effect of flux additions on the viscosity, T_{liq} and T_{br}, surface tension, heat flux and the ability to absorb inclusions has been studied in several of these studies on both F-free and low F powders [70–73, 75, 78, 81]; the results are summarised in Table 6.3. The effect of crystallisation on the viscosity, T_{liq}, T_{br} and f_{crys} of the remaining liquid for a F-free slag has also been determined [86].

Powders for casting billets
F-free mould powders have been developed for casting billets of plain C steels [27, 70, 74, 85]. In designing F-free mould slags, it is important that the values of the following properties should all match those of the original slag: (i) $\eta_{1300°C}$ (ii) T_{br} and (iii) degree of crystallinity. However, high-viscosity slags are frequently used in

Table 6.3 The effect of flux additions (in mass%) on melting temperature, T_{br}; viscosity, surface tension and heat flux measured in simulation experiments and effect on absorption of inclusions

Ref	Compound	$dT_{liq}/\%$ ($dT_{br}/\%$) °C $\%^{-1}$	($d\eta/\%$) dPas $\%^{-1}$	q_{hor} MWm^{-2}	q_{hor} MWm^{-2}	Absorption	Composition of base slag
Lu [81]	BaO	−15	−0.15[b]	−6.2[b]		↑	*C/S* = 0.8; 1.0;1.2; A4;L3;F2.3;B2; Ba1-5
Lu [78] Wen [72]	Li_2O	−33[c] −30	−0.35[c] −1	−9[c]	0	↑	*C/S* = 0.8; 1.0;1.2; A4;L2-10;F2.3 *C/S* = 0.95–1.15; N5-8;B4-8;;L1.5; Mn4; T4-7
Chen [73] Wen [72]	Na_2O	−2.5	−0.3	−17*	0 ([8] %)[e]		*C/S*1.05;M7; N6; Mn5 *Fl*2.6 *C/S* = 0.95–1.15; N5-8;B4-8;;L1.5; Mn4; T4-7
Lu [83]	K_2O	−59	−0.37	−4		↑	*C/S* = 0.8; 1.0;1.2; A4; L3;F2.3;B2; K1-5
Lu [75] Fox [70] Wen [72]	B_2O_3	−8[a] (−43) −19	−0.2[a] −0.2	a	+0.04	0	*C/S* = 0.8; 1.0;1.2; A4;L2;F2.3;B2-10 *C/S* = 0.95-1.15; N5-8;B4-8;;L1.5; Mn4;T4-7
Wen [72]	MgO	+5	−0.12		0		*C/S* = 0.95–1.15; N5-8;B4-8;;L1.5; Mn4;T4-7
Wen [72]	MnO	−7	0		−0.15 [4%][e]		*C/S* = 0.95–1.15; N5-8;B4-8;;L1.5; Mn4;T4-7
Wen [72]	TiO_2	+4.5	e				*C/S* = 0.95–1.15; N5-8;B4-8;;L1.5; Mn4;T4-7
Lu [84]	NaF	−7	d	−1.7	−0.1 [4%][e]	↑	*C/S* = 0.8; 1.0;1.2; A4;L2;F2.3;B2

() indicates T_{br}
cf-15 for CaF_2
[a]Varies with *C/S* ratio
[b]Mean of three slags with *C/S* ratios of 0.8; 1; 1.2
[c]Converted to Li_2O (additions of Li_2CO_3)
[d]Sharp drop from 3 to 6%, then steady
[e]Unaffected to 6% MnO or NaF, then sharp increase
M MgO, *N* Na_2O etc. in slag composition

billet casting and these slags have no break temperature since they form a super-cooled liquid slag and do not crystallise [70, 74]. This is due to the fact that in HC steels a thick, strong steel is required and so restriction of the heat extraction from the shell is not important. Thus, it is only necessary to match the viscosity (i.e. the lubrication) when designing F-free mould powders for billets [27, 70, 74]. Thus, in this case, 2.3% CaF_2 was replaced by 2%Na_2O and 1.1%B_2O_3. In plant trials it was found that [27, 70, 74]

(i) The F-free flux performed as well as the original F-containing powder.
(ii) There was no discernible pick-up of boron by the steel.

Powders for casting blooms and slabs

The criteria for designing F-free mould slags are much more stringent for casting slabs and blooms than for billet casting.

Wang et al. [87] pointed out that for a F-free flux to perform as well as a conventional slag, it must have similar values not only for (i) T_{liq} (ii) $\eta_{1300\ °C}$ (iii) T_{br} but also must satisfy the conditions for (iv) the ratio of crystalline phase in the slag film (R_p%, which is equivalent to 100 f_{crys}) (v) the crystallisation temperature (T_{crys}) and (vi) the nature of the η-T curve. They also carried out Raman spectroscopy on F-containing and F-free mould slags and found that F-free slag contained less Q^0 and Q^1 species and more Q^2 and Q^4 species (where superscript indicates the number of bonding O's, see Sect. 11.3.2) than the conventional slags (i.e. they contained more-polymerised species). A F-free flux for casting slabs was found to perform satisfactorily and they have also been used to cast blooms [88] slabs [71] and thin slabs [79]. One benefit of using F-free slags is that it results in significant decreases in SEN erosion rates [27, 87].

6.2.5.2 Replacement of Cuspidine with Other Crystalline Phase

Several investigators have designed F-free slags by replacing the *cuspidine* in the slag film with another crystalline phase; these are classified below in terms of the crystalline phase formed. The presence of crystalline phases in the slag film is essential to reduce the radiation conduction. Crystallisation of an initially glassy slag film causes (i) reflection and scattering of the IR radiation (thereby reducing the radiation conduction considerably) and (ii) shrinkage in the slag film, resulting in both an interfacial resistance ($R_{Cu/sl}$) and porosity in the slag film.

TiO_2 or $CaTiO_3$ (CT)

The replacement of *cuspidine* by adding TiO_2 to the powder has been studied by several workers [72, 73, 87, 89–91]; the CaO–SiO_2–TiO_2 (CST) system was studied as candidates for F-free fluxes [89]. Time–temperature–transformation (TTT) curves for slags with a (*C/S*) ratio of 0.8 and containing up to 17 mol% TiO_2 were constructed and it was found [89] that:

(i) The phase, *titanite* (CST) formed rapidly.
(ii) The CST phase did not increase the viscosity and melting temperature.
(iii) The precipitation of CST at higher temperatures was lower than that of *cuspidine*-forming (conventional) mould slags.
(iv) Na_2O additions probably reduce the time taken to crystallise (i.e. incubation time) as they did with *cuspidine* [92].

Wen et al. [72] studied the replacement of slags containing *ca.* 8%F and (*C/S*) ratios of 0.8–1.0 with TiO_2 additions of 1–9% (but typically 6%). They used blast furnace slag as a source of TiO_2 (and CaO, SiO_2 and Al_2O_3). They studied the effects of various oxide additions on T_{liq}; $\eta_{1300\,°C}$ and q_{hor} (in a simulation test) as shown in Table 6.3. A F-free powder was developed (*C/S* = 0.95–1.15; 5–8%N; 4–8% B; 2%L 4%Mn and 4–7% T). Plant trials were carried out on the casting of thin slabs and regular slabs for plain C and low-alloy steels. It was found that

(i) The crystalline phase formed in the slag film was *perovskite* (CT) and $f_{crys} = 0.4$.
(ii) The F-free slag had similar viscosities, melting temps and heat fluxes to the original, F-containing slag.
(iii) The surface quality of plain C and low-alloy slabs obtained with the F-free slag was satisfactory with fewer cracks produced than with the conventional slag.
(iv) There was a marked increase in q_{hor} when $Na_2O > 8\%$ and $MnO > 4\%$ (which may be due to the formation of more glassy phase).

It was subsequently reported that Na_2O reduced the incubation time for the crystallisation of *perovskite* and that the kinetics of crystallisation were similar to those for industrial mould slags, containing F (Fig. 6.14). Furthermore, it was shown that TiO_2 reduced the activation energies for the transitions (glass → crystal) and (liquid → crystal) [93].

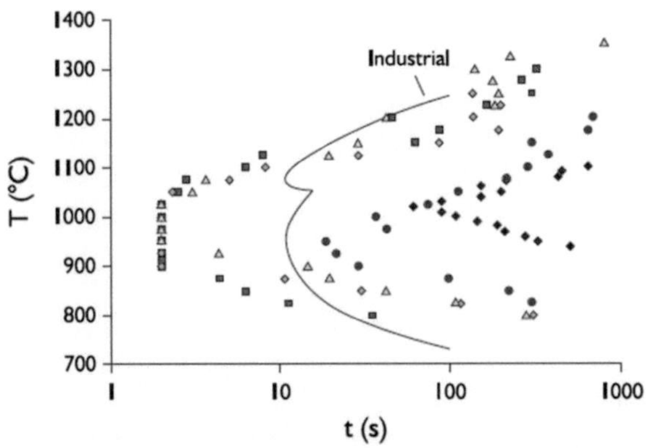

Fig. 6.14 Transformation-temperature-time (TTT) diagram for slags from the CaO–SiO_2–TiO_2–Na_2O system and for an industrial mould slag ♦ = CST; ● = CSNTA-3; ■ = CSTNA-1; Δ = CSTNA-4; ◊ = CSTNA-2; (permission granted, Wiley [85])

However, it was, subsequently, reported that the flux developed by Wen et al. [72] resulted in sticker break-outs when it was used on plant [15]. Thermodynamic calculations and laboratory experiments showed it was possible to form TiC or Ti (C/N) by the reaction of TiO_2 with carbon floating in the slag pool or in the slag bed (Eqs. 6.21 and 6.22)

$$TiO_2 + 3C = TiC + 2CO(g) \tag{6.21}$$

$$2TiO_2 + 4C + N_2(g) = 2TiN + 4CO(g) \tag{6.22}$$

Both TiN and TiC have low solubility in the molten slag and the undissolved particles agglomerate and block the infiltration of slag into the mould/strand gap (and also cause a marked increase in slag viscosity) [15]. These conditions lead to sticker break-outs [15].

Subsequently, Chen et al. [73] developed low F-fluxes, in which 7.5% F was reduced to 3%, by introducing 6%TiO_2 to form *perovskite*; no problems were reported when using these mould slags to cast peritectic, MC steel grades. However, *perovskite* has a high melting temperature which will restrict the liquid slag thickness and lubrication of the shell. These contradictions in casting behaviour when using mould slags with 6%TiO_2 need to be resolved.

6.2.5.3 NC_2S_3 ($Na_2O{\cdot}2CaO{\cdot}3SiO_2$)

The phase, NC_2S_3 was studied as a candidate to replace *cuspidine* as the crystalline phase formed in the slag film [94]. The phase diagram of the NCS system (shown in Fig. 6.15) contains both NC_2S_3 (T_{liq} = 1285 ± 5 °C) and $N_2C_2S_3$ (T_{liq} > 1400 °C).

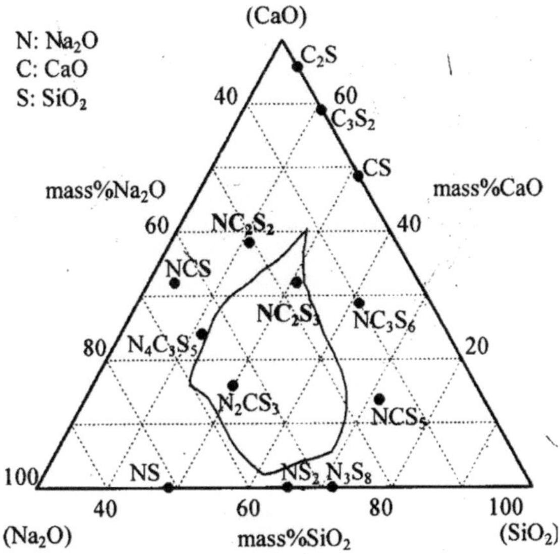

Fig. 6.15 Ternary phase diagram for the $Na_2O + CaO + SiO_2$ system; (permission granted, ISIJ, [94])

The effect of small additions of Li_2O, MgO, MnO and Al_2O_3 were investigated in order to reduce the melting temperature. The effects were quantified in terms of the composition length ($L = \{\Sigma(X_i - Y_i)^2\}^{0.5}$ where X_i and Y_i represent the compositions of the actual slag and NC_2S_3, respectively and i represents the individual constituents) [94]. A F-free flux ($C/S = 0.62$; $(Na_2O + Li_2O) = 22\%$; others 4%; $T_{sol} = 1100$ °C and $\eta_{1300\ °C} = 3.2$ dPas) was used to cast a HC (0.47%) steel slabs at $V_c = 0.7$ m min^{-1} [94]. The F-free flux performed as well as the original, F-containing powder. However, restriction of q_{hor} is not an issue when casting HC steels and its use when casting MC steels would be a more rigorous test of this type of mould slag.

6.2.5.4 Gehlenite (C_2AS) and Wollastonite (CS)

A fluoride-free flux (where 3% CaF_2 was replaced by 1.5%B_2O_3 + 2.5% Na_2O + 4%Al_2O_3 + 3%MgO) was used to cast HC billets [27]. The slag film contained a mixture of *gehlenite* (C_2AS) *pyrope* [M_3AS_3 and *wollastonite* (CS)] [27]. It performed well in plant trials and led to reduced SEN erosion rates.

6.2.5.5 Melilite

Melilite slags cover the range of solid solution between end-members, *akermanite* (C_2MS_2) and *gehlenite* (C_2AS). It can be seen from Fig. 6.16 that *melilite* slags tend to have higher viscosities than conventional mould slags (based on *cuspidine*). They have a reasonably high basicity and have a low oxygen potential.

Hanao et al. [20] pointed out that *melilite* slags retain some glass phase and thereby maintain some lubrication via the supercooled liquid formed above T_g (at *ca.* 600 °C). Since mould slags for casting billets frequently have high viscosities, *melilite* slags are suitable for casting billets and Hanao et al. [20] used these slags in the successful casting of round billets. *Melilite* slags have also been used in the casting of TRIP steels [20].

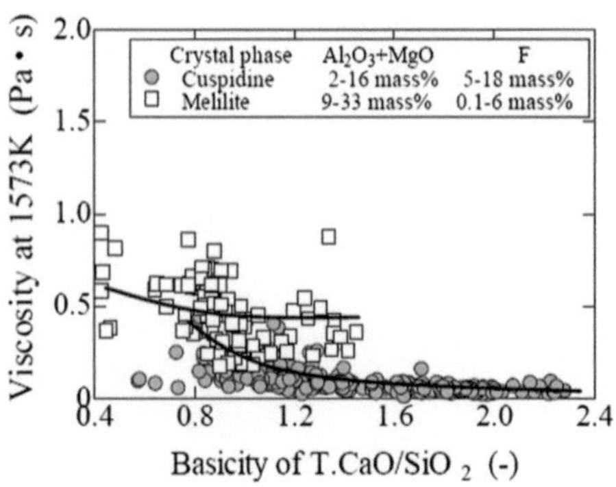

Fig. 6.16 Viscosity at 1300 °C as a function of basicity (*C/S*) ● = slags forming *cuspidine*; □ = slags forming *melilite*; (permission granted, ISIJ [20])

6.2.6 Reduced F-Powders

The first response by the powder manufacturers to the environmental and health and safety concerns posed by Fluorine emissions and the leaching of fluorides was to reduce the fluorine contents in mould powders. The advantage of this approach is that *cuspidine* is maintained as a crystalline phase, even if its concentration in the slag film is reduced [87]. Physical properties, simulation tests and plant trials were carried out on casting MC steels with a conventional powder (containing 7.5% F) and two candidate powders (containing 3.5% F and 6% TiO_2) with *C/S* ratios of 1.15 and 1.35 [73]. Some of the *cuspidine* in the slag film (formed with the conventional powder) is replaced by *perovskite* ($CaTiO_3$=CT) and Ca_2SiO_4 (C_2S) in the two candidate powders [73]. The plant trials showed that the candidate powder with *C/S* ratio = 1.35 performed as well as the conventional powder with similar heat fluxes and powder consumption but the slag film was thinner. No problems were reported with the formation of Ti(CN) and the sticker break-outs or false alarms.

6.2.7 C-Free Powders

The free carbon content of the casting powder is the main factor controlling the melting rate (Q_{MR} in kg min^{-1}) of the powder. The melting rate should be sufficient to match the required powder consumption. The carbon is non-wetting to the molten slag and prevents the slag globules from agglomerating (Fig. 6.17). However, the carbon is oxidised by any O_2(g) and CO_2(g) present in the bed. So the amount of carbon content decreases continually down the mould. Smaller particles of carbon provide more separation of the globules per unit mass of carbon. Thus the melting rate increases (Fig. 6.17) with (i) decreasing free carbon content (%C_{free})

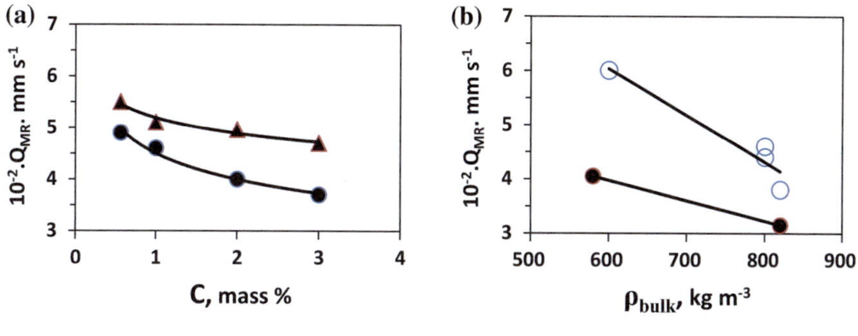

Fig. 6.17 Melting rate as functions of **a** Carbon content, ▲ = C black; ● = Coke; note $D_{C\ black} < D_{coke}$ so C black has lower melting rate and **b** bulk density ○ = 1% C; ● = 1.3% C; (permission granted, ISIJ, redrawn after [44])

Fig. 6.18 Photograph showing *Carbon smears* on the surface of thin slabs (courtesy of IMERYS Group [22])

(ii) with increasing particle size of the carbon (D_C) and (iii) with decreasing bulk density (ρ_{bulk}) of the powder as shown in Fig. 6.17b (i.e. $Q_{MR}\uparrow$ as $\%C_{free}\downarrow$ as $D_C\uparrow$ and as $\rho_{bulk}\downarrow$).

However, excess carbon tends to float near the surface of the slag pool (Fig. 6.1) and forms a layer of "*amorphous carbon*" [22]. Excess carbon can lead to

(i) Carbon pick-up by the steel, this is a particular problem when casting LC and ULC steels [95].
(ii) The formation of *Carbon smears* on the surface of thin slabs (Fig. 6.18 [22]) through contact between the shell and the amorphous carbon layer formed at the top of the slag pool.
(iii) The formation of depressions on the surface of slabs (see Sects. 11.8.1 and 11.8.2).
(iv) The formation of large slag rims or *ropes* which is promoted by both amorphous carbon and high Na_2O contents [22] and is prevalent when large amounts of Al_2O_3 are formed.

Carbon is not essential for the control of melting rate. Any material which (i) is non-wetting to molten slag and (ii) reacts with O_2 and CO_2 until it is consumed would be suitable for this task. Nitrides have been used for this purpose; BN has been used but is expensive and also poses a threat of Boron pick-up by the steel [23, 96]. However, other nitrides (e.g. Si_3N_4) have been used successfully to totally, or partially, replace the carbon in the casting powder [22–24, 96, 97].

6.2.8 *Powders for High-Speed Casting and Thin Slab Casting*

Thin slab- and high-speed casting (up to 8 m min^{-1}) pose difficult casting conditions, i.e.

(i) The high casting speed results in both a high heat flux and a short residence time in the mould; these conditions will tend to result, respectively, in an uneven shell in the meniscus region (prone to longitudinal cracking) and to a relatively thin shell, overall, at the mould exit (which make it prone to bulging and mould level variations).
(ii) The powder consumption tends to be low because of the high-casting speed and a mould powder with a low viscosity is needed to provide a reasonable Q_s value.
(iii) The high metal velocities can cause meniscus instability, slag entrapment and frequently EMBr is used to decelerate the flow velocity; EMBr tends to reduce the vertical heat flux which results in concomitant decrease in melting rate, slag pool depth and powder consumption.
(iv) In thin-slab casting there is little room to accommodate the SEN; consequently, misalignment of a high-conductivity, SEN can create "*cold spot*" in the mould resulting in variations in shell thickness (which can lead to longitudinal cracking).

Thus the principal problems are longitudinal cracking, poor powder consumption, slag entrapment and those issues associated with mould level variations. Several studies have been carried out to develop mould powders for these trying conditions [98–103].

The first thin-slab-casting slags tended to have high T_{br} values to control the horizontal heat flux; T_{br} values were *ca.* 100 °C higher than those calculated using either Eqs. 6.6–6.9 for the appropriate steel grade or derived from Fig. 6.6. However, more recent powders for thin-slab casting tend to have values for the required viscosity (η^{req}) and break *temperature* (T_{br}^{req}) that are consistent with values calculated from Eqs. 6.2–6.4 and 6.7–6.9, i.e. they follow the same empirical rules as those for conventional slab casting [104].

Specially designed moulds, e.g. "*football moulds*" are frequently used to accommodate the SEN. However, the sharp corners where the regular mould meets the football can also lead to cold spots and sequentially, to shells of variable thickness and to longitudinal cracking. Thus the tendency is to use a more crystalline slag with a high T_{br} to alleviate such problems. However, sharp corners should be avoided and the mould should be designed to reflect this.

The problems of low powder consumption values (about a half of the values of Q_s^{req} calculated with Eqs. 6.2–6.4) and low slag pool depths can be addressed by increasing the melting rate by reducing the % carbon and the carbon particle size. However, care must be taken to avoid both the % carbon and the carbon particle size being reduced simultaneously, since they have opposing effects on the melting rate. Precautions should be taken to avoid reducing the % carbon to <1% since the concomitant loss of energy from carbon consumption can lead to meniscus freezing (i.e. skull formation).

Bulging is a particular problem when casting MC-peritectic steels which are prone to longitudinal cracking where slags with high T_{br} and f_{crys} are used to reduce q_{hor}. This and the short residence time result in a thin shell at the mould exit which is prone to bulging and to mould level variations [105]. Mould level variations cause increases in (i) inclusions (ii) pinholes [106] (iii) large slag rims [104, 107] and an increased risk of sticker break-out [108]. In order to combat the problem of bulging, it has been proposed that the slag film should restrict heat transfer in the meniscus region and the upper mould but should promote heat transfer in the lower mould to create a thicker shell [104 p. 150]. Further work must be done if this target is to achieved.

6.2.9 *Powders for Casting Round Billets*

The major problem with casting round billets is that the shell shrinks during solidification and it is important that molten slag should fill any gap and help to support the billet. Slag films which crystallise develop "*surface roughness*" on the surface in contact with the mould and thus provide poor support for the billet. In order to provide the required support of the billet it is necessary to use a glassy slag film. The challenge is to provide both control of the horizontal heat flux whilst providing support to the billet. This was obtained in practice by using a powder which:

(i) Produced a glassy slag film (giving the necessary support).
(ii) Reducing the horizontal heat flux by creating a thick solid layer, d_s (via. a high solidification temperature) and by incorporating transition metal oxides (e.g. MnO) into the slag to absorb the IR radiation [40, 109, 110].

Subsequent work showed that slags forming *melilite* phase performed well [20].

6.2.10 *Powders for Casting Beam Blanks*

Longitudinal cracking is a problem in the casting of beam blanks (Fig. 6.19).

It was reported [111] that that the longitudinal cracking was worst

(i) At the regions denoted B and F in Fig. 6.19 where the horizontal heat flux is at its highest.
(ii) For mould slags with low solidification temperatures (T_{sol}).
(iii) For steels with higher S content.

Powders for casting beam blanks should provide a thick, crystalline slag film to minimise any temperature variations (i.e. shell thickness variations) which lead to longitudinal cracking.

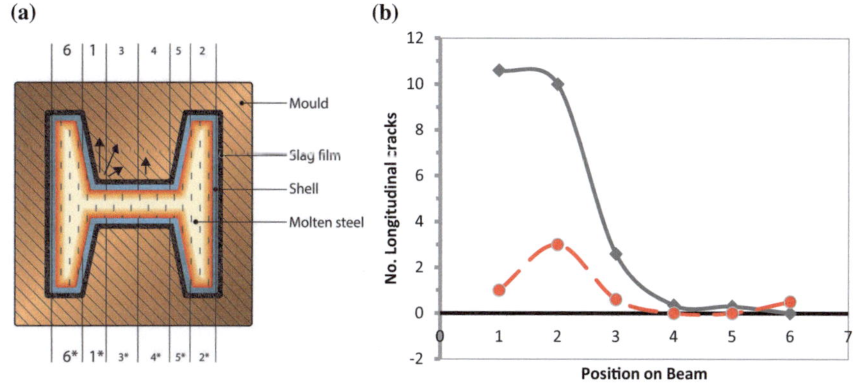

Fig. 6.19 **a** Schematic diagram showing the mould for casting H beams showing mould (*brown*) molten steel (*yellow*) solidified steel shell (*blue*) and slag film (*black*) and *black arrows* show heat flow paths (courtesy of Daniel Guerra). **b** Longitudinal cracking at different positions in H beam; –◆–sites 1, 2 etc. ● - -, 1*, 2* etc. (drawn from data reported by [111]); Note longitudinal cracking is prevalent at internal corners where heat flux density is high

6.2.11 Non-Newtonian Powders

The viscosities of most liquid slags are considered to be Newtonian since their viscosity is unaffected by the shear rate. In non-Newtonian melts the viscosity decreases with increasing shear rate. Non-Newtonian slags have been produced to reduce slag entrapment which occurs in the region of low stress near the slag/metal interface. The area of lubrication (the region where liquid slag infiltrates into the mould/shell gap) is considered to be a region of high stress. Thus a non-Newtonian slag will exhibit a high viscosity in the slag entrapment region and a lower viscosity in the lubrication region which will increase powder consumption.

Non-Newtonian slags were produced by adding Si_3N_4 to a conventional slag (with $C/S = 0.9$) at 1450 °C and holding it at this temperature for 1 h; the N content of the final slag was 0.2% [112, 113]. The viscosities of this flux are shown in Fig. 6.20a and it can be seen that the viscosity was 7 dPas at low shear rates but fell to 5 dPas at high shear rates. The viscosity of the conventional slag was 4.8 dPas for all shear rates. The N enters the silicate structure as a bonding N in the silicate tetrahedron and is also bonded to a Ca^{2+} ion which is also connected to an O in another tetrahedron (Fig. 6.20c); the N–Ca bond is weak and is severed as the shear rate is increased.

It was found that the non-Newtonian slag exhibited a lower contact angle on a plate of pure Fe than the conventional slag (Fig. 6.20b) and that the heat flux was greater for the non-Newtonian slag than for the conventional slag ($q_{hor}^{conv} < q_{hor}^{non-N}$) [112, 113].

Trials were carried out on a miniature continuous caster and it was found that the powder consumption (Q_s) was higher for the non-Newtonian flux than that for a Newtonian slag (η = 3.5 dPas) (Q_s = 0.5 and 0.35 kgm^{-2}, respectively) [112, 113].

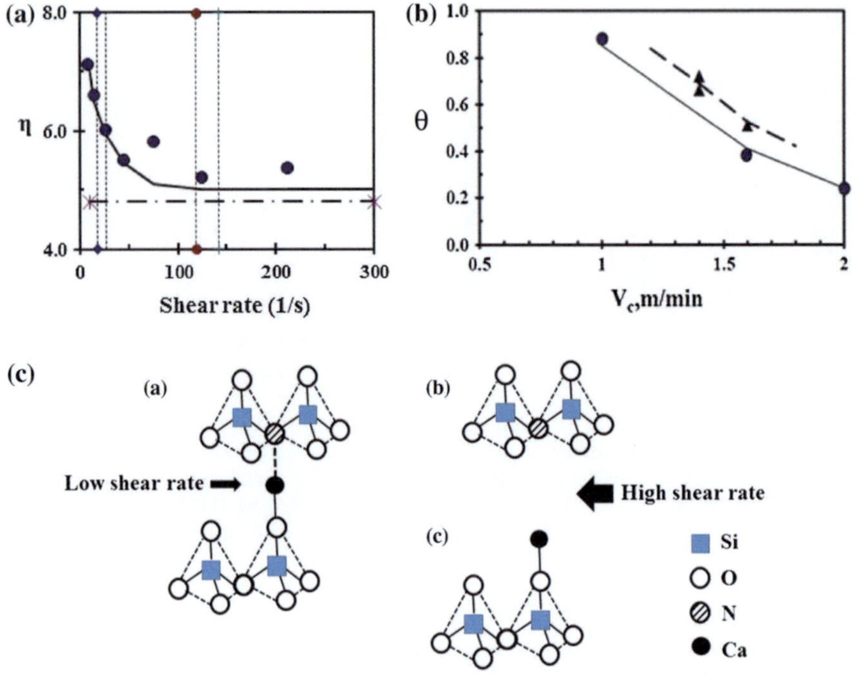

Fig. 6.20 Schematic diagrams showing **a** the viscosity of slag (at 1300 °C) as a function of shear rate o = Newtonian and ● = non-Newtonian behaviour **b** Contact angle as a function of temperature and **c** shows the change to the structure with increasing shear rate; (permission granted, ISIJ [113] re-drawn)

6.2.12 *Powders for Casting TRIP and TWIP Steels*

TRIP steels have both good ductility and high strength; they contain (i) retained *austenite* with good ductility and (ii) *martensite* formed during plastic deformation (of retained *austenite)* which provides the strength. TRIP stands for Transformation-induced plasticity. These steels tend to contain higher levels of C and Si and additions of Al (typically *ca.* 1%). The Al enhances the C content in the austenite and thereby promotes the stability of *austenite*; a further advantage is that it reduces the Si needed in the steel. Normally, steels with >0.1% Si cannot be galvannealed but the substitution of Si by 0.5–1%Al allows the steel to be galvannealed. A typical TRIP steel contains 0.14%C; 1.8%Mn; 0.6%Si; 1%Al.

TWIP steel stands for Twinning-induced plasticity steel. These steels have high Mn contents and the Al increases the stacking fault energy and stabilises the *austenite.*

6.2.12.1 Problems with Casting TRIP and TWIP Steels

TRIP and TWIP steels are difficult to cast because of the large amounts of Al_2O_3 formed by the reaction of $\underline{Al}$ in steel (denoted by underline) with SiO_2 in the slag.

$$2\underline{Al} + 3SiO_{2slag} = 3\underline{Si} + Al_2O_{3slag} \tag{6.23}$$

In fact, the reaction of the $\underline{Al}$ in the steel is not confined to SiO_2 since similar reactions will occur with any slag constituents above Al in the Ellingham diagram, e.g. TiO_2, MnO, etc.

Figure 6.21 shows the pick-up of Al_2O_3 when casting normal steels is frequently around 4% with *ca* 2% coming from steelmaking reactions and *ca.* 2% from the reaction, as shown in Eq. 6.23. [18]. When casting steels containing (0.4%Al) and (0.6–1%Al), the Al_2O_3 pick–up values increase to (14–18%) and (17–25%), respectively. Casting steels with Al contents >0.6% is difficult and frequently casting is limited to one ladle of steel. These large increases in the Al_2O_3 content result in significant increases in T_{liq}, T_{sol}, f_{crys} and viscosity with consequent, decreases in the powder consumption ($Q_{s\ req}^{slag} = 0.55/\eta^{0.5}\cdot V_c$ or $Q_{s\ req}{}^{powd} = 0.6/\eta\cdot V_c$) and the lubrication supplied. The changes in T_{br} and the crystal fraction in the slag film affect the horizontal heat transfer. Given these significant changes, it is not surprising that a high pick-up of Al_2O_3 by the slag is reported to result, sequentially, in erratic melting, fluctuating mould temperatures and false alarms on the sticker detection system [20, 114, 115]. It has been suggested that the Al_2O_3 formed tends

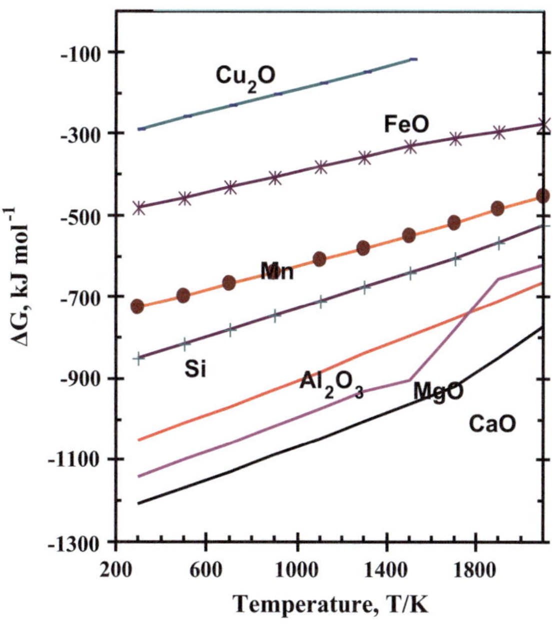

Fig. 6.21 Ellingham diagram showing Gibbs free energy of various oxides per mole of O_2 as a function of temperature (T_K) (Note the temperatures are in K.)

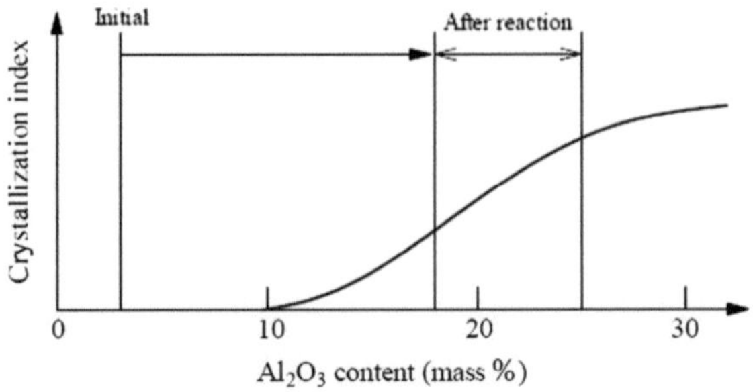

Fig. 6.22 Crystallisation index as a function of Al_2O_3 content of mould slag (permission granted, ISIJ [20])

to promote the formation of more crystalline phase at the expense of the glassy phase (Fig. 6.22) [20] resulting in a loss of lubrication [20, 115].

The large amounts of Al_2O_3 generated in casting high-Al steels are not easily dissolved in the slag pool and the particles tend to agglomerate and hinder the infiltration of molten slag. Furthermore, these agglomerated particles can be captured by the shell (forming depressions or grooves in the steel). Furthermore, the creation of large amounts of Al_2O_3 tend to result in large slag rims which have a tendency to fracture, which, in some cases, block slag infiltration and thereby, cause sticker break-outs.

6.2.12.2 Strategies to Minimise Al_2O_3 Pick-up

Several strategies have been adopted in designing powders for casting TRIP steels, namely

(i) To select a slag composition which remains in a low-melting region even with significant levels of Al_2O_3 pick-up.
(ii) To increase the FeO and MnO contents in order to minimise the pick-up of $\underline{Si}$ by the metal since the reaction (3 $MO_{slag} + 2\underline{Al} = 3\underline{M} + Al_2O_{3slag}$ where M=Fe or Mn) will occur in preference to Eq. 6.23.
(iii) To increase the volume of the slag pool to improve the dissolution of Al_2O_3 (this is usually carried out by including exothermic agents in the powder but the use of a thicker powder bed would also be beneficial).
(iv) To increase the driving force for Al_2O_3 dissolution (i.e. $C_{sat}–C_o$) by keeping the Al_2O_3 content in the powder (C_o) to low levels in slags based on calcium silicates.
(v) To minimise the formation of Al_2O_3 by keeping the SiO_2 content of the powder at a low level.

It is not possible to use all of these strategies simultaneously but some of them are incorporated into each of the approaches described below.

6.2.12.3 Different Approaches to Developing Powders for TRIP Steels

There have been two approaches taken in the development of powders for casting high-Al steels, namely

(i) To modify conventional casting powders.
(ii) To develop new powders based on calcium aluminates in order to minimise both the SiO_2 content of the flux and amount of Al_2O_3 formed by Eq. 6.23.

Conventional powders

Steels containing less than 0.6%Al can be cast using conventional powders forming slag films based on *cuspidine*. The slag pool is capable of dissolving the Al_2O_3 formed. However, difficulties are experienced when casting steels with higher Al contents.

The composition range of conventional "*cuspidine-forming* slags" is shown in Fig. 6.23 and the changes in composition (i.e. the gain in Al_2O_3 and losses in SiO_2, FeO and MnO) are shown by the direction of the arrow. The Al pick-up by the slag is cited as % Al_2O_3 on the basis of chemical analysis but some of the Al could be present as AlN [116].

Low-basicity powders [116–121]

At first sight, it would seem illogical to use low-basicity powders because this entails high SiO_2 contents which would promote the reaction $\underline{Al}$ and SiO_2 shown in Eq 6.23. However, careful examination of Fig. 6.24 indicates that the slag remains longer in the molten range when using a slag with *C/S* ratio of *ca.* 0.6.

Plant trials performed on a TRIP (1%Al) steel with a low-basicity (*C/S* = 0.55; N10%; *F10%*; Mn 6.5%) powder and the results were compared with those obtained when casting LC steels; it was found that [117]:

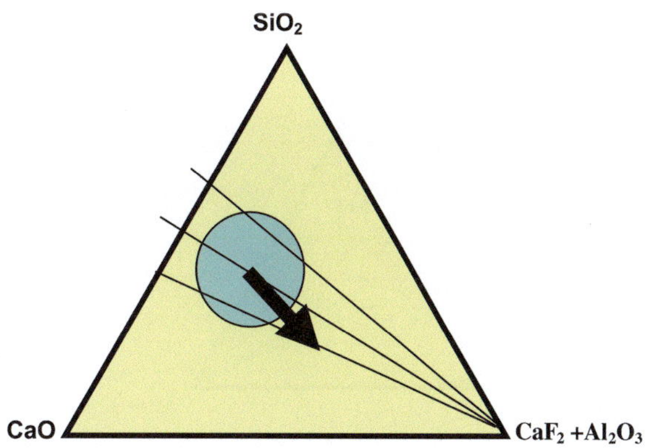

Fig. 6.23 Ternary diagram showing the approximate compositional bounds for "*cuspidine*-forming" powders; CaO = % CaO + %MgO + %Na_2O; ● = range of *cuspidine*-forming fluxes; *lines* represent *C/S* ratios, from top, of 0.667; 1.0 and 1.5; *arrow* shows the change in composition through reaction of $\underline{Al}$ and oxides

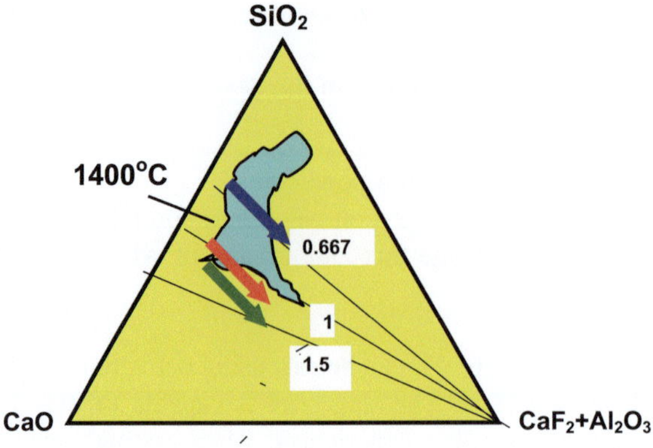

Fig. 6.24 Ternary diagram CaO + SiO_2 + (CaF_2 + Al_2O_3) where CaO = %CaO + %MgO + % Na_2O; the thin lines represent (*C/S*) ratios of 0.667, 1 and 1.5; the irregular shape is the low-melting region and the arrows show the change in composition for (*C/S*) ratios of 0.6 (*blue*) 1.0 (*red*) and 1.3 (*green*)

(i) The Al_2O_3 pick-up was 25% with losses of 16% SiO_2 and 5% MnO in the slag.
(ii) Different crystalline phases were formed in the TRIP steel casting (CaF_2 + $NaCaAlSi_2O_7$ + Al_2O_3) *cf. cuspidine* in LC steel casting.
(iii) There was little change in the horizontal heat flux.

High basicity powders

High-basicity slags would appear to be a more logical selection for casting steels containing Al because they contain (i) high concentrations of CaO to react with the Al_2O_3 formed and (ii) lower concentrations of SiO_2 to react with the $\underline{Al}$ in the metal. However, it can be seen from Fig. 6.25 that the Al_2O_3 generated will cause

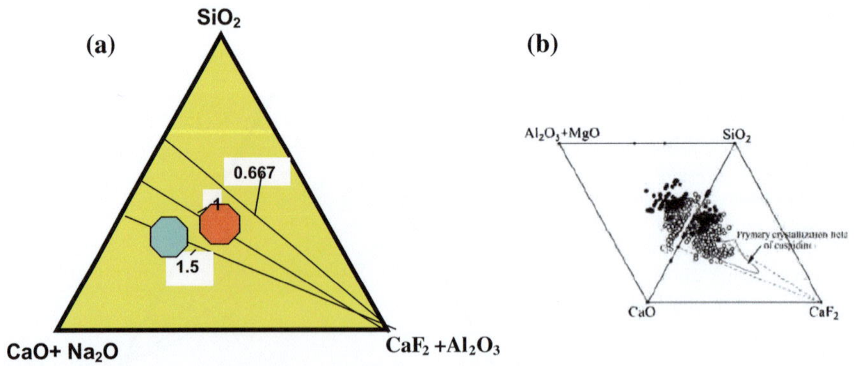

Fig. 6.25 Comparison of *cuspidine* and *mellite*—forming slags in **a** Ternary diagram showing positions of *cuspidine*—forming slags (●) and *melilite*-forming slags (●); *lines* represent different *C/S* ratios and **b** as a quaternary diagram *melilite* on *left*; *cuspidine* on *right*; (permission granted for **b** ISIJ [20])

the melting temperature of the slag to increase, so high concentrations of fluxes will be needed to lower the melting temperature.

Melilite—forming powders are high-basicity fluxes [20, 120]. *Melilite* consists of a solid solution ranging between end-members *akermanite* ($C_2M\ S_2$) and *gehlenite* (C_2AS). It can be seen that *gehlenite* contains more Al_2O_3 than *akermanite*, so these slags offer a way of accommodating the Al_2O_3 into a solid solution. *Melilite* has a melting temperature of 1280 °C which can be reduced by additions of fluxes. Hanao et al. [20] pointed out that *melilite* slags retain some glassy phase despite the accommodation of Al_2O_3 and this provides some measure of lubrication. Plant trials were carried out using a *melilite* slag when casting steels containing Al and gave no problems with either sticker break-outs or false sticker-alarms and provided improved surface quality [20]. Two examples of *melilite*-type mould slag have (*C/S*) ratios of 1.3 and 1.8 with additions of 8%Na_2O and MnO and F 8–13%.

Plant trials carried out on low-basicity (*C/S* = 0.55; N = 11%; L = 2%; Mn = 6%; F = 10%) powders resulted in increases in Al_2O_3 of 5–20% and SiO_2 losses of 10% [115]. It was found that the surface of the cast product contained [115]

(i) No "*drag marks*" indicating that there was sufficient lubrication provided by the slag.
(ii) Many horizontal and vertical depressions which contained open cracks (due to high heat flux in the meniscus region).

Powders based on calcium aluminates [114, 115, 122–124]

Most mould powders are based on the CaO–SiO_2 system with added fluxes. An alternative approach is to base the mould powders on the CaO–Al_2O_3 system with added fluxes. The CaO–Al_2O_3 (CA) system contains a eutectic at a composition corresponding to $C_{12}A_7$ (*ca.* 50% CaO) with T_{liq} = 1413 °C to which fluxes (e.g. Na_2O, Li_2O, CaF_2) are added to reduce T_{liq} further. The replacement of SiO_2 by Al_2O_3 results in a large reduction in the Al_2O_3 formed. In practice, these CA slags tend to contain 5–10% SiO_2 so some Al_2O_3 is still formed by reactions of $\underline{\text{Al}}$ with SiO_2, FeO, MnO and B_2O_3 but it is at a manageable level.

It has been reported that Al_2O_3 formed during the cast tends to get incorporated into the AlO_4 networks rather than the SiO_4 networks via Eq. 6.24 [123].

$$\{\text{Si}-\text{O}-\text{Si}\}+\{\text{Al}-\text{NBO}\}=\{\text{Al}-\text{O}-\text{Si}\}+\{\text{Si}-\text{NBO}\} \tag{6.24}$$

With regard to the fluxes, CaF_2 additions reduce the viscosity but do not break any Al–O bonds, so this viscosity decrease has been attributed to the liberation of silicate units (Q^0_{Si}) from the aluminate network [123]. It was suggested that CaF_2 additions had little effect on the overall polymerisation (i.e. the distribution of Q^n species) or on the number of BO_{Al} units but CaF_2 does tend to increase the number of depolymerised (Q^0_{Al}) units [123]. Additions of TiO_2 reduce the viscosity and have been reported to reduce the degree of polymerisation in the AlO_4 network [125, 126].

The evolution of crystal phases formed in these CA-type slags has been investigated [127, 128] it was found [127] that the size of crystals increased with increasing basicity, (*C/S*) and that B_2O_3 additions reduced crystallisation (it tends to promote glass formation).

The principal problems in casting these steels are (i) a loss of glassy phase leading to lubrication problems (ii) erratic melting with its knock-on effects of fluctuating mould temperatures and false alarms on the sticker detection system and (iii) the formation of large slag rims which are fragile and which can fracture and cause sticker break-outs.

Plant trials carried out on several candidate mould powders (in %; 30C;26A; 2–10S;10 N; 10–15*Fl*) indicated that the slab surface contained [115];

(i) Very few longitudinal and transverse depressions and surface cracks (*cf.* low-basicity powders) due to the lower horizontal heat flux.
(ii) *Drag marks* due to inadequate lubrication.
(iii) Less distinct oscillation marks (*cf.* those obtained with a low-basicity powder).

Consequently three further candidate compositions were developed [115] [(in %; 38–42C;12A; 9–12S;9–9N; 5–6L;9F; 10–16B) by optimising the following (i) the CaO/Al_2O_3 ratio (ii) the Na_2O substituted by Li_2O and (iii) the $\%B_2O_3$ (which will be reduced by $\underline{Al}$). The (crystal/glass) ratio and f_{crys} are reduced with increasing B_2O_3 since it promotes glass formation (and lubrication) and Na_2O and Li_2O which retard crystallisation; these oxides therefore affect the horizontal heat flux through their influence on crystallisation. It was found from the subsequent plant trials that [115]:

(i) The reduction in $\%SiO_2$ was <3% but $\%Al_2O_3$ was greater than expected due to the reduction of B_2O_3.
(ii) Drag marks were obtained when using slags providing insufficient glass (and lubrication) and which led to slag film breakage in the lower mould.
(iii) Depressions and cracks occurred in slags containing too much glass and too few crystals to reduce the heat flux to an acceptable level.

Overall, the surface quality of TRIP steel slabs was better when using calcium aluminate slags *cf.* high basicity calcium silicate slags [115] but further work is required to optimise the (C/A) ratio and the levels of B_2O_3, Na_2O and Li_2O to provide sufficient glass for good lubrication and sufficient crystalline content to reduce the heat flux.

Plant trials were also carried out on a calcium aluminate mould powder [(C/A) = 1.5; in mass % 10S; 13.5N; 6M; 2L; 3Mn; 14 B] and it was found [124] that

(i) "*Flaring*" occurred in the mould which was attributed to hydration of the powder in the spray drying process ($CA + xH_2O \rightarrow CAxH_2O$); it was found that baking at 250 °C before use, eliminated this problem.

(ii) Large slag rims were formed and the properties differed from the expected values, e.g. $\eta_{1300°C}$ = 3.5 dPas and d_s = 3.6 mm).
(iii) Most of the B_2O_3 reacted with the $\underline{Al}$ in the metal ($B_2O_{3\ sl}$ + 3$\underline{Al}$ = $Al_2O_{3\ sl}$ +3$\underline{B}$) with B_2O_3 in the slag decreasing from 14% to <4%.

Although the use of calcium aluminate slags has produced some encouraging results, there is still room for further improvement in the selection of the fluxing agents.

In summary, CA-type slags have proved to be reasonably effective but some development of the mould powders is still needed to address the poor lubrication provided by existing powders.

6.2.13 Powders for Casting Stainless Steels

The principal difficulties in casting stainless steels arise because of

(i) The low liquidus temperature of *austenitic* stainless steels which results, sequentially, in lower values for the vertical heat flux (q_{vert}) the melting rate, pool depth (d_{pool}) and lower powder consumption (Q_s).
(ii) Reactions occur between Ti in Ti-stabilised steels and the more reducible oxides (e.g. SiO_2) in the slag (Eqs. 6.24 and 6.25).
(iii) High melting oxides (e.g. $CaTiO_3$) can be formed by the reaction of Ti in the steel with high basicity mould slags and it is necessary to keep the basicity below 0.9 to avoid the formation of $CaTiO_3$ [42].
(iv) TiN (or Ti(CN)) can be formed in the mould [13–15], in these Ti-stabilised steels, and since TiN has a very low solubility (<0.5%) in mould slag, particles do not dissolve significantly, so that TiN particles tend to agglomerate (cluster) over time to form "*lumps*" and "*clogs*" [14, 129]. These agglomerates hinder the infiltration of liquid slag into the shell/mould channel and thus, reduce powder consumption; in extreme conditions, they can cause sticker break-outs.

$$\underline{Ti} + SiO_{2sl} = \underline{Si} + TiO_{2sl} \tag{6.25}$$

$$2TiO_2 + 4C + N_2(g) = 2TiN + 4CO(g) \tag{6.26}$$

In mould slags with high basicity, the above reactions can result in the formation of *skin laminations* (of $CaTiO_3$ or TiN see Fig. 11–83b) which have high melting points [42]. The replacement of SiO_2 by TiO_2 (up to 10%) was found to reduce the viscosity but increase T_{br} and T_{liq} and the $CaTiO_3$ formed tended to suppress the formation of *cuspidine* [130]. Solid particles also increase the slag viscosity.

6.2.14 Powders for Casting Steels with Rare Earths

Some pig iron emanating from China contains low concentrations of rare earths (RE's, e.g. La, Ce, etc.). It has been mooted to incorporate these low concentrations of rare earths into steels in order to de-oxidise and desulphurise the metal. The presence of RE's in the steel would have a similar effect to that of $\underline{Al}$, described in Sect. 6.2.12 above.

$$2\underline{RE} + 1.5\,SiO_2 = RE_2O_3 + 1.5\,\underline{Si} \tag{6.27}$$

Similar reactions could be expected with FeO and MnO leading to a build-up in RE_2O_3 in the mould slag and $\underline{Si}$ in the steel (Eq. 6.28 where M = Mn or Fe)

$$2\underline{RE} + 3MO = RE_2O_3 + 3\underline{M} \tag{6.28}$$

Additions of RE_2O_3 to $MgO–SiO_2$ slags resulted in (i) a decrease in viscosity and (ii) an increase in surface tension [131] because γ_{SiO_2} is low *cf.* $\gamma_{RE_2O_3}$. Additions of Re_2O_3 to mould slags with basicities (*C/S*) of 0.5, 0.9 and 1.2 were found to reduce the viscosity but increase T_{liq} and T_{br} [132]. This suggests that RE_2O_3 acts as a network breaker in the silicate network. Thus the build-up of RE_2O_3 in casting slags should not prove a problem with regard to the viscosity and lubrication. ($Q_s = 0.6/\eta V_c$).

6.3 Summary

The various types of mould powders used in continuous casting have been described from the viewpoint of the functions to be carried out and the underlying principles in their design and development. The underlying principles for the selection of chemical composition and properties have been established for most casting powders. There are two types of mould powders where development is still active, namely, (i) F-free and F-reduced powders and (ii) powders for casting TRIP and TWIP steel grades.

For F-free fluxes the primary difficulty encountered lies in replacing *cuspidine* in the slag film. The two principal candidates are *perovskite* ($CaTiO_3$) and *melilite* and there have been claims that *perovskite* forms Ti(C/N) which leads to sticker break-outs and counterclaims that it does not result in sticker break-outs. This must be sorted out. Further work is also needed on the suitability of *melilite*-forming slags.

There are three types of powder currently used for casting steels with high-Al contents, namely (i) high basicity calcium-silicates slags (ii) low-basicity calcium–silicate slags and (iii) calcium aluminates slags. At the present time it seems that there are fewer casting problems when using calcium aluminate slags but further work is still required to determine the optimum mixture of fluxes in these powders.

References

1. KC Mills, *A review of ECSC –funded research*. EUR 13177, (1990) (Europ. Comm, Sci. Tech. Publ., Luxembourg, 1991).
2. R Shah, JG Williams, G Hecko, Proc. ISS Tech., 2003, Conf., (ISS, Warrendale, PA, 2003) p. 555.
3. T Sakuraya, T Emi. T Imai, K Emoto, M Kodama, Tetsu-to- Hagane, **67**, 1220, (1981).
4. P Andersson. Ironmaking and Steelmaking, **42**, 465, (2015).
5. P Andersson, *Thermal conductivity of powders used in continuous casting of slabs and billets*. Ironmaking Steelmaking in press
6. S Diehl, JA Moore, RJ Phillips, Proc. 78th Steelmaking Conf., Nashville, TN, 1995, (ISS, Warrendale, PA, 1995) p. 351.
7. R Saraswat, PhD Thesis, *"Modelling the effect of mould flux on steel shell formation during continuous casting*"," Imperial College, London, (2006).
8. KC Mills, NPL Report DMA (D) 405, *Fundamental Study of the behaviour of casting powders, Final Report EUR 9560*, ECSC contract 7210 CA/131/ 311/ 810 (Europ. Sci. Tech. Publ., Luxembourg, 1985).
9. PE Ramirez-Lopez, KC Mills, PD Lee, B Santillana, Met. Mater. Trans. B, **43B**, (1), 109, (2012).
10. M. Wolf, Proc. METEC Congress'94, 2nd Europ. Conf. Continuous Casting, Dusseldorf, 1994, (VDEh, Dusseldorf, 1994) vol. 1, p. 78.
11. S Ogibayashi, K Yamaguchi, T Mukai, T Takahashi, Y Mimura, K Koyama. Y Nagano, T Nagano, Nippon Steel Technical Report, **34**, 1, (1987).
12. AB Fox, PhD Thesis, *Mould fluxes their properties and performance.* Dept. of Materials, Imperial College, London, (2003).
13. T Mukongo, C Pistorius, A Garbers-Craig, Ironmaking Steelmaking, **31**, 135, (2004).
14. RC Nunnington, N Sutcliffe, Proc. 59th Electric Furnace Conf., Pheonix, AZ, (2001), p. 361.
15. Q Wang, YJ Lu, SP He, KC Mills, Z Li, Ironmaking and Steelmaking, **38**, 297, (2011).
16. S Hiraki, K Nakajima, T Murakami, T Kanazawa, Proc. 77th Steelmaking Conf., Chicago, 1994 (ISS, Warrendale, PA, 1994) p. 397.
17. KC Mills, AB Fox, ISIJ Intl., **43**, 1479, (2003).
18. G Skoczylas, Proc. 79th Steelmaking Conf., 1996, (ISS, Warrendale, PA, 1996) p. 269.
19. F Neumann, J Neal, MA Pedroza, AH Castiliejos E, FA Acosta G, Proc. 79th Steelmaking Conf., 1996, (ISS, Warrendale, PA, 1996) p. 249.
20. M Hanao, Y Tsukaguchi, M Kawamoto, Proc. 4th Intl Cong. Sci. Technol, Gifu, Japan, 2008, p. 694.
21. H Nakato, S Takeuchi, T Fujii, T Nozaki, N Washio, Proc. 74th Steelmaking Conf., 1991, (ISS, Warrendale, PA, 1991), p. 639.
22. J Macho, G Hecko, B Golomowski, M Frazee, McMaster Iron & Steelmaking Symp. "*Thinner slab casting*", McMaster University, Hamilton, Ont. (2005) p. 131.
23. H Takeuchi, H Mori, T Nishida, T Yanai, K Mukunashi, Trans. ISIJ, **19**, 274, (1979).
24. C Lefebvre, JP Radot, JN Pontoire, Y Roux, Revue de Metallurgie, CIT, **94**, 489, (1997).
25. Y Nakamura,T Ando, K Kurata, M Ikeda, Trans. ISIJ, **26**, 1052, (1986).
26. N Tsukamoto, Taikabutsu (Refractories), **44** (5), 270, (1992).
27. MC Bezerra, CA Valadares, IP Rocha, JP Bulota, MC Carboni, IL Scripnic, CR Santos, K Mills, D Lever, Proc. 37th Steelmaking Seminar, Porto Allegre, RS- Brazil, 2005, (ABM, Sao Paulo, 2005) p. 190. http//www.carboox.com/pdf/fluxante_2007.pdf.
28. MM Wolf, Ironmaker & Steelmaker, **2000**, (Feb), 65, (2000).
29. MM Wolf, Ironmaker & Steelmaker, **2000**, (Mar), 69, (2000).
30. MM Wolf, Ironmaker & Steelmaker, **2000**, (Apr), 58, (2000).
31. MM Wolf, Ironmaker & Steelmaker, **2000**, (May), 78, (2000).
32. MM Wolf, Ironmaker & Steelmaker, **2000**, (June), 22, (2000).
33. MM Wolf, Ironmaker & Steelmaker, **2000**, (July), 63, (2000).

34. MM Wolf, Ironmaker & Steelmaker, **2000**, (Aug), 75, (2000).
35. MM Wolf, Ironmaker & Steelmaker, **2000**, (Sept.), 90, (2000).
36. MM Wolf, Ironmaker & Steelmaker, **2000**, (Oct), 114, (2000).
37. MM Wolf, Ironmaker & Steelmaker, **2000**, (Nov), 67, (2000).
38. MM Wolf, Ironmaker & Steelmaker, **2000**, (Dec), 45, (2000).
39. MM Wolf, Proc. 81st Steelmaking Conf., 1998, (ISS, Warrendale, PA, 1998) p. 53
40. M. Kawamoto: Paper presented at Intl. Workshop on "*Thermophysical data for the development of mathematical models of solidification*", Gifu City, Japan, Oct. (1995).
41. R Carli, V Ghilardi, Iron and Steelmaker**,** **1998**, (June), 43, (1998).
42. T Kishi, H Takeuchi, M Yamamiya, H Tsuboi, T Nakano, T Ando, Nippon Steel Tech. Report, **34**, 11, (1987).
43. JY Park, SM Jung, I Sohn, Met. Mater. Trans., B, **45B**, 329, (2014).
44. M Kawamoto, K Nakajima, T Kanazawa, K Nakai. ISIJ Intl., **34**, 593, (1994).
45. K Schwerdtfeger, A Jablonka, Steel Research Intl., **64**, 77, (1993).
46. SY Kim, J Choi, Proc. 4th Intl. Conf. Continuous Casting, Birmingham, 2002, (IOM, London, 2002) p. 594.
47. JN Pontoire, JP Radot, V Delvaux, E Dehaussy, Revue de Metallurgie, CIT. **93**, 1237, (1996).
48. Y Ueshima, T Mizoguchi, T Kajitani, Proc. 9th Intl. Conf. Molten slags, fluxes and salts (Molten 12), Beijing, 2012, (Chinese Metals Soc., 2012)
49. H Yamamura, T Kajitani, J Nakashima, M Yamasaki, S Mineta, Nippon Steel Technical Report, **104**, 54, (2013).
50. Q Liu, GH Wen, JZ Li, XJ Fu, P Tang, W Li, Ironmaking and Steelmaking, **41**, 292, (2014).
51. T Watanabe, H Fukuyama, M Susa, K Nagata, Met. Mater. Trans. B, **31B**, 1273 (2000) see also ISIJ Intl., **42**, 489, (2002).
52. M Hanao, M Kawamoto, M Hara, T Murakami, H Kikuchi, Tetsu-to Hagane, **88** (1), 23, (2002).
53. H Nakada, K Nagata, ISIJ Intl., **46**, 441, (2006).
54. M Hayashi, T Watanabe, H Nakada, K Nagata, ISIJ Intl., **46**, 1805, (2006).
55. P Grieveson, S Bagha, N Machingawuta, K Liddell, KC Mills, Ironmaking and Steelmaking, **15**, 181, (1988).
56. JH Park, JS Park, JY Park, Steel Times Intl., **1998**, (Sept), CC14 (1998).
57. PP Sahoo, A Dey, Proc. 9th Intl. Conf. Molten slags, fluxes and salts, Beijing, 2012, (Chinese Metals Soc. 2012).
58. SH Chang, IJ Lee, MR Kim, SM Yang, J Choi, JS Park, Proc. Conf. on Continuous Casting in developing countries, Beijing, 1993, p. 832.
59. K Tsukaguchi, S Ura, A Shiraishi, Y Hitomi, T Nagahata, Steel Technol. Intl., **1995/6**, 175, (1995/6).
60. LE From: *"Optimisation of mould powder performance in casting long products- Tech Rept No. 6"*. MEFOS Report: MEF99024 K: P3568: Konto 130410 (1999).
61. JW Cho, JK Park, KH Moon, SH Lee, KH Kim, HS Jeong, " *Characteristics of molten mold flux feeding technology.* Proc. AISTech. 2007, Indianapolis, IN, 2007 (ISS Warrendale, PA, 2007).
62. JK Park, JW Cho, KH Moon, SH Lee, KH Kim, HS Jeong, "*Study on the initial solidification behaviour under the new process of molten mold flux feeding technology in the continuous casting mould.*" Proc. 7th Intl. Conf on Clean Steel, Balatonfured, Hungary, 2007, (Hung. Min. Metall. Soc., Budapest, 2007), p. 264.
63. AI Zaitsev, AV Leites, AD Litvina, B Mogutov, Steel Research, **65**, 368, (1994).
64. F Shimizu, F Tokunaga, N Saito, K Nakashima, ISIJ Intl., **46**, 385, (2006).
65. K Shimizu, AW Cramb, Iron and Steel Maker, **2002**, (6), 43, (2002).
66. H Heimbach, K Schulz, J Markardt, HJ Ehrenberg, Stahl u Eisen, **117**, 105, (1997).
67. DE Sturgill, Proc. 79th Steelmaking Conf., Pittsburgh, 1996, (ISS, Warrendale, PA, 1996) p. 265.

68. RE Hargrave, DW Reichgott, Proc.78th Steelmaking Conf., 1995, (ISS, Warrendale, PA, 1996) p. 385.
69. H Abratis, F Hofer, M Junemann, H Sardmann, H Stoffel, Stahl u Eisen, **116**, 85, (1996).
70. AB Fox, KC Mills, D Lever, C Bezerra, C Valadares, I Unamono, J Laraudogoitia, J Gisby, ISIJ Intl., **45**, 1051, (2005).
71. C Zhang, DX Cai, Proc. 8th Europ, Conf. Continuous Casting, Graz, 2014, (Austrian Metals Soc. 2014).
72. GH Wen, S. Sridhar, P Tang, X Qi, YQ Liu, ISIJ Intl., **47**, 1117, (2007).
73. LY Chen, GH Wen, CL Yang, F Mei, CY Shi, P Tang, Ironmaking and Steelmaking, **42**, 105, (2015).
74. AB Fox, K Mills, D Lever, M C Bezerra, CA Valadares, I Unamono, J Laraudogoitia, J Gisby, Proc. 36th Steelmaking Seminar, Vittoria, ES, Brazil, (2005) p. 222 available on line at http:/www.carboox.com/pdf/paper2005sfs.pdf.
75. Y Lu, GD Zhang, MF Jiang, Advanced Mater. Res., **233/235**, 805, (2011).
76. Z Wang, Q Shu, KC Chou, Steel Research, **84**, 768, (2013).
77. PS Kharlashin, Azovetal **1999** (4), 34, (1999).
78. Y Lu, GD Zhang, Mater. Sci. Forum, **675/677**, 877, (2011).
79. P Tang, GH Wen, ZB Wang, SB Fan, ZB Zhang, SC Li, H Li, Proc. ISS Tech. 2003 Conf., (ISS, Warrendale, PA, 2003) p. 567.
80. SY Choi, DH Lee, DW Shin, JW Cho, JM Park, J Non-Cryst. Solids, **15**, (8), 157, (2004).
81. Y Lu, X Fang, GD Zhang, Advanced. Mater. Res., **287/290**, 1866, (2011).
82. QC Wei, YQ Ding, KD Peng, J Chongqing Univ., **4**, 110, (1995).
83. Y Lu, GD Zhang, X Yu, Appl. Mech. and Mater., **71/78**, 2899, (2011).
84. Y Lu, X Fang, XF Yu, Advanced Mater. Res., **455/456**, 134, (2012).
85. JL Klug, DR Silva, SL Freitas, MMSM Pereira, NC Heck, ACF Viela, D Jung, Steel Res. Intl., **83**, 791, (2012).
86. SP He, X Long, JF Xu, T Wu, Q Wang, Ironmaking and Steelmaking, **39**, 593, (2012).
87. Q. Wang, SP He, KC Mills: Proc. 4th Intl. Conf. Cont. Casting Steel in Developing Countries, Beijing, China (2008). (Chinese Soc. Metals, Beijing, 2008), p. 715.
88. YK Jung, SK Jung, JW Cho, JH Park, "*Commercial plant trial of Fluorine-free mold flux for continuous casting of blooms*". Proc. SEAISI Conf., SEAISI, Singapore, (2014)
89. H Nakada, K Nagata, ISIJ Intl., **46**, 441, (2006).
90. JL Klug, R Hagermann, NC Heck, JFC Viela, PR Scheller. Proc. 9th Intl. Conf. Molten slags, fluxes and salts, Beijing, 2012 (Chinese Metals Soc., 2012)
91. J Klug, N Heck, A Faria, Proc. 8th Intl Conf. Molten slags, fluxes and salts, Santiago, Chile, 2009 (GECAMIN, Santiago, Chile, 2009), p. 1053.
92. H Hashimoto, T Watanabe, K Nagata, CAMP-ISIJ, **17**, 849, (2004).
93. X Qi, GH Wen, P Tang, J Non Cryst. Solids, **354** (No 52/54), 5444, (2008).
94. N Takahira, M Hanao, Y Tsukaguchi, ISIJ Intl. **53**, p. 818, (2013)
95. A Kusano, N Sato, M Okimori, S Fukunaga, K Nishihara, M Sato, Y Minigawa, Proc. 74th Steelmaking Conf., 1991, (ISS, Warrendale, PA, 1991) p. 111.
96. S Terada, S Kaneo, T Ishikawa, Y Yoshida, Proc. 74th Steelmaking Conf.,Washington, DC, 1991 (ISS, Warrendale, PA, 1991) p. 635.
97. B. Debiesme, J Radot, D Coulombet, C Lefebre, Y Roux, C Demarval, US Patent No 5 876, 482 (1999) and US Patent No 6, 328, 781, (2001).
98. T Kanazawa, T Marukawa, K Nakai, T Yamada, K Nakajima, Steel Times Supplement, **222** (6), 16, (1994).
99. M Kawamoto, T Kanazaka, S Hiraki, S Kumakura, Proc. 5th Intl. Conf. Molten slags, fluxes and salts, Sydney, 1997, (ISS, Warrendale, PA, (1997) p. 777).
100. M Kawamoto, K Nakajima, T Kanazawa, K Nakai, ISIJ Intl., **34**, 593, (1994).
101. M Hanao, M Kawamoto, M Hara, T Murakami H Kikuchi, A Yamanaka,. Proc. 5th Europ. Conf. Cont, Casting, Nice,2005, (La Revue Metall., Paris, 2005) see also Tetsu-to Hagane, **88** (1), 23 (2002)

102. JA Kromhout, S Meltzer, RW Zinngrebe, AA Kamperman, R Boom, Proc. 9th Intl. Conf. Molten slags, fluxes and salts, Beijing, 2012, (Chinese Metals Soc.)
103. JA Kromhout, M Kawamoto, M Hanao, Y Tsukaguchi, E Dekker, R Boom, Steel Research Intl., **80**, 575, (2009).
104. JA Kromhout, PhD Thesis, "*Mould powders for the high speed continuous casting of steel*", Univ. Delft, (2011), p. 165.
105. T Matsumiya, ISIJ Intl., **46**, 1800, (2006).
106. S Chakraborty, W Hill, Proc. 77th Steelmaking Conf., 1994, (ISS, Warrendale, PA, 1994) p. 389.
107. J Kromhout, RS Schimmel, Proc 8th Europ. Conf. Continuous Casting, Graz, 2014, (Austrian Metals Soc.)
108. Y Liu, XD Wang, M Yao, AB Zhang, H Ma, Z Wang, JC Ma, X Wang, GQ Shi, Ironmaking and Steelmaking, **41**, 748, (2014).
109. M Kawamoto, CAMP-ISIJ, **6**, 1176, (1993).
110. S Umeda, Y Tsukaguchi, H Mikai, Y Hitomi, M Kawamoto, CAMP- ISIJ, **7**, 302, (1994).
111. R Nishimachi, Y Ogura, SEAISI Quarterly, **25**, (Oct), 44, (1996).
112. K Tsutsumi, K Watanabe, J Kubota, S Hatori, Y Miki, T Suzuki, T Omoto, Proc. 7th Europ. Cont. Casting Conf., Dusseldorf, 2011, (VDEh, Dusseldorf, 2011) Session p. 1.
113. K Watanabe, K Tsutsumi, M Suzuki, H Fujita, S Hatori, T Omoto, ISIJ Intl., **54**, 865, (2014).
114. K Blazek, H Yin, G Skoczylas, M McClymonds, MJ Frazee, AIST Tech. 2011 (AIST, Warrendale, PA, 2011) also Iron & Steel Technol., **8** (3), 232, (2011).
115. JW Cho, KE Blazek, MJ Frazee, HB Yin, JE Park, SW Moon, ISIJ Intl., **53**, 62, (2013).
116. H Yin, G Skoczylas, Proc. ATS Tech. Conf., Indianapolis, 2006 (AIST-ISS Warrendale, PA, 2006) p. 753.
117. WL Wang, K Blazek, AW Cramb, Met. Mater. Trans. B, **39 B**, 66, (2008).
118. X Yu, GH Wen, P Tang, H Wang, Ironmaking and Steelmaking, **36**, 623, (2009).
119. IH Jung, MA van Ende, Proc. 8th Europ. Conf. Continuous Casting, Graz, 2014, (Austrian Metals Soc. 2014).
120. Q Wang, JH Chi, B Xie, Y He, B Zhu, W Chen, B Xie, JH Chi, Proc. Conf. Continuous Casting in Developing Countries, Beijing, 1993 (SEAISI, Singapore, 1993) p 842.
121. Y Tsukaguchi, CAMP- ISIJ, **21**, 826, (2008).
122. K Blazek, HB Yin, G Skoczylas, M McClymonds, M Frazee, Iron and Steel Technology, **8**, (3), 232, (2011).
123. TS Kim, JH Park, ISIJ Intl., **54**, 2031, (2014).
124. Q Liu, GH Wen, JZ Li, XJ Fu, P Tang, W Li, Ironmaking and Steelmaking, **41**, 292, (2014).
125. JL Li, QF Shu, KC Chou, Can. Metall. Quart., **54**, 85, (2015).
126. QF Shu, Z Wang, J L Klug, K Chou, PR. Scheller, Steel Research, **84**, 1138 (2013).
127. MD Seo, CB Shi, JW Cho, SY Kim,. Proc. 8th Europ. Conf. Continuous Casting, Graz, 2014, (Austrian Metals Soc., 2014).
128. JW Cho, SY Kim, SC Moon, H Shibata, Proc. 6th Intl Conf. Molten Slags, Fluxes and Salts, Stockholm, Helsinki, June 2000, Paper 154, Filename 341.pdf
129. P Misra, S Sridhar, AW Cramb, Met. Mater. Trans. B, **32B**, 963, (2001).
130. Z Wang, QF Shu, XM Hou, KC Chou, Ironmaking and Steelmaking, **39**, 210, (2012).
131. F Shimizu, F Tokunaga, N Saito, K Nakashima, ISIJ Intl., **46**, 385, (2006).
132. DY Wang, M Jiang, CJ Liu, PY Shi, J North Eastern Univ. (Natural Sci.), 26, (11), 1082, (2005).

Chapter 7
Fluxes for Ingot Casting

Abstract Mould powders for ingot casting have, in many ways, the same function as mould powders for continuous casting. However, in some aspects the properties and compositions must meet requirements other than those for continuous casting. After the introduction of continuous casting on a bigger scale and adaptation of mould powder for this technology, the development of ingot casting powder has mainly been based on experience derived from the development of CC mould powders. Thus, there has been little development in casting powder research and ingot casting technology for more than 30 years. The remaining steel grades that are cast by the ingot process route today tend to be high-grade, high-quality steels and there is a new, re-awakened interest in the development of both ingot casting technology and mould powders for ingot casting. Today most ingot casting is carried out with killed steel cast into bottom end up (BEU) ingots. Thus, the most important properties of the mould powders are good flowability and good thermal insulation. The creation of a mould powder slag is important but the thickness and composition of this must be designed according to the steel grade being cast and the process conditions which require careful control of the composition and melting speed. Most ingot powders are still produced with fly ash as a base but the use of synthetic mould powders is growing because of its greater versatility in composition. Granulated powders for ingot casting are also attracting a growing interest, since they are more environmentally friendly, causing much less dust during casting. They also give better flowability, compared to loose synthetic powders, and open up the opportunity of continuous feeding during casting.

Symbols, Abbreviations and Units

a^* Capillary constant
B_i Basicity index
g Acceleration due to gravity (ms^{-2})

The original version of this chapter was revised: For detailed information please see correction. The correction to this chapter is available at https://doi.org/10.1007/978-3-319-53616-3_12

Within this book minor changes to some of the figure permissions and the readability of the chapters have been made and several figures have been replaced with new versions.

K.C. Mills and C.-Å. Däcker, *The Casting Powders Book*,
DOI 10.1007/978-3-319-53616-3_7

k Thermal conductivity (Wm^{-1} K^{-1})
α^* Absorption coefficient
ρ_{Fe} Density of liquid steel (kg m^{-3})
ρ_{sl} Density of liquid mould slag (kg m^{-3})
γ_m Surface tension between liquid steel and gas above it (mNm^{-1})
γ_{msl} Interfacial tension between liquid steel and slag (mNm^{-1})
T_{br} Break temperature (°C)
T_{liq} Melting temperature (°C)

Abbreviations, Subscripts and Superscripts

CC Continuous casting

7.1 The Ingot Casting Process

The dominant casting process today is continuous casting but still around 8% of the World's annual steel output consists of ingot cast steels. There are a number of different motives underlying the use of ingot casting (instead of continuous casting), these are:

- Lack of investment capital to build equipment for continuous casting which is much more expensive compared to ingot casting.
- Large products such as rolls, shafts, etc., which can be produced by hot forging.
- Product programme with high demands on reduction rate, such as thick heavy plate.
- A product programme with a great variety of different dimensions.
- Steel grades which are very sensitive to surface cracks.

Ingot casting is carried out in cast iron moulds with square, round or polygon cross-sections. Ingots with square cross section are used for rolling into billets, rails and other structural sections or for hot forging to their final shape. Ingots with rectangular cross section (also known as slabs), are mostly used for rolling into flat products but also for forging of flat products. Round ingots are used for tube making. Polygon ingots are used in the production of tyres, wheels, etc. Typically, ingots weigh 5–20 tons for rolling, but range from 100 up to 300 tons when used for forging.

Cast iron is used to fabricate the mould which consists of grey iron, ductile iron or malleable iron. The moulds are essentially of two types;

- Wide-end-up moulds are used to produce ingots of killed carbon or alloyed steels.
- Narrow-end-up moulds are commonly used to produce rimming and semi-killed steel ingots.

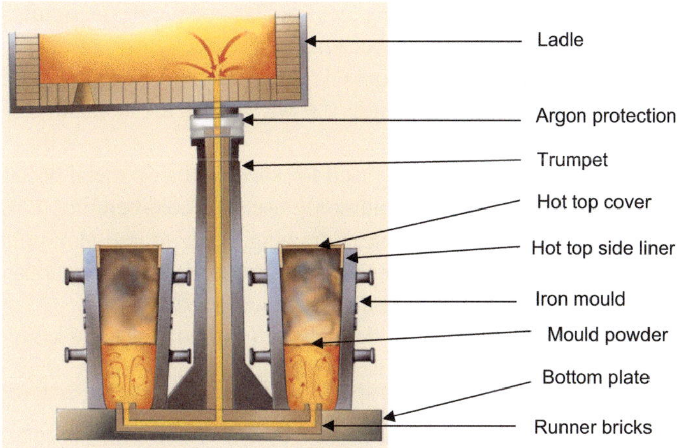

Fig. 7.1 Schematic diagram showing the bottom pouring of ingots (courtesy of Uddeholms AB)

The casting from the steel ladle can be made in two ways, either by top pouring or bottom pouring. In top pouring (or downhill teeming) the mould is simply filled directly from the ladle into the individual ingot. In bottom pouring (or uphill teeming) one or (mostly) more ingots are placed on a bottom plate (Fig. 7.1). The steel is poured via a sliding gate from the ladle into a trumpet where the open stream mostly is protected with an argon shield. The trumpet, made of iron, is thermally protected by a tube formed from bricks. This connects to a distributor at the bottom which distributes the steel stream to the mould inlet via the rectangular runner bricks. The steel is covered with mould powder which normally is added from paper sacks hanging about 30 cm from the bottom which open spontaneously when the paper burns in response to the radiation emitted from the incoming steel.

The thermal conductivity and thickness of the mould and bottom plate are sufficient to keep the surface temperature at a secure low temperature, and thereby avoid sticking between the steel and mould/bottom plate. So almost immediately a steel shell is formed when the steel comes in contact with the mould/bottom plate.

7.1.1 Classification of Ingot Cast Steels

The ingot casting procedure is classified according to the deoxidisation of the steel melt; thus, there are three groups, namely; rimmed, semi-killed, and killed steels.

Rimmed steel is also known as *drawing quality steel*, and has little, or no, deoxidising agent added to it during steelmaking which causes carbon monoxide to evolve rapidly from the ingot during solidification. This causes small blow holes to be formed at a certain distance from the steel surface which are later closed up during the hot rolling process. Another consequence is the segregation of elements where most of the carbon, phosphorus, and sulphur move to the centre of the ingot, leaving an almost perfect "rim" of pure steel on the outside of the ingot. This gives the ingot an excellent surface finish because of the steel rim, but also results in a strongly segregated ingot. Most rimmed steels have a carbon content below 0.25 wt-%, a manganese content below 0.6 wt-%, and are not alloyed with aluminium, silicon and titanium.

These types of steels are commonly used for sheet and strip metal because of their excellent surface condition and are commonly used for cold-bending, cold-forming, cold-heading and, as the name implies, drawing. Due to the non-uniformity of alloying elements they are not recommended for hot-working applications. The yield of rimmed steel is slightly better than that for semi-killed steel.

Semi-killed steel is incompletely deoxidised (with silicon) and contains a sufficient level of dissolved oxygen to react with the carbon to form carbon monoxide which creates blowhole-type porosity distributed throughout the ingot. The porosity eliminates the pipe found in killed steel and increases the yield to approximately 90% by weight. Semi-killed steel is commonly used for structural steel with carbon contents between 0.15 and 0.25 wt-% of carbon. Since pipe cavities are minimised, semi-killed steels are usually cast in big-end-down moulds without hot-tops. The quality of semi-killed steels is not as good as killed steels because of the large amount of manganese–silicates formed during steel-making. These do not readily separate from the metal and consequently, they tend to remain in the final ingot.

Killed steel is completely deoxidised by the addition of an agent in the ladle before casting; consequently, the level of oxygen is so low there is practically no evolution of gas during solidification. These steels are characterised by a high degree of chemical homogeneity and freedom from larger gas porosities. The steel is said to be "*killed*" because it will quietly solidify in the mould, with no gas bubbling. Common deoxidising agents include aluminium, ferrosilicon and manganese. Aluminium reacts with the dissolved oxygen to form aluminium oxide. Aluminium also has the added benefit of forming pin grain boundaries, which prevent grain growth during heat treatments. Killed steels suffer from deep pipe shrinkage defects. Thus, in order to minimise the amount of metal that must be discarded because of the shrinkage, the upper part of the ingot walls and top surface, must be provided with a so-called "*hot top*" for thermal insulation. A summary of the different steel types is made in Table 7.1.

Cross sections of ingots with different states of deoxidisation are schematically shown in Fig. 7.2 where it must be pointed out that the killed steel in (a) is cast without hot top insulation. The dots in the figures for the rimmed and semi-killed steels are carbon monoxide blow holes.

The introduction and expansion of the continuous casting process has, to a large extent, put an end to the casting of rimmed and semi-killed steels. This is exemplified by Sumitomo [1] who reduced the amount of rimmed steel for cold-rolled sheet production to 2% by the mid-80s and completely stopped the production in 1995 (Fig. 7.3). The high segregation in rimmed steels and the high amount of manganese–silicate inclusions in semi-killed steels have led to the situation where almost all ingot casting today is carried out using killed steels.

Table 7.1 Summary of steel types with different deoxidation treatment

	Rimming; no de-O	Semi-killed; de-O by Si	Killed: de-O by Al, Fe/Si, Mn
Steel composition	C < 0.25%; Mn < 0.6%	C = 0.15– 0.25%	Low alloyed and high alloyed carbon steels
Steel use	Sheet, strip	Structural	Tool steel, rolls, shafts, ball bearing etc.
Gas evolution	CO(g) → blowholes; closed in rolling	O % sufficient → CO(g) blowholes through ingot, closed in rolling	None
Type of casting	Top pouring; without hot top insulation	Top pouring; big-end-down; without hot top insulation	Bottom pouring for most cases Top pouring for very large ingots with multiple pouring Hot top insulation
Defects	Porosity in steel: S,P, C segregates to ingot centre	Blowholes Mn silicate inclusions	Deep pipe shrinkage-Minimized with hot top, surface cracks, macro segregates
Yield	Greater than semi-killed	>90%	80–85%
Steel quality		Inferior to killed steel	

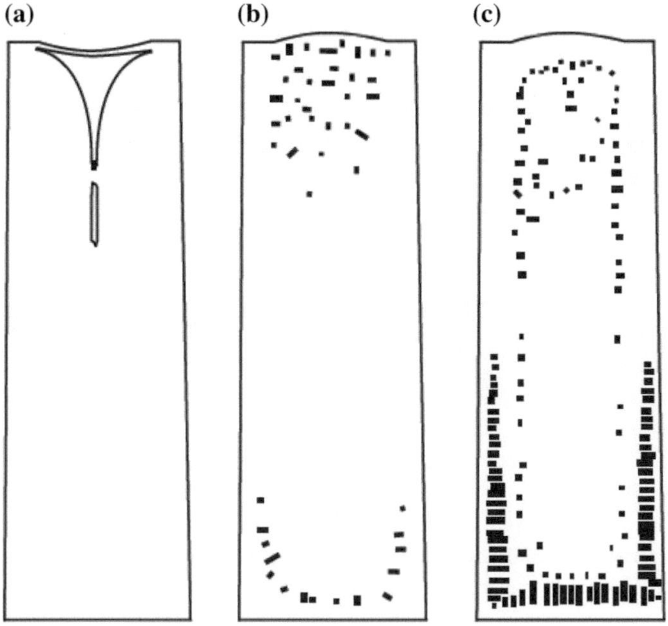

Fig. 7.2 Schematic figures showing cross sections of steel ingots for **a** killed steel, **b** semi-killed steel, **c** rimmed steel

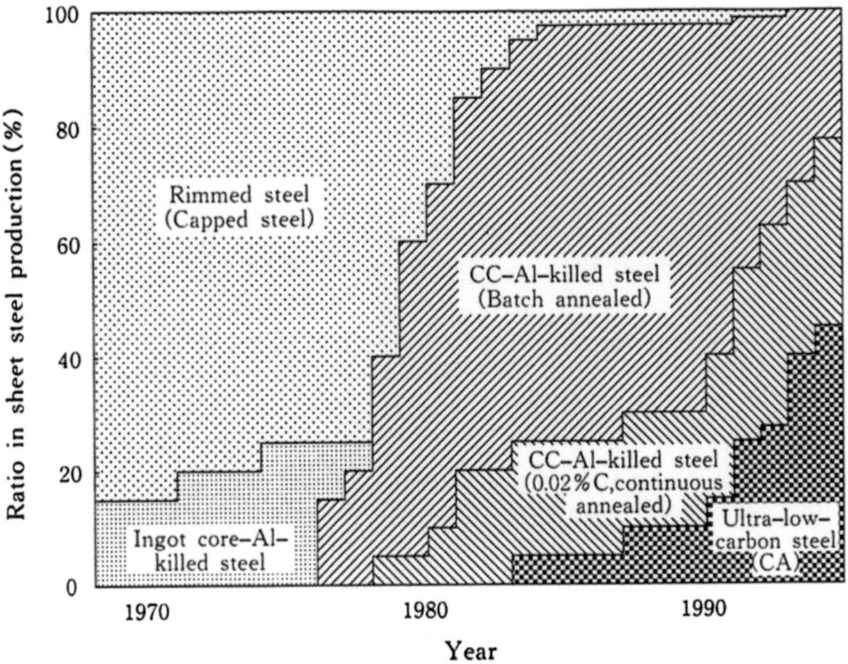

Fig. 7.3 Schematic diagram showing changes in steel production of cold-rolled sheet (Reproduced with permission from Nippon Steel & Sumitomo Metal Corporation [1])

7.1.2 Ingot Casting of Killed Steels

As mentioned above, when casting killed steels, a very deep pipe is formed (see Fig. 7.2a) in the upper part of the mould due to the solidification shrinkage caused by the density differences between liquid and solid steel. The density of ordinary carbon steel at room temperature is: $\rho = 7850$ kg/m^3 and as liquid at the melting temperature: $\rho = 7010$ kg/m^3. In order to eliminate the deep pipe, a technique to insulate the upper part of the mould has been developed, the so called "*hot top*". This consists of hot-top side liners of ceramic material, with good insulating properties, in the upper part of the mould. These are mounted in the mould in one piece for small ingots or in four or more sections for larger ingots. The top part of the ingot must also be thermally insulated, this is achieved with mould powder during the casting and normally, with exothermic powder, or board, at the end of casting. The hot top cover contains aluminium which makes it exothermic and also special graphite, which makes it expand and thereby improves its insulating properties. For larger ingots, with long solidification times, it is also common to further increase the insulation capability by adding insulating material, such as perlite or rice ash, to the top of the exothermic powder/board after it has burned through. A picture from a steel plant with hot top boards, where the ingot on the front right has just burned through, is shown in Fig. 7.4.

Fig. 7.4 Photograph showing hot top with exothermic boards; note that the board has just burnt through for ingot at *bottom right*

With a good insulated hot top, a reservoir of molten steel is created at the top of the ingot which feeds the ingot during its final solidification and shrinkage. This is shown in Fig. 7.5 from a simulation made with the software THERCAST® from Transvalor.

A properly installed hot top will eliminate the pipe and produce an almost flat top surface as shown in Fig. 7.6.

The yield of killed steels is normally around 80–85% and is mainly restricted in the upper part because of heavy macrosegregation especially for large ingots with a long solidification time. This is exemplified in Fig. 7.7 for a 65 ton steel ingot with a carbon content of 0.22 wt-% [2].

Traditionally, all ingots were top poured directly from the ladle into the mould, (so-called downhill casting) which is the cheapest way to produce ingots. When new, more advanced, steel grades were introduced, there was a need for better ingot quality which was achieved by the introduction of bottom pouring or uphill casting.

The pros (+) and cons (−) of uphill casting (cf. downhill casting) are:

\+ Better surface quality due to low amount of splashes during casting and the use of mould powders.

\+ Better steel quality and fewer inclusions since there is less re-oxidation (no open steel stream).

\+ Higher casting rate, and simultaneous casting of many ingots.

\+ Lower wear of moulds and bottom plates.

− Additional costs for extra material such as trumpets and refractory material.

− Extra working steps for preparation of casting and cleaning operation after casting.

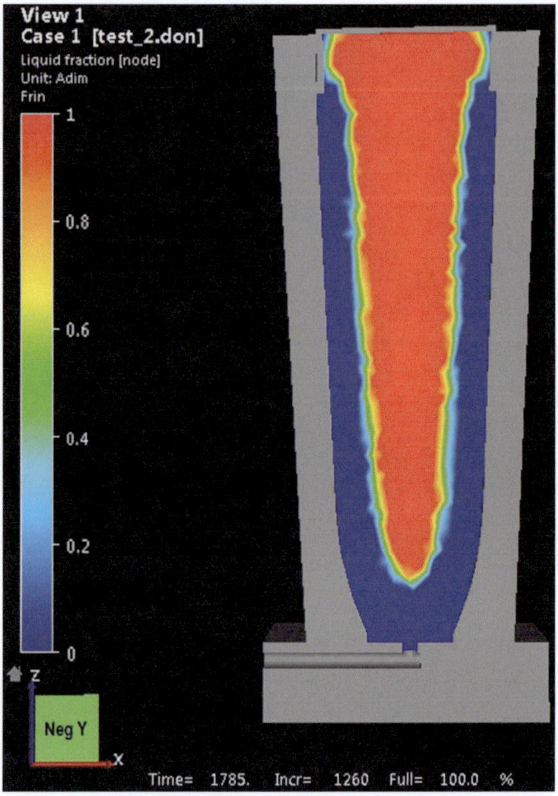

Fig. 7.5 Schematic drawing from a simulation of ingot solidification using THERCAST® from Transvalor, the colour scale represents the liquid fraction where red is fully liquid

Fig. 7.6 Photograph showing top surfaces from casting of killed steel with a well-functioning hot top

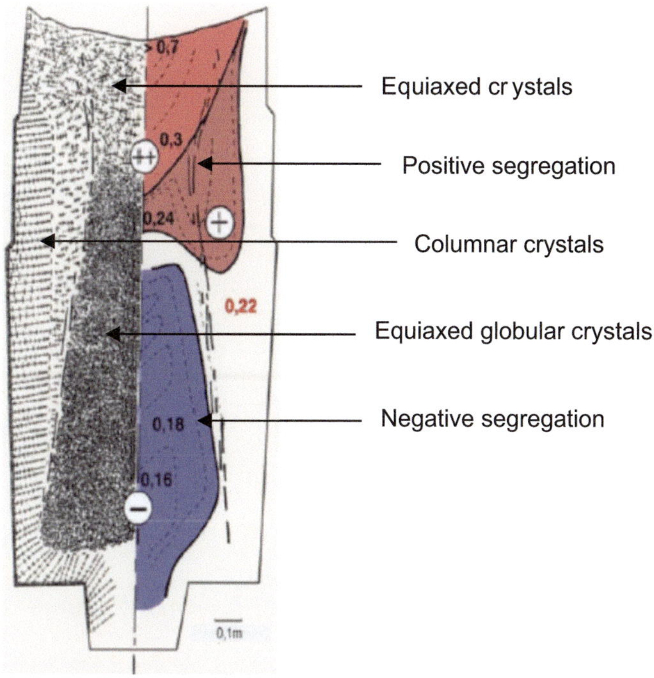

Fig. 7.7 Schematic drawing showing structure pattern (*left*) and a map of final average mass fraction of carbon (on *right* side of figure) (Reprinted from Materials Science and Engineering A 413–414, Macrosegregation in steel strands and ingots: Characterisation, formation and consequences, G. Lesoult, pp. 19–29, © 2005, with permission from Elsevier [2])

Today's demand for high-quality steel products has forced almost all ingot casting steel makers to use the uphill casting method.

7.2 Aspects of Importance for Ingot Casting Quality

More information on the ingot casting process is needed before any recommendation can be made on which mould powders to use. Thus, a more detailed description of the ingot casting process and its limitations are presented below to provide a better understanding of the use and composition of mould powders.

7.2.1 Surface Quality

To understand the function of mould powders and their effect on the surface quality, it is important to have an idea of the situation at the first stage of solidification in the meniscus area.

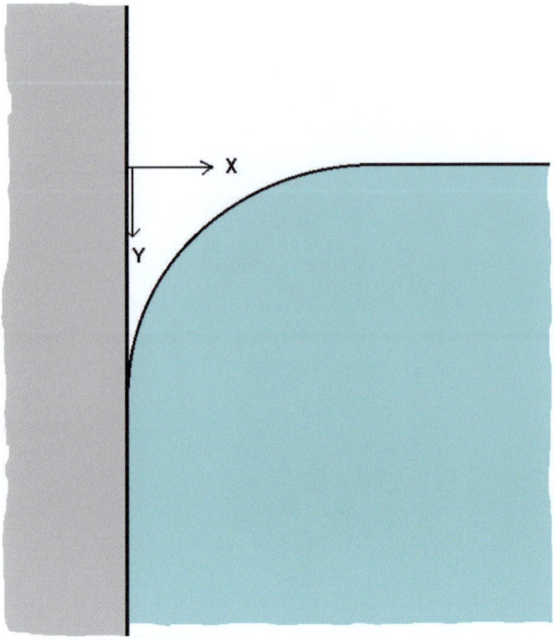

Fig. 7.8 Schematic drawing showing the liquid steel meniscus in the mould (Reproduced with permission from Nippon Steel & Sumitomo Metal Corporation [3])

This part is very well described by Tomono [3] who made in situ studies of solidification on both an organic melt (with analogous solidification behaviour to steel) and on steels (in where the meniscus was observed through a quartz window).

When the bottom of a mould is filled with liquid steel, a meniscus is formed in contact with the mould wall, as shown in Fig. 7.8.

For a straight wall, the shape of the meniscus can be calculated [4] according to the following equation (note this equation is not valid for corners or round format):

$$X = -2(a^{*2} - y^2)^{0.5} + \frac{a^*}{\sqrt{2}} \ln \frac{a^* \cdot \sqrt{2} + (2a^{*2} - y^2)^{0.5}}{y} + 0.3768\, a^*, \tag{7.1}$$

where x is the horizontal and y the vertical distance from meniscus.

a^* is the capillary constant defined as:

$$a^* = \frac{2\gamma_m}{g\,(\rho_{Fe})} \tag{7.2}$$

γ_m = surface tension between liquid steel and the gas above it
ρ_{Fe} = density of liquid steel
g = acceleration due to gravity

If the steel surface is covered by a mould powder slag, the capillary constant can be derived from the following expression:

$$a^{*2} = \frac{2\gamma_{\text{msl}}}{g\left(\rho_{\text{Fe}} - \rho_{\text{sl}}\right)} \tag{7.3}$$

γ_{msl} = interfacial tension between liquid steel and slag
ρ_{sl} = density of liquid mould powder slag.

Increasing interfacial tension leads to increases in both the radius of curvature and the total height of the meniscus.

If the meniscus shape is simplified to that of a quarter of a circle, the following simplified calculation of the radius can be derived [5]:

$$r = 1.7 \cdot \left[\gamma_{\text{m}} / \left(\rho_{\text{Fe}} - \rho_{\text{Sl}}\right) \cdot g)\right]^{0.5}. \tag{7.4}$$

Most mould powders used in uphill ingot casting tend to have a high carbon content which results in a low melting rate during the teeming operation. They function mostly as a heat insulating cover but also produce a reducing atmosphere to protect the steel surface from oxidation. The surface tension of liquid steel in contact with this type of gas (CO) is rather high, (1500–1850 mN/m) due to the low soluble oxygen content but is sensitive to the sulphur content.

During teeming the meniscus partially solidifies and forms a shell. This shell grows slowly in the meniscus region because of the surface/interfacial tension, which holds the shell away from direct contact with the mould. During the teeming process the first shell continues to increase in thickness, where the shell thickness achieved is dependent on the following variables:

- **Fluid flow**. Where a strong fluid flow leads to a thin solid shell.
- **Teeming speed**. Where a quick teeming leads to a thin solid shell.
- **Superheat**. Where high superheats also lead to a thin solid shell.
- **Interfacial tension**. A high interfacial tension leads to a long meniscus and a long, but weak, solid shell.
- **Steel grade**. This is dependent on both chemical composition (which determines the way of solidification) and the amount of micro segregating elements, which reduce the high temperature strength and thus, determine the strength of the steel.

The steel meniscus grows upward as the steel pours into the mould. This is different to continuous casting where the meniscus (i.e. shell) is both withdrawn downwards and subject to mould oscillation. This difference is reflected in the need for lubrication of the shell since lubrication is not important for gradual, upward movement in ingot casting but is essential in handling the stresses generated by the vigorous movement in the CC mould.

When the steel meniscus is covered by a thin pool of liquid slag, gravity feeding would be expected to cause some infiltration of liquid slag into the shell/mould gap. However, the high viscosity of the slag of most ingot casting mould powders would tend to restrict the slag infiltration.

Another difference between ingot and continuous casting lies in the horizontal heat transfer. In CC the thermal gradients across the slag film are large and the cooling rates are large because of a water-cooled copper mould. In ingot casting both the thermal gradients and the cooling rates are much smaller in the cast iron mould and hence the need to control heat extraction from the shell is much less demanding than in CC. However, in ingot casting control of the vertical heat flux is important to stop the steel meniscus from freezing during casting.

The shell forms periodic marks (so-called ripple marks) during the teeming. Two different types of ripple mark formations can be observed depending on the circumstances [3].

7.2.1.1 Ripple Mark Formation Due to Folding

This mechanism is schematically described in Fig. 7.9.

In this case, the steel shell formed at the meniscus is so weak that the "meniscus shell" is bent towards the mould surface by increasing static pressure due to the increase in the height of the bath (cycles 2–5). As the meniscus is pressed against the mould (cycle 6) the shell starts to solidify immediately. The next cycle of ripple mark starts to form when the shell has sufficient strength to withstand the ferrostatic pressure.

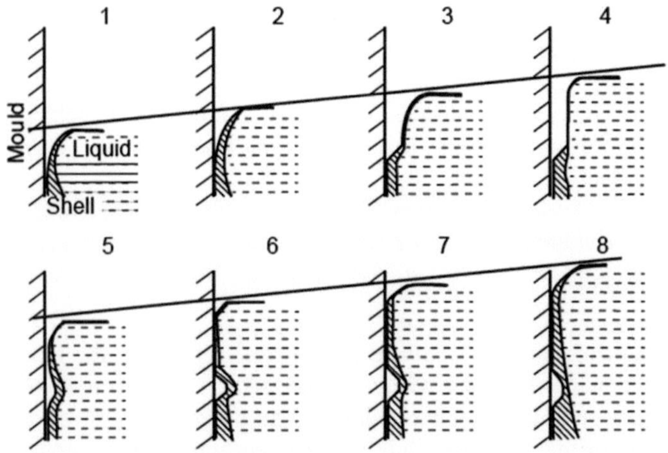

Fig. 7.9 Schematic diagram showing typical cycle of ripple mark formation due to folding (Reproduced with permission from Nippon Steel & Sumitomo Metal Corporation [3])

7.2.1.2 Ripple Mark Formation Due to Overflow

In this case, the solidified shell in the meniscus region is so strong that only the top part is bent towards the mould surface with increasing static pressure. If the bath level rises to a point which is higher than a certain critical position, the surface tension can no longer support the pressure and liquid steel overflows the solidified meniscus shell and fills the gap between mould surface and meniscus shell, thereby forming a "*hook*". This case is typical for casting at too low speed, low superheat or a steel meniscus with insufficient insulation or protection against oxidation. (Fig. 7.10).

In Tomono's [3] in situ studies of uphill-teemed steel, shell growth in the meniscus was varied by changing (i) the gas atmosphere (thus changing, simultaneously, both the surface tension and the heat transfer between steel and mould), (ii) mould material (thus changing the heat transfer) and (iii) the superheat of the melt. These changes affected both the pitch and the depth of the ripple mark, as shown in Fig. 7.11.

The experimental results can be summarised as:

- Increasing solidification in the meniscus area resulted in increases in both the depth and the pitch of the ripple mark.
- Both depth and pitch of folding marks (extent of solidification) increased with a lower surface temperature of the mould. In the above example, the mould temperatures were 190 °C for Copper (CU), 342 °C for Cast Iron and 615 °C for Stainless Steel (SUS).
- The depth and pitch decrease when the superheat in the mould is high (42 °C).
- Changing the atmosphere above the surface from air to hydrogen (H_2) caused two changes, which both influence the depth and pitch of the ripple mark:

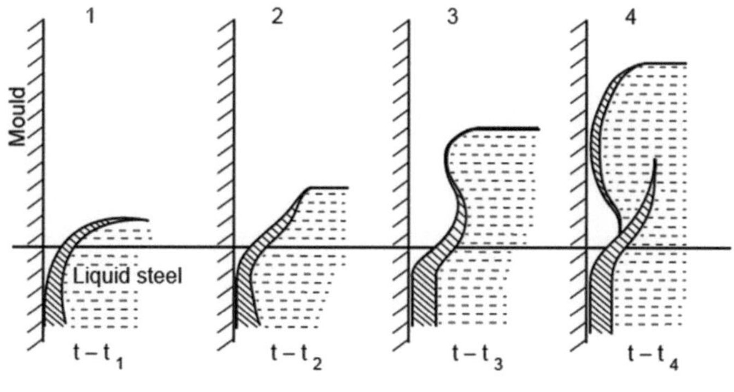

Fig. 7.10 Schematic drawing showing the formation of ripple marks by the overflow mechanism (Reproduced with permission from Nippon Steel & Sumitomo Metal Corporation [3])

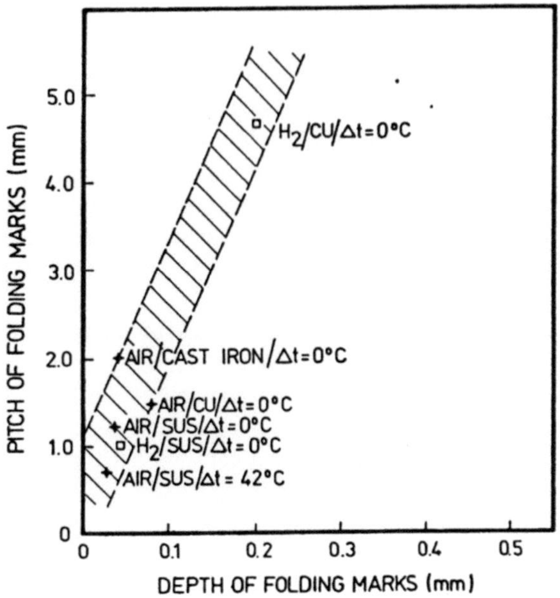

Fig. 7.11 Schematic diagram showing the relation between depth and pitch of folding marks. *SUS* Stainless steel mould, *CU* copper mould, H_2 and AIR is the gas atmosphere respectively, Δt superheat (Reproduced with permission from Nippon Steel & Sumitomo Metal Corporation [3])

- First and probably the most dominant variable is that the thermal conduction is changed by a factor of *ca.* 7 from a low value in air to a high value in hydrogen (see Table 7.2).
- Secondly the interfacial tension is changed from approx. 500–800 mN/m in air (dependent on the extent of absorption of oxygen and the formation of an oxide layer) to approximately 1500 mN/m for hydrogen; this will result in an elongation of the meniscus.

Similar results have been reported in another investigation [6] in which steel was uphill teemed in a mould consisting of three sides of ceramic material and one side of water-cooled copper. The mould was installed in a vacuum furnace, in which, vacuum or different gas atmospheres could be chosen. In Fig. 7.12 the surface temperature of the solidifying shell on the copper side is shown for casting trials with low carbon steel.

Table 7.2 Thermal conductivity of some gases from 27–337 °C at 1 bar pressure

	Thermal conductivity (°C), $mWm^{-1}K^{-1}$			
Gas	27 °C	127 °C	227 °C	337°C
Ar	17.9	22.6	26.8	30.6
O_2	26.3	32.3	39.2	45.7
H_2O (steam)	18.7	27.1	35.7	47.1
H_2	186.9	230.4		
H_2S	14.6	20.5	26.4	32.4
Air	26.2	33.3	39.7	45.7
CO	25.0	32.3	39.2	45.7

Comments:

- The casting trials were made without mould powder. This means that the heat flux from steel to the copper mould consists only of radiation and conduction.
- In the upper curve in which the casting is made in vacuum, the heat flux results solely from radiation.
- The other curves clearly reflect the differences in thermal conductivity for their respective gases in which they were cast according to Table 7.2.

In Fig. 7.12 it can be seen that the surface temperature decreases more rapidly for the P_{H_2O} curve than for the P_{O_2} curve despite the fact that $k_{O_2} > k_{H_2O}$. This anomaly is due to the oxidation of the steel which then provides a thermal resistance; the oxidation and thermal resistance will be greater for O_2 than H_2O thus the temperature decrease will be more rapid for H_2O than O_2.

From some of the castings, photographs were taken from the surface (shown in Fig. 7.13). The same pattern as that shown in Fig. 7.11 is also visible here. Casting in a H_2 atmosphere creates a large and strong solidification at the meniscus due to high values of the surface tension and thermal conductivity which, in this case, leads to deep ripple marks of the overflow type.

Casting in an oxygen atmosphere both lowers the surface tension but, above all, reduces the heat flux compared with that for a casting in a H_2 atmosphere. This results in shallow ripple marks probably of the folding type.

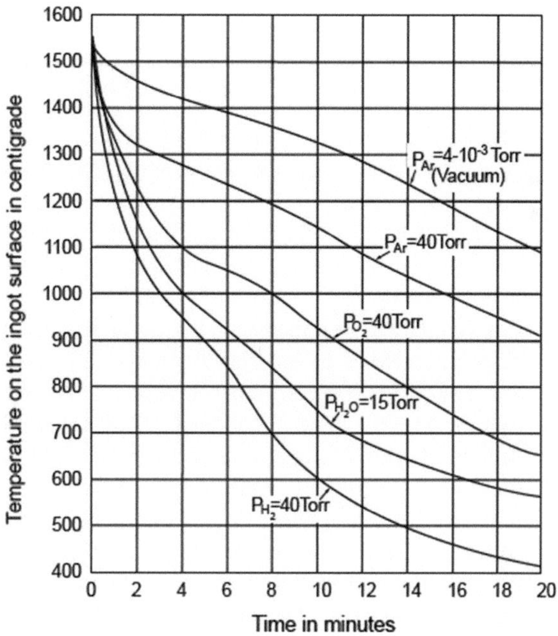

Fig. 7.12 Schematic diagram showing surface temperatures as a function of solidification time (Reproduced with permission from Verlag Stahleisen GmbH [6])

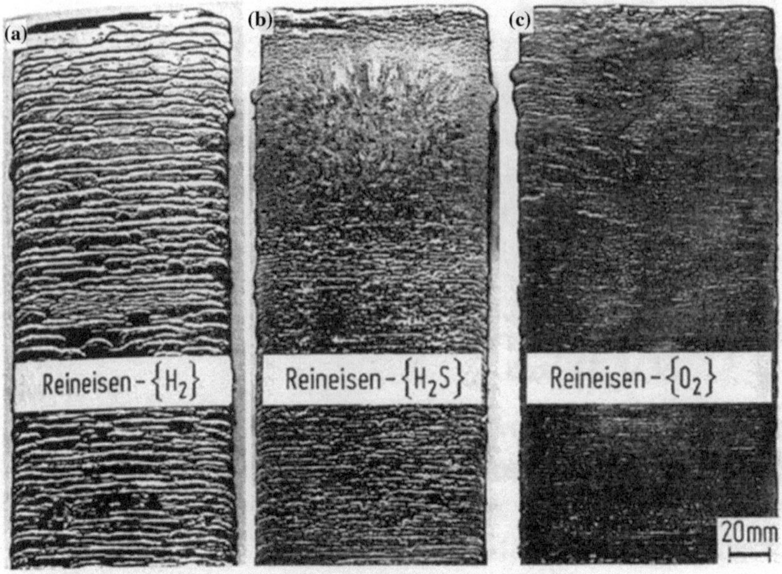

Fig. 7.13 Surface appearance for castings in different gas atmospheres: **a** H_2 **b** H_2S **c** O_2 (Reproduced with permission from Verlag Stahleisen GmbH [6])

7.2.1.3 Uphill Teeming with Mould Slag

If a molten slag is put on top of the steel bath, a number of things change:

- The interfacial tension is reduced by an amount, dependent on the chemical composition of the slag. Interfacial tension (γ_{msl}) values for mould powder slags lie in the range 900–1300 mN/m but are very dependent upon the S content of the steel and can be lowered by chemical reactions between steel and slag [7]. The shape of the meniscus will also change as a consequence of the density of the mould powder slag.
- If the mould powder has a high melting rate and/or a low viscosity, the mould slag can partially fill the gap between mould and ingot. The thermal conductivity of an amorphous mould slag is around 1.1 W/m, K at 300 °C compared to a value of 0.044 W/m, K for air or carbon monoxide at the same temperature. This will increase the thermal conductivity by a factor of *ca.* 25. In practice, the heat flux will be lower since some of the heat flux by radiation will be absorbed by the high FeO content in the mould slag [8]. This effect has an impact on the initial solidification when the distance between the steel shell and mould is very small and the gas conductivity has a major role and convection is suppressed by the thin gap. When the air gap (or gas gap) has developed, the contributions from radiation (along with some convection) to the heat flux (between steel surface and ingot) will be dominant. This is well illustrated in Fig. 7.14 [9] from

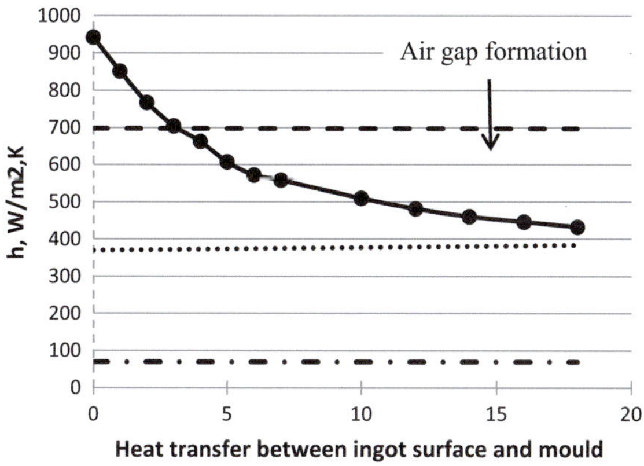

Fig. 7.14 Heat transfer between ingot and mould as function of solidification time. ● = measured values, ····· = calculated heat transfer by radiation, -·-·-·-· = calculated heat transfer by gas conduction with a gas gap of 1 mm. - - - - - = calculated heat transfer by gas conduction with a gas gap of 0.1 mm (Re-drawn with permission from Verlag Stahleisen GmbH [9])

measurements taken on a wide side of a mould where it can be concluded that, once a gas gap has developed, radiation is the dominant heat transfer mechanism, aided, to some extent, by gas convection.

The overall effect is that meniscus solidification, the pitch and depth of the ripple marks all increase (Fig. 7.15) with high rates of heat transfer.

Fig. 7.15 Example of an ingot with strong ripple marks due to mould slag entrainment between mould and ingot (Reproduced with permission from Verlag Stahleisen GmbH [10])

7.2.2 Inner Quality

The mechanical properties of high alloyed steel, produced by the ingot route, are controlled to a large extent, by the volume fraction, distribution, composition and morphology of inclusions and precipitates. Inclusions act as stress raisers with larger inclusions being much more detrimental than small ones. Ingot-cast steel can contain a great number of different types of inclusions; it is customary in steel terminology to classify inclusions by their origin [9, 11].

7.2.2.1 Exogenous

This type of inclusion arises from:

- Entrained slag (mould slag, slag from re-oxidation).
- Erosion products from ceramic lining (ladle, trumpet, runner).
- Entrapped hot top materials.
- Re-oxidation products.
- Entrained mould powder.
- Sand from the preparation of the runner bricks.

7.2.2.2 Endogenous

Inclusions of this type arise from:

- *A-segregates*, which arise due to the flow of solute-rich, interdendritic fluid via thermosolutal convection. They are characterised in the final solidified microstructure as channels of enriched solid, often with near-eutectic composition.
- *V-segregates* are found in the centre of the casting during the final stage of ingot solidification which is usually occupied by a network of loosely connected equiaxed grains. It is thought that *V*-segregates arise due to the fissuring of such networks under the combined actions of a metallostatic head (i.e. the weight of the material above) and the solidification shrinkage; this leads to the formation of open shear planes that can fill with any remaining liquid. This remaining liquid will have become enriched throughout solidification.
- Positive segregation.
- Negative segregation.
- Oxidation products formed in the micro segregates.

The mould powder does not affect all factors affecting the internal quality (e.g. *V*-segregates) but does influence the following:

- Entrained mould slag.
- Re-oxidation of mould slag.

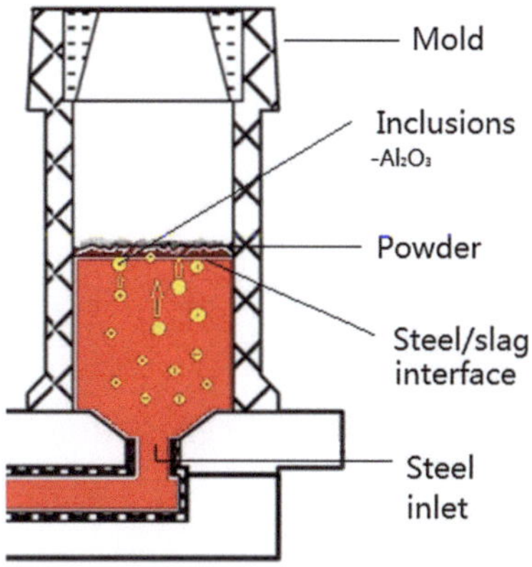

Fig. 7.16 Illustration of the flotation of Al_2O_3 inclusions to the steel/slag interface (Courtesy of Line Hallgren [12])

- Ability to absorb non-metallic inclusions of all origins that will rise to the steel surface during casting and solidification.

The principal non-metallic inclusion in ingot casting of aluminium-killed steel is alumina (Al_2O_3) which is formed by the de-oxidation of steel with Aluminium. It could also form as the result of the reaction between dissolved Al in the steel and reducible oxides (such as FeO, MnO and SiO_2) in the mould slag or by contact with the air from an unprotected steel surface.

For Al_2O_3 inclusions to be absorbed by the melted mould powder, the inclusions have to reach the steel/slag interface as shown in Fig. 7.16 [12].

The dissolution process starts once the inclusions reach the interface and have been absorbed by the slag. The absorption of Al_2O_3 inclusions in the mould slag phase results in property changes of the slag, as follows [12, 14]:

- The activity of Al_2O_3 will increase with absorption.
- Solid absorbed Al_2O_3 clusters will be dissolved and then diluted into the slag which will increase its viscosity.
- The accumulation of undissolved alumina clusters, as a second solid phase in the slag will increase its viscosity and will result in non-Newtonian behaviour.
- Once the Al_2O_3 has reached the saturation concentration in the slag, the thermodynamic driving force for the absorption will be zero.
- The heterogeneous Al_2O_3-saturated slag will prevent further dissolution of inclusions into the slag.

From a metallurgical point of view, an ideal mould powder slag with good absorption ability should have the following composition [12]:

- The basicity index, B_i (based on the oxygen-ion attractive forces and defined in Eq. 7.5 [13]) should be in the range of 1.5– 2.0.
- A viscosity in the range of 2–7 dPas at 1300 °C.
- The initial content of Al_2O_3 in the slag should be less than 15 wt-%.

– For Al-killed steels (especially when Si-content is low) a slag with a low SiO_2 content and a very low FeO content is preferred.

$$B_i = \frac{1.53(CaO + MgO + CaF_2) + 1.94Na_2O + 3.55Li_2O}{1.4SiO_2 + 0.1Al_2O_3} \quad (7.5)$$

7.2.3 Macro Segregation (Hot Top Insulation)

As shown in Fig. 7.2 the yield (from ingot to final product) for the top part of the cast is much restricted by the thermal efficiency of the hot top and the extent of the macro segregation. The mould powder influences the thermal efficiency of the hot top in two ways:

- Endothermic reactions of the remaining mould powder on the top of the ingot will increase the heat losses.
- Absorption of radiation improves the thermal insulation of the mould slag but is dependent upon its iron oxide content.

The heating, melting and the decomposition of carbonates in the mould powder are all endothermic reactions. The combustion of carbon is exothermic with the formation of $CO_2(g)$ providing almost 3 times the heat supplied by the formation of CO(g). Although the mould powder contains a large amount of carbon, most of the carbon forms a reducing atmosphere of CO(g), (C + O → CO) in the mould powder bed and it is only later that some of the CO is oxidised above the ingot surface to form CO_2. Thus the heat generation is limited and the carbon combustion is inefficient. Even with a high-carbon, fly-ash powder the overall effect of the powder is endothermic. This is clearly shown in an investigation where the thermal insulation of the exothermic hot top covering powders were studied [15] using the experimental equipment shown in Fig. 7.17. In these experiments a plate of SiC was heated to 1425 °C using super Kanthal elements while the plate was insulated with ceramic fibre. When the target temperature was reached and had stabilised, the

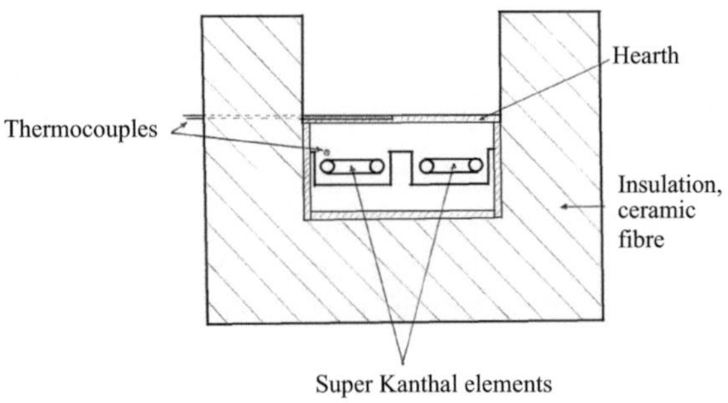

Fig. 7.17 Schematic drawing of the equipment for testing of thermal insulation of covering powders (Reproduced with permission from Swerea MEFOS [15])

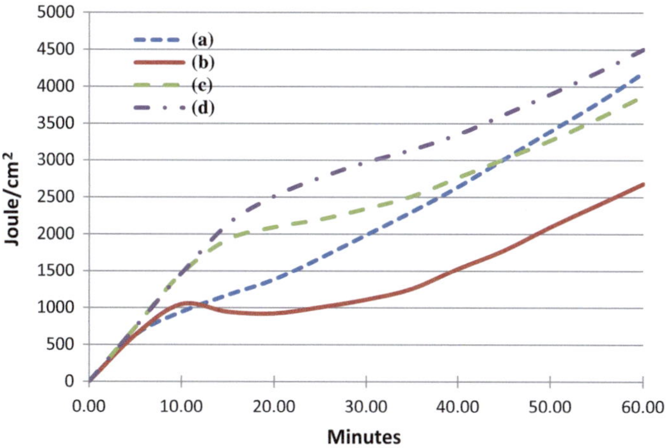

Fig. 7.18 Schematic diagram showing the accumulated heat losses from tests with exothermic covering powder. **a** 10 kg exothermic powder, **b** 20 kg exothermic powder, **c** 5 kg mould powder +20 kg exothermic powder, **d** 10 kg mould powder +20 kg exothermic powder (Re-drawn with permission from Swerea MEFOS [15])

insulation was taken off and the exothermic covering powder was applied. The SiC-plate, provided with thermocouples, was regulated to maintain the temperature. The additional energy required to maintain temperature was measured to monitor the thermal insulation provided by the powder.

The test equipment was designed to enable one to test a combination of mould powder and exothermic powder and the result from such a test is shown in Fig. 7.18 where the mould powder consisted of fly-ash powder.

The results given in Fig. 7.18 indicate the following:

- For the powder containing 10 kg exothermic powder, the exothermic reaction provides insufficient heat to compensate for the energy needed to heat the powder from room temperature to the test temperature, 1425 °C.
- Larger amounts of exothermic powder (20 kg) decrease the heat losses and have negative heat losses, during the period when the exothermic reactions occur.
- The combination of the fly-ash mould powder with exothermic powder increases the heat losses, especially during the first 40 min of the test.
- Increasing amounts of mould powder result in increasing heat losses.

When the mould powder finally melts, the molten slag cover affects the subsequent heat losses from the steel surface since the heat flux from the steel surface is influenced by the properties of the molten mould slag. Heat transfer through the slag involves three mechanisms, namely:

- Conduction.
- Convection.
- Radiation.

The conduction is rather low for mould powder slags so the largest contribution to the heat flux consists of convection and radiation. The convection contains two contributions (i) convection in the molten slag cover and (ii) gaseous conduction through the layer of any unmelted powder above the liquid pool. The convection in the liquid slag is dependent on the gradient ($T_{steel/slag}-T_{slag/powder})/d_{slag}$) and the slag viscosity, where ($T_{steel/slag}-T_{slag/powder}$) represents the temperature difference between the interfaces of the liquid slag with the liquid steel and the unmelted powder and d_{slag} represents the depth of the liquid slag layer. The gaseous conduction in the unmelted powder layer will be affected by both the temperature difference across the slag layers, the depth of the powder layer and the packing density (or permeability) of the layer.

Radiation conduction (k_R) across the liquid layer is high because it is proportional to T_K^3, where T_K is in K and is close to the liquidus temperature of the steel. The magnitude of k_R is inversely related to the absorption coefficient (α*) of the mould slag which increases significantly with % transition metal (e.g. FeO), as shown in Fig. 7.19 [16, 17]. The figure gives important information showing that the absorption of Fe_2O_3 is not so high but will increase strongly if the slag contains

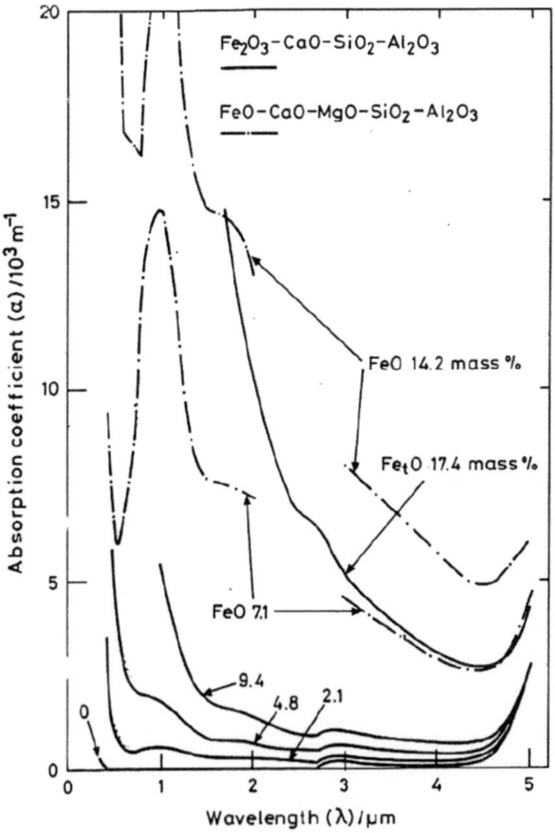

Fig. 7.19 Absorption coefficients for CaO–SiO_2–Al_2O_3 slag containing iron oxides obtained in study by M. Susa et al. Ironmaking and Steelmaking Vol. 20:5 (1993) (www.tandfonline.com). (Reproduced with permission from Taylor & Francis Ltd. [16])

FeO. The thickness of the liquid slag is also important. Values of k_R will be lower in the unmelted powder layer than in the liquid layer; this is due to the lower temperatures, the T_K^3 relation and the much higher absorption.

Fly-ash powders normally have a high FeO content. An exchange from a high-viscosity, fly-ash powder to high-viscosity mould powder with a low content of Fe_2O_3 (which will react during melting with the carbon in the powders: $Fe_2O_3 + C \rightarrow 2FeO + CO$) will therefore result in higher heat losses from the steel surface; this is due to less absorption of the radiation. If, on the other hand an exchange is made to a high basicity slag with low iron content this will crystallise on cooling leading to significant reflection and scattering of the irradiation.

7.3 History of the Development of Mould Powders for Ingot Casting (and CC)

In research and development it is important to know where it all started and the history of the first development; in the case of mould powder development, the first work was carried out for ingot casting, which is why a thorough description of that work is made in this chapter.

Traditionally ingot casting was made by top pouring of various (rimmed, semi-killed) steels and there were no need for mould powders. Splashing and bad surfaces on the bottom part of the ingot were general problems which were found to be reduced by surface coating of the mould surfaces with tar or other carbon additives. However, one attempt to add mould powder in connection with top pouring has been reported [18].

At the introduction of bottom pouring, surface cracks were a great problem because of the oxidation and the cooling of the unprotected steel surface (see Fig. 7.20). A laboratory scale, investigation from 1957 [19] revealed that surface dressings improved the surface quality of steel, especially if the dressings were volatile, since:

- They prevent re-oxidation of the steel meniscus.
- They keep surface films or slag away from the interface.
- They prevent excessive decarburisation of the ingot surface.
- They prevent sticking of splashed steel to the mould wall.
- They promote a smooth metal rise in the mould.

It was concluded from this study that bottom pouring required a steel surface covered with a non-oxidising atmosphere.

In the beginning of the 1960s more plants carried out tests on bottom pouring of killed steels and it was found that some sort of insulation was needed to protect the steel surface from freezing. From the inspired work of Thornton [19] it was obvious that this must contain re-oxidation material. However, the amount of steel being

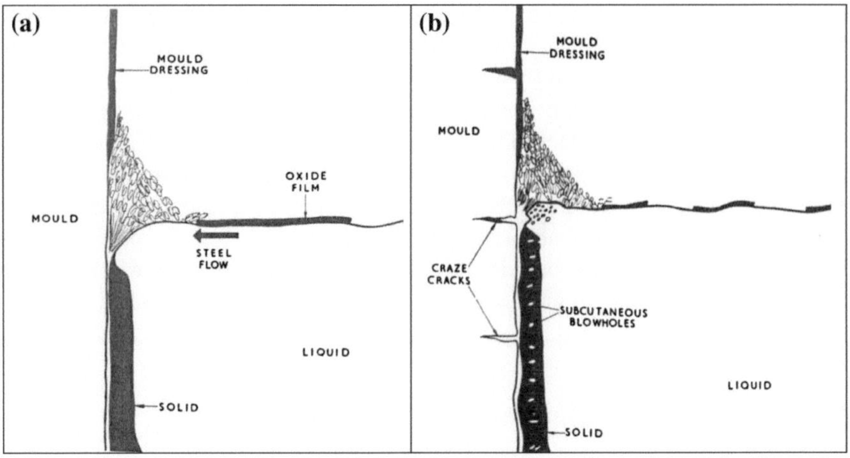

Fig. 7.20 Schematic illustration of **a** correct action of an active mould dressing **b** action of an active mould dressing on a crazed mould (Reproduced with permission from Steel Times International [20])

cast by bottom pouring in the mid-1960 was very low and only three steel plants in USA practised this technique.

In England they suffered from bad ingot surfaces with major defects such as hanger cracks, corner cracks and scabs. Thus, surface planing with an oxygen lance, or grinding, was necessary and could result in a yield loss of around 5%.

For many years, active dressings were applied in the manufacture of killed steels, to reduce these problems. Dehydrated gasworks tar was an effective dressing but it was difficult to apply an even coating of this material. Several problems were encountered as a result of its high viscosity at ordinary temperatures, sprays blockages and maintenance proved difficult. This led to the development of alternative lacquers of less viscous, hydrocarbon derivatives.

The study by Thornton [19] clearly showed that a correctly applied volatile hydrocarbon burns in advance of the steel meniscus, producing a slightly reducing atmosphere above the liquid steel surface. Furthermore, the current of hot gases prevented the oxide film from reaching and adhering to the mould (Fig. 7.20a).

This technique improved the surface quality of bottom-poured, killed steel but caused problems when older moulds with crazed cracks were used. Application of a fluid to a crazed mould wall resulted in the dressing being absorbed, by capillary attraction, into the fissures. Under these circumstances, the dressing is shielded from the heat of the rising metal and does not volatilise until the metal comes adjacent to the mould defect. The sudden evolution of gas caused turbulence within the mould and raised the hydrocarbon content of the rapidly solidifying, surface layer. This led to the formation of subcutaneous blowholes [20] as shown in Fig. 7.20b.

When production started of grain-size-controlled, killed steels containing aluminium, for all types of alloyed steels, the active dressings were unable to produce a sufficiently reducing atmosphere to prevent oxide layer formation. This layer eventually became entrapped, causing laps with associated hot tears, blowholes and subcutaneous refractory inclusions.

Thus, there was a need for the improvement and development of better surface dressings with the following requirements [20]:

- The dressing must prevent atmospheric oxidation of the steel and its alloys.
- It should contain no readily volatile matter which might get entrapped below the rising liquid level thereby causing blowholes.
- It should prevent teeming scum from becoming entrapped in the ingot surface.

In addition, the dressing should act as a good thermal insulator of the steel bath surface.

Rippon [20] was the first to describe the use of mould powders which were both insulating and which would partially flux at a temperature just below that of liquid steel. He commented that:

- The method, using mould powder, is simple to control, is reproducible and produces an ingot with fine surface quality when correct teeming rates are used on bottom-poured material.
- Perhaps the most remarkable effect of these powders, however, is the production of a good ingot surface when using crazed moulds; this is illustrated in Fig. 7.21.

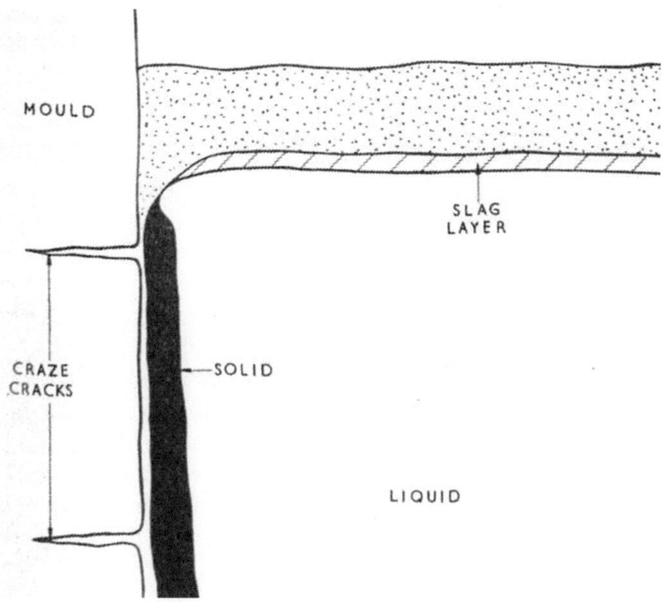

Fig. 7.21 Schematic illustration when a mould powder is used during teeming into a mould with a crazed surface (Reproduced with permission from Steel Times International [20])

The mould powder described by Rippon was based on fly ash, a by-product from coal combustion. This was revealed in an article by Alexander [21] where he also explained the problems that could be met by using this product. "Fly ashes vary considerably in composition, form, particle size and impurities, according to the fuel mix burnt, the efficiency and type of combustion furnace, the methods of separation from the flue gases by cyclone, electrostatic or washers and lagoons, and thus of 50 examined, only two or three are possible starters for this application".

Alexander [21] was very enthusiastic regarding the use of mould powders and put forward the following advantages in using this new technique:

(1) Elimination, or minimising, of subcutaneous cavities and inclusions.
(2) Elimination of cold laps and the improvement of the general surface smoothness.
(3) Minimisation of heat losses with the additional advantage that low tapping and pouring temperatures can be used whilst still producing first-class, quality ingots. It should be noted that in the early days of using these powders, it was common practice to leave an opening in the centre of the steel surface but subsequent work showed the importance of maintaining a complete cover of the steel.

In the USSR pure graphite powder were tested and used on a regular basis which was reported from a steel plant [22] with the following result:

- Treating killed steel with graphite powder while it was being bottom-cast in large plate, ingot moulds hinders the development of oxide films on the metal meniscus; this improves the surfaces of the ingots, reducing the amount of teeming laps and gas blowholes, also (to a lesser degree) the amount of cracks.
- The use of graphite during the teeming of killed steel at the Orsk-Khalilovo combine reduced the proportion of slabs and plates rejected and also the number of products with incorrect dimensions. Providing the amount of graphite does not exceed 250–350 g/ton of steel, the metal is not carburised.

Probably inspired by the work carried out in England, tests were also made with fly-ash powder in USSR [23]. It was observed that a normal fly ash, with low carbon content, fused rapidly during the casting process. This had a markedly adverse effect on the heat-insulating properties and gave rise to defects on the ingot surface.

To avoid this problem, they produced a mixture of fly-ash material and graphite with a ratio of ash (75–85%) to graphite (25–15%) controlled so as to produce a carbon content in the mixture of not less than 24% which was found to be optimum. If the carbon content was lower than 24%, it was difficult to obtain good flowability and spreading of the ash–graphite mixture; this made it difficult to ensure thermal insulation of the entire surface of the steel which rises in the big-end-up ingot moulds. A rim of molten metal incidentally appeared on the mould walls which could adversely affect the ingot surface quality.

Tests were carried out where a number of differently killed steel grades (conventional grades, low-alloyed steel, chromium and chromium–nickel steels) were cast with the ash-graphite powder. The results were then compared with those obtained using the old technique employing a mould dressed with tar. The following advantages were identified from two years' experience of using fly-ash–graphite powders:

- There was a great reduction of billet rejects due to surface defects.
- There was a reduction of top discard of 5–10 kg/ton steel.
- There was a reduction of metal rejected at customers' factories.
- There was an improvement in mould life by 1.5 kg/ton of steel.

Fly ash was accordingly established as a good material for mould powder production and is still the dominant raw material used for ingot-casting mould powders. It was soon recognised [23] that the fly ash could only be satisfactorily used in regular production if it was used as a base material into which other materials could be mixed to obtain the desired properties [24, 25]. Fly ash consists mainly of Al_2O_3, SiO_2 and FeO to which both carbon and sodium carbonate are added; the latter decreases the viscosity and increases the melting speed. It was found desirable to have a thin slag cover on the steel surface so as to avoid oxidation if open areas occurred, without powder cover, during the teeming operation. The slag layer counteracts carburisation of the steel by eliminating direct contact between the carbon-rich powder and the steel surface.

The reason for fly ash being so successful, as a base for mould powder production, lies not only in its low price [25] but also because it is a binary phase having the right chemicals and carbon structure. Adequate grain size, combined with the particle form, gave both good insulation and flowability. Only a few fluxing agents are necessary to achieve the right melting point.

Although, the first articles regarding mould powders were written in UK and USSR, the real pioneer in their development was Hans Joachim Eitel, the founder of Metallurgica. His PhD Thesis from 1990 [26] is a detailed description of the development of mould powders both for ingot and continuous casting and a brief summary of his work is worth bringing out into the light.

Eitel states that in the development of bottom-pouring, ingot casting, the introduction of mould dressing was not optimum because it introduced restrictions regarding the pouring rate. For a good casting, a high pouring rate was needed but that increased the risk for longitudinal cracks. If a low pouring rate was used the steel surface tended to freeze during the casting which lead to internal quality problems in the top section causing reduction in the yield.

To overcome this problem a number of different insulation materials, to be put on top of the steel surface, were tested such as:

- Wood shavings impregnated with water glass.
- Boards made of wood.
- Strips of corrugated cardboard.
- Rice ash.

None of these materials gave the desired result, so in 1957 the first trial was made with a pure fly-ash powder at the Belgium steel plant Cockerill Seraing. This showed good results and it was shown that it was possible to cast at a very low casting speed, which eliminated longitudinal cracks, without getting freezing of the top surface, due to the very good thermal insulation properties of the fly ash.

At the beginning, pure fly ash was used but at an early stage of the tests it was recognised that it was desirable to produce a molten slag layer to avoid fly-ash powder coming in direct contact with the steel surface (and thereby carburising it). As mentioned above, fly ash consists mainly of SiO_2, Al_2O_3 and FeO to which soda ash (Na_2CO_3) is added to lower the melting point of the powder, creating a high viscosity slag.

The first production of mould powder for ingot casting was started in February 1958 with fly ash from Kraftwerk Osterfeld and later also Kraftwerk Wehlheim and Schlägel & Eisen in Recklinghausen. Around 10–15 wt-% fine grained soda ash was added to the fly ash and then mixed. This was delivered to those steel plants in Germany using bottom pouring, i.e. Dillinger Hüttenwerke, Stahlwerk Phoenix-Rhein-Ruhr, Stahlwerk Südwestfalen and Edelstahlwerke Buderus.

The production of mould powder increased during the following years and it was found that the powder was not only a good product for ingot casting but could also be used as surface dressings for the ladle and tundish.

Different recipes were developed for mixtures of fly ash, soda ash and carbon additions (such as coke breeze) to meet the needs of different steel grades and casting practice. The need to add coke breeze was identified because of the great variations in carbon content in the fly ash. As the power plants became more efficient the carbon content fell and carbon additions became essential.

7.3.1 *Development of Mould Powders for Continuous Casting*

The development of mould powder for ingot casting was also the basis for the development of powders for the continuous casting process. This started on a small scale for billet casting in the 1950s and in 1961 mould powders were used for the first time in the slab machine at Dillinger Hüttenwerke. The introduction of the continuous casting technique was slow in the beginning due to technical problems, large investment costs and the need for higher education of both operators and maintenance staff. The percentage of world steel production being continuously cast rose sharply in the 1970s and 80s from 14.2% in 1975, to 30% in 1980 and 54.8% in 1987. However, in 1987 the CC-cast proportion in Germany was 87.9%, in France 93.1% and in Japan 93.3% [27].

When Dillinger Hüttenwerke started to operate their slab machine they needed lubricant to prevent sticking of the solidifying shell in the mould; this, traditionally, was performed by using vegetable oil. This technique, for slab casting had a number of drawbacks [38] such as:

- Unstable distribution.
- Hydrogen and carbon pickup.
- Large amount of breakouts.
- High depression frequency.

The much calmer and wider steel surface in a slab machine probably increased the problem (cf. billet casting) with its greater need for thermal insulation of the steel surface.

Due to these problems, as an initiative from Dillinger Hütte, a mould powder was developed in 1963 dedicated for continuous casting which was based on the experience obtained from ingot-casting powder development. However, in addition to providing good thermal insulation properties, the mould powder slag was required to act as a lubricant (i.e. to replace the vegetable oil).

This put new demands on both the development and quality assurance of the mould powder, since a poorly performing mould powder had a much greater financial impact, especially, if it resulted in a breakout. In the beginning, only silicon-killed steels (with low addition of aluminium) were continuously cast and it worked fairly well for these steels. A standard mould powder for ingot casting, containing a mixture of fly ash and soda was used. When they started to cast Al-killed, deep-drawing steels the situation changed dramatically and led to the introduction of the first, low-viscosity, mould powder for continuous casting which contained a mixture of; 10% fluorspar, 7.5% soda, 30% Portland cement and the rest fly ash.

The rapid spreading of the continuous casting technology, especially in Europe and Japan, went hand in hand with the development of mould powders for CC and became the prime interest. Further development of new ingot casting powders was, to a large extent, based on the experience derived from the development of powders for continuous casting.

7.3.2 Development of Synthetic Mould Powders

In the beginning it was considered that fly-ash powder contained all the necessary components needed for a mould powder, i.e. SiO_2, Al_2O_3, CaO, Fe_2O_3 and C. However, it was realised that this restricted range of components limited better design of the mould powders (e.g. the high FeO content of fly ash was undesirable when casting Al-killed steels). The increasing demand for better properties of CC powders led to the development of mould powders which excluded fly ash. These powders were a mixture of a number of different raw materials and are frequently referred to as "*synthetic mould powders*" to distinguish them from the fly ash-based powders. The raw materials and production technology of synthetic mould powders are described in Chap. 8.

7.3.3 Development of Granulated Powders

In 1976 the Rautaruukki steel plant in Finland required a CC mould powder which should be dust-free and thus was needed in an agglomerated form. To start this development, (led by Hans Joachim Eitel [26]), a list of demands were drawn up (Table 7.3).

With this background, different techniques for granulation were studied and two options were identified: i.e. roll granulation or spray-granulation.

Spray granulation was found to be the better method for two reasons:

- It resulted in a product with lower density.
- Lower water content in the granules and no need for subsequent drying.

At that time, spray granulation had been developed for the chemical- and food industry, so it seemed to offer a reasonably quick route in the development of mould powder granules (although it was an untried technology with no documented activities). With this background Eitel [26] in 1978 received a patent for granulation of mould powders with the title "*Verfahren zur Herstellung von Gießpulver*". Eitel was looking for a suitable factory which could be rented for trials with mould powder and found the company Dorst in Kochel am See which produced granules for the ceramic industry and in November 1978 the first trial was made to produce mould powder granules.

With optimisation of the spray technique regarding nozzle design, composition of the slurry and spray pressure, a reasonably good product could be manufactured. In the beginning, around 150 ton/month was produced for the first customers for the granules; i.e. the steel plants of Rautaruukki (Finland), Sollac (France) and BSC (UK).

The advantages of granules over loose powder were soon identified and the word was quickly spread among the continuous casting community. In 1983 the production was 1300 ton/month. With this success of the new product, a factory dedicated to the production of mould powder granules was built with a capacity of 1500 ton/month which was started in 1983. The increased demand for this product

Table 7.3 Demands for a new granulated powder generation

Mould powder	Mould slag
Good thermal insulation properties due to low density	Optimum viscosity depending on steel grade and process parameters of the casting.
Dust free	Protection of the steel surface from re-oxidation
Good flowability	Good absorption ability for non-metallic inclusions
Low water content	Good lubricating ability between steel shell and mould
Good homogeneity	Good and even heat transfer from steel surface to mould
Suitable melting properties	

lead to an increase in production and an investment in a second spray dryer in 1990. It also included sales of licenses of the technology to companies in Japan and USA.

7.3.4 *Today's Situation Regarding Mould Powders for Ingot Casting*

After the rapid increase in continuous casting and the development of mould powders specifically adapted for this technology, the development of ingot casting powder has mainly been based on experience from CC mould powder development. The literature of this subject has been sparse compared to the corresponding literature of continuous casting and were mostly published during the 1970s and 80s [24, 25, 28–37] with concerns to find the optimum mould powders for ingot casting. After that, for a period of more than 20 years, almost no research was published on mould powders for ingot casting; the only exceptions being [38, 39].

The remaining steel grades that are cast via the ingot process route today are high grade, high quality steels. This has sparked a new, awakened interest in the development of both ingot casting technology and mould powders for ingot casting. The need for more research was identified by Stahlinstitut, VDEh in Germany; consequently, they arranged the first international conference on ingot casting, rolling and forging (ICRF) 2012 in Aachen which was followed up by ICRF 2014 in Milan, Italy. These conferences resulted in a number of new proceedings concerning mould powders for ingot casting [12, 40–43].

7.4 Selection of Mould Powders for Ingot Casting

To understand the influence mould powders have on the quality of ingot cast steels one must distinguish between the two different materials involved.

- Mould powder.
- Mould slag.

7.4.1 *Important Properties of the Mould Powder*

The three key physical properties for ingot powders are:

- Melting behaviour.
- Flowability.
- Thermal insulation.

7.4.1.1 Melting Behaviour

A number of variables are known to influence the melting behaviour (see Chap. 4) such as:

Steel; Steel temperature, steel flow (pouring rate), ingot geometry.

Mould powder; Mineralogical composition, carbon content, carbonate content and effective thermal conductivity.

Addition of free carbon to the mould powder is the most efficient way to control the melting speed. While mould powders are composed of a mixture of mineral components with high melting points and fluxing agents, such as CaF_2 and Na_2O, carbon particles delay the contact and the sintering of oxide particles. The carbon also delays the agglomeration of molten slag droplets. The melting speed is controlled by both the amount and the size of the particles (where a finer carbon is much more effective (per unit mass of C) in reducing melting speed).

Mould slag irradiative absorptivity; This is the ability of the slag to absorb the IR radiation emitted by the steel shell through the molten mould slag to the covering, hot-top powder. The heat flux is influenced by the thickness of the slag and the concentration of transition metal oxides (e.g. FeO) in the slag. The slag viscosity also affects heat transfer from the steel bath to the powder by convection in the liquid slag since convection is inversely proportional to the slag viscosity.

7.4.1.2 Flowability

The ability of the mould powder to spread out over the steel surface is very important in order to ensure that the entire surface is covered with mould powder throughout the casting. It is important to prevent both the formation of an "*open eye*" and subsequent re-oxidation of the steel. Flowability is not so critical for smaller ingots but is important for larger and, especially, big-end-up (BEU) ingots where the negative ingot taper requires increased spreading of the powder as the pour proceeds. High-carbon, fly-ash, mould powders normally show a good flowability, whereas synthetic powders tend to exhibit poorer flowability [36].

Granulated mould powders normally have a good flowability because of their narrow particle size distribution.

7.4.1.3 Thermal Insulation Properties

Heat is lost from the molten steel by radiation, conduction and by liquid and gaseous convection. Radiation is reduced by ensuring the FeO content of the slag is high to absorb the infra-red radiation; alternatively, a high-basicity, crystalline slag could be used to reflect and scatter the radiation where high FeO contents are unacceptable (e.g. in Al-killed steels). Liquid convection is reduced by using a high-viscosity, fly-ash slag. Gaseous convection through the powder and sinter decreases with increasing bulk density; the latter, in turn, is increased by using

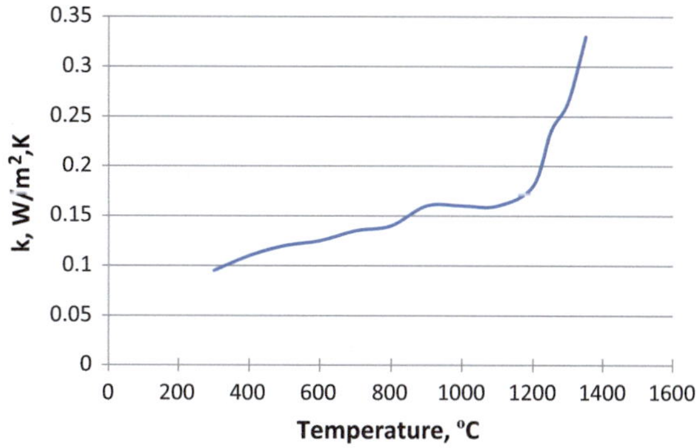

Fig. 7.22 Thermal conductivity of a granulated mould powder as function of temperature measured by the transient hot wire method [42]

granules of smaller size. The insulation properties of mould powders are normally very good as witnessed by the low thermal conductivity (Fig. 7.22 [40]) for a granulated mould powder. It must be noted that the thermal conductivity in reality is somewhat higher due to convection from the gases evolved during heating (described in more detail in Sect. 9.6.4).

7.4.2 Important Properties of the Mould Powder Slag

The most important properties for the mould powder slag are the following:

- Viscosity.
- Ability to absorb inclusions.
- Irradiative absorption properties.
- Interfacial tension.

7.4.2.1 Viscosity

The viscosity determines the infiltration of liquid slag into the gap between the solid steel and the mould and there is a large difference in viscosity for different oxide systems. The viscosity is very dependent upon the degree of polymerisation of the silicate or alumina–silicate network and upon temperature. A typical viscosity curve for a high viscosity, commercial soda-lime silica glass composed of 75 wt-% SiO_2, 15 wt-% Na_2O and 10 wt-% CaO is shown in Fig. 7.23 [44].

The viscosity curve is very steep and the dependence on temperature is strong due to the strong network bonding of SiO_4^{4-} tetrahedra in the glass.

Experimental, viscosity curves from high-basicity, blast furnace slags are shown as a comparison in Fig. 7.24 [45]. The compositions of these slags were 36–47 wt-% SiO_2, 32–43 wt-% CaO, 13.0 wt-% MgO and 8.0 wt-% Al_2O_3 and the silicate

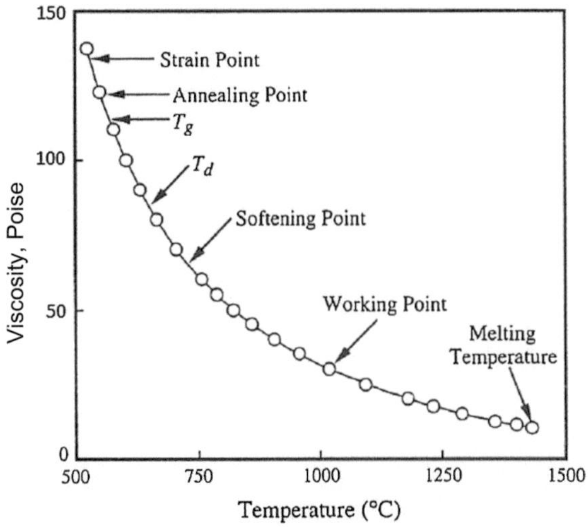

Fig. 7.23 Typical curve for viscosity as function of temperature for a soda-lime-silica melt showing characteristic points such as the glass transition temperature (T_g). (Reproduced by permission of The Royal Society of Chemistry [44])

network of the slag is much less polymerised than that of the glass shown in Fig. 7.23.

The temperature dependence is also much lower for the blast furnace slag compared to the glass because the melt is much more depolymerised due to a high concentration of CaO and MgO. The marked increase in viscosity at around 1375 °C is due to precipitation of crystals in the slags which is defined as the break temperature (T_{br}).

Due to a high amount of SiO_2 and Al_2O_3 in most fly-ash-based mould powders, the temperature dependency of the slag viscosity is similar to that shown for glassy slags in Fig. 7.23. In contrast, the viscosity–temperature relation for a typical high-basicity, synthetic mould slag has the same appearance as that for the blast furnace slag but has a higher degree of de-polymerisation which leads to:

- The viscosity is lower than the blast-furnace slag.
- The temperature dependence of viscosity is very low until the crystallisation (or break) temperature is reached.
- The break temperature is lower than that for the blast-furnace slag.

7.4.2.2 Ability to Absorb Inclusions

For clean steel production the mould slag can act as a flux to dissolve high-melting oxides and inclusions, (such as Al_2O_3, de-oxidation and re-oxidation products); this prevents these oxides from coalescing and possibly becoming trapped in the ingot surface. Emulsification and slag entrainment become easier with lower interfacial tension, lower slag viscosity and higher slag density [11].

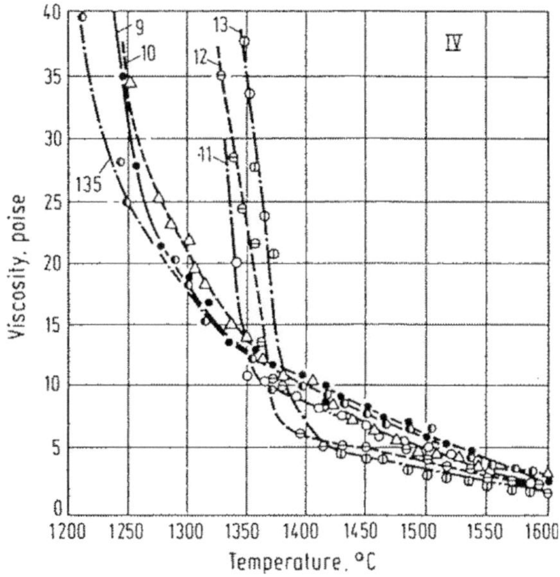

Fig. 7.24 Temperature dependence of viscosity (in dPas) of various blast furnace slags (Reproduced with permission from Verlag Stahleisen GmbH [45])

7.4.2.3 Irradiative Absorption Properties

A molten mould slag has a rather low thermal conductivity of around 0.4 W/m, K [8]. Of more importance is the ability to absorb radiation which is very much related to the content of FeO. Heat flux by radiation is mainly transferred within the infrared region 0.7–3.0 μm. The absorption coefficient of a mould slag at a wavelength of 2 μm can be calculated by the following formula, [8]:

$$\alpha^* = 30 + 910\,(\%FeO) + 5\,(\%MnO) + 410\,(\%NiO) + 390\,(\%Cr_2O_3) + 370\,(\%Cr_2O_3). \quad (7.6)$$

7.4.2.4 Interfacial Tension

The interfacial tension between mould powder slag and the liquid steel influences two important factors:

- The *size of the meniscus*, which, in turn, affects the size of ripple marks. A low interfacial tension results in a shorter meniscus, leading to shallow ripple marks, which normally is preferred and even to their disappearance.
- The *emulsification of the mould slag*. If the interfacial tension is low there is a risk that slag drops get dragged from the mould slag surface and can be trapped in the surface of the steel shell as macro inclusions. The risk for this to happen is greater when the fluid flow is high (as at the start of casting and for higher

pouring rates). On the other hand a mould slag with a low interfacial tension has a greater ability to absorb inclusions coming from the steel.

7.4.3 Selection of Mould Powders in Regard to Steel Grade

The principal tasks which the ingot powder is expected to perform are:

- To protect the steel surface from oxidation.
- To provide thermal insulation from the surface to prevent the steel from freezing during casting.
- To absorb inclusions from the steel during and after casting.

Unlike mould powders for continuous casting, ingot powders are not expected to:

- Lubricate the ingot/shell surface—so powder consumption is not an issue here except consumption must be sufficiently low to ensure that the thermal insulation remains intact for the entire cast.
- To provide controlled, horizontal heat transfer and solidification.

So what properties are desirable in an ingot mould powder?

- T_{liq}; ought to be low but not at the expense of producing a high melting rate.
- ***Viscosity***; should be low for good Al_2O_3 absorption but this will possibly lead to high consumption and a risk of mould slag floating in between the mould and ingot causing both increased heat transfer and slag patches on the ingot surfaces. So there is a balance and the viscosity must be chosen according to the steel grade being cast.
- ***FeO-content***; a high content will decrease thermal radiation during and after casting but means an increased risk of Al_2O_3 formation in Al-killed steel, so the trend today is to lower the FeO-content in mould powders for ingot casting.
- ***Interfacial tension***; a low interfacial tension provides better surface finish (less deep ripple marks) assisted by high Na_2O and a high S content in the steel. However, low interfacial tension leads to more slag entrapment and to scale formation during the rolling operation.
- ***Melting speed***; in most cases a low melting speed is desirable; this ensures that there is a good insulation of the steel surface during casting without much mould slag penetrating between the mould and the ingot.

Today there are two types of mould powders on the market, fly-ash-based and synthetic powders. Most powders are still, fly-ash, loose powders but granulated powders are being used to a higher extent mainly because:

- They are more environmentally-friendly with much lower emissions of dust during casting.

- The flowability of synthetic powders are normally inferior to the fly-ash powders but are improved with granulation, due to the narrower fraction distribution.

The selection of mould powders is to a large extent dependent upon the steel grade [33] but still fly-ash-based mould powders with high carbon content tend to dominate sales for the following reasons:

- Low price.
- Very good insulation properties.
- Low consumption rate which will secure a thick insulation blanket throughout the whole casting.
- High melting temperature which reduces the risk of slag entrapment during the early, turbulent, stage of the casting due to low slag formation at this stage.
- High irradiative absorption of the molten slag which reduces the heat losses from the top of the mould during solidification thus making the hot top more effective.

A typical composition of a fly-ash-based, high carbon, mould powder is shown in Table 7.4 and its properties are listed in Table 7.5.

Table 7.4 Chemical composition of a typical fly-ash mould powder

SiO_2 (Wt-%)	30.0–40.0
CaO+MgO (Wt-%)	2.0–8.0
Al_2O_3 (Wt-%)	15.5–21.5
Na_2O+K_2O (Wt-%)	7.5–10.5
FeO (Wt-%)	4.0–10.0
C_{free} (Wt-%)	19.0–21.0
CO_2 (Wt-%)	3.0–5.0
$\underline{C}_{total}$ (Wt-%)	20.0–22.0
$H_2O_{600°}$ (Wt-%)	<0.8
Basicity	<0.2

Table 7.5 Properties of a typical fly-ash mould powder; * denotes calculated value

Softening point, °C	1080 ± 30
Melting point, °C	1260 ± 30
Fluidity point, °C	1390 ± 20
$Viscosity_{1300\ °C}$, dPas*	Approx. 600
Absorption. Coeff., m^{-1}*	Approx. 5000–10,000
Sieve analysis, 500 μm	<1.0
Sieve analysis, 125 μm	12.0–18.0
Sieve analysis, 63 μm	26.0–36.0
Bulk density, kg/m^3	600–700

The drawbacks with the fly-ash powder are its low basicity and the high amount of iron oxide. The latter leads to a risk of alumina formation during the casting by the reaction:

$$3FeO + 2\underline{Al} \rightarrow Al_2O_3 + 3\underline{Fe} \quad (7.7)$$

The thickness of the liquid slag on top of the ingot will become very shallow because of the high melting point of the powder and the low melting rate (due to the high carbon content). It may be expected that the reaction rate for alumina generation will be reduced by the high slag viscosity; however, the ability to absorb inclusions will also be very low for the very same reasons.

The recommendations for selection of mould powders for ingot casting related to steel grades are as follows.

7.4.3.1 Low to Medium Carbon, Low Alloyed Steels

The primary functions of the mould powders are to protect the steel meniscus from re-oxidation and heat loss, and to provide a good cover during the whole teeming process. Fly-ash, high carbon, mould powders are capable of performing these tasks satisfactorily.

7.4.3.2 High Carbon (HC) Steels

It is possible to use a similar powder for high carbon steels, (>0.8 wt-%) to that used for low carbon, low alloyed, steels with one exception. Since HC steels have a much lower liquidus temperature and hence, a low pouring temperature, it is important to ensure that a liquid slag layer is formed to cover the steels. Thus, the melting properties must be improved by adding more fluidisers (i.e. soda or fluorspar) for casting HC steels.

7.4.3.3 Tool Steels

The demands on tool steel are very high, especially with regards to non-metallic inclusions; since the mechanical properties of the steel deteriorate with increases in both the number and size of the inclusions. Thus, a mould powder forming a mould slag with good capability to absorb inclusions is desirable. This calls for a high-basicity powder with a low FeO content.

The result is a low-viscosity mould powder slag and a mould powder with higher melting speed. The higher melting rate of the powder can lead to some drawbacks which must be taken into consideration when designing the powder in terms of carbon content and the magnitude of mould powder additions.

- A high consumption of mould powders can result in open steel areas without powder cover in the middle of the ingot ("*open eyes*") which poses a greater risk of creating alumina.
- Excessive melting can result in a very deep slag pool with slag entrainment between mould and ingot which can be detrimental to the surface quality and can also lead to problems in the heating furnace.
- A low-viscosity slag has greater ability to absorb non-metallic inclusions but it is also more reactive and if its sulphur content is too high there is a risk of increased sulphur pick-up by the steel and more sulphides in the final product.
- A mould powder slag with low iron content will have lower thermal absorptivity, thus, heat losses by radiation from the top surface during solidification will be larger which will have an impact on the hot top efficiency.

7.4.3.4 Ultralow Carbon Steels

For ultra-low carbon, (<0.02 wt-%,) steels, it is important to avoid carbon pickup during casting and total carbon of the mould powder should not exceed 1 wt-%. This, in turn, will increase the melting rate of the mould powder and ensure that a thick slag layer is formed, which will minimise the contact between mould powder and steel. To avoid excessive melting of the powder, with the risk of slag patches on the surface, the basicity and the amount of fluidiser ought to be lower than that used in the mould powder for tool steels.

7.4.3.5 Stainless and High-Alloy Steel

During casting of stainless and high-alloy steel grades, it is normal to generate a high concentration of refractory oxides. Absorption of non-metallic inclusions is therefore very important to avoid severe surface defects. A high-basicity powder is recommended with good melting properties entailing the addition of fluidisers such as Na_2O and CaF_2 in order to ensure a thick molten pool on the steel surface during casting. The carbon content should be low to avoid carbon pickup and to increase the melting rate.

An example of recommended mould powder selection from a large producer of mould powder for ingot casting is shown in Table 7.6.

Table 7.6 Typical properties of mould powders for ingot casting (courtesy of ALSICAL Hüttenwerkstechnik GmbH)

Brand name	PC10–PC27* GC10–GC27**	PC2 or GC2	PA1	PA3	PEx1
Range of application					
Steel grade	Unalloyed and alloyed carbon steels	Tool steels	Stainless steels Steels with Al and Ti	ELC-steels	Heating conductor material
Ingot size (ton)	All	0.5–2.0	All	All	0.5–2.0
Consumption (kg/ton)	1.5–1.8	2.0	2.5	2.0	6.0
Chemical composition, % (approx.)					
C_{free}	10–27	18	<4	<0.3	<0.5
SiO_2	33	23	43	50	25
CaO/MgO	8	30	30	40	25
Al_2O_3	18	9	9	4	20
Na_2O/K_2O	8	9	6	2	4
Fe_2O_3	2–5	3	2	1	15
H_2O	<0.6	<0.6	<0.6	<0.4	<0.4
Physical parameters					
Density (kg/dm^3)	0.6	0.7	0.6	0.4	1.0
Sintering point (°C)	1060	1030	1150	1100	Exothermic
Melting point (°C)	1260	1150	1200	1300	Exothermic
Fluidity point (°C)	1320	1170	1220	1320	Exothermic

Remarks *Carbon content depends on ingot size and steel grade. ***P* powder, *G* granule

7.5 Application Techniques for Mould Powders

Several techniques are available to apply the mould powders to the steel surface. Normally, mould powders are hung in paper bags around 300 mm from the bottom plate. This is shown in Fig. 7.25 where four 5 kg bags are hung from the hot-top tile wedges in a 10 ton ingot mould. When the steel enters the mould, the strong radiation from the steel surface sets the paper bags on fire and the mould powder pours out from the bags. It subsequently spreads over the steel surface.

Another way of applying the mould powder [25], is shown in Fig. 7.26 where the mould powder bag is placed on an expanded cardboard shape. This will eventually burn, which ensures that the mould powder reaches the steel surface when a certain amount of steel has been filled in the ingot.

Fig. 7.25 Photograph showing 10 ton ingot mould prepared with mould powders

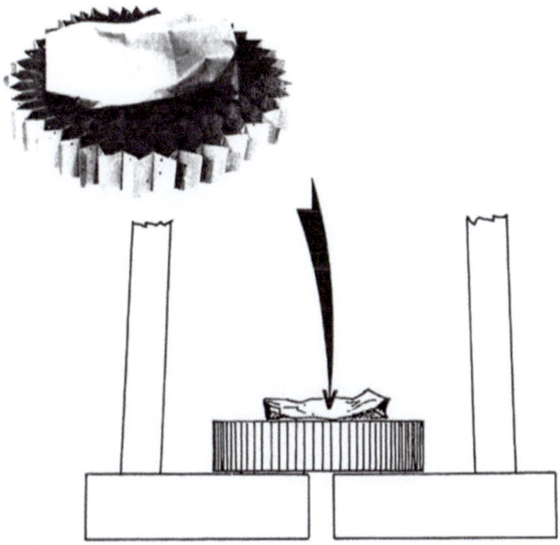

Fig. 7.26 Diagram showing mould powder on an expanded cardboard shape (Reproduced with permission from Association for Iron & Steel Technology [25])

Attempts have also been made to add the mould powder as a preformed board as shown in Fig. 7.27 [25]. This method has not become widespread probably because most ingots today are big end up (BEU) and this means that the board will not cover

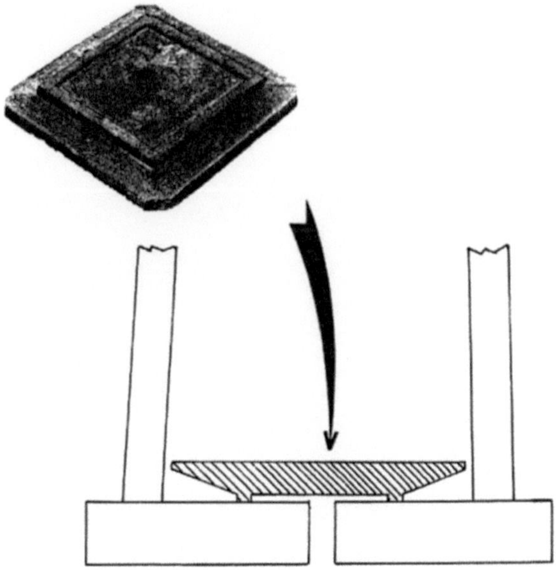

Fig. 7.27 Schematic drawing showing a pre-formed board of flux material (Reproduced with permission from Association for Iron & Steel Technology [25])

the full meniscus during the casting, resulting in unprotected areas on one, or more, sides of the ingot.

This problem was clearly experienced from work within a RFCS project regarding ingot casting, IPTINGOT (RFSR-CT-2011-00006) where pre-formed, mould powder boards were tested. From the plant trials it was shown that mould powder boards can have another positive effect when it comes to minimise the addition of mould powder.

In order to avoid "*open eyes*" with the increased risk of alumina creation, it is common practice to over-compensate and add more mould powder than necessary to ensure that the centre of the mould is covered throughout the whole teeming operation, especially for BEU ingots. This has a negative impact on the hot-top function because of the endothermic nature of the melting of mould powder (see Fig. 7.18). In some cases this problem is solved by manually adding small mould powder bags during the casting to cover the "*open eye*" but in many cases this is impossible because there is no manual access to the mould.

In these cases a pre-formed mould powder board could be a solution, since these can be hung over the middle of the mould before casting. An example of such a test, from the IPTINGOT project, is shown in Fig. 7.28.

If the board has a chemical composition with a high ability to absorb non-metallic inclusions, and has low sulphur content, this would be a way of achieving cleaner steel.

Another way to add mould powder is by continuous additions from a bin through a steel pipe where the addition rate can be regulated by a ball valve. This

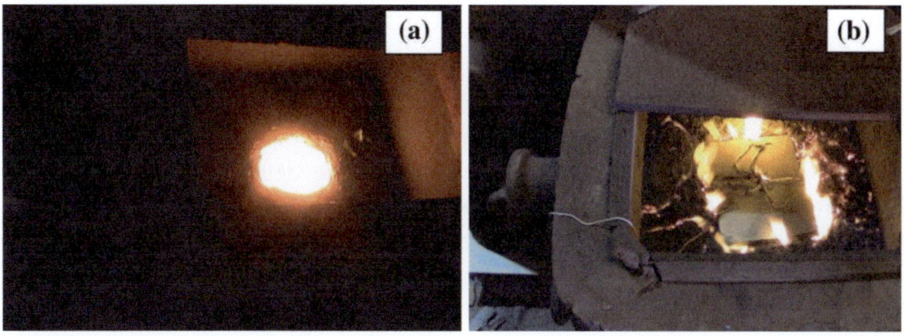

Fig. 7.28 Photographs showing. **a** An "*open eye*" in a 6 ton ingot. **b** Closing the "*open eye*" with a mould powder board

technique is suitable for granulated powders but loose powders cannot be applied in this way. Using this technique it is possible to avoid the splashes which normally emerge in the beginning of the casting, when the mould powder bags open; this results in a better surface in the bottom part of the ingot. The overcompensation of mould powder to avoid "*open eye*" can also be reduced by the continuous feeding in the centre of the ingot. An example of such equipment is shown in Fig. 7.29.

Fig. 7.29 Photograph of the equipment for continuous feeding of granulated mould powder

7.6 Use of Mould Powders to Minimise Defects and Process Problems

There are few publications providing practical advice on how to avoid defects in ingot casting and how mould powders can influence defect-formation. The catalogue of ingot defects, published by Stahlinstitut VDEh [10], provide some information. However, the best advice, so far, is that reported by McCauley and Paul [33] who describe the different defects that can emerge during ingot casting, and how to avoid them, these include:

- Laps and ripple marks.
- Entrapped oxides.
- Slag patches.
- Porosity.
- Cracks.
- Bottom-end defects.

The reader is recommended to study their paper [33] but a short summary is given in this chapter.

7.6.1 *Laps and Ripple Marks*

Laps form in two ways; through oxidation of the meniscus and by rapid solidification of the meniscus. If the meniscus is not protected from the atmosphere, a coherent oxide layer can form on the curved surface of the liquid steel. The oxide layer prevents the liquid steel from rising near the mould until a sufficient, ferrostatic head has been generated and the liquid metal can overflow the oxide layer. Similarly, if the heat loss from the meniscus is too rapid, a solidified shell of metal can form on the curved meniscus near the mould, having the same effect as an oxide layer.

To produce a lap-and ripple-free surface, a combination of higher teeming temperature and faster teeming with an insulating flux cover is desirable. However, excessive superheats can cause refractory erosion problems and surface crackings so the desired teeming rate will depend strongly on the steel grade being cast.

7.6.2 *Entrapped Oxides*

Entrapped oxide patches are caused by the coalescence of high-melting oxides, such as Al_2O_3, CeO_2, Cr_2O_3 and $MnTiO_3$. These can be trapped in the advancing, solidifying shell. The mould powder slag fluxes these high-melting oxides so they can be retained in a liquid flux pool on the meniscus or be consumed by infiltration between the mould and advancing steel shell.

Entrapped oxide-type defects can also be caused by the flux itself, if the melting characteristics of the flux are not properly controlled. If the flux sinters at the steel surface, a sintered mass of the flux can become trapped in the solidifying steel in the same way as high melting oxides. The lower viscosity fluxes used for stainless and high alloy steels are consumed rapidly; consequently, the flux layer must be allowed to melt faster to maintain a liquid flux pool on the meniscus.

7.6.3 Slag Patches

Intermittent slag patches are associated with a high-fluidity flux, or a flux which melts too fast, or unevenly. If the fused flux layer on the meniscus is too fluid, or the thickness increases rapidly, the molten flux will infiltrate between the mould and ingot shell. These patches are simply layers on the ingot surface and are not entrapped in the surface. Slag patches usually fall off during handling of the ingot or can be easily removed from the ingot leaving a surface which is smooth, shiny and free from oxidation.

7.6.4 Porosity

Surface porosity in bottom poured ingots is sometimes a problem. However, the variation in the nature and the inconsistent appearance of the porosity makes it difficult to pinpoint the origin and to find a solution to the problem.

In many cases, the source of porosity is the ingot mould. Wet moulds, for example, will increase the incidence of porosity. Also, dirty and cracked, or crazed, moulds provide places where moisture and gases are trapped and then absorbed by the steel. A routine of mould cleaning and repair will solve most porosity problems. Arranging production so that moulds are reused hot, rather than allowing them to cool to ambient temperature will also reduce porosity, caused by moisture on the mould.

The flux cannot be eliminated as a source of porosity because of the presence of CO and CO_2 from the oxidation of carbon; it can also arise from the decomposition of carbonates contained in the flux, and the small amount of surface moisture present with all materials. Results from steel grades exhibiting severe surface porosity problems have shown that a non-gas producing flux will significantly reduce the incidence of porosity.

7.6.5 *Cracks*

Longitudinal surface cracks are a problem associated with teeming conditions and steel grade. High superheats and teeming rates enhance the risk of cracking in crack-sensitive grades, especially, those containing 0.10–0.16 wt-% C and re-sulfurised steel grades. In the 1010-type and similar steels, the peritectic transformation causes major volume changes in the rapidly-solidifying, ingot shell. If solidification is too rapid, cracking can easily occur. Also, if the pouring rate is too fast, the build-up of ferrostatic pressure can cause cracking of a weak, insufficiently thick, ingot shell.

Mould flux has little direct influence on the cracking except to reduce the heat losses from the meniscus.

7.6.6 *Bottom-End Defects*

Bottom-end defects are the result of the initial addition of flux powder to the mould after teeming has begun. The rapid development of a large amount of flux and the turbulence associated with the initial pouring stream causes some of the flux powder to get trapped between the mould and ingot. Under these conditions there is insufficient time for the flux to establish equilibrium of the fused flux and the powder flux layers: the flux trapped between the mould and steel does not have time to melt. Bottom-end defects, therefore, are characterised by entrapped, sintered flux powder and roughness caused by any gas generated by the flux between the mould and semi-solid steel.

Multiple ingates and flared ingate geometry will reduce the incoming velocity and turbulence; this minimises the occurrence of bottom-end defects.

The logistics of a bottom-pour setup precludes direct operator involvement; hand additions at the start of teeming are not an acceptable solution. Manual additions of flux during the pour have caused the formation of severe ripple marks, suspending the bags of the flux too close to the bottom plate (i.e. <15 cm) and near the mould wall increase the probability of bottom-end defects. Lowering the flux melting point has proven successful in some situations.

References

1. A. Okamoto, Sumitomo Search, **59,** pp. 3–11, (1997)
2. G. Lesoult, Materials Science and Engineering A, **413–414**, pp. 19–29 (2005)
3. H. Tomono, PhD Thesis No. 330, Dépaŕtement des Materiaux Ecole polytechnique Federale de Lausanne (1979)
4. J.J. Bikerman, Physical Surfaces (Academic Press, New York and London, 1970), pp. 10–15

5. T.G. Gammal, U. Schoneberg, Stahl u. Eisen, **112** (1), (1992)
6. H. Jacobi, Arch. F. Eisenhüttenwesen, **47** (7), pp. 441–446 (1976)
7. H. Gaye, L.D. Lucas, M. Olette, P.V. Riboud, Can. Metall. Quart., **23** (2), 179, (1984)
8. M. Susa, K. Nagata, K.C. Mills, Ironmaking and Steelmaking, **20** (5), 372, (1993)
9. A.F. Diener, A. Drastik, W. Haumann, Arch. F. Eisenhüttenwesen, **43** (7), pp. 525–533 (1972)
10. Blockfehlerkatalog, Stahlinstitut VDEh (Verlag Stahleisen GmbH, Düsseldorf, 2003)
11. L. Zhang, B.G. Thomas, Met.Mater. Trans. B, **37B**, 733, (2006)
12. L. Hallgren, C-Å Däcker, L. Teng, Proc. 2nd Intl. Conf. Ingot Casting, Rolling and Forging, Milan, 2014, (Associazione Italiana di Metallurgia, 2014)
13. H. Nakato, S. Omiya, Y. Habu, T. Emi, K. Hamagami, T. Koshikawa, J. of Metals 1984, 44, (1984)
14. M. Valdez, G. S. Shannon, S. Sridhar, ISIJ Intl., **46** (3), (2006)
15. B. Sjöholm, report from Swerea MEFOS, MF79012 (1979)
16. M.Susa, K Nagata, K.C. Mills, Ironmaking and Steelmaking, **20** (5), 372, (1993)
17. H.A. Fine, T. Engh, J. F. Elliott, Met. Trans. B, **7B**, 277, (1976)
18. V.A. Efimov, V.I. Legenchuk, G.V. Sivtsov, I.M. Konovalov, G.D. Bykov, A.G. Tatyanshchikov, Stal, **12**, 927 (1962)
19. D.R. Thornton, JISI July, 300, (1956)
20. J.R. Rippon, STEEL TIMES, **12**, 369, (1965)
21. W.O. Alexander, STEEL TIMES, **25**, (1966), p. 251
22. G.A. Sedach, N.P. Izotov, G.N. Kamyshev, Stal, **1967** (Feb),124 (1967)
23. G.A. Panev, A.M. Ofengenden, G.G. Zhitnik, M.L. Plepler, L.I. Krupman, Steel in the USSR, **1972**, (Jan), 16, (1972)
24. R. Sato, Bull. Jap, Inst. Metals, **12** (6), 48, (1973)
25. R.H. Hammerton, A. Hettler, Proc. 62nd Steelmaking Conf., 1979 (ISS, Warrendale, PA, 1979), p. 68
26. H.J. Eitel, PhD Dissertation Fakultät für Bergbau, Hüttenwesen und Geowissenschaften, RWTH, Aachen, (1990)
27. Jahrbuch Stahl, Band 1, (Verlag Stahleisen 1988), p. 114
28. S.P. Bakumenko, V.M. Shatov, E.V. Verkhovtsev, N.A. Ponomarev, B.B. Gulyaev, Stal, **12,** 669, (1976)
29. R. Gray, H. Marston, Iron and Steel Society Reference Text 101 (1979), p. 93
30. R.L. Harvey, A.P. Banks, Foseco INC. Seminar, Cleveland, Ohio, 1983
31. M.W. Nichols, A.P. Lingras, D. Apelian, Proc. 2nd Intl. Symp. Metallurgical Slags and Fluxes, Lake Tahoe, 1984 ed. DR Gaskell and HA Fine, (TMS-AIME, Warrendale, PA, 1984), p. 235
32. R.W. Paul, W.L. Mc Cauley, Proc. 2nd Intl. Symp. Metallurgical Slags and Fluxes,Lake Tahoe, 1984 ed. DR Gaskell and HA Fine, (TMS-AIME, Warrendale, PA, 1984), p. 253
33. W.L. Mc Cauley, R.W Paul, Iron and Steel Engineer, **1985** (Sept.), 36, (1985)
34. K. Sumitomo, M. Hashio, T. Kishida, A. Kawami, Iron and Steel Engineer, 1985, (Mar), 54, (1985)
35. J.G. Bartholomew, R.L. Harvey, D.J., Hurtuk, Proc. 69th Steelmaking Conf., 1986 (ISS, Warrendale, PA,1986), p. 121
36. C.M. Loane, K.H. Hashimoto, Proc. 69th Steelmaking Conf., 1986 (ISS, Warrendale, PA, 1986), p. 129
37. H. Marston, Proc. 69th Steelmaking Conf., 1986, (ISS, Warrendale, PA,1986), p. 107
38. R. Carli, A. Del Moro, C. Righi, La Metallurgia Italiana, **2008**, (May), 13, (2008)
39. B.V.R. Raja, Steel World, **2008** (Dec.), 34, (2008)
40. J. Lönnqvist, C.-Å. Däcker, P. Andersson, 1st Intl Conf. Ingot Casting, Rolling and Forging, Aachen, 2012, (Steel Institute VDEh, 2012)
41. M. Alloni, R. Carli, A. Del Moro, 1st Intl Conf. Ingot Casting, Rolling and Forging, Aachen, 2012, (Steel Institute VDEh 2012)

42. C.-Å. Däcker, 1st Intl Conf. Ingot Casting, Rolling and Forging, Aachen, 2012, (Steel Institute VDEh 2012)
43. M. Svensson, E. Sjöqvist-Persson, L. Hallberg, C-Å Däcker, 2nd Intl. Conf. Ingot Casting, Rolling and Forging, Milan,2014, (Associazione Italiana di Metallurgia, 2014)
44. J.E. Shelby, *Introduction to Glass Science and Technology*, (Royal Society of Chemistry, New York 1997)
45. Slag Atlas, 2nd edition, (Verlag Stahleisen GmbH, Dusseldorf, 1995)

Chapter 8
Manufacture of Mould Fluxes

Abstract It is difficult to get a complete knowledge of the composition of mould powders because the source and mixture of the raw materials is considered by the manufacturers to be confidential. Most raw materials are open-pit-mined minerals and can vary considerably in composition from batch to batch. For that reason, the mould powder is composed of a number of different raw materials where price optimisation is also an important factor. This chapter provides information on how mould powders are produced, its constituents, and how the quality and properties are controlled by the steel plants in order to get a better understanding of the mould powder and how it affects the casting process. The data sheet supplied by the manufacturer gives little information of value when it comes to the performance of the powder in the mould and when the given chemical composition of the mould powder can be achieved in an endless number of combinations of raw materials.

8.1 Introduction

Mould powder in a liquid state (mould powder slag) consists of a mixture of oxides (mainly of silicon, alkali, alkali-earths) and fluorides. The liquid slags are ionic in nature. The building block of most mould powders is the SiO_4^{4-}-tetrahedron which has the ability to create a three-dimensional network, providing the SiO_2 content is high enough. The network is broken down with the addition of cations, such as Ca^{2+}, Mg^{2+}, Na^{+}, K^{+} and Li^{+}. Alumina forms anions and increases the polymerisation of the silicate network, resulting in increased viscosity. Fluorine decreases the binding force and consequently reduces the viscosity of the slag system.

The original version of this chapter was revised: For detailed information please see correction. The correction to this chapter is available at https://doi.org/10.1007/978-3-319-53616-3_12

Within this book minor changes to some of the figure permissions and the readability of the chapters have been made and several figures have been replaced with new versions.

K.C. Mills and C.-Å. Däcker, *The Casting Powders Book*,
DOI 10.1007/978-3-319-53616-3_8

8.2 Raw Materials

It is apparent that the desired properties of a mould powder can be achieved with several different combinations of chemical composition. Table 8.1 shows the ranges of different combinations that are commonly used in powders for continuous casting and in synthetic, ingot-casting powders [1].

Older generations of mould fluxes contained Fe_2O_3 or FeO as fluidizers and as wetting agents. However, these oxides are easily reduced by Al-killed steels and change the properties of the mould powder slag during casting and create alumina inclusions. For this reason, mould powders used in casting of these steels tend to have a low level of iron oxides.

However, FeO has the ability to absorb radiative heat and this property is used in high-viscosity, mould slags for billet casting when "*soft cooling*" is desired.

It is preferable to restrict the fluorine content. High amount of fluorine causes excessive erosion of the SEN and fluoride emissions are detrimental to the environment and pose a health hazard.

The raw material selection, for the most part, is driven by costs and environmental concerns. With the exception of a few synthetic products, such as cement and prefused materials, most raw materials are open-pit mined minerals and thus are subject to considerable variations in composition from batch to batch. For that reason, it is preferable that mould powders are composed of a number of different raw materials. In Table 8.2 some common raw material sources for mould powder production are listed.

Fly ash was the main component of mould powders in the past. However, the wide variations in composition and the high amount of iron oxide in fly ash have resulted in it being mostly replaced by other raw materials today; the principal exception is in ingot casting where it is still the dominant raw material for most mould powders [2] (see Chap. 7).

Table 8.1 Typical composition ranges for mould powders (wt%) (Reproduced with permission from Association for Iron & Steel Technology [1])

Network (glass) formers	SiO_2	17–56
	Al_2O_3	0–13
	TiO_2	0–5
	B_2O_3	0–19
Network breakers, basic oxides	CaO	22–45
	MgO	0–10
	BaO	0–10
Network breakers, alkali	Na_2O	0–20
	K_2O	0–5
	Li_2O	0–5
Network breakers, fluidizers	F	2–15
	MnO	0–5
	Fe_2O_3	0–10
Melting speed control	C	2–25

Table 8.2 Raw materials for mould powder production

SiO_2	Feldspar, wollastonite, perlite, spodumene, cullet (crushed glass), fly ash, quartz dust
CaO	Lime, limestone, calcite, wollastonite, cement, slags, gypsum, calcium silicate.
Al_2O_3	Mullite, corundum
Na_2O/K_2O	Natrite, potash, feldspar, sodium silicate, cullet (crushed glass)
F	Fluorspar, sodium fluoride, cryolite, sodium hexafluorosilicate
Li_2O	Spodumene, lithium carbonate
C	Carbon black, coke dust, fine graphite, fly ash

Fig. 8.1 Photograph of fly ash, spherical, smooth particles which mainly consist of a SiO_2–Al_2O_3 glass

Fly ash is a mineral product which originates from energy production of electricity and heat in carbon-fuelled power and heating plants. It consists of the non-combustible part of carbon particles which are burned in the combustion chamber of the energy plant. Each carbon particle becomes a fly ash particle (Fig. 8.1) which is separated in an electrostatic filter before the flue gas is emitted from the plant's chimney. Fly ash contains some carbon, the amount of which depends upon the efficiency of the combustion in the power plant.

It must be pointed out that most mould powder producers consider the choice of raw materials as confidential and information of how this is made is not normally provided. However, the chemical composition and the results of DSC–TG analysis provide a reasonable idea of how the mould powder has been composed. It is possible to get further understanding of the raw materials used in the formulation of the mould powder from X-ray analysis and some geological expertise.

It is considered good practice to use

- A combination of minerals with similar melting points to promote homogeneous melting of the mould powder and
- A minimum number of different mineral sources (i.e. to keep the recipe simple) which helps to provide good quality control.

8.2.1 Selection of Carbon Additions to Mould Powders

Addition of free carbon to the mould powder is the most efficient way to control the melting speed. Whilst mould powders are composed of a mixture of mineral components (with high melting points) and fluxing agents (for example CaF_2 and Na_2CO_3), the carbon particles delay the contact and sintering of these components. The carbon also delays the agglomeration of the slag droplets. The effect of carbon on melting speed is determined by both the amount and the size of the carbon particles, where finer carbon particles cause a greater reduction of the melting rate per unit mass of carbon. Thus melting speed is controlled by both the amount and particle size of the carbon. The particle size is dependent on carbon source which is illustrated in Table 8.3. As shown in Fig. 8.2 it can be seen that carbon black, because of its high surface area and high reactivity, has the lowest ignition temperature and graphite the highest.

Carbon black has the finest particle size and is believed to coat each particle of the mould powder. Sakuraya et al. [3] have shown that a minimum of 0.4 wt% carbon black is needed to coat a typical mould powder with a monolayer of carbon black.

Table 8.3 Typical particle size and burn temperatures for different carbon types (Reproduced with permission from Association for Iron & Steel Technology [1])

Type of carbon	Particle size (μm)	Burn temperature (°C)
Carbon black	0.028	386–522
Metallurgic coke	20	511–718
Graphite	74	613–897
Fine graphite	1.36	613–897

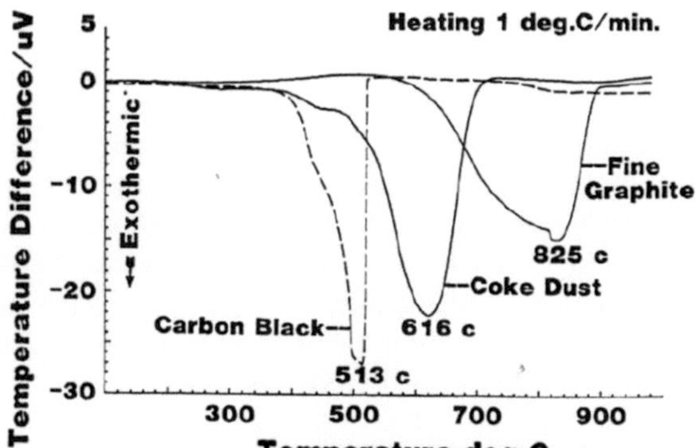

Fig. 8.2 Diagram showing thermal analysis curves for carbon black, metallurgical coke dust and graphite (heating rate 1 °C/min) (Reproduced with permission from Association for Iron & Steel Technology [1])

Carbon particles are non-wetting to molten slag and thus separate the molten globules and prevent them from agglomerating. For a given mass of carbon, the finer particle sizes will provide more separation than coarser particles. However, the carbon particles react with oxygen in the mould and gradually disappear; the oxygen sources are gaseous diffusion through the bed, CO_2 emitted during decomposition of carbonates and oxides like FeO in the powder. Finer particles have a greater (surface area/volume) and so will tend to combust more readily. The fact that the melting rate decreases with decreasing particle size indicates that the separation of globules determines the melting rate rather than the carbon combustion rate.

8.2.2 *Reactions During Melting and Cooling of Mould Powders*

A large amount of information regarding the mould powder can be gained from a combined DSC (Differential Scanning Calorimetry) and TG (Thermal Gravity) measurement. An example from a measurement from a commercial mould powder for continuous casting of peritectic slabs is shown in Fig. 8.3 [4].

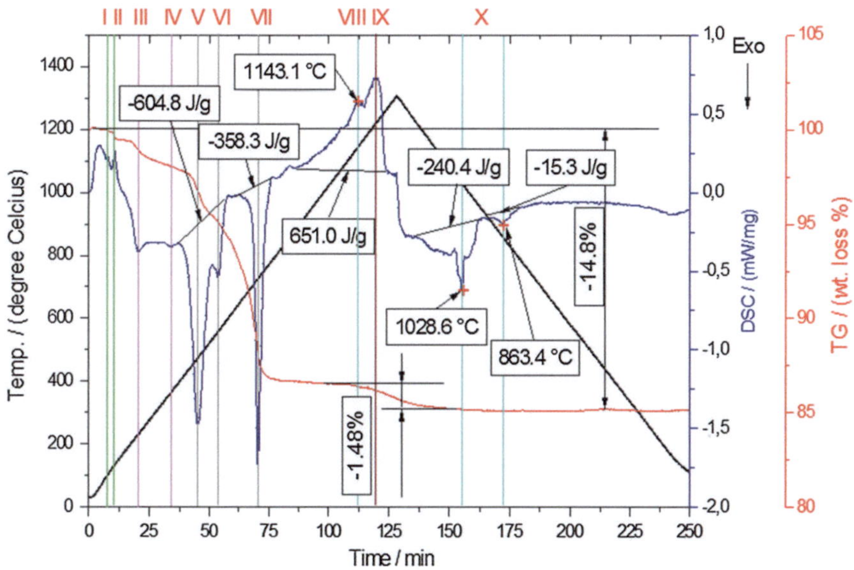

Fig. 8.3 Diagram showing DTA/TG examination of a mould powder used for peritectic steels. The temperature at any specific time is given from the curve shown in bold; *red* weight loss in %; *blue* DSC (mW/mg) (Reproduced with permission from Jernkontoret [4])

During the heating and cooling process, possibly, the following endothermic/exothermic reactions occur [4]:

I. Gas desorption (possibly) at around 95 °C. This is an endothermic reaction accompanied by a small weight loss.
II. Water evaporation at around 130 °C. This is an endothermic reaction accompanied by weight loss.
III. Ignition of organic substance (binding agents for granulation of powders) at around 225 °C. This is an exothermic reaction accompanied by weight loss.
IV. Ignition of sulphur at around 360 °C. This is an exothermic reaction accompanied by weight loss. The sulphur can come from the coke in the powders.
V. Combustion of carbon black at around 472 °C. This is an exothermic reaction accompanied by weight loss.
VI. Combustion of coke at around 554 °C. This is an exothermic reaction accompanied by weight loss.
VII. Combustion of graphite at around 722 °C. This is an exothermic reaction accompanied by weight loss. There is a small endothermic event before this peak at around 700 °C, which may be caused by the decomposition of sodium carbonates. These two phenomena are so close that the endothermic reaction is almost overlapped by the big exothermic peak.
VIII. Around 950 °C also a small endothermic peak is shown which reflects the decomposition of calcium carbonates (Na_2CO_3, K_2CO_3). The following peak at 1143 °C relates to the melting of the powder. This is an endothermic reaction; there was no significant weight change.
IX. Evaporation of Na_2O at around 1216 °C: this is an exothermic reaction accompanied by weight loss of *ca.* 1.5%.
X. Crystallisation of the slags: this is an exothermic reaction without any significant weight change. Normally, crystallisation occurs in two steps: the first stage at high temperature and second stage at a lower temperature, during which some latent heat due to crystallisation is released. The two stages correspond to the two peaks appearing in the DSC signal.

8.3 Manufacturing

The manufacturing process is shown schematically in Fig. 8.4.

The production steps are as follows:

- A large amount of different raw materials are used for the production of different mould powders; these arrive at the plant in a fine-grained condition (max. 70 mesh) either as bulk in a silo transport or in big bags.
- A rigid control of the incoming raw materials is exerted using chemical analysis.

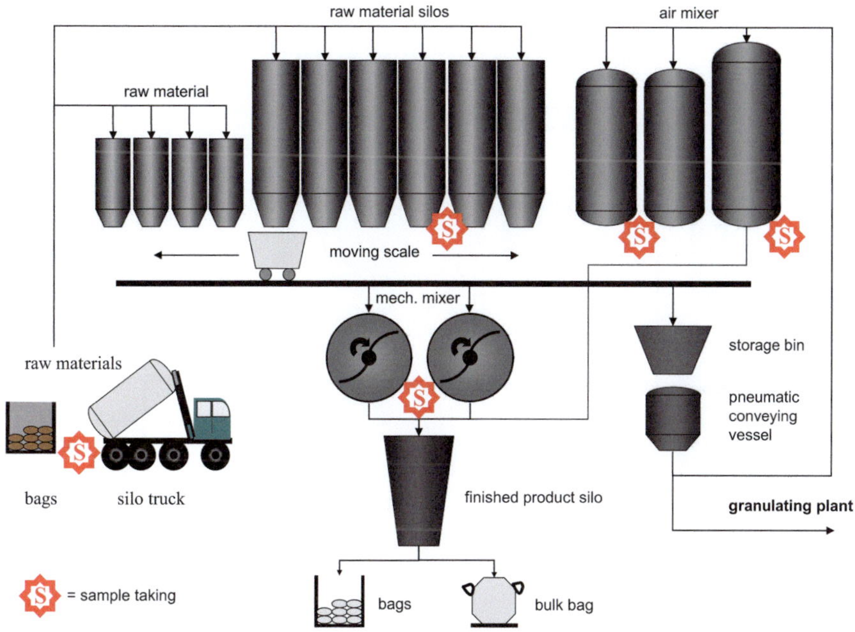

Fig. 8.4 Diagram showing material flow for production of loose mould powder (courtesy of IMERYS)

- The material is blended according to the desired formula. This is done either in a mechanical mixer or in an air mixer. The carbon materials are added at this stage.
- After mixing, the non-granulated (loose) powders are ready for delivery either in big bags or in small plastic bags for direct use.

For granulated powders, the mixed powder undergoes a second production process shown schematically in Fig. 8.5.

The production steps are as follows:

- Batches of *ca*. 10 tonnes powder are mixed with approx. 4 tonnes of water in a turbo mixer.
- The slurry is transported to a second tank where the binding agent is added. Previously, water glass was frequently used as binder but this has gradually changed to an organic material (for example, maize starch) which gives better melting properties for the powder.
- The slurry is transported through a tube filter to a spray dryer.
- In the spray dryer, hot air, heated with gas or oil to a temperature of around 600 °C, circulates through the dryer. The slurry is injected into the hot air through a nozzle consisting of a large number of small holes.
- When the slurry is pressed out of the nozzle, a mist of droplets is formed. The surfaces of the droplets dry quickly (due to the heat and the large amount of

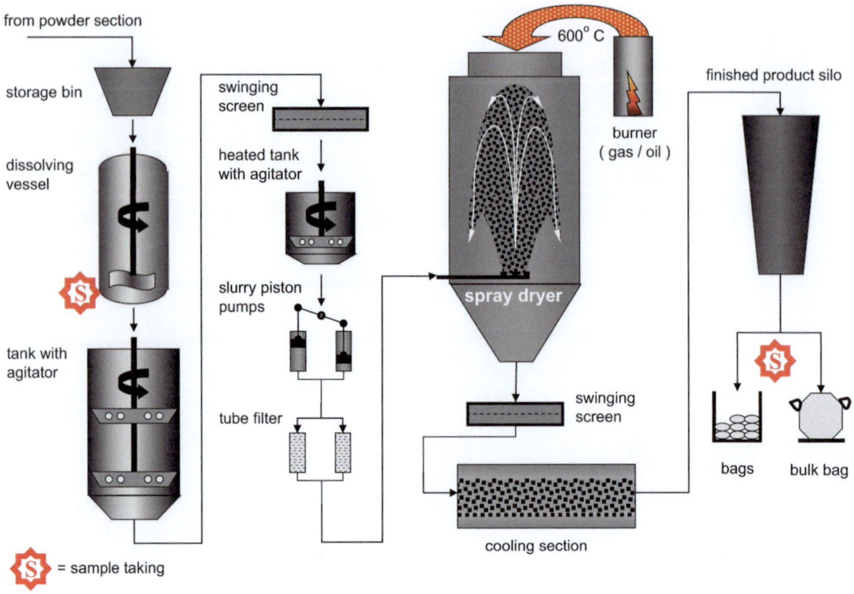

Fig. 8.5 Schematic diagram showing material flow for production of granulated mould powder (courtesy of IMERYS)

Fig. 8.6 Schematic diagrams showing the hollow appearance of granulated mould powder

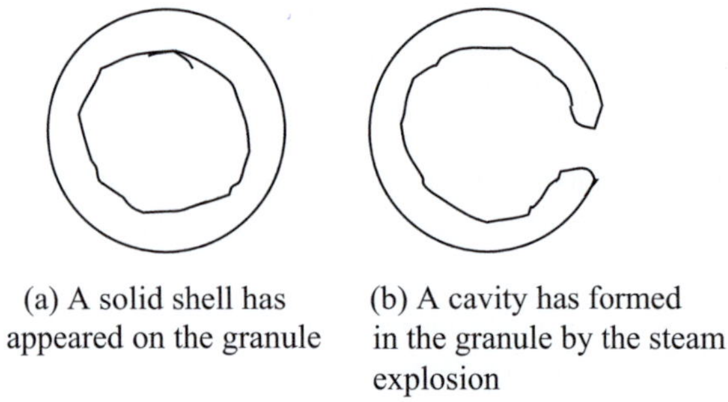

convection coming from the hot air) forming a shell of solid, dried material (Fig. 8.6a). With continued drying an overpressure is formed within the droplets because of its large water content. Finally, the granule explodes leaving a hollow granule as shown in Fig. 8.6b.

- The granule lands on the bottom of the dryer from where they are collected, transported, screened and packed into small bags or big bags.

Due to the cavity and a narrow particle size distribution, granulated mould powders possess a low density. A photograph of a granulated mould powder at high magnification is shown in Fig. 8.7.

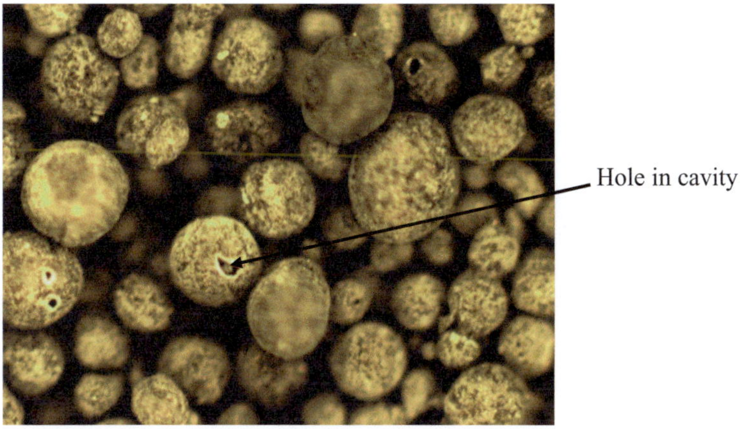

Fig. 8.7 Photograph of mould powder granules, magnification 40X

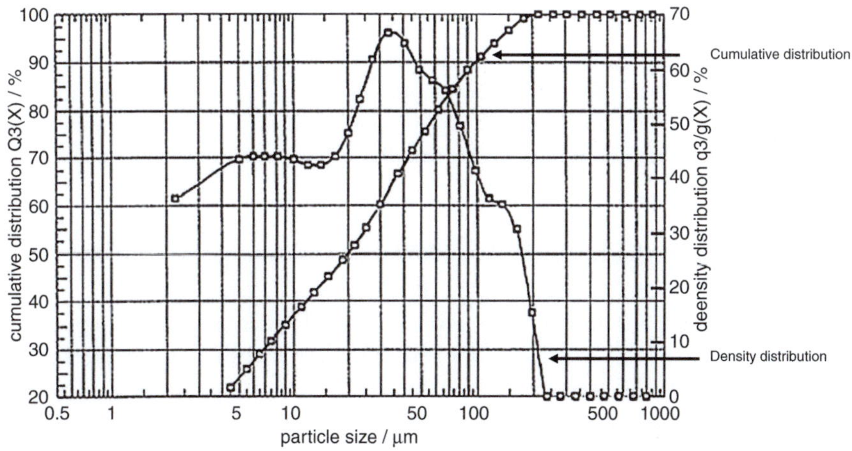

Fig. 8.8 Diagram showing particle size and cumulative distribution of standard, non-granulated, mould powder

The particle size distribution varies significantly between non-granulated and granulated powders as can be seen in Figs. 8.8 and 8.9, respectively.

Granulation of mould powders increases the price of the mould powders but has a number of advantages [5], such as

- The composition of the mould powder is uniform throughout the mould powder bed which makes its process behaviour both stable and reliable.
- The narrow particle distribution, with a low amount of fine particles, results in a low amount of dust during handling and casting; this makes them much more environmentally friendly to use.

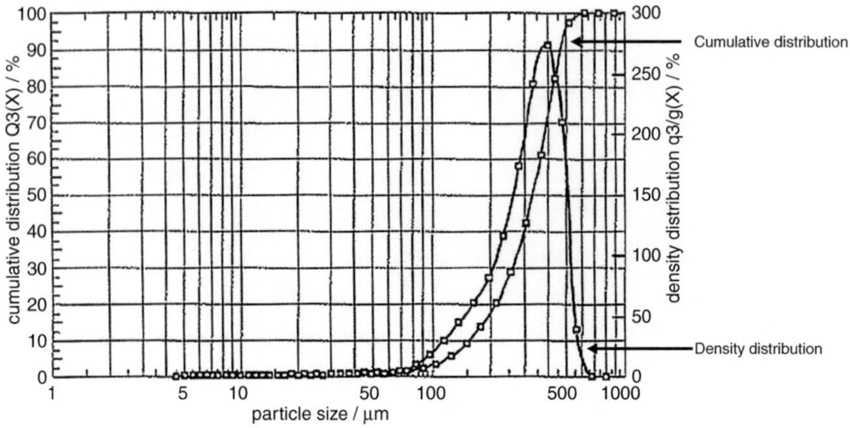

Fig. 8.9 Diagram showing particle size and cumulative distribution of standard, granulated, mould powder

- The granules will have a good flowability that makes it possible to use an automatic feeding system. This, in turn, leads to a secure feeding and control of the height of mould powder in the mould. The mould pool depth will be more even due to more uniform melting properties; this, in turn, ensures more homogeneous feeding of molten slag into the gap between steel shell and mould.

8.4 Quality Control at the Manufacturer

The raw materials for mould powders can vary a great deal in composition. Consequently, it is important that the manufacturer carries out regular control checks on both of the incoming raw material as well as the batches of mould powders before dispatch to the steel plants.

8.5 Information Provided by the Manufacturer

An example of a typical data sheet for mould powders from a manufacturer is shown in Table 8.4.

The following information is given:

- Chemical composition of the mould powder;
- Melting range as determined by DIN-standard 51730;
- Calculated viscosity using Riboud model [6] which is based on multiple statistical analyses from a large number of synthetic and commercial mould powder slags;
- Sieve analysis and
- Bulk density.

Table 8.4 Typical data sheet for a granulated mould powder

Batch number: 11/100		Packing date: 30-May-11	
Chemical analysis (%)		Melting Range (°C)	
LOI	10.62	Softening point	1155
C tot.	5.81	Melting point	1220
C free	4.05	Flowing point	1250
SiO_2	34.82		
Fe_2O_3	0.85		
Al_2O_3	3.85		
CaO	39.10	Viscosity (dPa s)	
MgO	1.90	(*IRSID Math. Method*)	
$Na_2O + K_2O$	4.82		
F-	7.22	1300 °C	1.7
MnO	0.00		
S	0.08		
H_2O (120 °C)	0.35	Sieve analysis	
		Mesh size (mm)	Residue (%wt)
		0.400	24.0
CaO/SiO_2	1.12	0.125	91.0
		0.063	98.5
Bulk Density	0.72 (kg/dm^3)		

Compared to most other products, the information is very sparse and does not give much information on how the mould powder will function in the steel plant. The missing information includes the following:

- Some information regarding the raw materials used to compose the mould powder;
- Measured viscosity with the rotary spindle method. Actually there is an ASTM standard for mould powders [7];
- Melting rate;
- Break or solidification temperature (determined with a viscometer with a small distance between crucible and spindle);
- Flowability, which is an important property especially for ingot casting of big end up, moulds (BEU). There is also an ASTM standard available [8] for this property;
- Interfacial tension between mould slag and steel;
- Mould slag irradiative absorptivity and
- Crystal phase (f_{crys}) in the slag film.

8.6 Delivery Control by the Steel Makers

Large steelmakers have their own production of mould powders and thereby exert full control of the product, including all necessary information on its raw materials.

Another way of maintaining full control of the raw materials is to use 100% prefused mould powder but this is very expensive and non-realistic but does reduce the fluorine emissions on the caster.

Most steel plants have to live with the sparse information provided by the manufacturer. In order to get more necessary information, ambitious steelmakers exert their own delivery control of the mould powder batches received from the manufacturer.

One example of such a delivery test is shown in Table 8.5. The following tests are performed by a Swedish steelmaker:

- Chemical composition;
- DTA to get a "*fingerprint*" of the mould powder behaviour and an idea of its components, and important information on the crystallisation temperatures;

Table 8.5 Information from a delivery control of mould powders A and B

Mould powder	A	B
SiO_2 (wt%)	32.9	32.7
CaO (wt%)	39.1	28.8
MgO (wt%)	0.37	1.77
Al_2O_3 (wt%)	6.5	4.7
TiO_2 (wt%)	0.03	0.11
Fe_2O_3 (wt%)	0.47	1.24
MnO (wt%)	<0.10	<0.10
Na_2O (wt%)	7.1	11.3
K_2O (wt%)	0.17	0.31
F (wt%)	8.3	9.4
C_{total} (wt%)	5.5	6.6
Loss on ignition (wt%)	10.1	14.1
H_2O (wt%)	0.3	0.3
Basicity (CaO/SiO_2)	1.19	0.88
Basicity ($CaO + MgO/SiO_2 + Al_2O_3$)	1.0	0.82
Melting time (s)	169	115
Viscosity 1300 °C (Poise)	2.72	2.51
Break temp. (°C)	1267	1165
DTA melting temp. 1 °C	1079	1223
DTA melting temp. 2 °C	1231	1146
DTA solidification temp. 1 °C	1250	1247
DTA solidification temp. 2 °C	–	–
Density (g/cm^3)	0.74	0.54

- Viscosity, to obtain both an experimental value of the viscosity and the break temperature (or solidification temperature). Observe that the determination of the break temperature is made possible by the use of a rotary viscometer with a narrow gap between rotor and crucible;
- Melting time which is determined from the time a certain amount of mould powder is melted in a tube-type furnace at 1500 °C and
- Bulk density, which is determined by weighing a certain volume of mould powder.

References

1. R. Bommaraju, Proc. 72nd Steelmaking Conf., 1991, (ISS, Warrendale, PA, 1991) p. 131.
2. M. Alloni, R. Carli, A. del Moro, 1st Intl. Conf. Ingot Casting, Rolling and Forging, Aachen, 2012 (Steel Institute VDEh 2012).
3. T. Sakuraya, T. Emi, K. Emoto, T. Koshikawa, Proc. 2nd Process Technology Conf., Chicago, 1981, (ISS, Warrendale, PA, 1981) p. 141.
4. T. Sun, X. Hu, L. Hallgren, L. Teng, Jernkontoret TO24-192 (2012).
5. H.J. Eitel, PhD Dissertation, Fakultät für Bergbau, Hüttenwesen und Geowissenschaften, RWTH, Aachen, 1990.
6. P.V. Riboud, Y. Roux, L-D. Lucas, H. Gaye, Fachberichte Hüttenpraxis Metallweiterverarbeitung, 19, 859, (1981).
7. ASTM Designation: C 1276 – 94, Standard Test Method for Measuring the Viscosity of Mold Powders Above Their Melting Point Using a Rotational Viscometer.
8. ASTM Designation: C 1444 – 99, Standard Test Method for Measuring the Angle of Repose of Free-Flowing Mold Powders.

Chapter 9
Properties of Mould Fluxes and Slag Films

Abstract In this chapter, the properties and the structure of the slags used in continuous casting are collated, analysed and discussed. The required property values of mould slags for successful casting are determined by the mould dimensions, the casting conditions and the steel grade being cast; these can be calculated using empirical rules. Required values of the powder consumption (Q_s^{req}) viscosity (η^{req}), break temperature (T_{br}^{req}) and the fraction crystal in the slag film (f_{crys}) are the key properties. However, other properties such as the interfacial tension are important when dealing with slag entrapment, thus a large amount of property data for mould slags are needed. Mould slags have been classified here into two groups (i) *Conventional* slags used in continuous casting and (ii) *Specialised* slags, which includes F-free slags, non-Newtonian slags, calcium aluminate (CA) type slags, etc. Some properties are very dependent on the structure of the slags. Thus, published data on slag structures for all types of slag have been collated and analysed. Routines, developed to estimate the various properties, are outlined and reviewed. The data for the following properties of the various mould slag families have been reviewed (i) *viscosity* (ii) *Liquidus* (T_{liq}), *Break* (T_{br}) and *Glass transition* (T_g) *temperatures* (iii) *Thermal conductivities* (iv) *Surface and Interfacial tensions* (v) *Densities and thermal expansion coefficients* (vi) *Heat capacity* and *enthalpy* and (vii) *Optical properties*. In addition, the crystallisation process in mould slags and the ability of the liquid slag to dissolve inclusions are reviewed and discussed. The requirements and the properties of slags used in continuous casting and ingot casting are compared and discussed.

Symbols, Abbreviations and Units

A	Surface area (m^2)
d	Thickness of slag (m)

The following abbreviations to chemical formulae have been used; A = Al_2O_3; B = BaO; C = CaO; F = FeO; *Fl* = CaF_2; L = Li_2O; M = MgO; Mn = MnO; N = Na_2O; S = SiO_2; T = TiO_2; e.g. 40C + 40S + 20A = 40% CaO + 40%SiO_2 + 20% Al_2O_3or C_3S_2Fl = $3CaO{\cdot}2SiO_2{\cdot}CaF_2$.

The original version of this chapter was revised: For detailed information please see correction. The correction to this chapter is available at https://doi.org/10.1007/978-3-319-53616-3_12

Within this book minor changes to some of the figure permissions and the readability of the chapters have been made and several figures have been replaced with new versions.

K.C. Mills and C.-Å. Däcker, *The Casting Powders Book*,
DOI 10.1007/978-3-319-53616-3_9

C_p	Heat capacity ($JK^{-1}\ kg^{-1}$)
D	Diameter (m)
f	Oscillation frequency (Hz, cycle min^{-1})
f^*	Fraction of flux-forming slag
k	Thermal conductivity ($Wm^{-1}K^{-1}$)
Q	Powder consumption (various)
q	Heat flux ($Js^{-1}K^{-1} = WK^{-1}$)
R	Thermal resistance (m^2KW^{-1})
R^*	(Surface area/volume) of mould
s	Stroke (m)
T	Temperature (°C or K)
T_{br}	Break temperature (°C or K)
t	Time or thickness of mould (m)
t_n	Negative strip time (s)
t_p	Positive strip time (s)
V_c	Casting speed ($m\ min^{-1}$ or $m\ s^{-1}$)
V_m	Velocity of mould ($m\ s^{-1}$)
w	Width of mould (m)
α	Thermal expansion coefficient $(°C, K)^{-1}$
α^*	Absorption coefficient
η	Viscosity (d Pas)
ρ	Density ($kg\ m^{-3}$)
CA	Calcium aluminate slags
CC	Continuous casting
CP	Carbon potential
(C/S)	Basicity
DSC	Differential scanning calorimetry
DTA	Differential thermal analysis
EMBr	Electromagnetic braking
EMS	Electromagnetic stirring
FF	Fluoride-free slags
FP	Ferrite Potential
HC	High-carbon steels
HiAl	High-Al steels (TRIP)
IC	Ingot casting
IR	Infrared
LC	Low-carbon steels
MC	Medium-carbon steels
NN	Non-Newtonian slags
OM	Oscillation mark
SEN	Submerged entry nozzle

Subscripts, Superscripts

crys Crystalline phase
gl Glass phase
l Liquid
m Metal
s Solid
sl Slag

9.1 Introduction

The structure and the thermophysical properties of continuous-casting slags are discussed in this chapter. It will be seen that the values of the properties of conventional slags for successful continuous casting (such as viscosity and T_{br}) are determined by the mould dimensions, casting conditions and the steel grade being cast. These *required* property values can be calculated by empirical rules. However, in recent years, special mould slags have been developed to cast specific steel grades or for certain casting conditions. These include: (i) fluoride-free slags (ii) carbon-free powders (iii) non-Newtonian slags and slags for casting (iv) high-Al steels (v) stainless steels (vi) round billets and (vii) high-speed thin slabs. Although, the required properties for these types of slags tend to conform with the values derived from empirical rules they do have to deal with other problems which must be met. Consequently, powders and slags are classified below as either "*Conventional*" or "*for Specialist casting*" (see Chap. 6). The structure and property data for all types of mould slag are collated and discussed below.

9.2 Structure of Slags

The performance of slags in the continuous-casting process depends largely on their properties (e.g. viscosity) and the properties are, in turn, very dependent upon the structure of the mould slag [1–3].

9.2.1 Effect of Individual Slag Components on Structure

Casting powders are essentially calcium silicate slags with added fluxes. The properties of slags are principally determined by the structure of the slag. It should be noted that slags are ionic in nature and that reactions, such as that shown in Eq. 9.1, are ionic (i.e. involve the transfer of electrons); thus they are best described in the form of Eq. 9.2 where the underline indicates an element in the metal phase:

$$4\underline{Al}_{steel} + 3\,SiO_{2slag} = 3\underline{Si}_{steel} + 2Al_2O_{3slag}. \tag{9.1}$$

$$4\underline{Al} + 3\,Si^{4+} = 3\underline{Si} + 4Al^{3+} \tag{9.2}$$

Pure SiO_2 has a three-dimensional structure based on covalent Si–O bonds. Each Si^{4+} ion is surrounded by four O^- ions in the form of a tetrahedron (Fig. 9.1a) and each O^{2-} connects two Si^{4+} tetrahedra. For pure SiO_2, this results in a three-dimensional polymerized network (Fig. 9.1b) [1, 2]. Thus, in mould slags, SiO_2 is considered to be a *network former*.

When cations, such as Na^+ or Ca^{2+}, are added (in the form of Na_2O or CaO) to SiO_2, they break some of the Si^{4+}–O^- bonds and replace them with an ionic Na^+–O^- bond (for Ca^{2+} the cation is bonded to two O^- bonds). Further addition of Na^+ ions results in progressively- more de-polymerisation of the melt; the cations (Na^+, Ca^{2+}) are referred to as "*network breakers*". Thus silicate slags contain both covalent and ionic bonds (Fig. 9.2) and the thermophysical properties are very dependent upon the level of polymerisation in the slag.

The different types of bonds are frequently classified in terms of the O bonds formed, namely, (i) *Bridging O's* (e.g. Si–O) denoted both as BO and O° (ii) *Non-bridging O's* (denoted NBO or O^-) and (iii) *Free O's* (i.e. bonded to cations but not to Si) which are denoted as FO or O^{2-}.

When Al_2O_3 is introduced into the silicate network the Al^{3+} ions can be accommodated into the silicate structure (i.e. Al^{3+} adopts 4–coordination like Si^{4+}). However, a Na^+ ion must be sited near the Al^{3+} to provide electrical charge balance [i.e. forms ($NaAl^{4+}$)] [1–3] (Fig. 9.3). If M^{2+} (e.g. Ca^{2+}) ions do the charge balancing, the M^{2+} ions must be placed between two Al^{3+} ions. It is considered that a Na^+ ion acting on charge-balancing duty cannot act as a network breaker. Thus, Al_2O_3 additions act principally as *network formers* but when large amounts of Al_2O_3 are added to the slag (e.g. during the casting of high-Al, Trip steels) the Al^{3+} ions can also act as *network breakers* (i.e. exhibit *six*—(sometimes denoted *VI*)—*fold coordination* and for this reason, Al_2O_3 is often referred to as "*amphoteric*". For most mould slags:

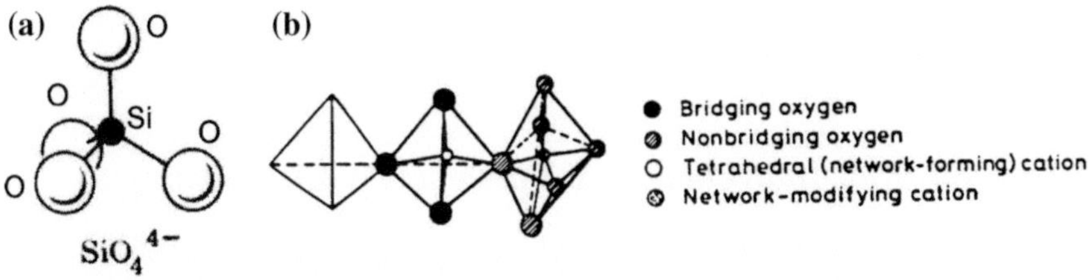

Fig. 9.1 Schematic diagrams showing **a** Si^{4+}–O^- bonds arranged in form of tetrahedron and **b** the creation of a network through the connection of each O^{2-} to two Si^{4+} ions (permission granted, VDEh, Stahleisen, [3])

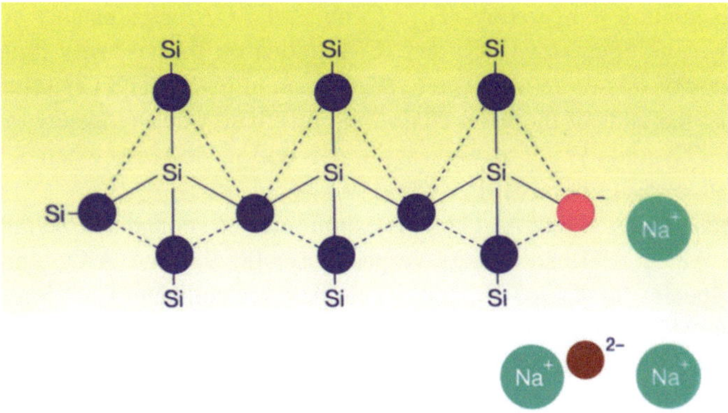

Fig. 9.2 Schematic drawings of silicate chain with Bridging O (O°) shown in *dark blue*, non-bridging O (O^-) in *pink* and free O^{2-} (shown as *inset* with free Oxygens)

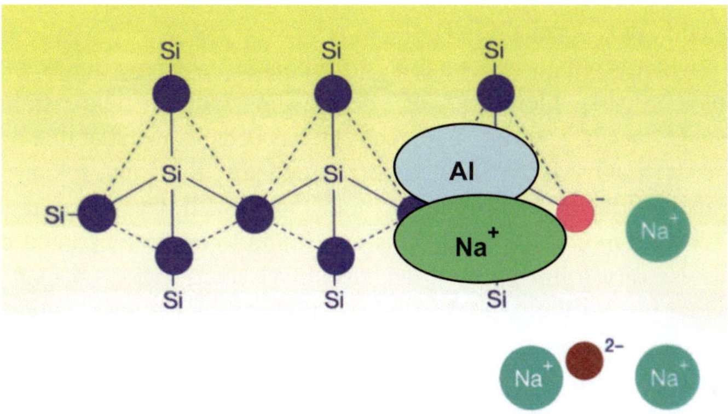

Fig. 9.3 Schematic drawing showing Al^{3+} incorporated into silicate (Si^{4+}) chain which requires cation (shown here as Na^+) to maintain charge balance

(i) Al_2O_3 contents are low (<7%) and thus Al^{3+} ions tend to be in tetrahedral (4 or IV) coordination.
(ii) Charge-balancing duties tend to be carried out by K^+ or Na^+ ions since they have lower field strength cf. Ca^{2+} ions. The effect of replacing Ca^{2+} by K^+ as charge-balancing cations and its effect on viscosity have been studied [4, 5].

Large amounts of Al_2O_3 are produced when casting high-Al (TRIP) steels. For low (C/S) slags used to cast these steels, increasing the (X_A/X_S) ratio increased the viscosity (presumably due to more cations being needed for charge balancing leaving fewer for network breaking) [6]. Calcium aluminate (CA) slags are also used to cast these steels. There is a eutectic in the CaO–Al_2O_3 system close to the $C_{12}A_7$ composition (corresponding to 48% CaO) to which fluxes are added to reduce the melting temperature (T_{liq}). Commercial CA slags usually contain (5–10%) SiO_2. The effects of (C/S) and (C/A) ratios on the viscosity indicated that (i) the viscosity was more sensitive to (C/A) than to (C/S) (ii) SiO_4 units tended to be placed at the ends of the AlO_4 chains and (iii) there was a tendency to segregate into Q^n_{Si} + Q^n_{Al} units [7]. The viscosity was found to decrease when K^+ replaced Ca^{2+} in cation–NBO bonds [5].

It has been reported that Al–O bonds exhibit four-, five- and sixfold coordination (i.e. form AlO_4, AlO_5 and AlO_6, respectively) [8, 9]. The AlO_4 units are the dominant species for $X_{CaO} > 0.5$, i.e. when there are sufficient Ca^{2+} ions to charge balance the AlO_4^{5-} ions; any excess Ca^{2+} cations are then free to depolymerise the AlO_4 network and reduce the viscosity. Thus viscosity decreases with increasing CaO (or CaO/Al_2O_3 ratio).

In the casting of high-Al steels, the Al_2O_3 concentration increases steadily as the cast proceeds and there may come a point where there are insufficient cations to charge balance the AlO_4 anions; under these circumstances some of the Al ions must take on the charge-balancing duties. In these conditions, additions of Al_2O_3 to the slag result in decreases in the number of AlO_4 units and to gradual increases in both AlO_5 and AlO_6 species [7–9].

Mould powders contain various fluxes, the charge-balancing duties will be done by cations in the hierarchy of decreasing cation field strength (i.e. $K^+ > Na^+ > Li^+ > Ca^{2+} > Mg^{2+}$). Recently, it has been shown that that the addition of K_2O to a $CaO–Al_2O_3–SiO_2$ slag resulted in K^+ ions taking over the charge-balancing duties performed by Ca^{2+} ions with the released Ca^{2+} ions gradually forming more NBOs [5]. This was found to be accompanied by a slight increase in viscosity followed by a decrease in viscosity when $(X_K/X_A) > 0.9$ [5]. This behaviour was attributed to the competition between the average bond strength of the NBOs (i.e. O–Ca vs. O–K) and the number of NBOs [5].

Conventional mould powders contain calcium fluoride (CaF_2) since it fluxes the slag and aids the formation of *cuspidine* ($3CaO{\cdot}2\ SiO_2{\cdot}CaF_2$ or C_3S_2Fl) crystals in the slag film.

When F^- ions are added to a slag, few Si–F bonds are created and although some Al–F bonds may be formed, most of the F^- ions bond with Ca^{2+} and Mg^{2+} cations [10]. Recent work [11–14] has indicated that when Fluorine is added to a silicate melt, the F^- ions bond almost exclusively with Ca^{2+} ions to form CaF_2 which then acts principally as a diluent to the silicate structure. It has been proposed that Ca^{2+} and F^- ions take up preferred positions; this allows them to form CaF_2 clusters [15]. The CaF_2 formed does not interfere with the concentrations of the various silicate ions formed (i.e. the various Q^n units remain unaltered). Thus the CaF_2 acts merely as a *diluent* or "*thinner*".

Fluorine additions can be made by either replacing CaO with CaF_2 or by adding CaF_2 to the slag. For the case, where CaO is replaced by CaF_2, it has been suggested that the F^- additions promote polymerisation of the melt [11–14, 16]. In contrast, CaF_2 additions serve only to dilute the polymerised slag.

The mechanism accounting for how CaF_2 modifies the network has been studied by a number of investigators [1, 2, 11, 12, 14, 16–18]. Basically, there models fall

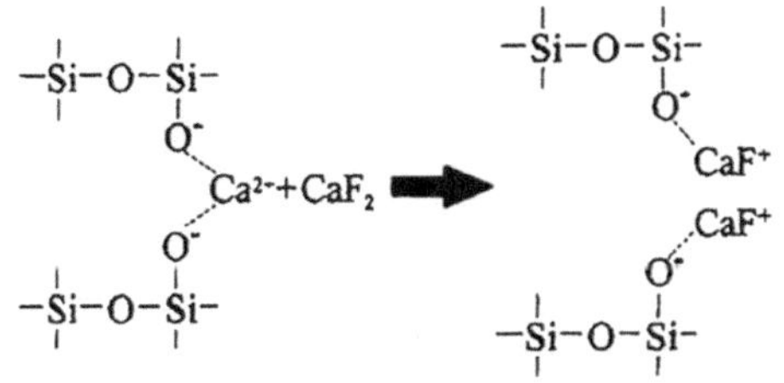

Fig. 9.4 Proposed mechanism for the addition of CaF_2 to silicate slag (permission granted, ISIJ, [16])

into two types (the Baak-Bills and Sasaki models [11, 19, 20]. The mechanism shown in Fig. 9.4 is similar to the Sasaki mechanism [11].

TiO_2 additions would be expected to behave like SiO_2 and further polymerise the melt. However, TiO_2 has been shown to decrease the viscosities of slags [21–24, 26]; this suggests that TiO_2 may act as a weak network breaker. There are three possible ways that TiO_2 can modify the silicate structure, namely. (i) Ti^{4+} is incorporated into the SiO_4 network (ii) Ti acts as a network breaker (by adopting 5- or 6- coordination) and (iii) by forming TiO_2-like structures (clusters) [25]. There is some dispute which is the dominant mechanism in the geological literature and recent publications on slags are equally inconclusive. The results reported by Li et al. [26] for CA-based slags indicate the Ti prefers to form TiO_2-like clusters and there is a small increase in polymerisation (the decrease in viscosity arises because the Ti–O bond is weaker than the Si–O bond). Alternatively, it has been suggested that that TiO_2 acts as a "*thinne*r" [27]. However, the spectroscopic results reported by Zhang et al. [25] support the formation of TiO_6 octahedra (i.e., Ti acts as a network breaker).

There use of B_2O_3 in mould powders as a flux has been limited by the fear that it could lead to B-pick-up by the steel. However, it has been used quite widely used as a flux in slags usd to cast high-Al steels. B_2O_3 is considered to be a network former in slags and is known to form complex borosilicate structures. However, B_2O_3 additions frequently result in a decrease in viscosity [28–31]. In pure borax (B_2O_3) the structure is made up of 3-coordinated BO_3 units arranged in the form of 2-dim. rings. However, additions of CaO or Na_2O result in the formation of structural units with both 3- and 4-coordination (i.e. three-dimensional BO_4 units) respectively [29] as shown in Eqs. 9.3 and 9.4. It has been reported that the BO_3 ring species are dominant [29] in calcium aluminate mould slags. It has also been pointed out [30] that the viscosity is determined by the strength of the bond of the flow unit; since the B–O bond is weaker than the Si–O bond, B_2O_3 additions result in a decrease in viscosity.

$$2BO_3 + CaO = 2\,BO_4^- + Ca^{2+} \tag{9.3}$$

$$2BO_3 + CaO = 2\,BO_3^- + Ca^{2+} \tag{9.4}$$

It might also be anticipated that Fe_2O_3 and Cr_2O_3 would act like Al_2O_3 in slags and result in the accomodation of Fe^{3+} and Cr^{3+} into the silicate network. This is largely true in the case of Cr_2O_3 (except when the slag contains a significant Al_2O_3 content) but Fe_2O_3 appears to act as both *network breaker* and *network former* [32, 33]. However, the reducing conditions in the mould would appear to rule out the formation of both Fe_2O_3 and Cr_2O_3 in the mould slag.

In summary, the various constituents of mould fluxes can be classified as follows:

Network formers: SiO_2, Al_2O_3

Network breakers: CaO, MgO, BaO; FeO, MnO, CrO, Na_2O, Li_2O and K_2O

Fluxes: CaF_2 and B_2O_3.

9.2.2 Parameters to Represent the Structure of Slags

It is important to have a parameter to represent the structure of the silicate slag. Several parameters have been used and are outlined below (Sect. 9.2.2).

9.2.2.1 Basicity

Various basicity indices (e.g. B = %CaO/ %SiO_2) have been used to represent slag structure in the calculation of viscosities [34]. The major problem with this approach lies in dealing with oxides which can act as both a *networker former* and *network breaker* (e.g. Fe_2O_3, Al_2O_3).

9.2.2.2 NBO/T and Q

The ratio of (non-bridging O/Tetragonal O) ratio, which is denoted (NBO/T) is a *measure of the de-polymerisation* of the melt [1, 2]. It is the ratio of the mole fraction of *available* network-breaking oxides (where *available* indicates corrected for charge balancing) divided by the mole fraction of the network-forming oxides. Thus (NBO/T) ratio is a kind of basicity index which allows for corrections for cations acting on charge-balancing duties [33]. There has been a long-standing problem on how to deal with the CaF_2 when calculating (NBO/T) for mould fluxes. It was suggested [3] that CaF_2 acts as (i) a *network breaker* in polymerised slags and (ii) a *diluent* in basic slags. Recent structural studies indicate that when fluorine is added the F^- ions bond to Ca^{2+} ions and thereupon, remain apart from the slag [11–14, 16]. This would suggest that mould fluxes can be regarded as (*remaining slag* + CaF_2). Thus, the CaF_2 concentration has been ignored when calculating the NBO/T ratio for the remaining slag (Eq. 9.5 where X is the mole fraction and where $X_{MO} = X_{MgO} + X_{CaO} + X_{BaO} + X_{FeO} + X_{MnO} +..$ and $X_{M_2O} = X_{Li_2O} + X_{Na_2O} + X_{K_2O}$).

$$NBO/T = 2(\Sigma X_{MO} + \Sigma X_{M_2O} - X_{Al_2O_3})/(X_{SiO_2} + 2X_{Al_2O_3}) \quad (9.5)$$

Since it is not known whether TiO_2 should be regarded as a *network breaker* or *network former* it is also suggested that TiO_2 and ZrO_2 concentrations should be ignored when calculating the (NBO/T) for the slag component and to determine any effects on property values as a function of X_{TiO_2} or X_{ZrO_2}.

Many people find it easier to visualise polymerisation than de-polymerisation, and thus, the parameter, Q, which is a *measure of the polymerisation* of the melt is preferred here; it can be calculated from (NBO/T) using Eq. 9.6.

$$Q = 4 - (NBO/T) \quad (9.6)$$

Typical, "*remaining slag*", Q values range from Q = ca. 2 (for high-speed casting with $\eta_{1300\,°C}$ = 0.6 dPas) to Q = ca. 3 (for high-viscosity billet fluxes, $\eta_{1300\,°C}$ = >15 d Pas). (Note Q-values for the remaining slag are calculated by ignoring X_{CaF_2}, X_{TiO_2} and X_{ZrO_2}).

9.2.2.3 Optical Basicity (Λ)

The major problem when using (NBO/T) or Q, lies in the fact that it does not differentiate between different cations e.g. Na^+ or Ca^{2+}. The *optical basicity* was introduced to partially resolve this problem [35–37]. The *optical basicity* (Λ) is a measure of the *electron donor power* of different ions relative to that of CaO [38, 39]. Measurements were derived from the shift in the frequency absorption band of the UV region (associated with 6s → 6p transition) which is related to the basicity of the slag. It was subsequently shown that it could be calculated from Pauling electron negativity values [39].

It was also used as a measure of the de-polymerisation of the melt [35–37] and can be calculated by Eq. 9.7 where m is the number of O atoms, e.g. 1 for CaO and 2 for SiO_2 and Λ_I is the optical basicity value for oxide 1.

$$\Lambda = \Sigma(X_1 m_1 \Lambda_1 + X_2 m_2 \Lambda_2 + X_3 m_3 \Lambda_3 + \ldots)/\Sigma(X_1 m_1 + X_2 m_2 + X_3 m_3 + ..) \quad (9.7)$$

The composition can be corrected to allow for the cations on charge-balancing duties [35] by subtracting $2X_{Al_2O_3}$ from $\Sigma\ (X_1 m_1 \Lambda_1) + \Sigma\ (X_2 m_2 \Lambda_2)$ where 1 and 2 refer to MO and M_2O, respectively; this can be done by assuming that charge-balancing duties are done (i) preferentially by the largest cations or (ii) according to the concentrations (i.e. randomly). Optical basicities for the various constituents of casting powders are given in Table 9.1.

The principal disadvantage to the use of optical basicity results from:

- Uncertainties in the Λ values for some components, e.g. transition metal oxides. FeO, and, more importantly for mould fluxes, CaF_2.

Table 9.1 Values of optical basicity cited in Slag Atlas [3]

	CaO	SiO_2	Al_2O_3	MgO	BaO	Na_2O	Li_2O	K_2O	FeO	MnO	CaF_2	B_2O_3	TiO_2
UV shift	1.0	0.48	0.605	0.78	1.15	1.15	1.0					0.42	
Pauling *e. n*	1.0		0.60	0.78	1.15	1.15	1.0	1.4	0.51	0.59	0.43, 0.67	0.42	0.61
Electron densities	1.0	0.47	0.66	0.92	1.08	1.11	1.06	1.16	0.94	0.95		0.42	0.65

Values derived from (a) measurements of UV shift (b) Pauling electron negativities (b) and (c) Electron densities [38, 39]

- The optical basicity is a reasonable measure of the M–O bond strength [35] but does not differentiate between the size of cations which is important in electrical resistivity and diffusion.

The corrected optical basicity has been used to estimate the following properties: viscosity [36, 41], density [42], electrical conductivity and resistivity [43] and thermal conductivity of molten slags [35] but has not been widely used for mould fluxes because of the uncertainty in the value for Λ_{CaF_2}.

9.2.2.4 Concentrations of O°; O⁻ and O²⁻

The structure can be represented in terms of the concentrations in the slag of *bridging* O's (O°) *non-bridging* O's (O^-) and *free* O's (O^{2-}). The principal limitation to this approach has been that it requires a mathematical model [44–46] to determine these concentrations. However, recently a method has been developed to calculate these concentrations without the aid of a mathematical model [40, 47–49] and this would appear to be exceedingly useful in predicting properties since it can differentiate between Si–O and Al–O bonds and their effects on specific properties.

Recently, the method has been applied to the prediction of viscosities of slags containing CaF_2 [49] but the model, at the present time, has not been extended to slags containing B_2O_3 which is used as a flux, particularly in calcium aluminate mould slags used for casting high-Al steels.

9.2.3 Effect of Cations

Although the degree of polymerisation in the silicate network is the dominant factor affecting the properties of slags, the nature of the cations (e.g. Ca^{2+} or Na^+) can also affect the properties [50, 51]. Cations can affect properties of slags (i) *directly* (e.g. the effect of cation size on electrical conductivity) or (ii) *indirectly* by affecting the slag structure. The properties are affected by the following.

9.2.3.1 The Field Strength of the Ionic M^+–O^- Bond

Which is frequently represented by either the parameter (z/r^2) where z is the charge and r is radius of the cation (M^+) or by the optical basicity, Λ (Fig. 9.5). The bond strength increases as (z/r^2) increases and as Λ decreases; The optical basicity (Λ) has been reported to have a linear and inverse relationship with (z/r^2) [35].

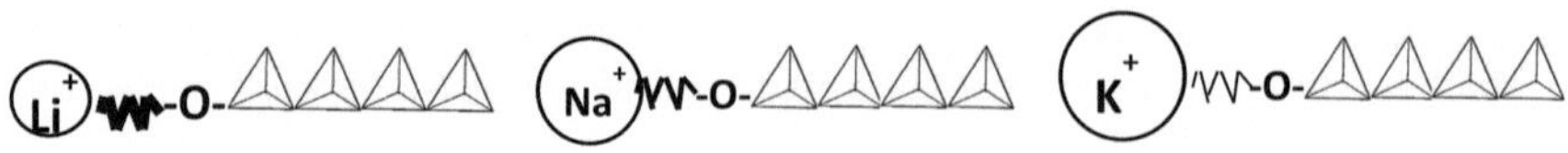

Fig. 9.5 Schematic diagram showing relative strengths (denoted by thickness of bond) of $M^{+}–O^{-}$ bonds for Li^{+}, Na^{+} and K^{+} [50]

9.2.3.2 The Fraction of Ionic Bonding in the M–O Bond

M–O bonds are mostly ionic but also contain some covalence (e.g. the fraction of ionic bonding (expressed as %) is 78, 48 and 59% for CaO, FeO and MnO, respectively [40]); oxides with significant levels of covalent bonding tend to promote the formation of the glass phase.

9.2.3.3 The Size of the Cations

Since larger cations tend to (i) hinder the motion of one layer of molecules over another layer and thereby increase viscosity and (ii) find it difficult to pass through the "holes" in the silicate network which thereby cause a decrease in electrical conductivity and diffusion (Fig. 9.6). The field strength (z/r^2) is also affected by cation size and large cations (e.g. K^{+}) form weaker NBOs and thus are more easily broken than those formed with Na^{+} and Li^{+} ions; this should aid the electrical conduction and diffusion. However, any advantage for the K^{+} must be offset against the more sluggish movement of these large ions.

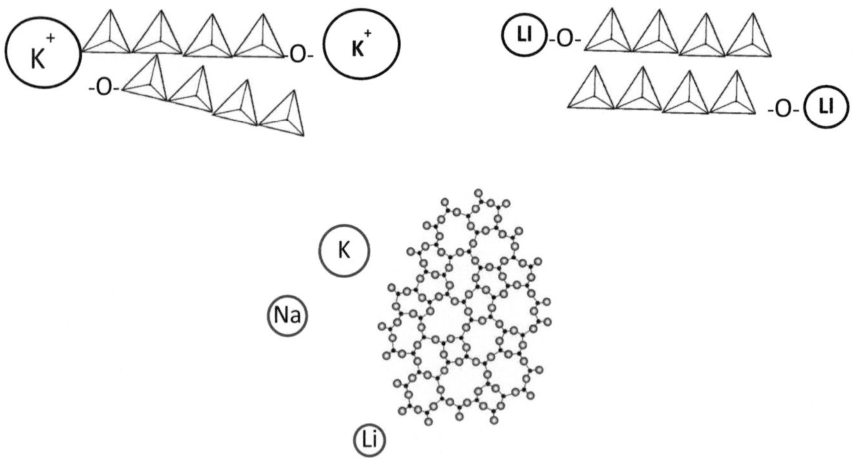

Fig. 9.6 Schematic diagram showing hindrance caused by large cations in (*top*) viscosity and (*bottom*) electrical conductivity [50]

9.2.3.4 The Bridging of Cations

Bridging occurs when divalent cations (M^{2+}, e.g. Ca^{2+}) carry out network breaking or charge-balancing duties; this effect is absent with monovalent cations (M^{+}, e.g. Na^{+}) and could lead to increases in viscosity and electrical resistivity (Fig. 9.7).

9.2.3.5 The Number of Cations

CaO gives rise to one Ca^{2+} ion, whereas, Na_2O gives two Na^{+} ions; the number of cations available to move is an important parameter in electrical conductivity and diffusion [50, 51].

9.2.3.6 The Mixed Alkali Effect

The mixed alkali effect occurs when the slag contains two (or more) cations with very different radii (e.g. when K_2O is added to $Li_2O–SiO_2$, the decrease in electrical conductivity is much greater than that predicted from linear mixing rules). It has been attributed to the cations occupying foreign sites [52].

9.2.4 *Effect of Temperature on Properties*

Slag properties are very sensitive to temperature and this sensitivity increases with increasing polymerisation of the melt. Increasing temperature tends to loosen the silicate structure (akin to the de-polymerisation caused by addition of cations). Thus the effect of increasing temperature on various properties can be predicted since it is equivalent to de-polymerisation (i.e. it causes decreasing viscosity, electrical resistivity and thermal conductivity).

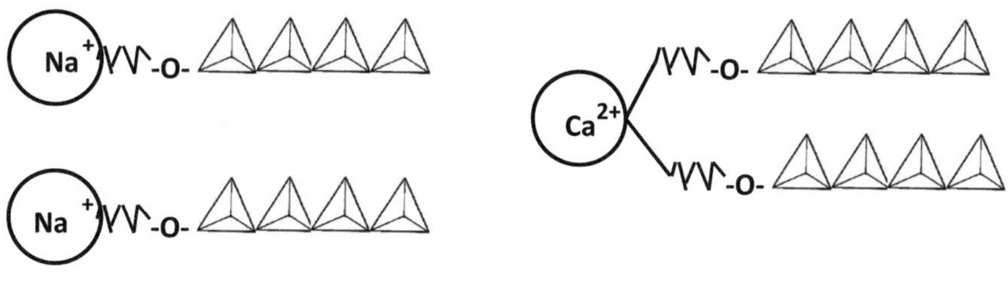

Fig. 9.7 Schematic representation of "*bridging*" with Ca^{2+} ions: note bridging is absent with Na^{+} ions [50]

9.2.4.1 Solid Slags

Most slag films consist of a mixture of glass and crystalline phases. The ratio of glass and crystalline phases in the slag film is dependent upon both composition (glasses are promoted by high Q value) and the thermal history of the sample.

Most mould slags can be produced in the form of a glass by rapid quenching of the molten slag. If the heat capacity of this glassy mould slag is measured during heating, the following sequence of events occurs:

- There is a smooth increase in C_p until the temperature exceeds the glass transition temperature (T_g)
- At T_g there is a "*step increase*" in C_p (Figs. 9.8a and 9.9 [53]) and a threefold increase in thermal expansion coefficient (α) (Fig. 9.8b); these events are associated with the transition of the glass to a supercooled liquid (*scl*).

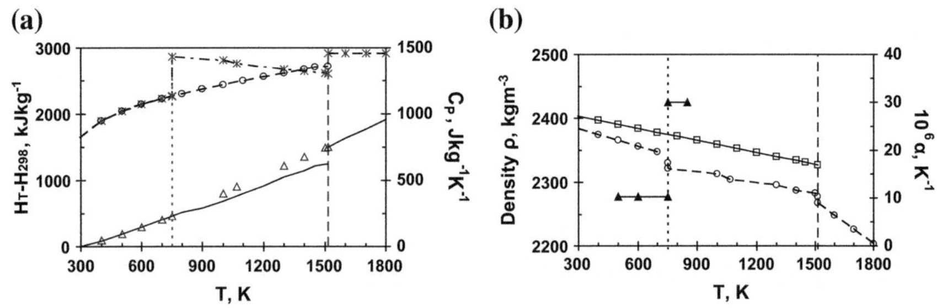

Fig. 9.8 Schematic diagrams showing the effects of the glass transition on **a** C_p (*right* axis) and enthalpy (H_T–H_{298}) (*left* axis); C_p : - .-. -, and * = *scl* and liquid phases; o and *solid line* = glass and crystalline phase and (H_T–H_{298}): *solid line* = glass, crystalline and liquid phases; Δ = *scl*; and **b** density and thermal expansion coefficient, α (*right* axis) Density (*left* axis), □ = crystalline phase o; glass and *scl* and liquid phases; α; ▲ = glass and *scl* phases

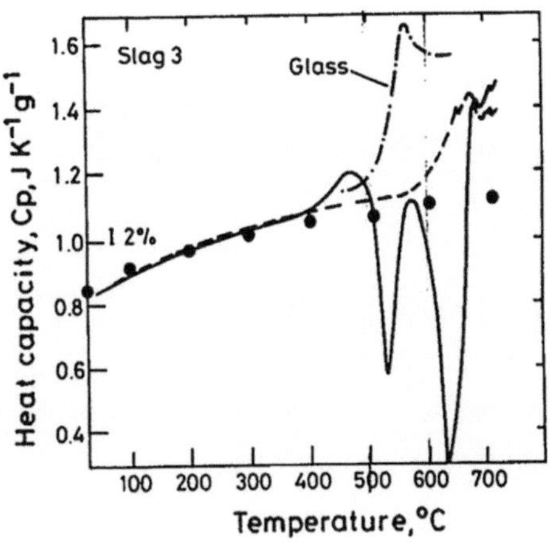

Fig. 9.9 Heat capacity of a mould slag as a function of temperature showing both the step-increase in C_p at T_g for the glass and the slag film and the apparent decrease in C_p accompanying crystallisation (*solid line*) (permission granted, IOM/Taylor and Francis [53])

- In the case of some mould slags, an inverted peak in C_p is observed to start at about 100 °C above T_g; this is due to partial crystallisation of the mould slag (Fig. 9.9). (Note crystallisation is a first-order, exothermic transition and in DSC measurements the exothermic, enthalpy of transition manifests itself as an *apparent* decrease in C_p).
- The mixture of (glass+crystalline) phases undergoes an endothermic transition at the solidus temperature of the slag (T_{sol}) when the crystal fraction starts to melt and this process (crystal → liquid) is complete when the temperature reaches T_{liq}; note the *scl* undergoes a smooth transition to liquid (Fig. 9.8a).

The (crystal → liquid) transition is accompanied by an enthalpy of fusion and a change in density (or volume), whereas the equivalent values for the (*scl* → liquid) transition are very low or are zero.

9.2.4.2 Liquid Slags

When a molten slag is cooled below T_{liq} it forms either crystallites or a supercooled liquid (*scl*). If viscosities are measured in a cooling cycle, when solid particles (crystallites) are precipitated (Fig. 9.10a), there is a sharp increase in viscosity and the slag exhibits non-Newtonian behaviour (i.e. measured viscosities differ for different rotation speeds). In contrast, when a liquid slag transforms to a *scl* it is accompanied by a smooth increase in viscosity (Fig. 9.10b) until the temperature reaches T_g ($\log_{10} \eta$ (dPas) = 13.4 at T_g).

9.3 Crystallisation in Mould Fluxes

9.3.1 *Importance of Crystallisation to the Process*

The first slag to form in the (shell/mould) channel is probably glassy because of the high cooling rates involved. However, crystallisation of the slag film takes place as the cast proceeds. The slag film controls the heat transfer from the shell to the mould.

Crystallisation of the slag film is very important to the continuous-casting process. The three principal factors affecting the heat transfer across the slag film are (i) the *slag film thickness* (which is related to the break temperature, d_s↑ as T_{br}↑)

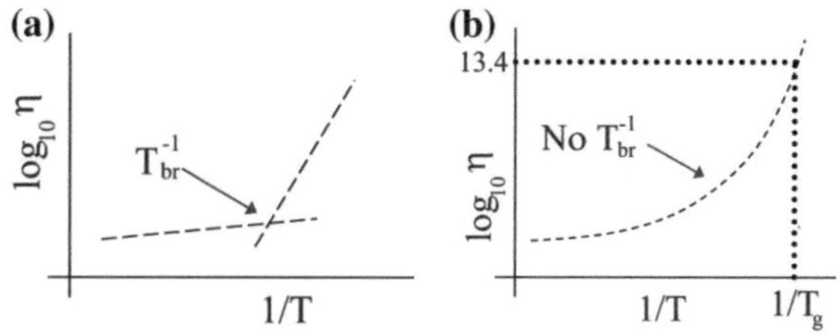

Fig. 9.10 Plots of $\log_{10}$ viscosity as functions of reciprocal temperature (T_K^{-1}) for **a** a slag where crystallites are precipitated (Q = 2) and **b** glassy slag forming a *scl* (Q > 2.5)

(ii) the *degree of* crys*tallisation* of the slag film (or f_{crys}) and (iii) the *size of the* crys*tallites* (D_{crys}). Increased crystallisation in the slag film results in:

- Both *scattering and reflection* of the IR radiation (thereby, decreasing the effective thermal conductivity) ($q_{hor} \downarrow$ as $f_{crys} \uparrow$) [54, 55]; the reflection and scattering of IR radiation increases with increasing crystal size (D_{crys}) with the extinction coefficient of the slag, increasing with increasing D_{crys} [56, 57].
- The creation of both *porosity* and an interfacial resistance ($R_{Cu/sl}$) due to the formation of a "*gas gap*" (or "*surface roughness*") at the mould/slag interface [58] caused by the shrinkage resulting from the (glass → crystal) transition ($\rho_{crys} > \rho_{glass}$); the porosity and $R_{Cu/sl}$ both reduce the horizontal heat flux ($q_{hor} \downarrow$ as $f_{crys} \uparrow$) [59].
- An increase in lattice thermal conductivity (k_{lat} with $k_{lat\ crys} \approx 2k_{lat\ gl}$) [60–62]; thus ($q_{hor} \uparrow$ as $f_{crys} \uparrow$).

Since the overall effect of crystallisation is to reduce the horizontal heat transfer, it follows that the combined effects of the reduction in radiation conduction along with the creation of the interfacial resistance ($R_{Cu/sl}$) are more important than the increase in lattice thermal conductivity resulting from crystallisation. There has been some dispute concerning the relative magnitudes of the decrease in heat flux due to scattering of radiant energy and the interfacial resistance. However, the analysis of heat flux data obtained in plant trials [63, 64] identified the scattering of radiant energy as the more important process. As mentioned above, scattering of IR radiation is also affected by the size of the crystals in the slag film with scattering (i) increasing with increasing size and (ii) being at an optimum level for D_{crys} = 2–3 μm [56]. The crystal size has been reported to increase with increasing slag basicity (C/S) [65].

The formation of a slag film with a well-developed crystalline fraction is essential in minimising longitudinal cracking in peritectic MC steels. In contrast, for HC steels a thin, slag film, composed of glass (with no crystalline phase) is important to promote horizontal heat transfer to create a thick and stronger shell to avoid sticker breakouts. This is especially important for steel grades which result in a weak shell, e.g. HC steels, in ULC steels where *ferrite* is formed and in *ferritic* stainless steels. The soft, weak shell needs strong cooling to avoid sticker breakouts.

9.3.2 Crystalline Phases Formed in Slag Films

In conventional mould slags, the primary crystallising phase is *cuspidine*. However, other primary crystalline phases are formed in other, specialised, mould slags (e.g. F-free and CA-type slags used to cast high-Al steels); these include the phases, CT (*perovskite*) [66] CST [67], NC_2S_3 [68] *melilite* [69]. Consequently, the phases formed in conventional and fluoride-free mould slags are discussed individually below.

9.3.2.1 Conventional (F-Containing) Powders

Grieveson et al. [70] produced a phase diagram for F-containing, mould slags with (%CaO/%SiO_2) ≈ 1.0. It was found that *cuspidine* was always formed, but the nature of the other phases (both glassy and crystalline) formed was affected by their chemical composition. The phase diagram for the $3CaO \cdot 2SiO_2 \cdot CaF_2 - CaF_2$ system has been investigated [57, 70] and the effects of Na_2O on the *cuspidine* phase relations have also been reported [69, 72].

Certain oxides are reported to promote the formation of crystalline phases, whereas others promote the formation of glassy phases.

Crystalline phase formation is promoted by increases in concentrations of the following:

(i) CaF_2 (for *cuspidine* formation);
(ii) High basicity (C/S) [73] or decreasing SiO_2 content; both the amount of *cuspidine* and its crystal size (D_{crys}) in the slag film increase with increases in basicity (C/S) [65] from 0.8 to >1 [65, 74].
(iii) Li_2O
(iv) Particles of ZrO_2 [77] TiO_2 [75] and Ti(CN) [25, 27, 75, 76] are reported to act as nucleating agents and evidence has been obtained for *cuspidine* nucleating on ZrO_2 particles [77].
(v) The application of an electric field to the slag affects crystallisation [78, 79] with both AC and DC increasing f_{crys} with crystal size (D_{crys}) increasing with DC and decreasing with AC [79].

Glass phase formation is reported to be promoted by additions of the following oxides:

(i) SiO_2
(ii) B_2O_3 [79, 80]
(iii) Na_2O [81]; MnO [82, 83] (with >4%MnO and >7%MgO) and FeO [56] which all tend to retard the crystallisation process [82, 83];

In F-containing powders (denoted here as conventional powders) the principal crystalline phase is *cuspidine* ($3CaO \cdot 2SiO_2 \cdot CaF_2$ which is sometimes abbreviated to C_3S_2Fl).

Cuspidine is usually the first phase to crystallise; consequently, cuspidine precipitation is usually referred to as primary crystallisation. Since it is formed at an early stage, *cuspidine* is found as small crystals on the mould-side of the slag film (i.e. the cold region) [84, 85]. The Gibbs free energy of formation of *cuspidine* has been determined as ΔG_f (kJmol^{-1}) = −5188 + 0.82 T_K [86] and as $\Delta G_f = -RT_K \cdot \log p_{SiF_4} = -5253 + 0.864\ T_K$ [87] where T_K is in *K*.

The kinetics of crystallisation are usually carried out by determining Transformation–time–temperature (TTT) curves [88, 89] for the mould slags (see Sect. 9.2.4). It was found that crystallisation times increased with increasing SiO_2 content in the slag.

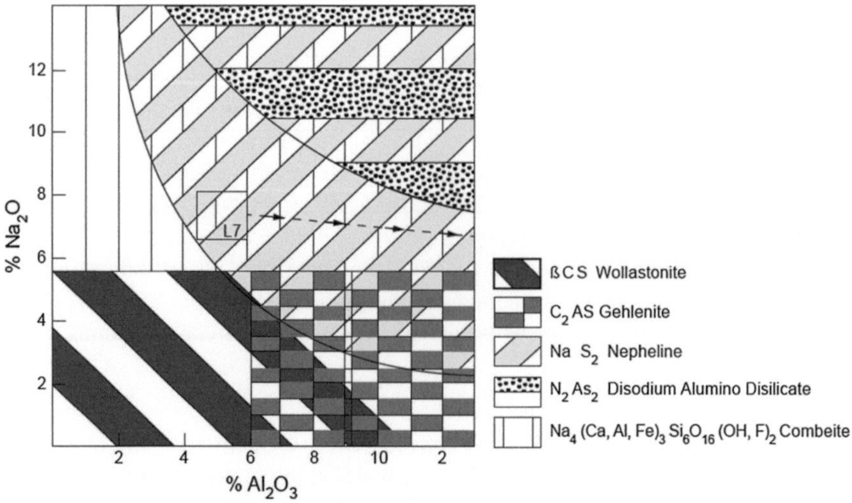

Fig. 9.11 Phases formed on solidification by molten casting fluxes (*cuspidine* is not shown but is always present) (permission granted, IOM/Taylor and Francis [70]; re-drawn Swerea KIMAB)

Secondary crystallisation consists in the precipitation of other phases (e.g. *nepheline* or *gehlenite*) and tends to be found towards the shell side of the slag film. The mineralogical nature of these late-forming phases for powders with a C/S ratio close to 1.0 is shown as a phase diagram in Fig. 9.11 [70].

Agitation was found to have significant affect on primary crystallisation but had a much smaller effect on secondary crystallisation [90].

A study of the effect of Na_2O on the formation of *cuspidine* (C_3S_2Fl) crystals found that (i) F^- ions tend to bond with Ca^{2+} ions (ii) the maximum amount of *cuspidine* after annealing was obtained with a sample containing 4% Na_2O [72]. The latter finding was attributed to the effect of Na_2O on both the nucleation and growth rate (via a lower viscosity). Most mould slag compositions lie outside the phase field for *cuspidine* and it has been reported that moving the slag chemical composition into the *cuspidine*-field provides slag films with reduced, horizontal, heat fluxes [69].

Low-basicity (C/S) mould slags have been used to cast high-Al steels and CaF_2 was identified in the mould slag [91]. More recently, thermodynamic databases have been used to identify the crystalline phases formed in these low (C/S) mould slags [92].

9.3.2.2 Fluoride-Free Fluxes

In recent years, there has been a drive to reduce the magnitude of fluorine emissions in steelmaking operations. In F-free fluxes, the absence of F means that no *cuspidine* can be formed to scatter the IR radiation and thereby reduce the horizontal heat flux.

Successful F-free powders have been developed for billet casting but the development of a successful mould powder for slab casting remains an ongoing challenge at the present time. Investigations into possible replacements for CaF_2 tend to fall into one of two classes:

(i) Those where CaF_2 is replaced by other fluxes, e.g. Li_2O [93–95] Na_2O [96, 97]:K_2O [98] BaO [99, 100] B_2O_3 [28, 101, 102] TiO_2 [66, 67, 102, 103].
(ii) Those where *cuspidine* is replaced by specific, crystalline phases (e.g. *perovskite*, $CaTiO_3$ or CT) [66, 103] or *melilite* (solid solution ranging from *akermanite* (C_2MS_2) to *gehlenite* (C_2AS) [69] or the phase NC_2S_3 [68].

It should be noted that the nature of the crystalline phase is not an issue, providing (i) the crystalline phase brings about the necessary reduction in horizontal heat flux and (ii) it causes no collateral problems. One example of the latter powders where *cuspidine* is replaced by $CaTiO_3$ [66] reputably, resulted in a high incidence of sticker breakouts; this was attributed to the formation of Ti(CN) (formed by reactions of the TiO_2 and C and N_2 in the mould) [104]. However, this view has been disputed recently [102].

9.3.2.3 Calcium Aluminate Slags

The effect of the (C/A) ratio on the phases formed in CA-type slags has been studied [105]. When the (C/A) ratio was less than 0.9, CaF_2 and lithium aluminate, LA, were the principal phases formed. With (C/A) = 1, no CaF_2 remained and the phases, LA, C_3A, $C_{11}NA_4$ and $C_{3\text{-}x}N_xA$ but when (C/A) = 2.3 the LA disappeared and was replaced by CaO [105]. The LA phase formed at high alumina comcentrations has a high melting temperature. In CA slags containing B_2O_3 the phases (C_9B_3Fl and CB+ a large amount of C_3S_2Fl) were detected [106].

9.3.3 Crystallisation Process

It has been reported above that primary crystallisation results in the formation of *cuspidine* and other phases (e.g. *carnegieite*, NAS_2) are formed in secondary crystallisation [70]. Homogeneous nucleation tends to be diffusion-controlled [77] with outward diffusion of MgO, Al_2O_3 and Na_2O and inward diffusion of CaO, SiO_2 and CaF_2 [84]. However, heterogeneous nucleation has been observed to occur on ZrO_2 particles [77]. It has also been suggested that TiO_2 and ZrO_2 particles [77] promote crystallisation principally by acting as nucleating agents, (ZrO_2 has low solubility, ca. 2%, in mould slags). It is known that ZrO_2 and TiO_2 tend to increase the break temperature and thereby create thicker slag films.

Scheller [111] observed tiny (5 μm) droplets of metals (e.g. Na) resulting from redox reactions in the mould slag; it was suggested that these droplets could also act as nucleation sites for crystallisation.

It has also been suggested [107] that pores in the slag film can act as nucleating agents (Fig. 9.12a, b). However, this is a "*Chicken and egg situation*" and it is debatable whether the pores are located near the crystals because of their nucleation of crystals or, alternatively, the pores are the result of the shrinkage accompanying the crystallisation process. Figure 9.12c shows the number of pores as a function of pore diameter and it can be seen that there is a large number of fine pores [107]. It was suggested that in the liquid layer, bubbles can move, collide and grow and thus larger bubbles are found in the liquid layer. However, once the crystallisation is initiated there is rapid dendritic growth and the bubbles cannot escape into the liquid and consequently, get trapped in the slag film [107].

Slag components which promote the glassy phase (e.g. B_2O_3; Na_2O, Li_2O [108, 109] FeO [110] and high amounts of MnO and MgO [83] reduce the f_{crys} in the slag film. However, components which tend to promote *cuspidine* formation (e.g. CaF_2 and the (C/S) ratio [69] also increase f_{crys}. The absorption of large amounts of alumina results in increased crystallisation [69]. Components, such as TiO_2, result in the formation of *perovskite* ($CaTiO_3$) and thereby increase f_{crys}.

The application of an electrical potential has also been found to affect the crystallisation [78, 79] with increasing voltage causing increases in f_{crys} with both AC and DC and an increase in D_{crys} for DC and a decrease in D_{crys} for AC.

It has been reported that the crystal grain size for the slag film in a simulation test was greater in the lower mould than in the upper mould.

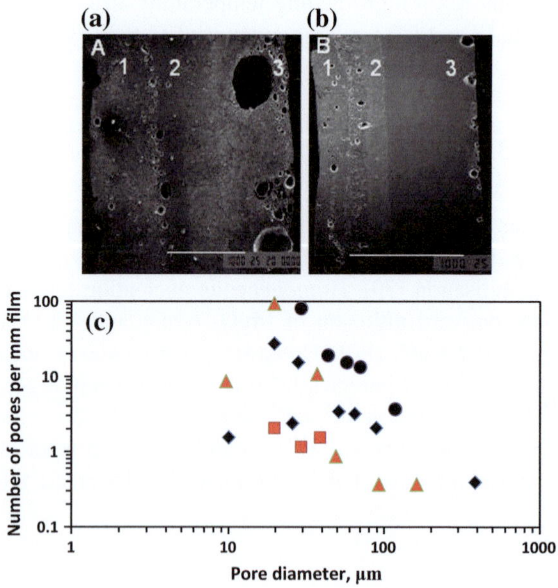

Fig. 9.12 **a**, **b** Photographs of slag films *A* (f_{crys} = 0.9) and *B* (f_{crys} = 0.3) with mould on left and shell on the right; showing *1* = fine crystals *2* = dendritic crystals and *3* = glassy (former liquid) [107] and **c** number of pores as a function of pore diameter, *A* = ●, ♦; *B* = ▲, ■; [107] (courtesy of, Li and Carboox [107])

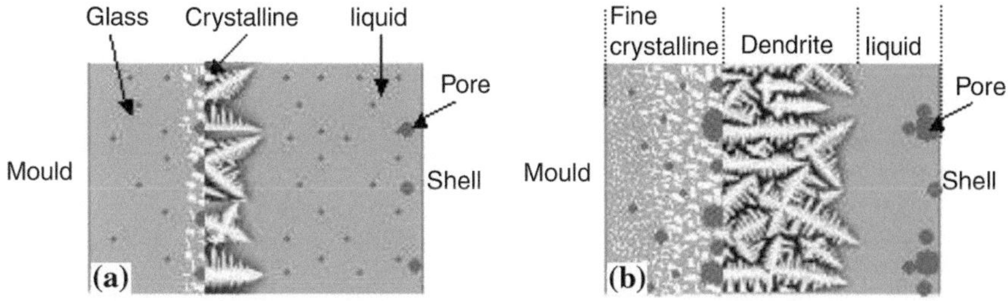

Fig. 9.13 **a**, **b** Schematic drawing showing the development of the crystalline nature of the slag film during the casting showing dendritic growth in the warmer parts of the slag film and fine crystals formed in cold regions of slag film where there is significant undercooling (courtesy of Li and Carboox [107])

9.3.4 Crystallisation Kinetics

The first slag to penetrate into the mould/shell gap probably forms a glassy slag film because of the high cooling rate involved. This glassy, slag film crystallises over time. The rate of crystallisation of the slag film is dependent upon both the time and temperature experienced. The kinetics of crystallisation is usually represented in the form of a Time–Temperature–Transformation (T–T–T diagram which plots time and temperature and shows contours to reach a certain % of transformation. A typical example is shown in Fig. 9.14a. Such diagrams are usually constructed from measurements from the double hot thermocouple (DHTC) method [88, 89, 111] or other methods such as DTA [112] high-temperature optical microscopy or by quenching in sealed capsules [77, 113].

The kinetics of crystallisation involves two processes, nucleation and growth. Temperature has opposing effects on these processes; *nucleation* is promoted by undercooling (i.e. decreasing temperature) whereas *growth* increases with increasing temperature. Take a case where the TTT curves exhibit a single nose indicating the precipitation of a single phase. There are two temperature regimes [114]; at high temperatures (just below T_{liq}) there is no undercooling but as the melt cools as the undercooling/nucleation increases and once a crystal is nucleated, the high temperatures promote rapid crystal growth.

At lower temperatures, nucleation is easy but the growth process is slow (due to the high melt viscosity) and so fine crystals are formed.

9.3.4.1 Conventional Fluxes

It is generally assumed that the first slag film formed is glassy because of the high cooling rates involved. Over time, this glassy slag film can crystallise, the extent to which it crystallises is dependent upon its composition (especially the C/S ratio) the annealing time and the local temperature. Longitudinal cracking tends to be more prevalent in the early part of the casting. In these early parts of the cast the

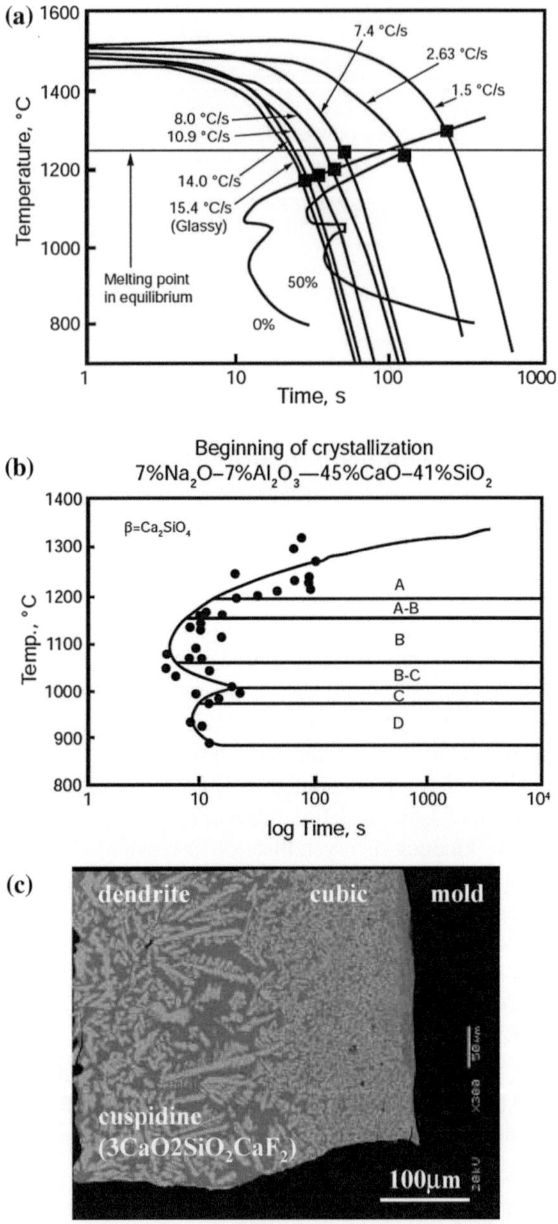

Fig. 9.14 **a** T–T–T diagram for a mould slag with composition 45%CaO + 41% SiO_2 + 7%Al_2O_3 + 7% Na_2O showing the onset of crystallisation (*curve* and points) and **b** resultant morphologies; *A* = Equiaxed crystals formed at high temperature; *B* = Columnar crystals; *C* = faceted crystals and *D* = very fine crystals formed at high undercooling and **c** Photograph of section of slag film showing dendritic growth of crystals on shell side and fine cubic crystals on mould side (permission granted, (**a**, **b**) ISS, re-drawn Swerea KIMAB after [88] (**c**) ISIJ [67])

crystalline layer of the slag film is not fully developed (Fig. 9.13a) and this results in high heat fluxes and to longitudinal cracking. However, the cracking will tend to decrease as the crystalline layer of the slag film (Fig. 9.13b) develops and this layer reflects and scatters the IR radiation.

The primary crystallisation (of *cuspidine*) occurs by inward diffusion of CaF_2, CaO and SiO_2 and by the outward diffusion of Na_2O, Al_2O_3 and MgO [84]; it would appear that MnO aids this process [82] whereas TiO_2 additions result in the formation of *perovskite* (CT) which suppresses the formation of *cuspidine* [30].

A typical example of transformation–time–temperature (T–T–T) plot is shown in Fig. 9.14a [88]. The various smooth curves represent different cooling rates where the two irregular curves show the contours for 0 and 50% transformation. It can be seen from Fig. 9.14a that:

- The cooling rates increase as we move from right to left and the sample tends to become more glassy, as we move from right to left.
- When annealing at 900 °C, the sample remains totally glassy for just over 10 s and is 50% crystalline at 50 s.
- The rate of crystallisation decreases at higher and lower temperatures (e.g. 800 and 1200 °C, respectively).

It can be seen from Fig. 9.14b that large equiaxed crystals are formed at high temperature and that fine crystals are produced at lower temperature; thus, crystal size tends to increase with increasing local temperature in the slag film. These features can also be seen in the slag film shown in Fig. 9.14c [67]. The time taken for crystallisation to occur is referred to as the "*incubation time*". It has been reported that Na_2O reduces the incubation time and is thus beneficial for faster crystallisation of the slag film [109].

Rocabois et al. [77] examined the theory regarding the temperature of the "*nose*" in Fig. 9.14a and suggested a value of $T_{nose} = 0.77\ T_{liq}$ for most slag systems, However, they found values for mould slags tended to follow $T_{nose} = 0.86\ (\pm 0.06)\ T_{liq}$. They also looked at the critical cooling rate $(dT/dt)_{crit} = (T_{liq} - T_{nose})/t_{nose}$ which is the cooling rate necessary to produce a glass.

Inspection of TTT curves for various mould fluxes shows the transformation contours tend to move to right with increasing SiO_2 content (or increasing polymerisation, Q), i.e. they require longer crystallisation times. Some mould fluxes with high SiO_2 contents may not crystallise at all.

It can be seen from Fig. 9.14a that the morphologies of the crystals are also dependent on time and temperature with large dendritic crystals being formed in high-temperature regions and progressively finer crystals being formed with increasing supercoiling (i.e. lower temperatures) [88].

The effects of the following oxides on the crystallisation kinetics have been investigated: Na_2O [57, 67, 72], Li_2O [57, 65], K_2O [57], MnO [82], B_2O_3 [65].

9.3.4.2 F-Free Fluxes

TTT diagrams have also been developed for F-free fluxes containing TiO_2 [81] and it was found that TiO_2 reduced the activation energy for crystallisation and that Na_2O also reduced the incubation time. However, it was also reported that the horizontal heat flux increased when $Na_2O > 7\%$ and $MnO > 4\%$ [81] to these oxides promoting the glass phase. This latter finding is consistent with observations of Susa et al. [56, 110] who observed that FeO (another transition metal oxide) additions (up to 2%) reduced the apparent reflectivity of the crystals. This is a surprising result since FeO, MnO, NiO, etc. are added to increase the absorption of IR radiation in mould fluxes where no crystalline phases are formed (i.e. $q_{hor}\downarrow$ as % FeO $\uparrow$). Thus for slag films where crystals are formed, the total heat flux increases with increasing FeO since the FeO reduces both the reflectivity (R') of the crystals [110] and the grain size (D_{crys}) of the crystals [56] (i.e. $q_{hor}\uparrow$ as % FeO $\uparrow$ since $R'\downarrow$ and $D_{crys}\downarrow$). The reflectivity was found to increase with increasing grain size and it has been calculated that control of grain size could give a 7–8% decrease in the horizontal heat flux [56] and manifests itself as an increase in Extinction coefficient [57].

9.3.4.3 Calcium Aluminate Slags

The effect of (C/A) ratio on the incubation times of CA-type slags has been investigated [105]. It was found that the incubation times were longest (i.e. crystallisation was slowest) when (C/A) = 1.5 [105].

9.3.5 *Effects of Crystallisation*

Crystallisation causes changes in the properties of mould slags; these are detailed below.

9.3.5.1 Density

The crystalline phase has a greater density than the glass ($\rho_{crys} > \rho_{glass}$) because of the better packing in the crystal; thus, shrinkage occurs during crystallisation. Slag films contain some porosity and this porosity lowers the density of the slag film. At temperatures above T_g, the supercooled liquid (*scl*) formed from the glass has a much higher thermal expansion coefficient than that of the crystalline phase.

9.3.5.2 Transport of Heat

It was pointed out that the shrinkage accompanying crystallisation in the slag film has the following effects on the horizontal heat flux (q_{hor}).

(i) The tiny gas bubbles in the crystallised region (Figs. 9.12 and 9.15) reduce both the effective thermal conductivity of the slag film and the heat flux (q_{hor}) [107].
(ii) The creation of an interfacial resistance ($R_{Cu/sl}$) or "*gas gap*" at the mould/slag interface also reduces q_{hor} [115].

(iii) The crystallites scatter the IR radiation, reducing both k_R and q_{hor} [56, 110].
(iv) The lattice thermal conductivity is increased (i.e. $k_{cryst} \approx 2k_{glass}$) [60, 61].

The overall effect of crystallisation is to reduce q_{hor}.

9.3.5.3 Hydrogen in Slag Film

Significant levels of porosity have been reported to occur when the slag film contains high levels of H_2 and this has been linked to incidences of sticker breakout [116, 117].

Presumably, the porous slag film has a low thermal conductivity which reduces the horizontal heat flux and results in a thin, weak shell. Possible sources of H_2 are (i) high moisture levels in casting powders and (ii) water leaks in the continuous-casting process.

9.3.5.4 Heat Capacity

The heat capacity of the *scl* formed from the glass phase at $T > T_g$ is higher than that of the crystalline phase for the temperature range between T_g and the solidus temperature of the slag, as can be seen in Fig. 9.9.

9.3.5.5 Friction Forces and Lubrication of the Shell

Hering et al. [118] reported that the liquid frictional forces are dependent upon the mineralogical phases formed (Figs. 9.15 and 9.16); the friction force for *nepheline* being greater than those where *combeite*; *wollastonite* and *gehlenite* are formed which may be related to the T_{liq} values of the phases.

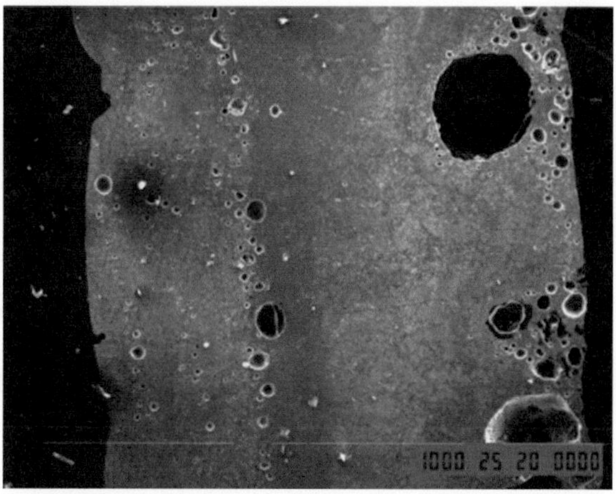

Fig. 9.15 Photograph showing fine pores associated with crystallisation (*left* = mould; *right* = shell) (courtesy of Carboox [107])

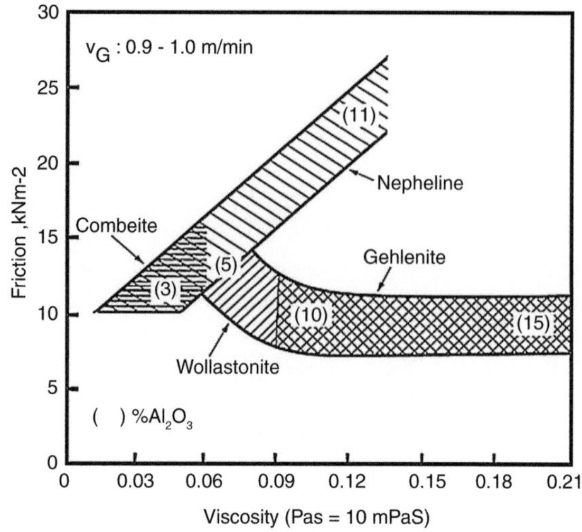

Fig. 9.16 Friction force as a function of slag viscosity (in Pas = 10 d Pas) showing the effect of mineralogical phase on the friction (Reproduced with permission from Stahlundeisen, re-drawn Swerea KIMAB after [118])

There is little, or no, crystallisation in some high-viscosity billet slags. These slags form a supercooled liquid (*scl*) at temperatures above T_g. Although, the slag film has a high viscosity, it will tend to move when subjected to stress from the moving shell. In thermal expansion measurements at $T > T_g$, the slag sample cannot support its own weight, and so the sample collapses due to the pressure of the probes, (see Fig. 9.41a). Thus, a glassy slag film (*scl*) will provide some measure of lubrication to the shell. In contrast, crystalline phases tend to be more rigid and are prone to fracture [119]. The fracture of the slag film can occur near the bottom of the mould and results in variable heat transfer and in "*star and spongy cracking*" [120, 121].

9.3.6 Methods of Determining Fraction of Crystalline Phase in Slag Films

The fraction of crystalline phase (f_{crys}) in the slag film is an important process variable. Yet, hitherto, it has not been treated with such reverence, as say, the slag viscosity or the break temperature. A possible reason for this is that it f_{crys} has proved difficult to quantify and the measurements obtained tend to show considerable variation according to the method used.

Experimental Methods

Several methods have been proposed to determine the fraction of crystalline phases (f_{crys}) formed in the slag film under steady state conditions. The various techniques used are summarised below. There is an urgent need for an inter-comparison of the results for f_{crys} obtained with these various methods. The development of a standard test and recommended procedures for determining f_{crys} are urgently needed.

9.3.6.1 Metallographic Method

In this method, the slag film is mounted so that its cross section can be examined and then polished to 1 μm and etched in 2.5% aqueous HF solution [83, 107]. The sample is then examined in an optical microscope and the relative amounts of crystalline and glassy phase are determined. The method is slow and time consuming and since the observation area is small and a large number of observations are needed to obtain a representative value for the whole slag film [107, 122].

9.3.6.2 X-Ray Diffraction Methods

The slag film is ground to a powder and then subjected to X-ray diffraction (XRD). A typical XRD pattern is shown in Fig. 9.17; the crystalline phases are characterised by sharp peaks and the glassy phases appear as a diffuse "*hump*", as shown in the lowest pattern in Fig. 9.17. Measurements can be made of (i) the intensities (I) of specific peaks (corresponding to crystallographic plane) [54, 55, 123] or (ii) to the area under the peaks [61, 62, 123].

9.3.6.3 Measurements of Intensities

These experiments are usually comparative methods where the intensities of a specific peak are compared with those of standard samples with known f_{crys} values. The intensities of peaks from the slag film have been compared with the following:

(i) *Cuspidine*/α-Al_2O_3 [54, 55].
(ii) *Cuspidine*/glass mixtures [123].

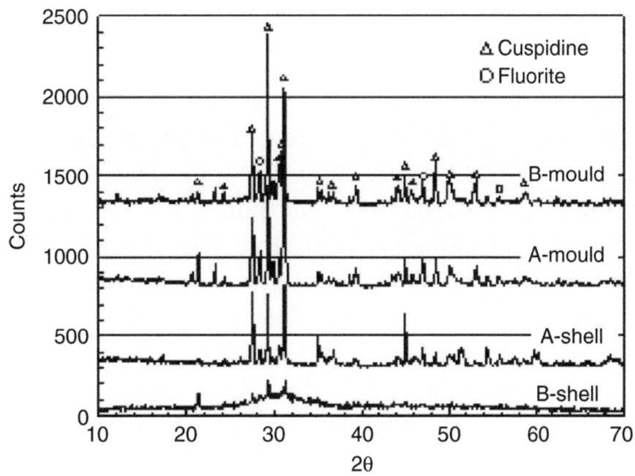

Fig. 9.17 Typical X-ray diffraction (XRD) patterns of two slag films (*A* (f_{cryst} = 0.9) and *B* (f_{crys} = 0.3)) showing sharp peaks associated with crystalline phases and the "hump" due to glassy phase (bottom pattern) (courtesy of S Riaz [123])

In these experiments, the ground slag film is mixed with known weights of the Al_2O_3 or glass and the samples homogenised. The intensities of specific peaks in the XRD pattern are then compared with those from standard samples (e.g. samples of 100/0, 75/25, 50/50, 25/75, 0/100, respectively, of *cuspidine* and α-Al_2O_3 [54, 55]. The XRD patterns of these various mixtures are then measured and the ratio of integrated intensities (e.g. $I_{cuspidine\ (222)}/I_{Al_2O_3(012)}$) determined (where $_{()}$ denotes the crystallographic plane of the peak [54, 55]. In these experiments the slag film is ground and mixed (in the case of the Al_2O_3 experiments) with a known amount of Al_2O_3 and the intensity ratio determined by XRD. The intensities are then compared with those for the standard mixtures. Since the samples are small at least five measurements should be made to obtain a representative value. One major problem with this method is that pure *cuspidine* is extremely difficult to obtain.

9.3.6.4 Measurements of Areas

It can be seen from Fig. 9.17 that the glass appears as a "*hump*" and the crystalline phases as sharp peaks. Thus the total area under the peaks (A_{total}) represents (glass + crystalline) and the area under the hump (A_{hump}) represents the glass phase. The crystal phase is represented by (A_{total} – A_{hump}). Standard mixtures of glass and *cuspidine* were made up and it was observed that f_{gl} and f_{crys} could be determined from Eq. 9.8 [123].

$$f_{gl} = 1 - f_{crys} = \left(A_{hump}/A_{total}\right) \tag{9.8}$$

A similar method has been used by Andersson [61, 62]. Since the sample size is small, a number of determinations should be made to ensure the f_{crys} is representative of the whole slag film.

9.3.6.5 Measurement of Physical Properties

Measurements of heat capacity (C_p) and thermal expansion coefficient can be used to measure f_{crys}. When a glass is heated through the glass transition temperature (T_g, which occurs around 600 °C, for most mould slags) it transforms from a solid into a supercooled liquid (*scl*). This transition is accompanied by (i) a large-step change in heat capacity (ΔC_p^{Tg}) and (ii) a threefold increase in thermal expansion coefficient (α) [124, 125]. The method consists in determining the property changes at T_g for (a) the glass prepared from the slag film and (b) the slag film [124, 125]. Values of f_{crys} can then be calculated by Eq. 9.9. It was found difficult to use measurements of the thermal expansion to determine f_{crys} since the sample (*scl*) collapses about 650 °C which severely restricts the temperature range of the measurements [124].

$$f_{gl} = (1 - f_{crys}) = (\Delta C_p^{Tg})_{slagfilm} / (\Delta C_p^{Tg})_{glass}. \quad (9.9)$$

Measurements of thermal conductivity/diffusivity of the slag film at room temperature would appear to offer considerable potential. Measurements on both a slag film and a glass (made from the slag film) can provide a sensitive scale for measuring f_{crys} since $k_{cryst} \approx 2k_{glass.}$

9.3.7 *Tests to Simulate* f_{crys} *Formed in Slag Film*

Several tests have been reported to simulate the f_{crys} formed in the slag film with the de-carburised mould powder. These tests are briefly described in Table 9.2.

9.3.8 *Empirical Rules to Calculate the Crystal Fraction in Slag Films*

The fraction of crystalline material (f_{crys}) will be dependent upon both the composition and the thermal history of the slag film (time and temperature). Equations correlating f_{crys} with compositional parameters have been reported by several investigators [80, 83, 108]. In the relation proposed by Sakai [108] f_{crys} is shown as a graphical function of (X_{CaO} + X_{MgO} + X_{CaF_2}). In the Sankaranarayanan relation [80] f_{crys} is represented as function of the de-polymerisation index (DI) but does not contain the component CaF_2 which is essential for *cupidine* formation in mould slags.

Table 9.2 Summary of published tests using mould fluxes to determine the fraction crystalline phase (f_{crys}) developed in slag films

Test/reference	Details of method
Lidefelt [126]	Triangular—(side 30 × 30 mm) × 45 mm long hole machined in refractory brick and Cu plate inserted. Molten flux at 1500 °C poured into brick. Cut down the middle and examined metallographically to determine f_{crys}
Sakai [108]	Hole (30 × 30 × 5 mm wide) machined in stainless steel block. Molten mould flux poured into mould which was held at 300 °C. f_{crys} determined metallographically
Stollberg [127]	25 g molten mould flux poured into a copper mould held in cold water. The sample was then removed and examined metallographically to determine f_{crys}
Li [107]	The following procedure was found to provide best match to f_{crys} for slag films. 20 g of molten decarburised mould powder is heated to 1300 °C and held there for 20 min. It is then poured into a stainless steel crucible at 610 °C and then held at 610 °C for 20 min. The f_{crys} of the sample was determined metallographically

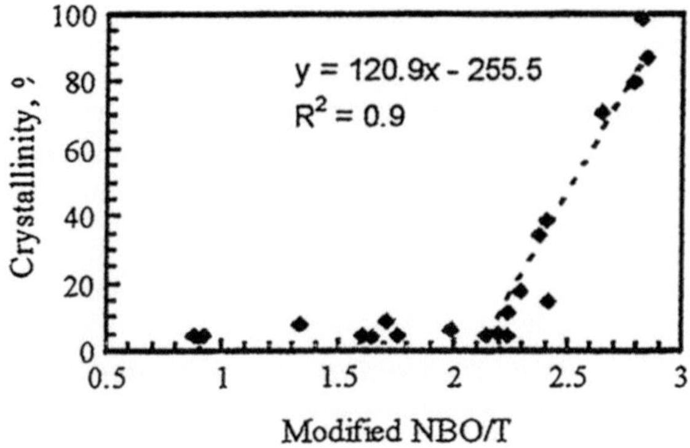

Fig. 9.18 Fraction crystalline phase (f_{crys} expressed as %) as a function of modified (NBO/T)* ratio (permission granted, SAIMM, [83])

Li et al. [83] measured the fraction of crystalline phase (f_{crys}) in slag films taken from the mould and tested various parameters (i.e. modified O/Si ratio, modified mole ratio and modified NBO/T* ratio and maximum *cuspidine* formed) to provide a fit with the experimental f_{crys} values (Fig. 9.18). It was found that the *modified* (NBO/T)* provided the best fit (Eqs. 9.10 and 9.11) where *X is* the mole fraction of a component (e.g. CaO). Note CaF_2 is treated as a network breaker in the calculation of the modified (NBO/T)* in Eq. 9.9. This equation presumably pertains to the time when the crystallisation has achieved a steady state.

$$\%\text{crys} = 100 f_{\text{crys}} = -284 + 141.1(NBO/T))^* \tag{9.10}$$

$$\begin{aligned}\text{Modified } (\text{NBO/T})^* = 2\{&X_{\text{CaO}} + X_{\text{FeO}} + X_{\text{Na}_2\text{O}} + X_{\text{K}_2\text{O}} \\ &+ X_{\text{CaF}_2} + \{X_{\text{MgO}} + X_{\text{MnO}}\} - X_{\text{Al}_2\text{O}_3}) \\ &/ (X_{\text{SiO}_2} + 2X_{\text{Al}_2\text{O}_3} + \{X_{\text{MgO}} + X_{\text{MnO}}\})\end{aligned} \tag{9.11}$$

Where the contents in the curly bracket { }, are included in the numerator if MgO is <7% and MnO is <4% and in the denominator when they exceed these values. It can be seen from Eq. 9.8 and Fig. 9.18 that $f_{crys} = 0$ (i.e. slag film is totally glassy) for mould slags with (NBO/T)* values of 2.0 or less and $f_{crys} = 1$ (i.e. slag film is totally crystalline) for slags with (NBO/T)* > 2.75.

9.3.9 *Data for* f_{crys}

Measurements of f_{crys} have been carried out by several workers [54, 61, 62, 83, 110, 123, 125, 128, 129], with f_{crys} increasing as (i) Li_2O decreased [108] (ii) as (C/S) or NBO/T increases [130]. The evolution of f_{crys} with time has also been investigated [57, 71].

9.4 Physical Properties of Mould Slags

The performance of the mould slag in the continuous-casting process is greatly affected by its physical properties. The various physical properties are discussed below. In the following text, mould slags used for continuous casting are divided into two categories, namely, (i) conventional slags (containing CaF_2) and (ii) mould slags for specialist casting (see Chap. 6) which may, or may not, contain CaF_2. Although the required property values are largely determined by the mould dimensions, casting conditions, steel grade, etc., the compositions of the various slags must be adjusted to meet the required values.

Some of the required tasks carried out by continuous-casting mould powders are identical to those performed by powders in Ingot casting, whereas others vary in importance. Powders for ingot casting are described in detail in Chap. 7 but the properties of powders for continuous casting and ingot casting are compared in Sect. 9.7.

9.4.1 Thermodynamic Properties and Liquidus Temperatures (T_{liq})

The Gibbs free energy for *cuspidine* formation has been determined in two studies and the ΔG_f values were found to be in good agreement [86, 87].

A thermodynamic database covering some mould powder compositions has been developed by Jung and van Ende [131] and has been used to predict the pick-up of Al_2O_3 when casting a high-Al steel with a low basicity mould slag containing CaF_2.

9.4.1.1 Importance to the Process

The *liquidus temperature* (T_{liq}) is the temperature where the casting powder becomes completely molten. Knowledge of the T_{liq} of the casting powder helps us to define the upper limit of the molten slag pool. A decrease in the liquidus temperature would be expected to increase the depth of the molten slag pool.

9.4.1.2 Measurement Methods

The melting range of a casting flux can be measured by a variety of techniques outlined in Table 9.3.

Table 9.3 Methods used to determine liquidus temperature

Method	Details of method	Comments
Boat tests	Known mass of decarburised sample placed in combustion boats and placed in muffle furnace (for say, 7 min) to come to thermal equilibrium at various set temperatures and then quenched. Examined metallographically for evidence of melting	
Quench tests [113]	As above, but mould powder sealed into Pt tubes	
DTA/DSC [132]	There are 2 pans, one holding the sample and the other a reference material. The pans sit in a metal block and are heated at a set heating rate and the temperature difference between the two pans is monitored continuously (Fig. 9.19a). When sample undergoes an exothermic or endothermic event there is departure from the baseline. For accurate temperatures, measurements should be carried out with (i) calibrants to check the temperature scale and (ii) corrected for heating rate. A typical DTA/DSC output is shown in Fig. 9.19b	
Hot stage microscope SHTC DHTC [133, 134, 136]	The melted sample is placed in a loop in the thermocouple (which also acts as heating element) and allowed to solidify. The sample is then heated at 5 °C min^{-1} and the specimen observed through a microscope for melting. The sample is then cooled at 15 °C min^{-1} and monitored for signs of precipitation. Double hot thermocouple technique contains 2 thermocouples and a thermal gradient can be imposed across specimen. The temperature field in DHTC has been modelled [136]	Tendency for undercooling
Leitz microscope test [135] DIN 51740 [135]	The shape of a pressed cylinder of de-carburised powder is monitored as it is heated at a set heating rate. The temperatures where the sample takes up a set shape is determined $T_{softening}$, $T_{hemisphere}$ and $T_{fluidity}$ are measured to define the melting range (Fig. 9.20)	T_{liq} values tend to differ from T_{liq} values measured by DTA/DSC

9.4.1.3 Liquidus Temperature Data

Values of the liquidus temperatures have been reported by several groups [17, 66, 95, 98–100, 102, 113, 132]. The effects of various constituents on the liquidus temperatures for various types of mould slags have also been reported: for conventional fluxes; Al_2O_3 [138, 139], Li_2O [95, 139], B_2O_3 [100], MnO [139, 140],

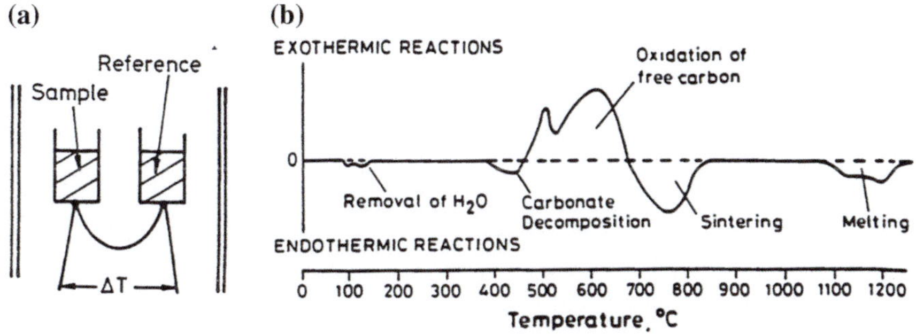

Fig. 9.19 Schematic diagram showing **a** DTA apparatus and **b** Typical DTA/DSC trace for a casting powder during heating (permission granted, Europ. Sci. Tech. Publ. [137])

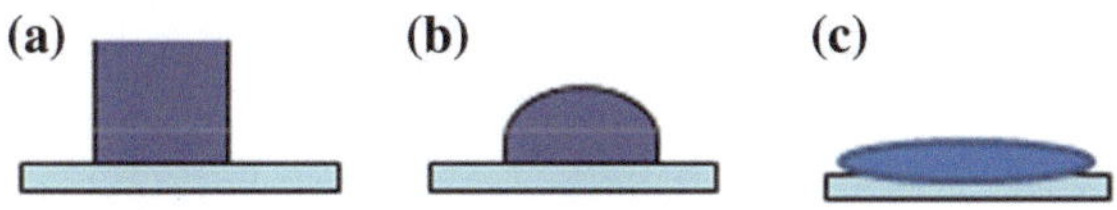

Fig. 9.20 Schematic drawings **a–c** showing typical shapes corresponding to $T_{softening}$, $T_{hemisphere}$ and $T_{fluidity}$, respectively

Na_2O on *cuspidine* phase [69, 141]: F-Free:Na_2O [68]:Hi-Al low (C/S) [142]; CA-based: B_2O_3 [29]. The T_{liq} and T_{br} of the remaining liquid have been determined after precipitation of the crystallising phase [128].

9.4.1.4 Methods to Calculate T_{liq}

Commercial thermodynamic software (e.g. Factsage, MTDATA, Thermocalc) are capable of providing reliable values for T_{liq} of casting slags.

Less accurate, values of T_{liq} for most mould slags can be calculated to ±30 °C using Eqs. 9.12 and 9.13 [83, 143]; these were derived using numerical "*best fit*" analysis of measured T_{liq} values and compositions of mould fluxes. This method identifies the changes in T_{liq} resulting from the addition of specific oxides (or fluorides) on the slag and provides a mean value (or constant) to represent this trend. When this method is applied to slags in general, T_{liq} values are subject to much greater uncertainty (>100 °C) due to the effect of the phase field on T_{liq} [144]. The smaller uncertainty value associated with mould flux data is due to the fact that mould fluxes tend to have similar compositions and tend to lie in the same phase field.

$$T_{liq}\,(^\circ C) = 1191 + 11.4\%SiO_2 - 11\%CaO + 4.2\%\,Al_2O_3 + 5.7\%MgO - 10.1\%Na_2O - 15.8\%K_2O + 1.9\%F + 8.3\%Fe_2O_3 + 11.6\%MnO \quad (9.12)$$

$$T_{liq}\,(^\circ C) = 1200 - 1.518\%SiO_2 + 2.59\%CaO + 1.56\%Al_2O_3 - 17.1\%MgO - 9.06\%Na_2O - \%K_2O + \%Li_2O + 4.8\%\,F - 9.87\%FeO - 2.12\%MnO \quad (9.13)$$

9.4.2 *Break Temperature* (T_{br})

The *break (or solidification) temperature* is the temperature where solids are produced on cooling. The term, Break temperature, T_{br}, is defined as the temperature where the viscosity increases dramatically during cooling. Values of T_{br} tend to decrease with increasing cooling rate and the cooling rates in the mould are vey high. It should be noted that T_{br} is not an equilibrium value like T_{liq} and that T_{br} decreases with increasing cooling rate and thus most cited values relate to a specified cooling rate (usually 10 K min^{-1}). However, at high cooling rates it has been observed the T_{br} tends to come to a fixed value [145].

9.4.2.1 Importance to the Process

The break temperature is important because it determines the boundary between solid and liquid layers in the slag film and thus the magnitude of the solid and liquid film thicknesses (i.e. d_s and d_l). Since the horizontal heat flux, q_{hor}, is inversely dependent upon d_s, any relation between q_{hor} and d_s (which is difficult to measure) can be represented in terms of q_{hor} and T_{br}. The friction forces exerted on the shell are dependent upon $(d_l)^{-1}$; thus, T_{br} also affects the lubrication supplied to the shell by the mould slag.

9.4.2.2 Measurement Methods

The *break temperature* is derived from viscosity measurements on liquid mould slags when cooling at a set rate. It is detected by a sudden departure from the baseline (Fig. 9.21) and the appearance of non-Newtonian behaviour (due to presence of solid particles). It should be noted that it is difficult to detect T_{br} if the distance separating the two concentric cylinders is too large (say 10 mm) and a value, of say, <6 mm is needed to detect T_{br}.

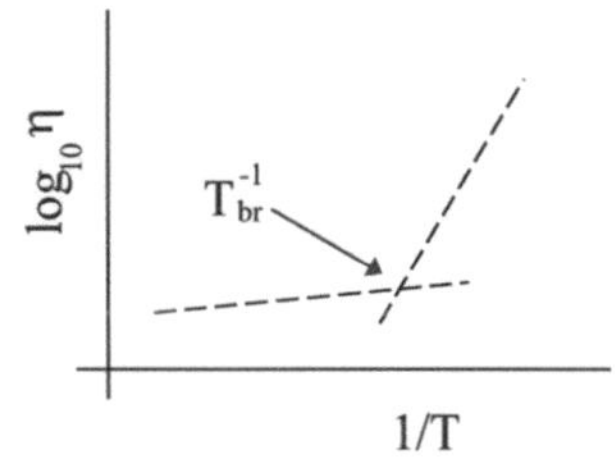

Fig. 9.21 Schematic diagram of $\log_{10} \eta$ as a function of reciprocal temperature (K^{-1}) showing the break temperature

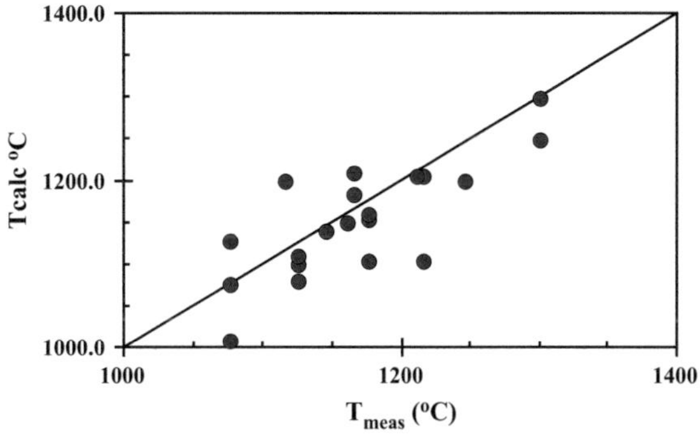

Fig. 9.22 Plot of measured values versus calculated values for T_{br} (permission granted, ISS/AIST, re-drawn from [74])

Kawamoto et al. [74] defined T_{br} as the point where the fraction solid in the (solid+liquid) exceeds a value of 0.49 and calculated this temperature using a thermodynamic software package. The calculated and measured T_{br} showed reasonable agreement [74] (Fig. 9.22).

The *solidification temperature* is measured by DTA/DSC or by hot thermocouple method when cooled at a set rate.

Some workers use the term, crys*tallisation temperature*. Feldbauer and Cramb [146] have described this term as confusing since T_{crys} will depend upon on TTT curves and thus will be dependent upon the method and procedures used.

9.4.2.3 Data for T_{br}, T_{sol}

Values of T_{br} have been obtained from viscosity measurements e.g. [147–150] and the effect of individual components on the T_{br} or T_{sol} has been studied by a number of workers: Al_2O_3 [151], ZrO_2 [151, 152], TiO_2 [27, 76, 109, 151, 153], B_2O_3 [28, 29], MnO [140], RE O [154], Na_2O [130], BaO, Li_2O [155], Basicity [130].

9.4.2.4 Methods to Calculate T_{br}

Numerical analysis of experimental values of break temperature derived during viscosity measurements yielded Eq. 9.14 [147, 156].

$$\begin{aligned} T_{crys} = T_{br}(^{\circ}C) &= 1241.6 - 0.0215X_{MgO} - 0.0141X_{Al_2O_3} \\ &- 0.0449X_{Na_2O} - 0.0855X_{CaF_2} \\ &- 0.0641X_{Li_2O} - 0.1528X_{B_2O_3}. \end{aligned} \tag{9.14}$$

Values of T_{br} were also estimated using the following relation [157] which was derived from regression analysis (i.e. "*best fit*") of experimental data for T_{br} and mould slag compositions. The uncertainty associated with this equation is ±25 °C.

$$\begin{aligned} T_{br}(^{\circ}C) = 1120 &- 3.3\%SiO_2 - 8.43\%\,Al_2O_3 + 8.65\%CaO - 13.86\%MgO \\ &- 3.3\%Na_2O - 18.4\%FeO - 3.2\%MnO - 2.2\%K_2O - 6.6\%Li_2O \\ &- 6.47\%F \end{aligned} \tag{9.15}$$

The performances of some of the above equations have been checked against T_{br} values obtained during viscosity measurements [148].

9.4.3 *Glass Transition Temperatures* (T_g)

The *glass transition temperature* (T_g) is the temperature, on heating, where a glass is transformed into a supercooled liquid (*scl*) and this transition is accompanied by a marked change in some properties (e.g. in C_p and thermal expansion coefficient). Values of T_g [53, 132, 158] for most mould slags lie close to 600 °C (i.e. fall in the range, 570–630 °C). A regression analysis was carried out on T_g data obtained in DSC and DTA studies which yielded Eq. 9.16 [143]; the uncertainty in the calculated T_g values is ±30 °C.

$$\begin{aligned} T_g(K) = 906 &- 330.5X_{SiO_2} + 190X_{CaO} + 440X_{Al_2O_3} \\ &- 449X_{N_2O} + K_2O - 11X_{MgO} + 154X_{CaF_2} \\ &- 309X_{MnO} - 1391X_{FeO} \end{aligned} \tag{9.16}$$

9.4.4 Viscosities (η)

9.4.4.1 Importance to the Process

The slag viscosity is important to the continuous-casting performance because it affects:

(i) The lubrication supplied to the mould (which correlates with $1/\eta$).
(ii) The SEN erosion rate which is proportional to $(1/\eta)$ and
(iii) slag entrapment can be reduced by increasing the viscosity of the mould slag (but at the expense of the powder consumption [159]).

9.4.4.2 Factors Affecting Viscosities of Mould Slags

The viscosity is very dependent upon the degree of polymerisation of the slag. In fact, the viscosity has been used by some workers to represent the slag structure.

The viscosity increases with increasing concentrations of (i) *network formers* (e.g. SiO_2, Al_2O_3) (ii) decreasing concentrations of *network breakers* (e.g. CaO, MgO, MnO and FeO) and (iii) *fluxes* (e.g. Na_2O, K_2O, Li_2O, CaF_2, B_2O_3).

It is customary to represent the temperature dependence of viscosity by using either the Arrhenius or the Weymann relations (denoted as subscripts *A* and *W* in Eqs. 9.17 and 9.18, respectively)

$$\eta = A_A exp\,(B_A/T_K) \tag{9.17}$$

$$\eta/T_K = A_W exp\,(B_W/T_K) \tag{9.18}$$

where η is in dPas, T_K is in K and A_A and A_W are the pre-exponential terms and B_A and B_W represent the activation energy (E) term ($B = E/8.314$).The parameters, B_A and B_W increase with increasing polymerisation (i.e. increasing Q).

It is customary to cite the viscosity at 1300 °C (1573 K) since this approximates to the mean temperature in the liquid layer of the slag film.

9.4.4.3 Measurement Methods

Several methods have been used to measure slag viscosities, these are detailed below. Some of the classical methods of measuring viscosity (e.g. capillary viscometry) are rarely used because of the difficulty in obtaining a uniform temperature zone of sufficient length at the required temperatures.

Concentric Cylinder Method

These viscometers consist of two concentric cylinders namely, the crucible (holding the molten sample) and the bob (Fig. 9.23). When one of these cylinders is rotated

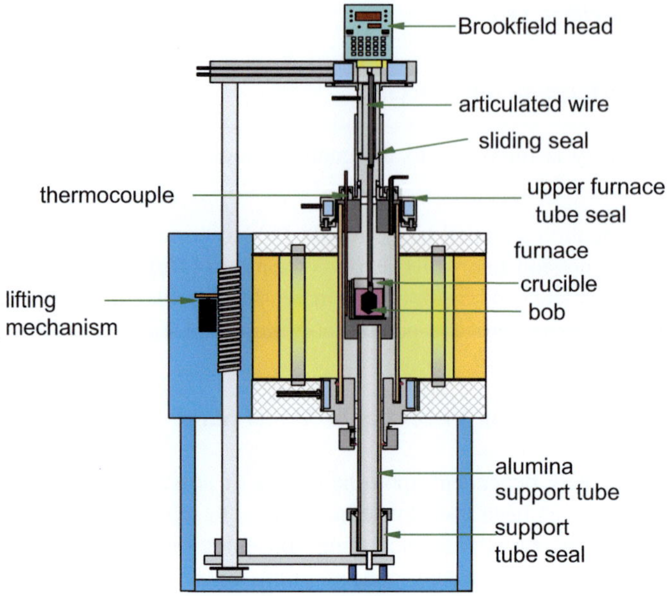

Fig. 9.23 Schematic diagram showing the rotating bob (or *cylinder*) method for measuring viscosity (courtesy of RF Brooks, NPL, Teddington, UK)

at constant speed, the viscosity can be determined from measurements of the torque developed [3, 159, 160]. It is difficult to obtain the accurate alignment of the two cylinders but the viscosity of mould fluxes is usually high enough to ensure self-centring of the bob; it is for this reason (i.e. ease of alignment) that the rotating bob (or cylinder) method (denoted RCyl) has become the more popular method.

Fig. 9.24 Schematic drawing showing the oscillating plate method [162, 163] (courtesy of RF Brooks, NPL, Teddington, UK)

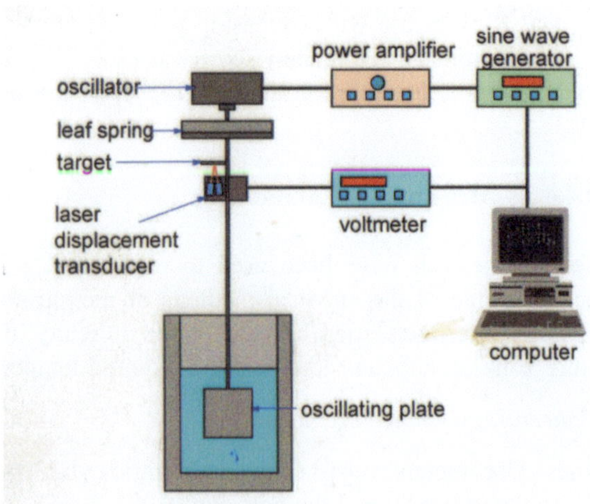

The various sources of uncertainty in the rotating cylinder method have been analysed by Wu et al. [161]. Ideally, the distance between the cylinders should be small to ensure the stress gradient is linear.

Oscillating Plate Method

The oscillating plate method has been used to measure the viscosities of mould fluxes [162, 163]. A plate is immersed in the melt and undergoes vertical oscillations in the fluid. When steady state has been established the amplitudes of the oscillations are determined in both the fluid (φ) and in the gaseous atmosphere (φ_{atm}) (Fig. 9.24). Care must be taken to ensure "*backwash*" from the crucible walls does not affect the results. The method gives values for the product ($\eta \cdot \rho$) and so values of the slag density (ρ) are needed to calculate the viscosity (η).

$$\eta \cdot \rho = G\{(\varphi_{atm}/\varphi) - 1\} \tag{9.19}$$

where G is a cell constant which is determined in calibration experiments with liquids of known viscosity.

Falling Ball Method

In this method, the time for a sphere to travel through a known distance in the slag is measured; this can be done by either letting a ball fall or by dragging a wire attached to the sphere vertically through the melt [3]. The principal problem lies in the difficulty of obtaining a zone of uniform temperature of sufficient length. The viscosity is calculated from the following equation where g is the gravitational constant, r_{sph} and v_{sph} are the radius and velocity of the sphere, ρ is the viscosity and subscript, *fl* refers to the liquid slag sample.

$$\eta = 2\, g\, r_{\text{sph}}^2 \left(\rho_{\text{sph}} - \rho_{fl}\right)/v_{\text{sph}}. \tag{9.20}$$

More Recent Methods

The draining crucible (DC) method [164] and the SLLS (Surface Laser Light Scattering) methods have not been applied yet to slag systems but would appear to be viable methods for these systems.

In the ***draining crucible technique***, the sample is allowed to drain through an orifice of known diameter in the base of the crucible and the rate of mass loss is recorded. The viscosity, surface tension and density can be derived from the hydrodynamic analysis of the data [164].

In the ***Surface laser light scattering*** **(SLLS)** method "*ripplons*" are monitored [165–167]. The surface of a melt may appear smooth but it is being continually deformed by thermal fluctuations of the molecules. Capillary waves (*ripplons*) have small amplitudes (*ca.* 1 nm) and a wavelength of *ca.* 100 μm which is dependent upon the frequency. Ripplon action relies upon surface tension for restoration and the kinematic viscosity ($\nu = \eta/\rho$) for oscillation damping. The spectrum of the

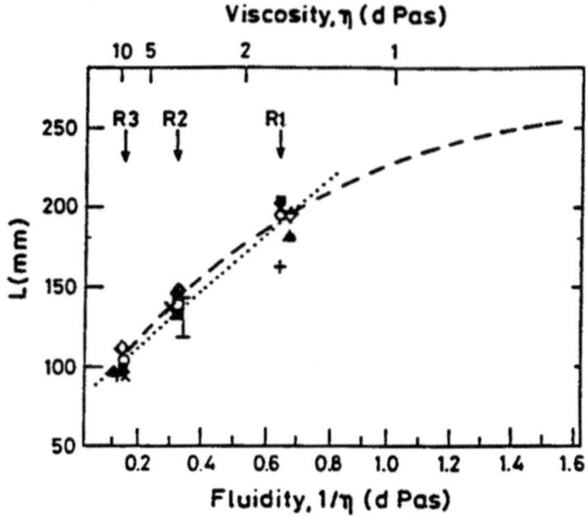

Fig. 9.25 The length of the slag ribbon (L) as a function of the fluidity ($1/\eta$) for Inclined Plane Test; R_1, R_2, R_3, refer to reference materials with known viscosity (permission granted, ISS/AIST, [168])

ripplons is derived using a Fourier spectrum analyser; this allows the surface tension and the viscosity to be determined, simultaneously. The method has been successfully used for measurements on liquid silicon [166] and $LiNbO_3$ up to 1480 °C. It has been estimated that it can be used for liquids with viscosities in the range 0.005–10 dPas [165] which covers most of the mould flux range. However, a value of the density is required to calculate the dynamic viscosity (η).

The advantages of these methods are that they also provide simultaneous values for both viscosity and other properties (density and surface tension with DC method and surface tension with SLLS method).

Inclined Plane Method

This simple test has been used for quality assurance tests on mould powder supplies. In this test 10 g of de-carburised mould flux is weighed into a carbon crucible and this is then placed in a muffle furnace at the required temperature for about 15 min to allow temperature equilibration [168]. It is then removed from the furnace and poured onto a V-shaped inclined plane (usually an incline of 9° is adopted) where it forms a slag ribbon. The length (L) of the solidified slag ribbon is then measured and is related to ($1/\eta$). It is possible to obtain viscosities of reasonable accuracy with this method by taking suitable precautions and using improved procedures [168] (Fig. 9.25).

9.4.4.4 Viscosity Data for Mould Slags

There are a large number of investigations of viscosities of mould fluxes; for convenience, these have been divided into two tables covering conventional mould fluxes (i.e. containing CaF_2) in Table 9.4a and mould slags for specialist casting in

Table 9.4a Summary of viscosity measurements on mould fluxes

References	Method	Viscosity, η_{1300} dPas	Slag type	Comments
Esaulov [17]	OscV	0.5–100	Conv	10 mould slags
McCauley [176]	RCyl	1–14	Synth	13 synthetic mould slags; effect of (C/S) = 0.8–1.2; Slags 1–6 similar composition, η varies ±25% around mean: B_2O_3 slag
Riboud [176, 177]	RCyl	1–12	Synth	16 synthetic slags from CSNFl system
Zhuk [178]	RCyl		Conv	Effect of B_2O_3 addition
Lanyi [179]	RCyl	0.3–10	Conv	18 slags; Effect of Al_2O_3 B_2O_3 additions; Activation energies
Kim [147]	RCyl	0.4–3	Conv	35 slags: Effect of F, Li_2O, Na_2O, B_2O_3 and Al_2O_3 on η
Chang [180, 181]	RCyl	3–20	Conv	(32–37)C +(39–46)S + 12A + (1–5)N + (0–12)Fl
Mills [132, 168]	RCyl	2–6	Conv	4+3slags; post-measurement composition
Kawamoto [182]	O Plate	1	Conv	Effect of Al_2O_3, SiO_2, B_2O_3, MgO, BaO, CaO, K_2O, CaF_2, Na_2O, Li_2O on η
Lopez [183]	RCyl	8.5–12.4	Conv	Slags for billet casting
Elahipanah [148]	RCyl	1–3	Conv	9 mould slags;
Park [14, 184]	RCyl		Conv	Effect of CaF_2 on viscosity
Wang [154]		1–2	Conv	Effect of RE oxides $\eta\downarrow$ as %RE $\uparrow$; also B_2O_3; Li_2O added
Lu [95, 98, 100, 102, 185, 186]	RCyl	1.5–4	Conv	No temps given: 1300 °C? (C/S) = 0.8, 1 and 1.2; Effect of Na_3AlF_6; B_2O_3; BaO; K_2O; NaF; Li_2O on η of slag;
Persson [149]	RCyl	0.7–4.3	Conv	7 slags + additions of Al_2O_3; post-measurement slag composition
Marschall [150]	RCyl	0.6–1.1	Conv	2slags; Effect of Al_2O_3 and TiO_2 on viscosity
Shahbazian [187, 188]	RCyl		Synth	CaO + SiO_2 + (13–40)CaF_2

In chemical formulae note: *A* Al_2O_3, *Fl* CaF_2, etc. and 5M = 5% MgO; Mould powders are mostly for conventional slag, *Conv* Conventional; *Synth* Synthetic; *RCyl* Rotating cylinder method; *OscV* Oscillating Viscometer

Table 9.4b. The effects of individual components on the viscosities of various mould slags has been studied by a number of workers: e.g. Al_2O_3; [5, 96, 168, 170]: TiO_2 [30, 76, 82, 133, 168–172]; ZrO_2 [96, 152] MnO [140, 147]. The effect of precipitates on the slag viscosities have been reported after crystallisation [128, 173] and after TiC precipitation [142]. The effect of applying an electric potential on the measured viscosity has been studied and it was found that increasing DC voltage caused a

Table 9.4b Viscosities for specialist mould powders: η_T is for 1300 °C unless stated otherwise

References	Method	η_T dPas	Slag type	Comments
			FF-flux	
Fox [28, 101]	RCyl	1	FF*-flux	Effect of B_2O_3 on η
Wen [66]	RCyl	2–9	FF-CT	F-free powders; Effect of (C/S); TiO_2; MnO; Na_2O; Li_2O; MgO; B_2O_3 on η (36C, 38S, 6A, 11N, 9%F)
Wu [161]	RCyl	2–6	FF-CT	Effect of TiO_2, Na_2O and Al_2O_3 on η
Chen [102]	RCyl	0.7–1	FF-CT	CaF_2 replaced by TiO_2
Wang [189]	RCyl		FF-flux	Effect of B_2O_3 on η
Klug [190]	RCyl	17	FF*-flux	Effect of Na_2O on η of slags for billet casting
Wang [172]	RCyl	4–7	FF-CT	Effect of B_2O_3; TiO_2 on η
Zhang [191]				
			NN	
Tsutsumi [192, 193]	RCyl	4	NN	Non-Newtonian flux; effect of different shear rates on η
			HiAl	
Li [6]	RCyl	$\eta_{1500,}$ =2–3 =6–8	low (C/S)	Effect of (A/S) on η: 25C + (20–45)S + (35–10)A + 20Fl; (1350–1500 °C); η_T at minimum at ca. (A/S) = 1.0
Zhang [168]	RCyl	3–14	low (C/S)	Effect of (A/S) on η
Yu [170]	RCyl	1–5	low (C/S)	Effect of (A/S) on η
Higo [5]	RCyl	$\eta_{1500,}$ 24–36	low (C/S)	Effect of (A/S) on η
Hanao [69]	Osc P	3–5 *cf* 2	*Melilite*	Melilite slag cf conventional slag; Also used in round billets
			CA	
Kim' Park [7]	RCyl	η_{1500} = 1–2	CA	CA + (7–12S) +5 M + (0–15)Fl; (1500–1600 °C); Q^n values $\eta_T\downarrow$ as (C/S)↑ as (C/A)↑ and CaF_2↑
Li, Shu [26]	RCyl	η_{1500} = 0.7–1	CA	CA + 5M + 5L + 8N + 10Fl + 0 – 10T; (1450–1700 °C) $\eta_T\downarrow$ as %T↑
Huang [29]	RCyl	η_{1673} = 3–27	CA	Effect of B_2O_3 on η
Q Liu [194]	RCyl		CA	
			SS-Ti st	

(continued)

Table 9.4b (continued)

References	Method	η_T dPas	Slag type	Comments
Kusano [195]	RCyl			Effect of TiO_2 on η
Wang [30]	RCyl		Conv	30C + (28–38)S + 6A + 1.5M + 10N + 14Fl + T (0–10) Replace SiO_2 by TiO_2 causes↓ η
Zhang [196, 197]	RCyl	$\eta_{1500} = 1.5$	Conv	Effect of TiC and TiN particles on η; non-Newtonian
Mukongo [27, 76]	RCyl	1.0	Conv	Effect of TiO_2 on η
Kishi [133]	RCyl	1–2.5	Conv	TiO_2 lowers—when (C/S) < 1
Chen [102]	RCyl	0.7–1		CaF_2 mostly replaced by TiO_2
			HSTS	V_c up to 8 m min^{-1}
Kromhout [198]	RCyl			Viscosity fits with empirical rules for slag selection
Kawamoto [199]	Osc P			

FF Fluoride free; *NN* non-Newtonian; fluxes for casting steels; *Hi Al* High-Al slag (3 types, denoted as (i) low (C/S) (ii) *Melilite* and (iii) CA = calcium aluminate slag); *HSTS* High-speed thin slabs; *SS* Stainless steel; *for casting billets: *Osc P* Oscillating plate method

decrease in viscosity [174, 175] due to the decrease in metal/slag contact angle which, in turn increased "*slip*" between the slag and the rotor (and the crucible).

9.4.4.5 Methods to Calculate Viscosity

Several models have been reported to estimate viscosities of mould fluxes from their chemical compositions and these are summarised in Table 9.5 where these estimated values are, for the most part, subject to uncertainties of *ca.* ±25%. This uncertainty is high because the uncertainty associated with the experimental viscosities in the databases is also high (±10–25%).

9.4.5 Thermal Conductivities

9.4.5.1 Importance to the Process

The occurrence of both longitudinal cracking and sticker breakouts are associated with the control of horizontal heat transfer; this involves both the thermal conductivity and thickness of the slag film.

Table 9.5 Summary of models used to calculate viscosities of mould slags from chemical composition

Model/references	Details of model	Uncertainty
Riboud [200] Modified const [6, 29, 168, 170]	A_W; B_W functions of 5 groups "CaO" + "SiO_2" + "Al_2O_3" + "CaF_2" + "Na_2O" where $X_{\text{"CaO"}}$ contains X_{CaO} +X_{MgO} + X_{FeO} + X_{MnO} etc.; ln $A = -19.81 + 1.73\ X_{CaO} + 3.58\ X_{CaF2} + 7.02\ X_{Na2O} - 35.76\ X_{Al_2O_3}$ $B = 31140 - 23896\ X_{CaO} - 46356\ X_{CaF_2} - 39159\ X_{Na_2O} - 68833\ X_{Al_2O_3}$ For calculation of η for fluxes for High-Al steels-low(C/S) [6, 168] CA–type [6, 170] +FF [29]	±30%
Iida [34]	η(Pas) = $A\eta_0$ exp (E/Bi) where η_0 = viscosity of hypothetical network-forming melt and Bi = basicity index $E = 11.11 - 3.65 \times 10^{-3}\ T$ and $A = 1.745 - 1.962 \times 10^{-3}T + 7 \times 10^{-7}\ T^2$ and $Bi = \Sigma(\alpha i\ \%i)_B/\Sigma(\alpha_i\ \%_I)_A$ where A = acid oxides and B = basic oxides or fluorides: $\eta = 1.8 \times 10^{-7}\ (M = i\ T^m)^{0.5}$ exp $(H_i/RT^m)/V_m^{0.667}$ exp (H_i/RT^m) where $H_i = 5.1\ T^m$ and values of α given	±25%
Kim [147]	$ln\ A = -2.307 - 0.046\ X_{SiO_2} - 0.07\ X_{CaO} - 0.041\ X_{MgO} - 0.185\ X_{Al_2O_3} + 0.035\ X_{CaF_2} - 0.095\ X_{B_2O_3}$ $B = 6807 + 70.7\ X_{SiO_2} + 32.58\ X_{CaO} + 312.7\ X_{Al_2O_3} - 34.8\ X_{Na_2O} - 176.1\ X_{CaF_2} - 167.4\ X_{Li_2O} + 59.7\ X_{B_2O_3}$ where X in mole %	±45%
Hanao [201]	Neural network model; available commercially	
Zhang [44–46]	Calculates BOs, NBOs and FOs and distinguishes between Al–O and Si–O bonds	
Gupta [37]	3 groups "Acidic": Yx = (%SiO_2/60) + α(%Al_2O_3/102); "Basic" Y_O = (%CaO/56) + (%MgO/40) + (% Na_2O/63) +. (%FeO/72. "Fluorides" Y_F = (%F/19): $N_F = Y_F/(Y_X + Y_O + Y_F)$: $N_X = Y_X/(Y_X + Y_O + Y_F)$: $N_O = Y_O/(Y_X + Y_O + Y_F)$; Na^+/Ca^{2+}affected η: Values of a_0 to a_5 given for 1573 and 1673 K; $\eta = a_0 + a_1(M^+/M^{2+}) + a_2N_X + a_3Nx\ (M^+/M^2) + a_4(M^+/M^2)^2 + a_5N_X^2$	±35%
Du, Seetha raman [202] Hayashi [203]	$\eta = hN_A\rho/M$ exp $(\Delta G_\eta/\text{RT})$ where h = Planck constant, N_A = Avogadro No Structure accounted for—thermodynamics:$\Delta G^*_\eta = \Sigma\Delta G^*_\eta$ (oxides) + ΔG_η^{mix} + $3R^*T\ X_1X_2$:ΔG_η^{mix} for interactions of cations only Works well for synthetic slags (±15%)	
Mills [33, 144, 204]	Two different models to calculate viscosity of wide compositional range of slags	
GH Zhang [47–49]	Calculates BOs, NBOs and FOs and distinguishes between Al–O and Si–O bonds	
Mills [143]	For cont. casting slags only; Values of X_{CaF_2} and X_{TiO_2} calculated; Q_{rem} calculated for remaining slag. Note T in K: $ln\ \eta_{1573\ K,\ ref} = 0.1227$ exp $(1.1454\ Q_{rem}) - 9\ X_{CaF_2}$; $B_\eta = 795830.8 + 8.476 \times 10^{-6}$exp $(Q/0.2611) + 795841.4$ exp $(Q/211587.4)$; $ln\ \eta_T = ln\ \eta_{1573} + (B/T) - (B/1573)$	

(continued)

Table 9.5 (continued)

Model/references	Details of model	Uncertainty
Elahipanah [148, 205]	Reviews performances of several models	
Shu [205]	Arrhenius equn; CMAS system; takes charge balance into account	±20%

Subscripts: *A* Arrhenius; *W* Weymann equations

9.4.5.2 Factors Affecting Thermal Conductivities of Slag Films

Horizontal heat transfer from the shell to the mould involves several heat transfer mechanisms, namely:

- Lattice (or phonon) thermal conduction.
- Radiation conduction.
- Convection in the liquid layer (aided by movement of the shell and mould oscillation).

As mentioned above, the effective thermal conductivity of the slag film (k_{eff}) has a significant effect on the horizontal heat flux (q_{hor}) and k_{eff} is largely determined by the physical state of the slag film. The magnitude of k_{eff} is affected by:

- The fraction of crystalline phase (f_{crys}) present in the slag film since (a) $k_{crys} \approx 2\, k_{gl}$ and (b) crystalline grains reflect IR radiation back towards the shell [56, 110] and thereby reduces radiation conduction (k_R) and (c) the shrinkage associated with crystallisation results in the formation of both pores and an interfacial resistance ($R_{Cu/sl}$). The overall effect is that $(k_{eff})_{crys} < (k_{eff})_{gl}$ and for most slag films it is thought the k_R contribution to k_{eff} is less than 20% of the total [115].
- The presence of gas pores in the slag film reduces k_{eff} (see Sect. 9.3.3).
- The presence of transition metal oxides in the slag film (e.g. FeO, NiO) reduces k_R through the absorption of infrared radiation [53, 54, 206] but this must be offset against a reduction in reflected heat transfer associated with the promotion of more glass phase by FeO, MnO, etc.

9.4.5.3 Measurement Problems

There are two major problems encountered when measuring thermal conductivities of slags, namely, how to deal with both *convection* and *radiation conduction*.

With convection, the usual approach taken is to determine a value of the thermal conductivity of the liquid slag (free from convection) and then to calculate the convective contribution using turbulence models (e.g. k-ε model). The main difficulty lies in determining the value of the thermal conductivity of liquid slag, free from convective contributions. This is usually achieved by using a *transient* method in which the measurements are completed before convection has time to develop (typically around 1 s). *Steady state* methods are unsuitable for these measurements. The two transient techniques in common use are the Laser Pulse (LP) and the Transient Hot Wire (THW, sometimes referred to as the line source) method.

Slags are semi-transparent media and thermal conductivity measurements on both solid and liquid slags at high temperatures are known to contain contributions from both lattice (k_{lat}) and radiation conduction (k_R, described in Chap. 3, Sect. 3.2.1) as shown in Eq. 9.21.

$$k_{eff} = k_{lat} + k_R \tag{9.21}$$

Values of k_R have been found to increase with increasing sample thickness, *d*, [207] until a critical point is reached, beyond which, k_R remains constant. Samples below this critical point are described as *optically thin* and those above, as *optically thick*. Optical thickness occurs when $\alpha * d \geq 3$, where α^* is the absorption coefficient and *d* is the sample thickness. Values of k_R can be calculated for *optically thick* samples by use of the following equation where σ = Stefan–Boltzman constant, n = refractive index and T_K is the temperature in K. Contributions from k_R increase dramatically with increasing temperature (since k is proportional to T_K^3)

$$k_R = 16 \cdot \sigma \cdot n^2 T_K^3 / 3\alpha^* \tag{9.22}$$

In partially crystalline samples, IR radiation can be both absorbed and reflected and it is not possible to replace α^* by the Extinction coefficient (E) in Eq. 9.22 without knowledge of the *albedo* (or reflectivity coefficient). There is little information on the albedo but a method for calculating this parameter has been reported [208].

9.4.5.4 Measurement Methods

Laser Pulse (LP) Method [209–213]

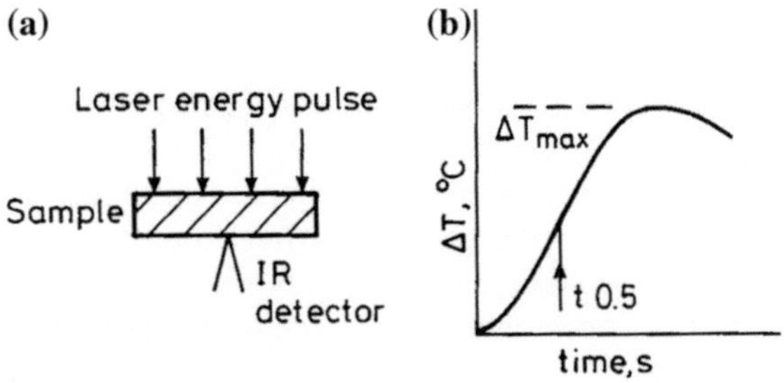

Fig. 9.26 Laser pulse method used to determine thermal diffusivities of solids **a** specimen assembly and **b** typical temperature transient (permission granted, Verlag Stahleisen GmbH [3])

The laser pulse provides values for the thermal diffusivity (a); the thermal conductivity (k) is derived from the thermal diffusivity by using the following relation, where C_p and ρ are the heat capacity and the density of the slag, respectively.

$$k_{\text{eff}} = a_{\text{eff}} \cdot C_p, \rho \tag{9.23}$$

The sample is usually in the form of a disc (10–15 mm diam. × 1–2 mm thick) which may be coated on both planar surfaces. Experiments have been carried out in two ways. In the first method (Fig. 9.26a), a laser beam is directed onto the front face of the specimen and the temperature of the back-face is monitored continuously. The thermal diffusivity is calculated using the relation, $a = 1.37\ d^2/\pi^2\ t_{0.5}$

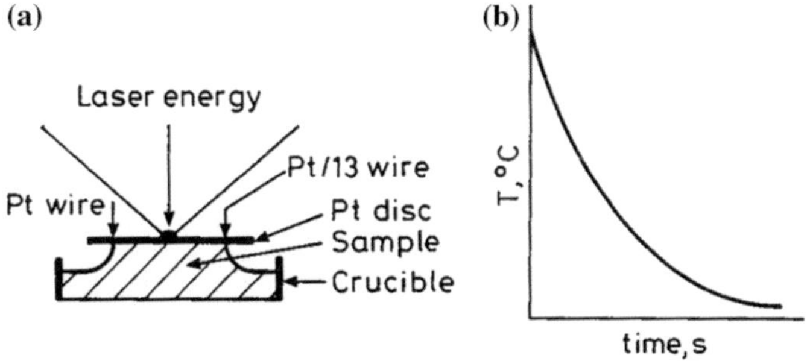

Fig. 9.27 Laser pulse method used to determine thermal diffusivities of liquid and solids **a** specimen assembly and **b** typical temperature transient (permission granted, Verlag Stahleisen GmbH [3])

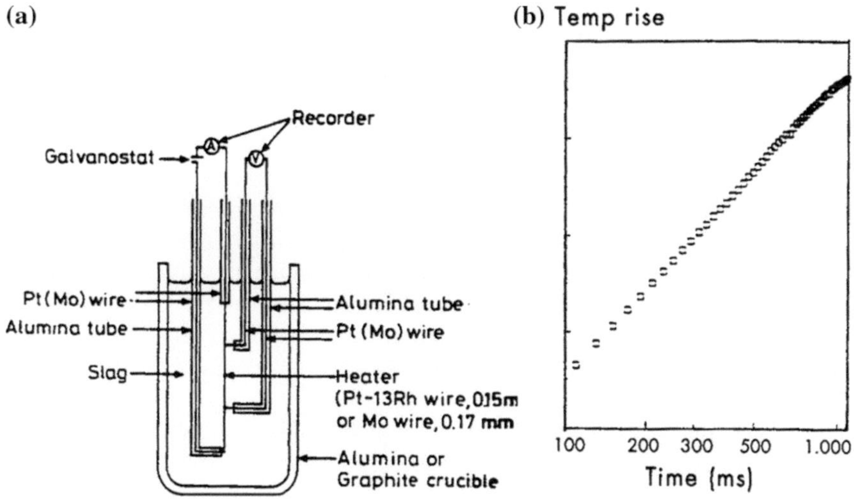

Fig. 9.28 Transient hot wire method **a** schematic drawing of probe and **b** temperature transient (permission granted, Verlag Stahleisen GmbH, [3])

where $t_{0.5}$ is the time required to reach $0.5\Delta T_{max}$ (see Fig. 9.26b) and d is the sample thickness. Corrections are usually made to account for radiation losses. This method is used only for measurements on solid samples.

The second method can be used for solid and liquid samples (Fig. 9.27). The sample is in good contact with a Pt disc (Fig. 9.27a) and the laser pulse is directed onto the Pt disc and the temperature decay of the Pt disc is monitored continuously. Numerical methods have been used to disentangle k_R and k_c contributions [214, 215].

Transient Hot Wire (THW) Method [54, 55, 61, 62, 216, 217].

In this method the sample is placed in a crucible and melted. Then a long, thin wire (typically, 0.1 mm diam.) is immersed in the molten slag. The thin wire acts as both a heating element and as a temperature sensor (Fig. 9.28a). When thermal equilibrium is attained, a current is applied for a short duration (ca. 1 s). The temperature of the wire is monitored continuously. The thermal conductivity (k) can be calculated from the following equation where q = heat input per unit length of wire, r = radius of wire, ΔT = temperature rise and $C = exp\ \gamma^*$ where γ^* = Euler's constant.

$$\Delta T = (q/4\pi k)\, ln\left(at/r^2C\right) \tag{9.24}$$

However, k is usually determined from the slope (=1/k) of the plot of the temperature rise (ΔT) as a function of ln time (Fig. 9.28b). The use of graphite components in the THW method is not recommended since reactions between FeO, MnO, etc. (in the mould flux) and graphite produce CO (g) bubbles. The bubbles increase porosity in the sample which lowers the thermal conductivity [62].

The hot wire method can also be used to determine the effective thermal conductivity of powders [61, 218, 219].

9.4.5.5 Thermal Conductivity Data for Mould Slags

The various measurements of thermal conductivity (k) and diffusivity (a) recorded for mould slags and powders are summarised in Table 9.6; the entries are grouped according to specimen type. The various factors affecting heat transfer have been discussed above. Although k_{lat} for crystals is approximately twice that of for the glass phase, the total thermal heat flux (q_{eff}) is lower in the crystalline phase than in the glass. This is due to the scattering of IR radiation by the crystals and the subsequent reduction in k_R. The k_R contributions can also be reduced by introducing transition metal oxides (e.g. FeO, NiO) into the glass which absorb the IR radiation [206].

Most slag films contain both glassy and crystalline phases. It should be noted that crystallisation results in some porosity in the slag film which will tend to reduce k_{eff}.

Table 9.6 Summary of thermal conductivity (*k*) and diffusivity (a) on mould fluxes

Specimen	Reference	Method	Values	(dk/dT) or {da/dt}	Comments
Liquid	Bagha [220]	Sim	k = ca. 0.3		Analysis of heat flux data; Cu cold finger
	Nagata [216]	THW	k_T = ca. 0.2		
	Powell [221]	THW	k_T = ca. 0.2		
	Shibata [211] Cho [59]	LP	$10^7 a = (4–6.5)$		Several liquid slags
	Hayashi [222]	LP	$10^7 a = (4.65)$		
	Gornerup [223]	LP	$10^7 a = (4)$		
	Monaghan [224]	LP	$10^7 a = (4)$		Equivalent to $k \approx 1.5\ Wm^{-1}K^{-1}$
	Waseda [210] Ohta [212, 225]	LP	$10^7 a = (4)$	{0}	(1100–1300 °C) Studied effects of TiO_2, ZrO_2, HfO_2 on k_R contribution
	Andersson [61]	THW	k_T = ca. 0.2		Powder sample heated into liquid phase
	Kim [226]	THW	k_{1500} = 0.2–0.3		Liquid for CSB system
Slag film					
	Susa [53, 227]	LP	$10^7 a = (6–8)$		5 slag films
	Andersson [62]	THW	k_{25} = 0.75–1.2		3 slag films from C-steel billet casting; Slag films exhibit increasing porosity with crystallis' *n*
Crystal					
	Shibata [211]	LP	$10^7 a = (4–5)$		3 part crystalline slags; 25–800 °C
	El Gammal [228]	RHF	k = 1.2–1.7		Probably contains k_R contributions
	Hayashi [222]	LP			Effect of f_{crys}
	Ozawa [54]	THW	k_{25} = 1.5–1.8	(2×10^{-4})	k increased with increasing f_{crys} but *(dk/dT)* changed from positive to negative
	Andersson [61, 62]	THW	k_{25} = 1.2–1.8		Partially cryst. samples; $k_{eff,\ 25} = 1.07 + 0.7\ f_{crys}$: k_{767} = 1.65; (dk/dT) calculated from k_{25} and k_{767}

(continued)

Table 9.6 (continued)

Specimen	Reference	Method	Values	(dk/dT) or {da/dt}	Comments
Glass					
	Taylor [209]	LP	$10^7 a = (4–5)$	{ca. 0}	10 different samples (200–900 °C) $10^7\ a = 4.5 \pm 0.5$
	Shibata [211]	LP	$10^7 a = (4–5)$		2 glassy mould slags (25–800 °C)
	Hayashi [222]	LP	$10^7 a = (4.6)$		
	Ozawa [54]	THW	$k_{25} = 1.1 \pm 0.03$		
	Andersson [61, 62]	THW	$k_{25} = 1.07 \pm 0.03$	(0.78×10^{-3})	(25–767 °C); $k_{gl\ T} = 1.07 + 0.78 \times 10^{-3}(T–25\ °C)$ k_T decreases sharply at 767 ± 10 °C
Bed					
Powder	Taylor [209]	LP	$k_{200} = 0.17–0.35$	(3×10^{-4})	(200–700 °C) in vacuo k calculated from a values
	Nagata [216]	THW	$k_T = 0.2–0.4$		Goes through maximum at ca 1000 °C
	Goto [219]	THW	$k_T = 0.12$		4 MPs values very similar
	Neumann [229]	THW	$k_{25} = 0.12–0.22$	$(7 \times 10^{-5}*)$ $(11 \times 10^{-5})^G$	* = powder; ^{G} = granules
	Andersson [61]	THW	$k_{25} = 0.12–0.16$	(9×10^{-5})	$k_{25} = 0.010 + 1.69 \times 10^{-4}\ \rho_{bulk}$
	Bratchikov [230]		k_{100} = ca. 0.3*		Compressed *powder* $\rho_{bulk} = 1555$ kg m^{-3}; 10–1000 °C
	Ichihashi [231]				
Powder bed	Macho [218]	THW tests	$k_{25} = 0.06–0.1$	(94×10^{-5})	
	Diehl [232]	Hot-plate	k_{25} = 1*–1.4**		Probably contains contributions from gas convection
	Ichihashi [231]	Sim	$k_{20} = 0.5$		$T_{crit} = 1030$ °C
Sinter	Taylor [209]	LP			k increases as $\rho_{bulk}\downarrow$ Suggests k_{conv} contribution
	Andersson [61]	THW	0.4		For powder sample which sintered at high temp.

Note k values in Wm^{-1}K^{-1}; $10^7\ a$ values are denoted by parenthesis e.g. (4) = 4×10^{-7} m^{2}s^{-1}; *Sim* Simulation experiments; *THW* Transient Hot Wire, *LP* Lase Pulse; *powder; *G* Granules

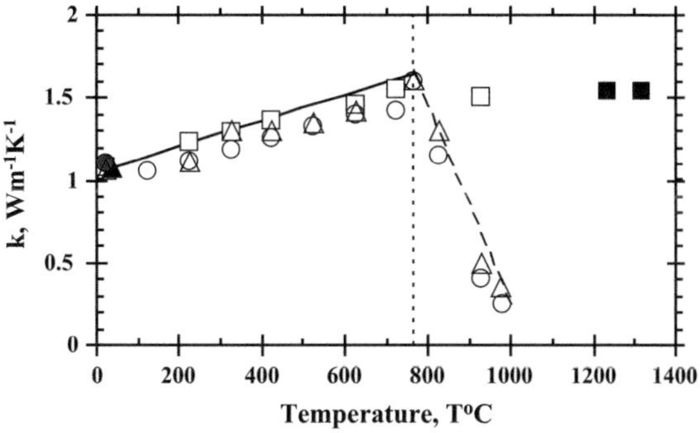

Fig. 9.29 The thermal conductivity of glassy, mould slags as a function of temperature; o, Δ = [61]; *solid line* = Eq. 9.25; *dashed line* = values above T_{crit} [61]; ●▲ = Ozawa [54]; □, ■ = *k* values calculated from $10^7 a = 4.5$ and 4 m^2s^{-1}, respectively, for glass and liquid; vertical *dashed line* = T_{crit}

9.4.5.6 Glassy Slags

It can be seen from Table 9.6 and Fig. 9.29 that thermal conductivity values have been measured using the THW method for glassy, mould slags [54, 61, 62]. Measurements have also been reported for a powder sample heated into the liquid which then formed a glass on cooling [216]. The thermal conductivities at 25 °C for glassy mould slags were found to lie between 1.04 and 1.1 $Wm^{-1}K^{-1}$ [54, 61, 62].

When a glass is heated, it undergoes a transition from a frozen glass to a supercooled liquid at the glass transition temperature (T_g); T_g occurs at *ca.* 600 °C for most mould slags. However, the slag crystallises at a temperature about 50 °C above T_g (see Fig. 9.9); this results in increases in both the fraction of crystalline phase (f_{crys}) and in k_T.

Andersson [61, 62] found that k_T for all glassy samples increased with increasing temperature up to a critical point (T_{crit} or the deformation temperature) beyond which, there was a marked decrease in k_T with increasing temperature (Fig. 9.29).

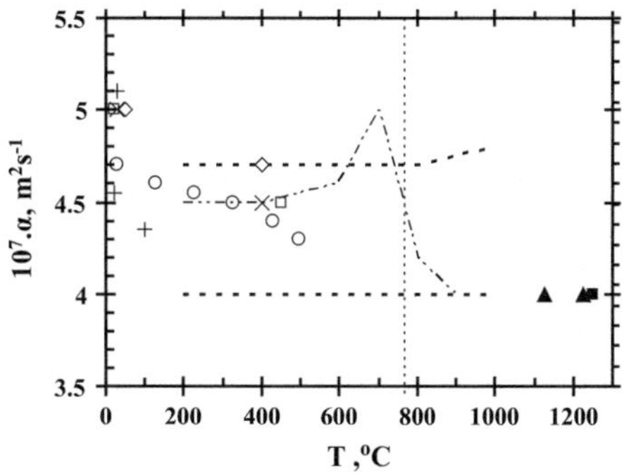

Fig. 9.30 The thermal diffusivity ($10^7.a$) as function of temperature for glass and liquid mould slags; o = mean of 3 samples studied by Shibata [211]; *dashed lines* = high and low values of 10 slags [211]; *dotted line* = sample THA [211] showing sharp drop near T_{crit}; += Values obtained by TPS method [61]; x, ◇, □ = [222]; Liquid ▲ = Ohta [212, 225] values; ■ = Gonerup [223].; vertical *dotted line* = T_{crit} (Reproduced with permission from Taylor and Francis, www.tandfonline.com [61])

Thus the temperature dependence of k_{gl} for glassy mould slags can be described by Eq. 9.25. Values of T_{crit} were found to occur at 767 ± 10 °C for mould slags [61] and measured values for k_{gl} at T_{crit} had values of 1.65 ± 0.05 $Wm^{-1}K^{-1}$ (Fig. 9.29). It should be noted that T_{crit} is ca. 50–100 °C above the point where crystallisation is usually initiated but on occasion, crystallisation can occur below T_g (for instance see Fig. 9.9).

$$(25-767\,°\mathrm{C})\quad k_{gl}\left(\mathrm{Wm^{-1}K^{-1}}\right)=1.07+0.78\times10^{-3}(T-25\,°\mathrm{C}) \qquad (9.25)$$

Measurements of absorption coefficients of glassy mould slags indicate that a 2-mm thick, initially glassy slag film would be optically thin.

Thermal diffusivity values have also been reported for glassy mould slags by a number of investigators, most of which were obtained with the LP method [209, 211, 222]; these values are shown in Fig. 9.30. The results show that the thermal diffusivity of all the glassy mould slags were found to lie between 4 and 5 × 10^{-7} m^2s^{-1} and values were virtually independent of temperature (Fig. 9.30). These values of a_{LP} were converted into k_{LP} values using estimated C_p and density values and are compared with the k_{THW} values in Fig. 9.30. It can be seen that k_{THW} values are in good agreement with k_{LP} values for temperatures up to T_{crit} but for $T > T_{crit}$, k_{THW} values decrease markedly with temperature, whereas k_{LP} values, shown as squares, increase slightly with temperature (Figs. 9.29 and 9.30). Similar behaviour has been identified [204] with k_{LP} and k_{THW} measurements for CaO–Na_2O–SiO_2 slags of similar composition [233, 234]. Three propositions have been put forward to account for the divergence:

1. k_{LP} values contain significant contributions from k_R since the surface area (A) of the heat source is 10× greater in the LP (*cf.* THW) experiments (A_{LP} = *10*A_{THW}).
2. k_{THW} values are low because of electrical leakage from the hot wire into the melt.
3. Crystallisation results in shrinkage of the slag and the formation of an air gap at the slag/heater interface.

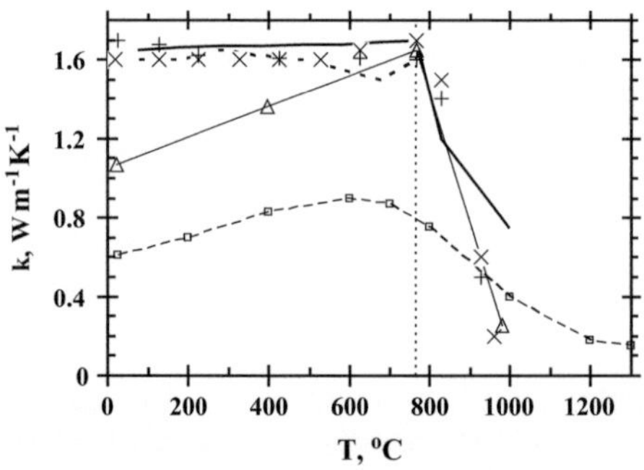

Fig. 9.31 Thermal conductivity of mould slags as a function of temperature; Glass: denoted Δ and faint line [61]; Partially crystalline; 4 samples denoted by x, +, *bold line*, *dotted lines* [61]; □ = values for a melted powder on cooling [216]; vertical *dotted line* = T_{crit} (Reproduced with permission from Taylor and Francis, www.tandfonline.com [61])

At this point of time, the problem is unresolved and no values can be recommended for k for temperatures above 760 °C. However, it has been suggested that the rapid decrease in k_{THW} above T_{crit} is due to the loss in rigidity of the silicate network between T_{soft} and T_{flow} [50] and that T_{crit} corresponds with the temperature where the viscosity had a value of 10^6 dPas [204] i.e. T_{crit} is equivalent to the deformation temperature of the slag.

9.4.5.7 Slag Films and Partially Crystalline Samples

It was noted that when glass samples were heated to high temperatures that the values recorded on the cooling cycle were significantly higher than those measured for the glass in the heating cycle, as shown in Fig. 9.31 [61, 62]. This enhancement of k values is due to the crystallisation of the sample at higher temperatures and these high k values are maintained, or increased, during subsequent measurement cycles. Thus the thermal conductivity increases with increasing crystallisation.

Thermal conductivity and diffusivity values for crystalline samples of mould fluxes are significantly higher than those for glassy samples [61, 209, 211]; this is in line with measurements on other slags where $k_{crys} \approx 2k_{glass}$ [60]. The effect of the fraction crystalline phase (f_{crys}) on the thermal conductivity can be clearly seen in Fig. 9.32a, b [54, 55, 61]. Equation 9.26 was obtained from the data given in Fig. 9.32b.

$$k_{25}(\mathrm{Wm^{-1}K^{-1}}) = 1.07 + 0.7 f_{crys}. \tag{9.26}$$

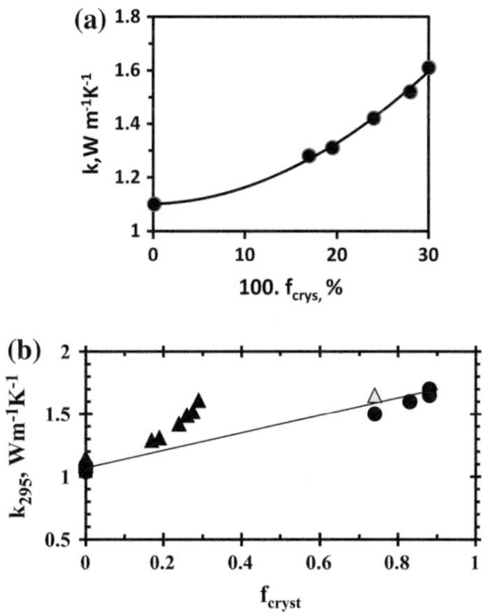

Fig. 9.32 Thermal conductivity of mould slag **a** as function of fraction crystalline phase (f_{crys}) expressed as % and **b** k_{25} as a function of fraction crystalline phase; ● [61]; ▲, Δ [54, 55] for fluxes 1 and 3, respectively; (permission granted, Taylor and Francis, www.tandfonline.com, using data from [54, 55, 61])

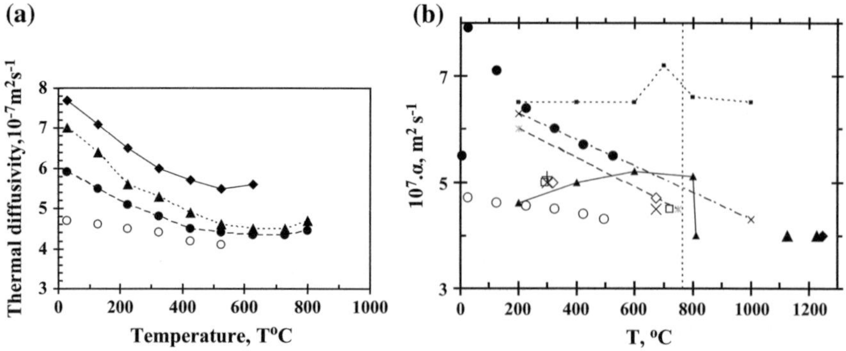

Fig. 9.33 Comparison of thermal diffusivity values for slag films, partially crystalline samples and liquid; **a** for partially crystalline (●▲♦) and glassy (o) mould slags with C/S values of 1.07 = ●; 1.16 = ▲ and 1.29 = ♦ [211] **b** slag films = various curves [53]; ● = partially crystalline (C/S = 1.29) and o = glass of same slag; liquid, ▲ = Ohta [212, 225]; ♦ = Gonerup [223] (**b** permission granted, Taylor and Francis, www.tandfonline.com [61])

Since values of $k_{25} > 1.8$ $\mathrm{Wm^{-1}K^{-1}}$ have been recorded and the partially crystallised samples contain some porosity, Eq. 9.26 may provide an underestimate of k_{25} for fully—crystalline samples.

It can be seen from Fig. 9.32 that the k_{THW} values for various partially crystalline mould slags (like the glass phase) attain a value of $k_{THW} = 1.65 \pm 0.05$ $\mathrm{Wm^{-1}K^{-1}}$ at T_{crit} [61]. It is interesting to note that k_{THW} values for partially crystalline samples show the same rapid decrease for temperatures above T_{crit} that the glassy phase samples display. This may be due to the rapid deterioration in k_{gl} for $T > T_{crit}$ for the glassy phase.

The temperature coefficient (dk/dT) for temperatures between 25 and 767 °C for partially crystalline samples can be positive or negative depending on whether $k_{25} < k_{Tcrit}$ or $k_{25} > k_{Tcrit}$, respectively.

Thermal diffusivity values for partially crystalline samples and slag films are shown in Fig. 9.33. It can be seen that (i) the values for the slag films have similar values to those of the partially crystalline samples (ii) the values are significantly higher than those of the glass and (iii) there is some variation in values, presumably due to differences in f_{crys}. It is also interesting to note the a_{LP} value recorded for one slag film shows a drop near T_{crit}.

It has been calculated that in partially crystalline samples, k_R contributions constitute <20% of the effective thermal conductivity (k_{eff}). This suggests that k_{eff} values derived on slag films with a significant crystalline fraction are unlikely to contain large contributions from k_R. The k_{THW} values for partially crystalline samples indicate a decrease in k for $T > 767$ °C; this could indicate that f_{crys} is not high enough to provide sufficient rigidity in the sample. However, it is difficult to make recommendations for k values for $T > 600$ °C until the discrepancy between k_{LP} and k_{THW} is resolved.

9.4.5.8 Powder Bed

The effective thermal conductivity (k_{eff}) of mould powders and the powder bed are affected by a number of factors:

(i) The porosity of the sample with k_{eff} increasing as the porosity decreases.
(ii) The magnitude of k_{eff} increases as the number of points of contact between particles ($N_{contact}$) increases (Note small particles have a much greater number of points of contact than large particles per unit mass, thus k_{eff} increases with decreasing mean particle size (D_{mean}). It should also be noted that compression of the powder increases $N_{contact}$ and thus k_{eff} increases with increasing bulk density (ρ_{bulk}).
(iii) Contributions to k_{eff} from gaseous convection ($k_{gasconv}$) with $k_{gasconv}$ increasing as the permeability (P) of the powder bed increases (Note P increases as D_{mean} increases and thus $k_{gasconv}$ increases as D_{mean} increases).
(iv) Exothermic or endothermic reactions (e.g. carbonate decomposition, sintering, carbon combustion) will affect the local temperature gradients in the bed and hence the local heat flux.
(v) The thermal conductivity of the powders increases with increasing temperature; the mean temperature coefficient (dk/dT) is $+9 \times 10^{-5}$ $Wm^{-1}K^{-2}$ [61].

It should be noted that $k_{gasconv}$ will be affected by the nature of the experiment; it will be large for experiments where there is a large temperature gradient across the sample and where the sample is heated from below. Furthermore, decomposition of carbonates, the oxidation of carbon particles and argon injection will all lead to enhanced gaseous convection in the bed.

Powders have a smaller diameter and higher bulk density than granules and consequently, have lower porosity and more points of contact, so $k_{powd} > k_{gran}$. The following relation has been reported [61] and the results are shown in Fig. 9.34.

$$k_{25}\left(Wm^{-1}K^{-1}\right) = 0.010 + 1.69 \times 10^{-4}\rho_{bulk} \quad (9.27)$$

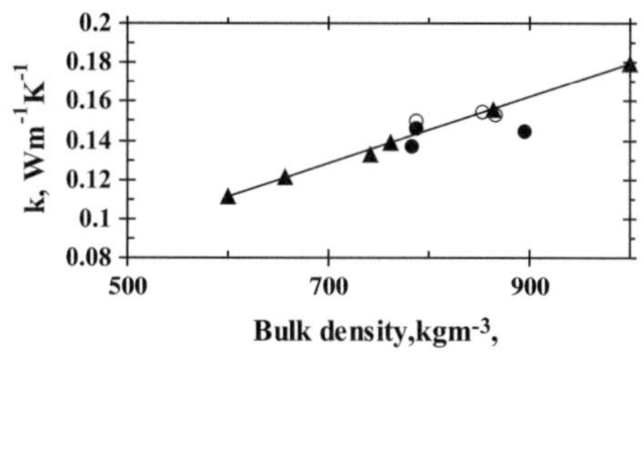

Fig. 9.34 Effective thermal conductivity of mould powders as a function of bulk density, using data from [61, 62] obtained using the THW method

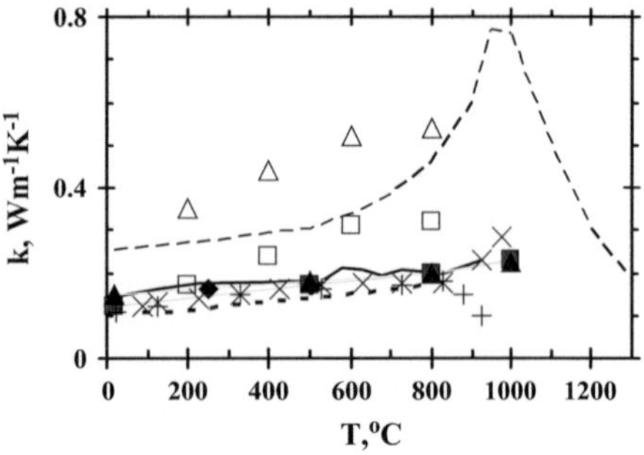

Fig. 9.35 Thermal conductivity of powders; *solid bold line*; *dotted line*, +, X = Andersson [62]; ♦ = k THW due to Macho [218]; *dashed line* = Nagata [216] values; □, Δ, O = values due to Taylor [209]; ▲, ■ = Neumann [229] granules and powder, respectively; (permission granted, Taylor and Francis, www.tandfonline.com [61])

The results reported for the thermal conductivity of powders are shown in Fig. 9.35 and Table 9.6. Inspection of these results shows that measurements of k_{eff} fall into three bands:

(i) k_{eff} = *ca* 0.12 $Wm^{-1}K^{-1}$ with the values reported by Macho [218] Neumann [229] and Anderson [61] in good agreement.
(ii) k_{eff} = 0.1–0.35 $Wm^{-1}K^{-1}$ and 0.3–0.5 $Wm^{-1}K^{-1}$ at 200 and 800 °C, respectively, [209, 216]; the higher k_{eff} values probably resulted from higher bulk densities.
(iii) k_{eff} = *ca* 1–1.4 $Wm^{-1}K^{-1}$ [232] derived in thermal insulation tests where the enhanced k_{eff} values probably arise from large $k_{gasconv}$ contributions.

The divergence in these values (0.1 to >1.0 $Wm^{-1}K^{-1}$) raises the question as to which value to use. When the gaseous convection contribution is calculated independently, the value of $k_{powd} \approx$ *ca* 0.12 $Wm^{-1}K^{-1}$ would be appropriate. However, the value k_{eff} = *ca* 1–1.4 $Wm^{-1}K^{-1}$ obtained in thermal insulation tests contains contributions from both k_{powd} and $k_{gasconv}$ and should be used (cautiously) when $k_{gasconv}$ is not calculated independently. It should be noted that these findings indicate that $k_{gasconv} \approx 10\ k_{powd}$ in these experiments which aim to simulate the heat transfer in the powder bed.

Values have also been measured for the sinter portion of the bed. A value of k = ca. 0.4 $Wm^{-1}K^{-1}$ was derived for a powder sample cooled after heating to high temperature; it was suggested that this may constitute a representative value for the sinter layer [61]. Much higher values of a_{LP} (*ca* $7 \times 10^{-7}\ m^2s^{-1}$ for a fully dense sample) and k_{LP} (1–2 $Wm^{-1}K^{-1}$) were derived [209] but these were for dense (compressed) de-carburised powder samples.

Table 9.7 Summary of published models to calculate the thermal conductivity of slag films and liquid mould fluxes

Model/Reference	Details of model
Mills [144]	Calculates, $ln\ k$ at 25 °C and T_g; $ln\ k = -a + b \times 10^{-5}\exp(Q/c) + dX_{Li_2O}$: $ln\ k_{25}\{ln\ k_{Tg}\}$; $a = 0.424\{0.435\}$, $b = 2\ \{5\}$, $c = 0.299\{0.332\}$; $d = 3.2\ \{3.0\}$
Mills [204]	Accepts k_{THW} data: $k\ (Wm^{-1}K^{-1}) = a + b\ \exp\ (Q/c) + d\ \exp\ (Q/e) + f(X\ (z/r^2))$; k_{25}:$\{k_{Tcrit}\}$ $a = 7.94\ \{-0.85\}$; $b = 7.22\ \{0.987\}$; $c = 199.3\ \{1.006\}$; $d = 0.109\ \{-0.8\}$, $e = 3.634\ \{0.965\}$. $f = 0.61\ \{1.62\}$
Fluegel [235]	Glass samples-only covers low levels of F. Numerical fit of thermal conductivity data for glasses at ambient temperatures
Mills [143]	Accepts k_{THW} data Glass; $k_{25} = 1.07\ Wm^{-1}K^{-1}$. Partially crys: Calculate f_{crys} from composition. $k_{25} = 1.07 + 0.7\ f_{crys}$. $k_{767} = 1.65$; $k_{Tliq} = 0.0278\ \exp\ (1.0242Q)$

crys crystalline

9.4.5.9 Liquid Slags

There is a large discrepancy in the values measured with the THW and LP methods with $k_{LP} \approx 1.5\ Wm^{-1}K^{-1}$ and $k_{THW} \approx 0.2\ Wm^{-1}K^{-1}$ [216, 221] and it is difficult to recommend a value at this stage.

9.4.5.10 Calculation of Thermal Conductivity

Few models have been reported to calculate the thermal conductivity of mould fluxes as can be seen from Table 9.7.

9.4.6 *Interfacial Tension (γ_{msl}) and Surface Tension (γ_s)*

9.4.6.1 Importance of Interfacial Tension to the Process

The interfacial tension, γ_{msl}, (between the metal and slag) affects the following:

- *Slag entrapment* with entrapment decreasing as γ_{msl} increases [146, 236].
- The *shape of the steel meniscus* is affected by by γ_{msl}.
- It has also been proposed that the *adhesion of scale and scum* to the steel surface is dependent on interfacial tension [237].

Values of the surface tension of the slag (γ_{sl}) are needed to calculate the interfacial tension (γ_{msl}) as can be seen from Eq. 9.28.

9.4.6.2 Factors Affecting Surface and Interfacial Tensions of Mould Slags

The interfacial tension (γ_{msl}) is given by the Good–Girifalco equation [238] where γ_m and γ_{sl} are the surface tensions of the metal and slag phases, respectively, and φ is an interaction coefficient.

$$\gamma_{\mathrm{msl}} = \gamma_m + \gamma_{\mathrm{sl}} - 2\varphi(\gamma_m \gamma_{\mathrm{sl}})^{0.5} \tag{9.28}$$

The parameter, φ, was found to have a value of 0.5 for slags free of FeO but increased with FeO additions [239]. It was proposed that the following equation could be applied

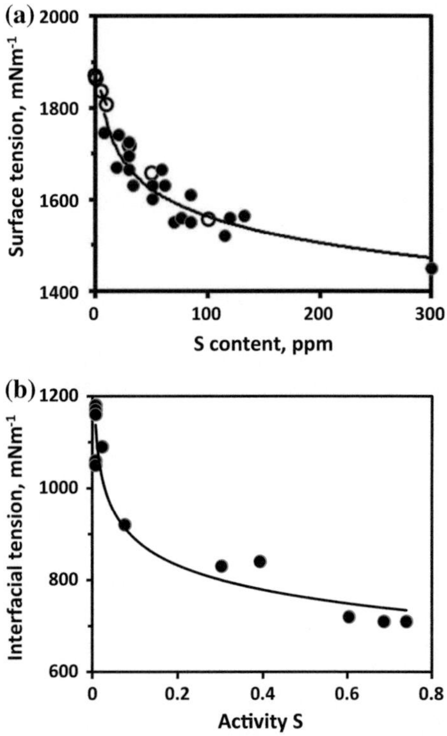

Fig. 9.36 The effect of S on **a** surface tension (γ_m) of pure iron at 1550 °C (o) and steel (●, using data from [242]) and **b** interfacial tension of mould slag and steel (permission granted, ISS/AIST [146])

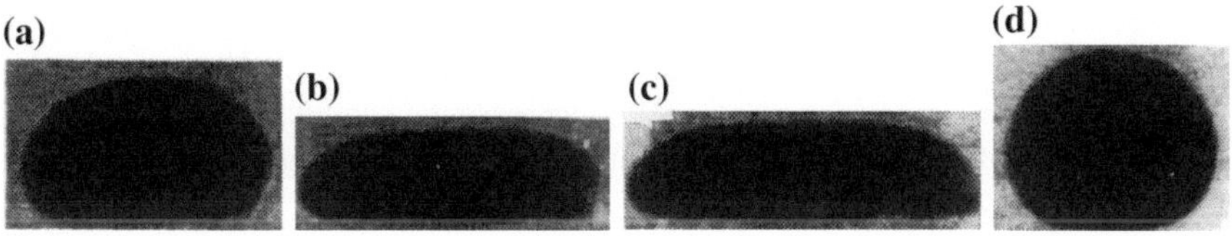

Fig. 9.37 Transient sessile drop profiles during metal/slag reactions **a** at start of experiment **b**, **c** during the experiment and **d** at end of the experiment, when slag/metal reactions are complete (permission granted, IOM/Taylor and Francis) [244]

$$\phi = 0.5 + 0.3X_{FeO} \tag{9.29}$$

Alternatively, Chung [238, 240] suggested that the interaction coefficient (ϕ) could be calculated from values of the Gibbs energies (ΔG_i) of formation the various liquid components of the mould slag.

The surface tensions of both steels and slags are dependent upon the concentrations of *surfactants* present. Surfactants tend to be materials with low surface tension and the surface layer of a liquid contains a high concentration of surfactants. That is why ppm levels of surfactants can have a significant effect on surface tension. In steels, the principal surfactants are *soluble* sulphur and oxygen. In contrast, the principal surfactants in slags are B_2O_3, K_2O and Na_2O and CaF_2; however, their effect on γ_{sl} is much less dramatic than the effect of S and O on γ_m of iron.

Surface tensions of the steel (γ_m) tend to have values between 1500 and 1850 mNm^{-1}, whereas the surface tension (γ_{sl}) of most slags have values in the range, 300–450 mNm^{-1}; thus $\gamma_m \approx 4\ \gamma_{sl}$. Consequently, the most important term in Eq. 9.28 is γ_m and this is very dependent upon the *soluble* S- and O-contents of the steel [241] (Fig. 9.36a). In most steels, the Al content is sufficiently high to hold the *soluble* $\underline{O}$ content down to ca. 4 ppm but Al has little effect on the *soluble*

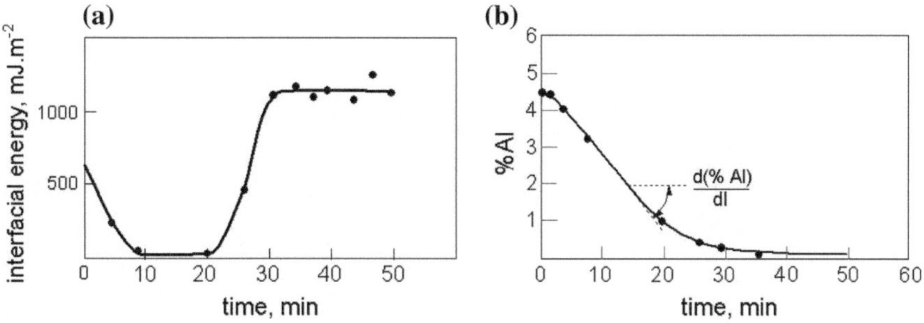

Fig. 9.38 Transient measurements of **a** Interfacial tension and **b** mass transfer rate of Al (*via* Eqs. 9.1 and 9.2) (permission granted, Taylor and Francis [243])

$\underline{S}$ content. Thus, γ_m is largely determined by the S content of the steel and it can be seen from Fig. 9.36a that 50 ppm S causes a 25% decrease in γ_m. However, FeO in the mould slag and metal/slag reactions will increase the *soluble* $\underline{O}$ in the steel and it is probably the latter which causes the apparent increase in the interaction coefficient (ϕ) (Eq. 9.29) and a decrease in γ_{msl}.

It has been reported [242–245] that reactions between slag and metal (e.g. S-transfer or Al transfer, Eqs. 9.1 and 9.2), the vigorous mass transfer (Figs. 9.37 and 9.38) apparently results in a dramatic reduction in the interfacial tension (Fig. 9.38a, b [243]). The interfacial tension remains low while there is vigorous mass transfer but increases sharply when most of the slag/metal reaction is complete (Fig. 9.38). It has been suggested (on the basis of observations on organic systems) that this effect is related to the large differences in volume of the slag and metal ($V_{sl} > V_m$) and that there would be a smooth change in interfacial tension if the metal and slag phases had similar volumes [246] i.e. $V_{sl}/V_m \leq 1$.

The interfacial tension has been reported to increase as (i) Al_2O_3 increases [245] and (ii) the contents of Na_2O, CaF_2, FeO and MnO decreased in the mould flux [245]. The interfacial tension is usually increased by reducing the Na_2O content of the flux [237].

Measured values of γ_{msl} tend to be in the range 1100–1400 mNm^{-1} and values of the surface tensions of mould fluxes tend to be in the range 300–450 mNm^{-1}.

9.4.6.3 Measurement Methods

Surface tension (γ_{sl})

There are several methods available to measure the surface tensions of molten mould fluxes, these are detailed below.

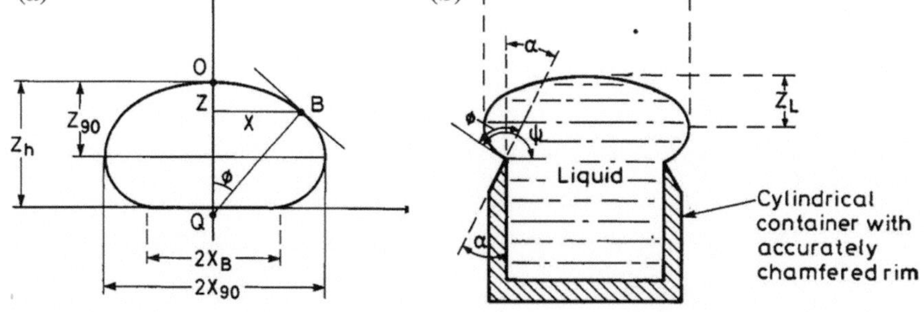

Fig. 9.39 Schematic drawings showing **a** sessile drop and **b** big drop (BD) methods (permission granted, Verlag Stahleisen GmbH, [3])

Maximum bubble pressure (MBP) method

In this method a gas/slag interface (i.e. a gas bubble) is formed at the tip of a metal capillary by gradually increasing the gas pressure. A transducer records the gas pressure continuously. The maximum bubble pressure (MBP or P_{MBP}) corresponds to the point where the bubble attains a hemispherical profile. The surface tension of the slag is calculated from the Laplace equation where h is the depth of immersion of the capillary in the slag and ρ is the density of the slag. It is customary to carry out measurements at various depths to derive the density of the slag [3]. It is also customary to chamfer the capillary to produce a knife-edge tip and to use a slow rate of bubble evolution [3].

$$P_{\mathrm{MBP}} = 2/r + \rho g h \quad (9.30)$$

Sessile drop (SD) and Big drop (BD) methods

The shape adopted by a sessile drop (Fig. 9.39a) represents the balance of surface and gravitational forces. These forces involve the surface tension and the density, respectively, and both properties can be derived by this method. A molten slag droplet is sited on a metallic plaque and the dimensions of the drop are measured accurately; several methods have been used to derive the surface tension and density from the dimensions [3]. In recent years, the accuracy of surface tension and density values has been improved by using software which calculates "*best fit*" values for the parameters affecting the drop profile.

The Big Drop (BD) method is a variant of the sessile drop technique (Fig. 9.39b) where the slag is held in a crucible and the surface profile of the drop is determined.

Pendent drop (PD) and Drop weight (DW) method

A pendent drop also represents the balance of surface and gravitational forces [3]. Frequently, when this method is applied at high temperatures, the sample is in rod form and the tip of the rod is heated by electron beam or laser heating. Thus surface tension values can only be determined for the liquidus temperature. Software similar to that used in the sessile drop method is used to provide a "*best fit*" of the measured drop profile.

The pendent drop method is often used in combination with the drop weight method; since the pendent drop eventually falls and forms a spherical drop which can be collected. The solidified drops are weighed and the mass (W_{ideal}) determined. The surface tension is calculated from the following equation where r = the radius of the tube or rod.

$$W_{\mathrm{ideal}} = 2\pi\gamma r \quad (9.31)$$

However, the measured weight (W_{meas}) of the drops is less than W_{ideal} and so a correction term (f_c) is usually applied and f_c is determined in preliminary experiments using liquids of known surface tension [3, 247].

$$W_{ideal} = 2\pi\gamma r f_c \quad (9.32)$$

Detachment methods

In this technique, a tube (or rod, plate or ring) is located just below the surface of the liquid. The probe is attached to a balance and then the mass is measured continuously as the probe is slowly pulled from the melt [3]. The force acting on the probe is measured; this force is the resultant of surface tension and mass. This force goes through a maximum at the point where the liquid film at the meniscus is about to break and thus, the measured weight also exhibits a maximum (W_{max}). The surface tension is calculated by Eq. 9.33 where r = radius of tube or probe and G is a correction factor

$$\gamma = (W_{max} g)G/4\pi r \quad (9.33)$$

Other methods

As mentioned above (Sect. 9.4.4.3) the draining crucible method is also capable of measuring the surface tension. A crucible has a small orifice machined in its base and the orifice is blocked with a stopper rod. Slag is placed in the crucible, melted and allowed to drain and the mass flow rate is monitored. The surface tension is derived from a hydrodynamic analysis of the data [164].

In the *Surface laser light scattering (SLLS) method* "*ripplons*" are monitored [165–167]. Ripplons are Capillary waves caused by thermal fluctuations of the surface. Ripplon action is initiated by surface tension and is damped by the kinematic viscosity. A Fourier spectrum analyser is used to determine the spectrum of the ripplons and the surface tension is derived from the data.

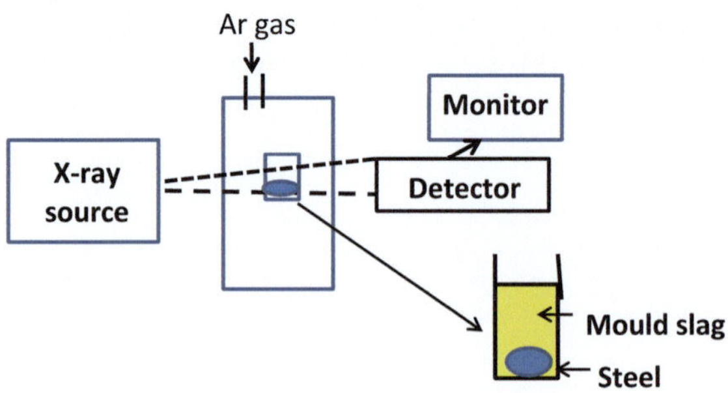

Fig. 9.40 Schematic drawing showing the X-ray sessile drop method

Table 9.8 Summary of surface and interfacial tension studies carried out on mould fluxes

Property	Reference	Method	γ Values mNm^{-1}	Comments
Surface tension, γ_{sl}	Yavoisky [248] Yakushev [249]		310–320 200–320	
	Krusina [250]		360–380	Slag:40C + 45S +10A + 5N + (0–15) Fl replacing S or C
	Elfsburg [251]	SD	430–550	Continual gas evolution from drop
	Kusano [252]	Det*	250–320	Det* = Detachmant method
	Lu [185]	RD	320–340	No temps given; (C/S) = 0.8; 1 and 1.2; 3–15% $Na_3Al\ F_6$ added ($d\gamma$/d% $Na_3Al\ F_6$) = −2.5
	Lu [102]	RD	305–355	No temps given; (C/S) = 0.8; 1 and 1.2; 2–10%B_2O_3 added ($d\gamma$/d%B_2O_3) = −6.5
	Lu [100]	RD	320–370	No temps given; (C/S) = 0.8; 1 and 1.2; 1–5%BaO added ($d\gamma$/d% BaO) = −4
	Lu [98]	RD	320–355	No temps given; (C/S) = 0.8; 1 and 1.2; 1–5% K_2O added ($d\gamma$/d% K_2O) = −5
	Lu [186]	RD	300–340	No temps given; (C/S) = 0.8; 1 1.2; 3–15%NaF added ($d\gamma$/d% NaF) = −2.5
	Lu [95]	RD	300–340	No temps given; (C/S) = 0.8; 1 and 1.2; 2–10%Li_2O added ($d\gamma$/d% Li_2O) = −3
	Monaghan [224]	MBP	320	(1200–1400 °C); $d\gamma/dT$ = ca. *0*
	Cheng [253]	RD	$\gamma_{1400} = 400$	$\gamma_{1400} = 400 + 560$ (*C/S*−1.0); $\gamma_{1400} = 415 + 8(\%Al_2O_3)$; $\gamma_{1400} = 400 - 15\%$F; $\gamma_{1400} = 420 - 10\%$ Na_2O
	Duberstein [254]	MBP	$\gamma_{1400} = 370 - 310$	($d\gamma_{1400}$/d%Na_2O) = −*6* $mNm^{-1}\ \%^{-1}$
	Konovalov [255]	MBP	$\gamma_{1550} = 300 - 350$	(C/S) = 0.5−2.28
	Gonerup [223]	XRSD	320−500	
Contact angle (θ)	Feldbauer [146] Nakato [237] Tsutsumi [192, 193]		$\theta_{1300\ °C} = 55$ $\theta_{1300\ °C} = 75$	Mould slag on LC steel: $\theta_{1300\ °C} = 55°$ Slag on steel: slag on solid Fe: $\theta_{1300\ °C}$: Conv $\theta = 75°$ NNslag; $\theta_{1300\ °C} = 65°$
Interfacial tension γ_{ms}	Gornerup [223]	XRSD	A- 1150 B- 910	

(continued)

Table 9.8 (continued)

Property	Reference	Method	γ Values mNm^{-1}	Comments
				2 mould slags: A = 37C + 29S + 7A + 7 N +6Fl + 0.8F Slag B – 29C + 33S + 5[a] + 11N + 9Fl +1.2F
	Elfsberg [251]	XRSD	800–1400	4 steels; 4 mould slags;
	Cramb [245] Feldbauer [146]	XRSD	1000–1350	γ_{msl} decreases↓- as soluble O and S in steel↑; -as slag- FeO↑; as Al_2O_3↓, as Na_2O↑; as CaF_2↓
	El Gammal [227]		1400–800	Effect of TiO_2 and S on γ_{msl} at 1550 °C of 40C + 40S + 20A
	Konovalov [255]		1000–1100	Sessile drop values for steel and MBP for slag
	Hagemann [256]	Drop wt	820–1060	Low (C/S) slag to cast highAl steel: 2 steels with 100 and 500 ppm S: ↑Al_2O_3 and MgO → γ_{ms}↑: ↑TiO_2 → γ_{ms}↓

γ and γ_{msl} are mNm^{-1} units.; *RD* Ring detachment method; *Drop wt* Drop weight method; *Det** Detachment

9.4.6.4 Methods to Measure Interfacial Tension (γ_{msl})

Two methods have been used to measure metal/slag interfacial tension.

X-ray sessile drop method

A cylinder of steel is placed in a MgO crucible and then covered with de-carburised mould flux and is then heated in a tube furnace under an inert atmosphere [3]. When both slag and metal have melted, a sessile drop of steel in molten mould flux is formed and observed by passing X-rays through the crucible (Fig. 9.40 [244]). The slag is transparent to X-rays whereas the steel is opaque and thus the image of the sessile drop can be observed. The contact angle and the dimensions of the drop are measured and the interfacial tension is measured using software packages based on the Laplace equation. Most of the published data have been obtained with this method.

Lens method

In this method mould slag is placed on the meniscus of molten steel. The slag melts and forms a lens-shaped drop [3]. The dimensions and contact angle are observed and measured.

Table 9.9 Summary of models available to calculate the surface tension of mould slags and interfacial tension

Model/references	Details of model	Uncertainties
Surface tension		
Mills [257]	Two different components; (i) Regular, e.g. CaO, SiO_2, Al_2O_3 and (ii) Surfactants B_2O_3, K_2O, CaF_2, Na_2O, Li_2O; Regular: $\gamma = \Sigma X_1 \gamma_1 + X_2 \gamma_2 + X_3 \gamma_3 +$; Surfactants- 2 regimes (a) $X_i < 0.12$;.$\gamma_{2i} = a + bX + cX^2$ and (b) $X_i > 0.12$; by $X_2 \gamma_2 = a' + b'X$; where, i = surfactant species Tends to over-emphasise effect of surfactant	±10%
Mills [143]	Above model modified; (1) surface active element are considered as regular components once surface active concentration ($X_{surf.} > 0.12$) (2) $d\gamma/dT = \Sigma(d\gamma/dT)_i$ for slag components	
Nakamoto [258]	Thermodynamic model involving cation ionic radii; ternary systems C + A + Fl; C + F+B_2O_3; and C + S+N	
Hanao [259]	Based on Tanaka model [239] using molar volumes of components oxides. Extended to 0.4C + 0.4S + 0.2Fl + additions of Al_2O_3 + MgO + Na_2O	±17.5%
Nakamoto [260]	Neural network model applied to ternary systems;	
Interfacial tension		
Chung [238, 240]	Cites values for ΔG: of constituents: ΔG_i^D. = = ΔG_{FeO}- ΣX_i ΔG_i $\varphi = 0.89 - 1.5 \times 10^{-3} \Delta G_i^D$	
Tanaka [239]	$\gamma_{msl} = \gamma_m + \gamma_{sl} + 2\,\phi\,(\gamma_m \cdot \gamma_{sl})^{0.5}$; $\phi = 0.5 + 0.3X_{FeO}$	
Mills [143, 261]	γ_{sl}; calculated as above; $\gamma_m = 1880 - 0.41(T - 1530\ °C) - 0.109\ ln\ (1 + e\{(194111/T) - 4.6849\}\ x\ \%S$ $\gamma_{msl} = \gamma_m + \gamma_{sl} + 2\,(\gamma_m \cdot \gamma_{sl})^{0.5}$;	

9.4.6.5 Surface and Interfacial Tension Data

The studies of the surface and interfacial tension involving mould slags are summarised in Table 9.8.

9.4.6.6 Methods to Calculate Surface (γ_{sl}) and Interfacial Tension (γ_{ms})

A few models have been reported to calculate the interfacial tension; these are summarised in Table 9.9.

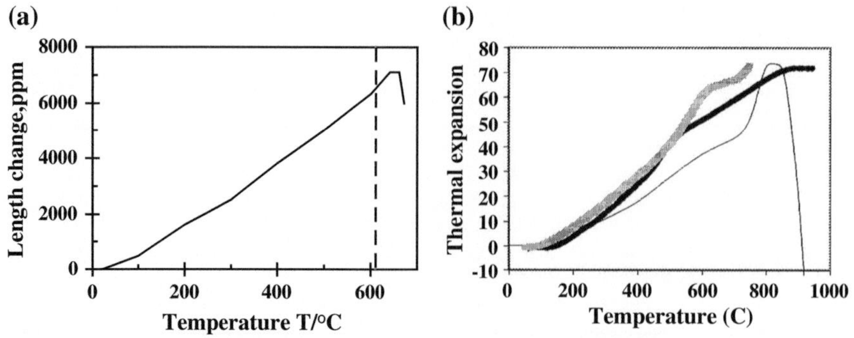

Fig. 9.41 Change in length as a function of temperature (°C) for **a** a glassy mould slag; dashed vertical line = T_g showing a sharp decrease at deformation temperature; **b** From top at 700 °C; slag film (*gray*) crystalline (*bold*) and glass (*faint line*) of same mould slag; (permission granted, Taylor and Francis, www.tandfonline.com [124])

9.4.7 Density (ρ) and Thermal Expansion Coefficient (α)

9.4.7.1 Importance of Density to the Process

The density of the mould slag is one of the factors affecting the rate of slag flow into the mould/strand channel (i.e. powder consumption). Density data are needed for the mathematical modelling of powder consumption.

9.4.7.2 Factors Affecting Density and Thermal Expansion

The molar volume (*V*) is only slightly affected by the polymerisation of the slag and this can be accommodated by expressing V_{SiO_2} (and $V_{Al_2O_3}$) as polynomials of (X_{SiO_2} V_{SiO_2}) where *X* is the mole fraction. The oxides with higher molecular weight (M) tend to have lower molar volumes (or higher densities) e.g. $V_{K_2O} < V_{Na_2O} < V_{Li_2O}$.

$$V\rho = M \tag{9.34}$$

The thermal expansion coefficient can be cited as either as the linear thermal expansion (α) or the volume thermal expansion (β) which are defined in Eqs. 9.35 and 9.36, respectively, where $\Delta T = (T - T_{ref})$ and T_{ref} = reference temperature (e.g. 20 °C) α and β values pertain to the mean temperature, i.e. 0.5 $(T + T_{ref})$.

$$\alpha = (L_T - L_{ref})/L_{ref}(T - T_{ref}) \tag{9.35}$$

$$\beta = (V_T - V_{ref})/V_{ref}(T - T_{ref}) \tag{9.36}$$

$$V = V_0(1 + \alpha\Delta T)^3 \tag{9.37}$$

The principal factors affecting the thermal expansion are (i) the *degree of polymerisation* (i.e. Q with α and β decreasing as Q increases, ($\alpha \downarrow$ as $Q\uparrow$)) and (ii) the *bond strength* of the M–O bonds with α decreasing as field strength (z/r^2) of the cation increases, where z = cation charge (e.g. 1 for Na^+ and 2 for Ca^{2+}).

The thermal expansion coefficient of a glass undergoes a threefold increase for $T > T_g$ when it transforms from a glass to a supercooled liquid (*scl*); this can be seen as an abrupt change of slope for the lowest curve in Fig. 9.41b. The sample collapses at the deformation temperature (which is ca. 50 °C above T_g) and no measurements are possible above this temperature. Crystalline samples do not undergo these changes but, instead, exhibit a sharp volume increase on fusion between T_{sol} and T_{liq}. Thus, the densities in the range between T_g and T_{sol} vary according to the degree of crystallisation developed in the sample (Fig. 9.8b).

9.4.7.3 Measurement Methods

Solid slags

Density

For solid slags at room temperature the density is usually measured by the *Archimedian* method. The slag sample, of known weight and volume, is suspended on a wire from a balance and the weight of the sample is determined in air (W_{atm}) and then in water (W_{fluid}) (or a more dense liquid, e.g. bromoform or acetylene tetrabromide, ρ_{25} = 2887 and 2953 kgm^{-3}, respectively). The density is then calculated from Eq. 9.38.

$$\rho = (W_{atm} - W_{fluid})/V \tag{9.38}$$

Densities at room temperature can also be determined by the Sink/Float method in which the sample either floats or sinks in standard liquids (like bromoform or acetylene tetrabromide) of known density.

Thermal expansion—Dilatometry

Thermal expansions of slags are usually measured by machining two parallel faces on the sample (usually *ca.* 30 mm long) and measuring the changes in length as the specimen is heated at a known heating rate. For glassy samples, the specimen collapses at the "*softening point*" which is *ca.* 50–100 °C above the glass transition temperature (T_g) and no further measurements can be made.

Liquid slags

The following methods are frequently used to measure the density of molten mould slags and fluxes; the experimental uncertainty is ca. 2–5%.

Archimedian (or buoyancy) method

A bob of known volume is suspended from a balance. The weight, before immersion (W_{atm}) and then after full immersion (W_{fluid}) in the molten slag, is measured. It is necessary to correct the density value for the effect of the surface tension forces acting on the wire [3].

Pyknometry (or Weighing) method

The slag is melted in a specially designed vessel of known volume and the mass of the slag is determined. Corrections must be made for the thermal expansion of the vessel.

Maximum bubble pressure (MBP) method [3, 254]

The MBP method has been described above (Sect. 9.4.6.3). The MBP (p_{MBP}) is determined at different depths of immersion in the molten slag (h_1, h_2, etc.) and the density is calculated by Eq. 9.39 or from the gradient of the plot of p_{MBP} as a function of h [3, 254].

$$\rho = (p_{MBP2} - p_{MBP1})/(h_2 - h_1) \tag{9.39}$$

Sessile drop method

The volume of a sessile drop of known mass is determined through measurements of the drop dimensions [3] Recent improvements in software to determine the profile of the drop have improved the accuracy of this method.

Draining crucible

This technique has not been used to date, on molten slags but would appear to be viable for the measurement of mould slag densities. The rate of drainage of slag from an orifice in the crucible is determined by continuous weighing of the crucible. The density is derived through hydrodynamic analysis of the measurements [164].

9.4.7.4 Density Data for Mould Slags

The reported density data for solid and liquid mould slags are given in Table 9.10 and for the thermal expansion coefficient in Table 9.11. For liquid mould fluxes, the density values lie mostly in the range ρ = (2500–2700) kgm^{-3}. The reported density and thermal expansion coefficient measurements are shown in Fig. 9.8b. There are few measurements of density and thermal expansion. This is quite surprising, given that these measurements are relatively easy to perform and are subject to low levels of experimental uncertainty.

It should be noted that Olivares et al. [262] annealed glass samples at 600 °C and found that the density values of annealed samples were lower than those for

Table 9.10 Density values reported for mould fluxes

Specimen	Reference	Method	Values	($d\rho/dT$)	Comments
Bed powder	Taylor [209]	V & m	ρ_{25}: 600–1000		
	Andersson [61, 62]	V & m	ρ_{bulk} = 700–1100		ρ_{bulk} increases with decreasing particle diameter
Glass					
	Taylor [209]	Arch	ρ_{25} = 2750 ± 50		
	Olivares [262]	SF	ρ_{25} = 2820–2920		2 mould slags + syn slags 39−46C + 38−48S + 6−16A + 6−16N + 4−16 Fl; 25 °C; Annealed 600 °C
Slag film					
	Susa [53, 227]	Arch	ρ_{25} = 2730–2940 ± 50		5 samples
Liquid					
	Monaghan [224]	MBP	ρ_{1300} = 2580 ± 50	–0.2	(1200–1450 °C)
	Matsushita [263]	LD-ESL	ρ_{1300} = 2580	−0.298	Supercooled (1150–900 °C)
	Matsushita [263]	SD	ρ_{1200} = 2850–2700		Seems high; Gas evolution- reaction with C plaque?
	Quested [264]	MBP	ρ_{1300} = 2650 ± 100		
	Konovalov [255]	MBP	ρ_{1550} = 2200–2400		(C/S) = 0.5–2.28
	Duberstein [254]	MBP	ρ_{1400} = 2740–2700		($d\rho_{1400}/d\%Na_2O$) = −8

V & m Volume and mass; *Arch* Archimedian; *SF* Sink of float; *LD* Levitated drop; *SD* Sessile drop

Table 9.11 Experimental values reported for the thermal expansion of mould fluxes

Specimen	Reference	Method	10^7 α; K^{-1} (25–500 °C)	Comments
Glass				
	Mills [132]	Dilatom	9–13	13 glassy mould slags
	Courtney [124]	Dilatom	6.5	Glassy mould slag
Slag film				
	Courtney [124]	Dilatom	8.5 and 11	$\{(dL/dT)_{T<Tg}/(dL/dT)_{T>Tg}\} = 27/12 = 2.3$
Part-crystalline				
	Mills [132]	Dilatom		α is *ca.* 10% higher than that for glass
	Courtney [124]	Dilatom	8.5	1 crystallised sample annealed 15 h at 900 °C

Dilatom. Dilatometry

un-annealed samples; this may be due to densification but could also be due to crystallisation of the sample. It should also be noted that crystallisation results in some porosity and thus slag films will have density values below that of the theoretical density.

9.4.7.5 Calculation of Densities and Thermal Expansion Coefficients of Mould Slags

The models reported for the calculation of density and thermal expansion from chemical composition are given in Table 9.12.

9.4.8 Heat Capacity (C_p) and Enthalpy (H_T–H_{298})

9.4.8.1 Importance of C_p and Enthalpy to the Process

Heat capacity data are needed for carrying out heat balance calculations in the mould.

Table 9.12 Outlines of models to calculate density and thermal expansion coefficient of mould slags

Model/references	Details of model	Uncertainties
Density		
Mills [143, 257]	Partial molar approach, $V_{1773} = \Sigma\ X_1V_1 + X_2V_2 + X_3V_3 +$ ... at 1500 °C, V ($10^{-6}m^3mo^{l-1}$) values for: CaO = 20.7; FeO = 15.8; Fe_2O_3 = 38.4: MnO = 15.6; MgO = 16.1; Na_2O = 33; K_2O = 51.8: TiO_2 = 24; P_2O_5 = 65.7; $SiO_2 = (19.55 + 7.97\ X_{SiO_2})$ $Al_2O_3 = (28.3 + 32X_{Al_2O_3} - 31.45X^2_{Al_2O_3})$ $(dV/dT) = +\ 0.01\%K^{-1}$	±2%
Olivares [262]	Unannealed : ρ_{25} (kgm^{-3}) = 2930 +4.1%C − 4.37%S − 2.99%A − 4.75%N + 29.9%Fl Annealed: ρ_{25} (kgm^{-3}) = 2930 +3.7%C − 3.4%S − 2.2%A − 3.5%N + 29%Fl	
Bottinga [257, 265]	Partial molar deals with slags with high Al_2O_3 contents	
Persson [266, 267]	Relates ρ to thermodynamic enthalpy of mixtures; Applied to binary systems and then to ternary and higher systems. CaF_2 not included.	
Fluegel [268, 269]	Numerical analysis of glass ρ_{25} database; only low concentrations of Fluorine covered	±0.5–3%
Fluegel [270]	Numerical analysis of ρ_T database for liquid phase glasses, 1200–1400 °C; no F % included	
Priven [271]	Based on MDL SciGlass database for glasses-does not include F%.	
Tokuda [163]	Data given to calculate V and (dV/dT) and hence ρ_T	
Thermal expans. coefficient		
Mills 15 [143]	Solid $10^6\alpha\ (20 - T_g) = 10$; Liquid; $(1/\alpha') = 0.25\ Q + 0.4\ (z/r^2)$ where $\alpha' = 10^6\ \alpha$	
Fluegel [269]	Numerical analysis of TEC data at 210 °C for glasses; no CaF_2 included.	
Fluegel [270]	Numerical analysis of TEC database for liquid phase glasses, 1200−1400 °C; no F % included	
Stebbins [272]	Partial molar approach: (dV/dT) for liquid	

Units ρ: kgm^{-3}: α: K^{-1}

9.4.8.2 Factors Affecting C_p and Thermal Enthalpy

The C_p of glassy samples of mould fluxes exhibit a "*step-like*" increase at T_g which is associated with the transformation of a frozen glass to a supercooled liquid (*scl*); this can be seen in the dashed line in Fig. 9.42a [53]. The twin valleys in the apparent C_p values in Fig. 9.42b indicate that further crystallisation of the sample occurs above 500 °C in this sample. The decrease in C_p is not a true effect but is due to the enthalpy of crystallisation (ΔH_{crys}) which is exothermic but which

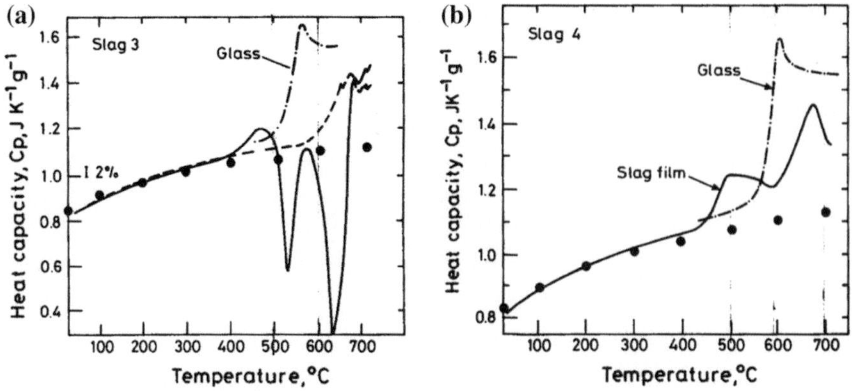

Fig. 9.42 Apparent heat capacity as a function of temperature for slag film and glassy specimen derived from slag film; ——slag film; -·-·-·- = glassy specimen; - - - - = extrapolation for slag film; ● = estimate **a** slag 4 showing step-increase in C_p of, glass at T_g; and **b** slag 3 showing an apparent decrease in C_p resulting from crystallisation (permission granted, Taylor and Francis, www.tandfonline.com [53])

manifests itself as an apparent decrease in C_p; Values for ΔH_{crys} can be derived by integrating the area under the curve.

These effects result in different C_p–T curves for glasses and crystalline specimens in the region between T_g and T_{sol} and the actual C_p and enthalpy values will fall between the two curves and will depend upon the fraction of the slag which has crystallised (see Fig. 9.8a).

9.4.8.3 Measurement Methods

Values of C_p and enthalpy can be determined using a variety of calorimeters but many of the traditional, calorimetric methods have been replaced by more rapid techniques, especially for measurements at high temperatures. However, in practice, the principal methods employed are differential scanning calorimetry and drop calorimetry [273], or a combination of both. Experimental uncertainty is about ±2% with these techniques.

Differential Scanning Calorimetry [132, 274]

The principles underlying DSC are similar to those given above (in Fig. 9.19 [132, 274]). The sample (in the form of a disc or loose powder) is placed in one crucible and the other crucible is kept empty. The instrument is then heated at a set heating rate. The sample requires more energy than the empty crucible and thus the temperature of the sample crucible lags behind that of the empty crucible. This temperature difference is monitored. Three sets of measurement runs are made with the sample crucible (i) empty (ii) filled with a known mass of sample (m_{sample}) and (iii) with a sapphire (Al_2O_3) disc of known mass ($m_{Al_2O_3}$). The $C_{p\ sample}$ can be

Table 9.13 Details of investigations to measure C_p and enthalpies of mould fluxes

Specimen	Reference	Method	T °C	C_p Values $JK^{-1}\ kg^{-1}$	Comments
Glass	Mills [132]	DPSC	25 T_g	(800–880) {1050–1150}	10 samples (25–700 °C)
	Susa [53]	DPSC	T_g	{1100–1200}	5 samples made from slag films
	Mills [227]	DSC	T_g	{1250}	1 sample (500−720°C)
	Courtney [124]	DSC	T_g	{1100}	1 sample (450–720 °C)
Slag film	Susa [53]	DSC	25 T_g	(800–850) ca. {1100}	5 slag films
	Courtney [124]	DSC	T_g	{1100}	$C_p(scl)$ = ca. 1.5 $JK^{-1}\ kg^{-1}$
	Mills [125]	DSC	T_g	{1250}	1 slag film (500–720 °C); $C_p(scl)$ = ca. 1.5 $JK^{-1}\ kg^{-1}$
Liquid					No data available

() = $C_{p25\ °C}$. {} = C_{pTg}

calculated from the difference in signals (S_{sample}'–S_{empty}') and for ($S_{Al_2O_3}'$–S_{empty}') and m_{sample} and $m_{Al_2O_3}$ [274]. This type of DSC is usually denoted as differential *temperature* scanning calorimetry (DTSC). However, there is a second type of DSC in which the power required to keep the two crucibles at the same temperature is monitored (differential *power* scanning calorimetry, DPSC) [273].

Drop calorimetry [273]

The sample, of known weight, is placed in a Pt crucible and heated to the required temperature and held there until the temperature has equilibrated. The gate separating the furnace from the calorimeter is then opened and the sample is dropped into a massive, silver (or copper) calorimeter held at constant temperature. The temperature rise (<5 °C) is then measured accurately and converted into enthalpy using the mass and heat capacitance of the calorimeter. A series of blank experiments are carried out with the empty crucible and the enthalpy of the blank for the measuring temperature is then subtracted from that for the (crucible+sample). Values of the C_p of the sample at 25 °C are needed to differentiate the enthalpy values to obtain C_p values; the most accurate data are obtained by combining the drop-calorimetry data with C_p values obtained by DSC. Rapid quenching produces a glassy specimen which is in a metastable state; corrections are needed to correct the measured values, so they pertain to the reference (crystalline) state.

9.4.8.4 C_p and Enthalpy Data for Mould Slags

Values of the C_p have been reported for mould slags at temperatures between (25–720 °C); these are given in Table 9.13 and details of the investigations are given in

Table 9.14 Details of published models to calculate C_p and enthalpies (H_T–H_{25}) of mould fluxes

Model/reference	Details of model	Uncertainties
Partial molar model mills [33, 143, 144]	$C_p = a' + b'T + c'/T^2$. where, a', b' and c' are constants and are calculated by $a' = \Sigma (X_{CaO}\, a'_{CaO}) + (X_{SiO_2}\, a'_{SiO_2}) + (X_{Al_2O_3}\, a'_{Al_2O_3}) + (X_{CaF_2}\, a'_{CaF_2}) +$ Similarly with b' and c'.··· C_p of liquid by $C_{p\,liq} = \Sigma (X_{CaO}\, C_{p\,liqCaO}) + (X_{SiO_2}\, C_{pliqSiO_2}) + (X_{Al_2O_3}\, C_{p\,liq\,Al_2O_3}) + (X_{CaF_2}\, C_{p\,liq\,CaF_2}) +$ ···· $(H_T–H_{298}) = \int_{298}^{T} C_p\, dT = a(T–298) + 0.5bT^2 - 0.5b(298)^2 + (c/T) - (c/298)$ Enthalpy of fusion (ΔH^{fus}) calculated from entropy of fusion (ΔS^{fus}) where T is in K $\Delta S^{fus} = \Sigma(X_{CaO}\, \Delta S^{fus}_{CaO}) + (X_{SiO_2}\, \Delta S^{fus}_{SiO_2}) + (X_{Al_2O_3}\, \Delta S^{fus}_{Al_2O_3}) + (X_{CaF_2}\, \Delta S^{fus}_{CaF_2}) +$ ···· $\Delta H^{fus} = T_{liq} \Delta s^{fus}$	±2%
Stebbins [272]	Partial molar approach to C_p and ΔS^{fus}; Based on mineral and glass database; no CaF_2.	
Thermodynamic models [275–277]	Commercial thermodynamic models, such as Factsage; MTDATA and Thermocalc, provide accurate thermodynamic data for slag systems	

Table 9.14. The C_p values for a range of mould slags show little variation; C_p values at 25 °C and T_g (*ca.* 600 °C) are typically $C_{p\,25\,°C} = 840 \pm 40$ and $C_{pTg} = 1150 \pm 50$ JK^{-1} kg^{-1}. The C_p–T relationships are complex for temperatures above 550 °C since (i) the sample transforms to a supercooled liquid (*scl*) at T_g which usually occurs at 600 ± 30 °C and (ii) crystallisation occurs, usually at about 50 °C above T_g but which can occur as low as 550 °C for some slag compositions. Crystallisation is accompanied by an apparent sharp decrease in C_p which is not a true C_p value since the enthalpy of crystallisation manifests itself as an apparent decrease in C_p; the enthalpy of crystallisation can be obtained by integrating the area below the C_p–T baseline. No values were found for the liquid phase. However, estimated values for the solid phase were found to be in good agreement with the measured values for the solid phase and it is known that Kopp's Law provides good estimates for C_p of slags, in general.

9.4.8.5 Calculation of C_p and Enthalpy

The models reported for estimating heat capacity (C_p) and enthalpy (H_T–H_{25}) are given in Table 9.14. The predicted C_p values are compared against experimental values in Fig. 9.42.

Table 9.15 Summary of the investigations of optical properties for mould fluxes and slag films

Property/reference	Phase	Method	Values	Comments
Refractive index (n)	*Glass*		*n*	
Susa [53]		Abbe R	1.57–1.60	
Ozawa [55]		Abbe R	1.55–1.57	3 slags; 25 °C; λ = 633 nm; $n\downarrow$ with $\uparrow$% CaF_2 and (dn/dT) is negative
	Liquid			
Firoz [278]				42C + 42S + 6A + 5N + (0–12.5% CaF_2) (1127–1400 °C) (dn/dT) = ca. −0.0008; $n\downarrow$ as % $CaF_2\uparrow$
	Slag film			
Susa [206]		Abbe R	1.57–1.61	
Susa [53]		Abbe R	1.57–1.60	5 slag films; λ = 0.546 μm
Holzhauser [279]			1.60	Slag film used in simulation experiment
Yamauchi [282]				
Absorption coeff. (α)*	*Glass*		α*	
Mills [132]		Spect	100–500	7 glassy slags; α*at 25 °C;
Susa [206]		Spect	550–1800	2 slag films; (0.5–5 μm) Effect of various oxides; $\Delta\alpha^*$ (m^{-1}) = k%; k(FeO) = 910; k(NiO) = 410; k(MnO) = 5;
Ohta [212] Waseda [210]		Spect	200–2000	At 250C; λ = (2–5 μm) Effect of TiO_2, $ZriO_2$ $HfiO_2$, FeO
Diao [280] Chen [281]				Effect on α*; of FeO; MnO: TiO_2 [280]; Co_2O_3 [281]
Shibata [211] Cho [59]		Spect	500–1000	7 slag samples, (25–580 °C); λ = (2–5 μm); α* increases as Temperature increases
Ozawa [55]		Spect	200–500	3 slag samples; (25 °C); λ = (0–2500 μm)
Extinction coeff (E)	*Partially crystalline*		*E*	
Susa [53]		Spect.	2000–15000	5 slag films and glasses; Effect of Fe_2O_3 on *E* & α*
Susa [206]			2000–5000	
Susa [56]				λ = (0–2.5 μm) A' unaffected by D_{crys}, but $A'\downarrow$ with $D_{crys}\uparrow$ for D_{crys} > 2 μm
Shibata [211]		Spect	15000–35000	7 slags; 25–580 °C; λ = (2–5 μm); *E* i$\uparrow$ as $T\uparrow$

(continued)

Table 9.15 (continued)

Property/reference	Phase	Method	Values	Comments
Ozawa [55]		Spect	20000–40000	3slags: λ = (0–2500 μm); annealed at varius temps
Reflectivity (R′)	*Partially crystalline*		R'	
Susa [56]		Spect-IS		λ = (0–2.5 μm) Effect of crystal diam (D_{crys}); T′↓ as D_{crys}↑; D_{crys} = 1.2–2.2μ m
Susa [110]		Spect-IS		λ = (0–2500 μm); effect of 0–2% Fe_2O_3 and f_{crys} on R' R'↑ as f_{crys}↑ and Fe_2O_3↓; Effect↑ with λ↑
Transmissivity (T ′)	*Partially crystalline*		T'	
Susa [56]		Spect-IS		λ = (0–2.5 μm) Effect of crystal diam (D_{crys}) T'↓ as D_{crys}↑; D_{crys} = 1.2–2.2 μm
Susa [110]		Spect-IS		λ = (0-2500 μm); effect of 0-2% Fe_2O_3 and f_{crys} on T' T'↓ as f_{crys}↑ and Fe_2O_3↓; Effect↑ with λ↑
Emissivity (ε)	*Liquid*		ε	
Keene in [132]		cfBB	0.89	1 slag (1280–1450 °C)
	Solid			
Keene in [132]		cfBB	0.91	2 slags (1080–1260 °C)
Seko [283]		Spect-IS	0.38–0.60	Crystalline slag

λ wavelength; *Abbe R* Abbe refractometer; *cfBB* Comparison with black body; *Elip* Elipsometry; *IS* Integrating sphere; *Spect* Spectrophotometer

9.5 Optical Properties of Mould Slags

Radiation conduction is a major contributor to the heat flux at high temperatures. The radiation conductivity, k_R can be calculated for *optical thick conditions* ($\alpha^* d > 3$, where d = sample thickness, α^* = absorption coefficient (m^{-1}), σ = Stefan–Boltzmann coefficient and n = refractive index and T_K is in K).

$$k_R = 16\sigma n^2 T_K^3 / 3\alpha^* \tag{9.40}$$

Values for the various optical constants have been determined by various investigators and details of these studies are given in Table 9.15.

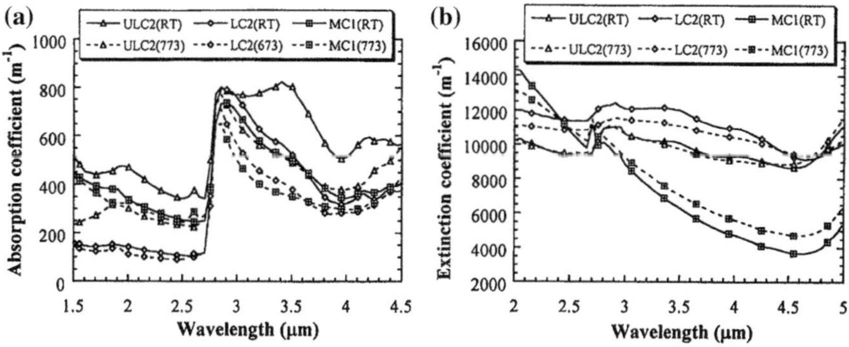

Fig. 9.43 Values of **a** Absorption coefficient and **b** Extinction coefficient at 25 °C and 500 °C, respectively, of various mould slags as function of wavelength; *LC* Low C; *MC* Medium C and *ULC* Ultra low C steels (permission granted, ISIJ [59])

9.5.1 Refractive Indices (n) [53, 55, 206, 278, 279]

Values of the refractive index have been reported for the glass, slag film and the liquid phases of mould slags (Table 9.15). The reported values range from 1.55 to 1.61 and a mean value, $n = 1.58 \pm 0.03$ is recommended.

9.5.2 Absorption Coefficients (α*) [53, 55, 56, 59, 110, 206, 211, 212, 280, 281]

The magnitude of k_R can be reduced in slag films by introducing either transition metal oxides to absorb, or crystals to scatter, IR radiation. Values of the absorption coefficient, α^*, of glasses and glassy mould slags are significantly increased by the presence of transition metal oxides in the slag (α^* (m^{-1}) = α^0 + K%MO where α^0 = absorption coefficient with 0% transition metal oxide (MO) where K has values of 910, 5, 410 m^{-1} for FeO, MnO and NiO, respectively, and $K_{Cr_2O_3} > K_{FeO}$ [53, 206]. The effects of other transition metal oxides on α^* for mould slags has also been studied, FeO, MnO and TiO_2 [212] and Co_2O_3 [281].

The absorption coefficient pertains to liquid and glassy phases but for crystalline or partially crystalline materials it is necessary to use the extinction coefficient, E (=α^* + s, where s = scattering coefficient). It is not possible to use Eq. 9.40 for partially crystalline materials since the scattering mechanism is different to that for absorption. In order to use Eq. 9.40 for crystalline slags it is necessary to have values for the *albedo* (*or reflection coefficient*) which is used to link scattering to absorption. Unfortunately, we have little data available for the albedo.

When IR radiation strikes a partially crystalline slag film, only a fraction of the energy is transmitted, since energy is also absorbed and reflected; Eq. 9.41 applies where A', R' and T' are absorptivity, reflectivity and transmissivity, respectively.

$$A' + R' + T' = 1 \tag{9.41}$$

From the Lambert–Beer Law, the transmissivity is given by the following equation, where d = sample thickness.

$$T' = \exp\{-(\alpha^* + s)d\} \tag{9.42}$$

Susa et al. [56, 110] measured A', R' and T' for two mould slags and reported that:

- Transmissivity decreased and reflectivity increased as the fraction of crystalline phase (f_{crys}) increased [110] (R' ↑and T'↓ as f_{crys} ↑).
- The addition of 1–2% FeO_x causes A' to increase but R' to decrease [110] due to the promotion of the glassy phase by FeO_x.
- R' increases and T' decreases with increasing grain size [56] (R' ↑and T'↓ as D_{crys} ↑); a grain size of 2–3 μm is recommended for the lowest q_{hor} [56].
- For casting MC, peritectic steels, the use of a mould flux free of FeO_x is recommended since it leads to a lower q_{hor} [56].

A summary of investigations of optical properties for mould fluxes and slag films is given in Table 9.15.

Typical absorption coefficients and extinction coefficients are shown in Fig. 9.43a, b, respectively, for mould slags used in the casting of different steel grades [53, 211]; it can be seen that the Extinction coefficients:

(i) Are considerably higher than the absorption coefficients.
(ii) Are particularly high for the mould slags used for casting medium carbon (MC) steels which tend to have a high basicity and which results in high values of f_{crys}.
(iii) Calculations using the absorption coefficients and assuming a slag film of 2 mm, indicate that the initial glassy slag film would be *optically thin* [284] i.e. the k_R value would be lower than that calculated from Eq. 9.40.

9.5.3 Reflectivity, Transmissivity and Emissivity

The published investigations for these properties are summarised in Table 9.16. It can be seen that the reflectivity (R') increases with:

(i) increasing fraction crystalline (f_{crys}) and increasing crystal diameter (D_{crys})
(ii) decreasing FeO content, since FeO promotes glass formation which tends to offset any gains from FeO increasing the absorption coefficient.

Table 9.16 Models and routines to calculate optical properties

Property/reference	Details of model	Uncert
Ref index, *n*		
Susa [56]	$\{(n^{-1})/10^{-3}\ \rho\} = 0.001$ k%; *k* values; CaO = 2.25; MgO = 2.0; K_2O = 1.89; Na_2O = 1.81; FeO = 1.87; MnO = 2.24; Cr_2O_3 = 0.22; Al_2O_3 = 2.14; SiO_2 = 2.07; TiO_2 = 3.97	±1%
Priven [271]	Model for glasses	
Matsushita [2, 287]	Review of models for calculating optical properties	
Abs, coeff. α^*		
Susa [56]	$\Delta\alpha^*$ (m^{-1}) = k %; *k* values: FeO = 910; NiO = 410; MnO = 5; $\Delta\alpha^* = 390(\%Cr_2O_3) + 370(\%Cr_2O_3)^2$	

Increased reflectivity reduces the Transmissivity.

Several models, or routines, have been reported to estimate the optical properties of mould slags; these are summarised in Table 9.16.

9.6 Thermomechanical Properties of Mould Slags

Slag rims and slag films can crack [198] and they must be repaired or replaced by the infiltration and solidification of molten slag. Slag films tend to crack in the lower half of the mould when liquid lubrication is low. The replacement process can take up to 20 min because of the low flow rates of slag infiltration. Slag films taken from the mould were found to show signs of fracture "*along a line of bubbles*" in the crystalline regions of the slag film [198]. Since crystallisation is accompanied by the creation of pores, it seems reasonable to assume that the "*line of bubbles*" was formed during crystallisation. Glassy slag films are less likely to fracture than crystalline slag films.

9.6.1 Thermomechanical Tests

The thermomechanical properties of molten slag have been measured in a test, shown in Fig. 9.44, where the mould powder is sandwiched between two crucibles and the displacement is measured as the upper crucible is lowered [285]. On heating, the rod was displaced gradually with increasing temperature; the displacement was observed to increase more rapidly when the temperature exceeded the deformation temperature (or softening point) temperature and increased sharply when $T > T_{liq}$.

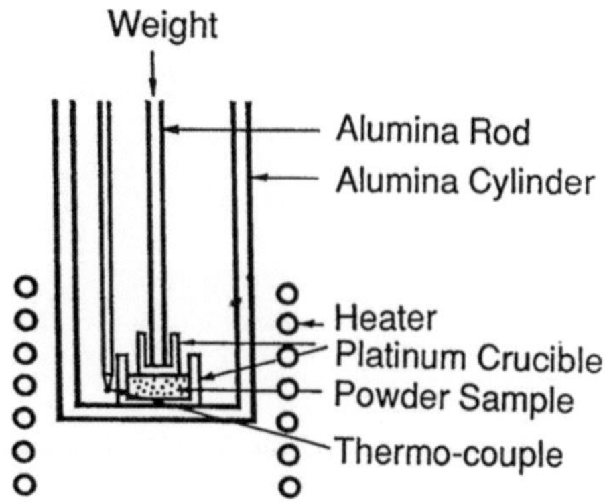

Fig. 9.44 Schematic diagram of the thermo-mechanical analyser (Permission granted, ISIJ [285])

A finite-element model has been developed to predict (a) shear stress (b) friction (c) slip and (d) fracture of slag film [286]; the crystalline phase was predicted to fracture in the meniscus region and the glass phase near the mould exit [286].

9.6.2 *Stress Relaxation*

The molten slag in the mouth of the infiltration channels is subjected to compression as a result of the ferrostatic pressure. This could affect the lubricating properties. Consequently, the stress relaxation of four mould slags were investigated [287] in which a sessile drop of liquid mould slag was compressed between two Pt plates, in an Instron machine, to 50% compression and the force measured. The compression was then terminated and the slag allowed to undergo stress relaxation. The stress attenuated until a constant stress value was attained; the time taken to reach this point, the relaxation time, was measured. It was found that the relaxation time increased with increasing polymerisation (or viscosity) of the slag [287].

9.7 Dissolution of Oxides, Nitrides and Carbides in Mould Slags

Non-metallic inclusions (such as Al_2O_3 or TiO_2) have a detrimental effect on the high-temperature, mechanical properties of steels. This loss in mechanical strength increases with increasing concentration and size of the inclusions. Thus, steelmakers go to considerable length to minimise inclusion concentrations; the continuous-casting mould provides one last chance to remove inclusions.

9.7.1 Origin of Inclusions

There are two types of inclusions found in steel:

- Oxides, sulphides, nitrides and carbides formed in steelmaking and from metal/slag reactions.
- Entrapped mould slag resulting from metal flow turbulence (Sect. 11.11).

The oxides and nitrides formed in steelmaking occur through reactions such as:

$$2\underline{\mathrm{Al}} + 3\underline{\mathrm{O}} = \mathrm{Al_2O_3}(s) \tag{9.43}$$

$$\underline{\mathrm{Ti}} + \underline{\mathrm{N}} = \underline{\mathrm{TiN}}(s) \tag{9.44}$$

$$\underline{\mathrm{Mn}} + \underline{\mathrm{S}} = \mathrm{MnS}(s) \tag{9.45}$$

where the underline indicates it is dissolved in the steel.

Oxides can also be formed by reactions between metal and slag (in the slag pool), e.g.

$$4\underline{\mathrm{Al}} + 3\,\mathrm{SiO_{2sl}} = 3\underline{\mathrm{Si}} + 2\,\mathrm{Al_2O_{3sl}} \tag{9.46}$$

where the subscript *sl* indicates that it is in the slag phase. The above reaction is not restricted to SiO_2, since similar reactions occur with FeO, MnO, NiO, CrO, B_2O_3 etc.

Ehrenberg [288] proposed the following reaction:

$$3\mathrm{CaO_{sl}} + 3\underline{\mathrm{S}} + 2\underline{\mathrm{Al}} = 3\mathrm{CaS(s)} + \mathrm{Al_2O_3}(s) \tag{9.47}$$

Typically, the pick-up of Al_2O_3 is 4%. It can be seen from Fig. 9.45 that the pick-up arises from both steelmaking reactions and metal/slag reactions [289]. However, much higher Al_2O_3 pick-up occurs (up to 35%) when casting High-Al (Trip and Twip) steels containing ca. 1% Al. It can be seen that the Al_2O_3 pick- up due to slag/metal reactions is variable; this is due to variations in (i) the Al content

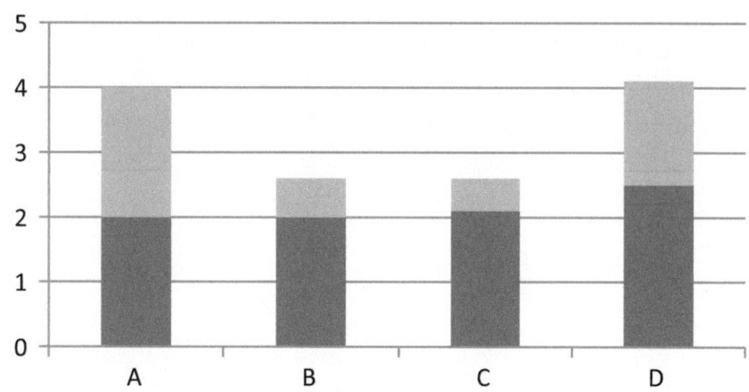

Fig. 9.45 Alumina pick-up (mass%) resulting from absorption of steelmaking reactions (*light grey*) and metal/slag reactions (*dark*) for four mould powders (permission granted, ISS/AIST re-drawn after [289])

of the steel and (ii) the concentrations of the more easily reduced oxides in the mould slag (e.g. FeO, MnO, SiO_2).

Kiyose et al. [290] developed a mathematical model which calculated the compositional changes associated with metal/slag reactions (Fig. 9.46) for steels containing Ti which reacts with SiO_2 in slag; they claim that these changes can result in the formation of cracks.

9.7.2 *Mechanism of Inclusion Removal*

For an inclusion to be successfully dissolved by the slag phase, several stages are involved:

1. Transport of inclusion to the slag/metal interface.
2. It must satisfy the interfacial requirements for passage into the slag phase.
3. Dissolve in the slag phase.
4. It must be transported away from dissolution site.

9.7.3 *Transport of Inclusions to the Slag/Metal Interface*

Most inclusions (e.g. Al_2O_3) have lower densities than molten steel and so can be removed by flotation. The velocity of the inclusion, (V_I) as it moves towards the slag pool, is given by Stokes Law (Eq. 9.48) where r = inclusion radius, g = gravitational constant, ρ = density and η = viscosity (in Pas) and subscripts, m and I denote the metal and inclusions, respectively.

$$V_I = 2\,r_I^2(\rho_I - \rho_m)/9\eta_m. \tag{9.48}$$

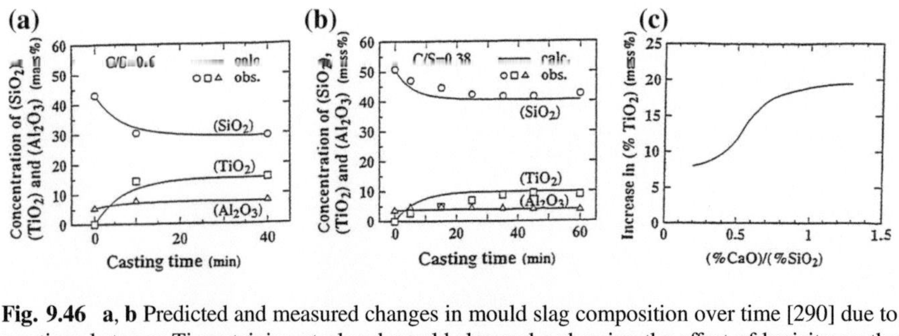

Fig. 9.46 **a**, **b** Predicted and measured changes in mould slag composition over time [290] due to reactions between Ti-containing steel and mould slag and **c** showing the effect of basicity on the the magnitude of TiO_2 pick-up (Permission granted, ISIJ [290])

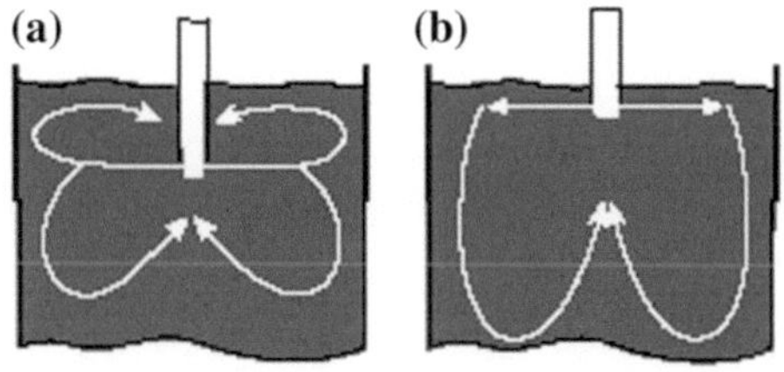

Fig. 9.47 Schematic diagrams showing metal flow in the mould **a** *double-roll* and **b** *single-roll* patterns; (permission granted, UNESID [291])

However, as we have seen in Sect. 5.3, the metal flow in the mould can be very vigorous and can move inclusions towards, or away, from the metal/slag interface. Furthermore, the flow pattern of metal flow is important; it can be seen that the "*single roll*" will initially tend to bring the inclusions towards the metal/slag interface before sweeping down the mould whereas the "*double roll*" (favoured in many continuous-casting operations) will tend to drag some of the inclusions away from the interface (Fig. 9.47).

The transport of inclusions to the interface is promoted by Argon bubbling and by the use of electromagnetic devices in the mould.

9.7.3.1 Argon Bubbling

Argon stirring in the mould is used to remove inclusions and to minimise clogging in the SEN. The attachment of inclusions to a bubble involves the interfacial properties of inclusion, metal and gas (denoted by subscripts *I*, *m* and *G*, respectfully). The following equation is derived from a balance of forces (Fig. 9.48) where θ is the contact angle.

$$\gamma_{Im} + \gamma_{mG} \cos\theta - \gamma_{IG} = 0 \tag{9.49}$$

Good flotation with the bubble occurs if the flotation coefficient (Δ, defined in Eq. 9.50) is positive and has a high value. Thus good inclusion removal occurs when (i) θ has a high value (i.e. is non-wetting) and (ii) the surface tension of the

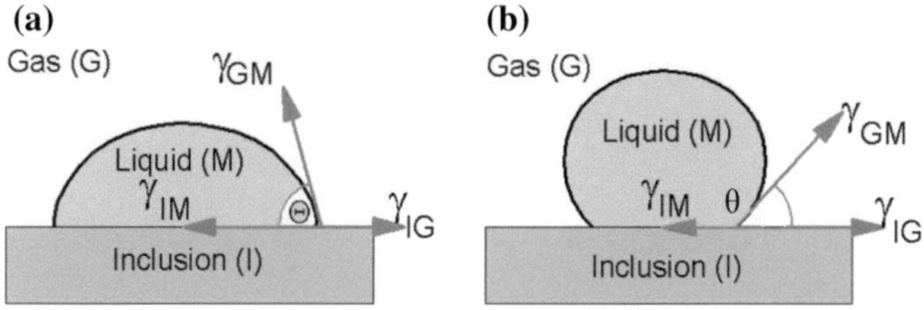

Fig. 9.48 Schematic diagram showing the balance of interfacial forces for **a** a wetting system and **b** a non-wetting system

metal (γ_{mG}) has a high value, which corresponds to steels with low sulphur contents.

$$\Delta = \gamma_{Im} + \gamma_{mG} - \gamma_{IG} \quad (9.50)$$

The contact angles (θ) for steel on most inclusions tend to be high (e.g. Al_2O_3, θ = ca. 130°; TiO_2, θ = ca. 80°; ZrO_2, θ = ca. 120°; TiN, θ = ca. 120°; [3, 292] Fig. 9.48). However, any reaction between the inclusions and the steel will cause a decrease in the contact angle (i.e. *reaction wetting*) [293].

9.7.3.2 Electromagnetic Devices

Several electromagnetic devices are used in continuous casting; these devices have different effects on steel cleanliness (i.e. inclusions) [294].

Electromagnetic stirring (EMS using AC* ca. 50 Hz*)

Inclusions and gas bubbles can be removed from the steel by using electromagnetic stirring (EMS). EMS of the steel is achieved by the interaction of a magnetic field (produced by a static induction coil) with the electrically conducting liquid metal. The magnetic field can be applied either to the mould (M-EMS) or to the strand (S-EMS). The principal benefits of EMS are that it provides (i) homogenisation of the melt (ii) a refined microstructure (iii) inclusion removal [294, 295] and (iv) has been reported to reduce the number of longitudinal cracks. The inclusions are removed because small particles, in a shear field, experience a (Saffman) lift force perpendicular to the direction of the flow. The inclusions are removed if the Saffman velocity exceeds the velocity of the solidification front [296]. The Saffman force increases as the size of the inclusion increases. When EMS is applied, the Saffman forces generated were found to successfully remove inclusions and gas bubbles with diameters greater than 100 μm [294, 295, 297].

Electromagnetic braking (EMBr)-DC

Electromagnetic braking (EMBr) exists in several forms. The original EMBr was developed to reduce the velocity of the metal flow (V_{metal}) in the SEN region by using a DC static magnetic field [294, 295, 298]. Initially EMBR was applied around the SEN region of the mould to reduce the velocity of the metal flow leaving the SEN; a reduction of 40% in V_{metal} has been reported [299].

Level magnetic field (LMF)-DC

However, another form of EMBr has been developed in which the DC magnetic field is applied across the whole width of the mould. This is known as *level magnetic field* (LMF) and it is effective in reducing metal flow penetration in the mould and in suppressing the mixing of different steel grades during ladle changes [294, 295]. The reduction in metal velocity and penetration promotes the flotation of inclusions and bubbles.

Flow control (FC)

EMBr has been developed to produce two magnetic fields across the full width of the mould where (i) the upper field reduces the meniscus turbulence and (ii) the lower field decelerates the metal flow, promotes Ar bubble and inclusion flotation [298] and increases the meniscus steel temperature by *ca.* 10 °C [300, 301]. This form of EMBr is often referred to as *Flow control* (FC) [298]. The reduction in metal flow velocity results in a 30% decrease in vertical heat flux and tends to produce shallow slag pools [302]. The velocity of the slag flow is also reduced as a consequence of the decrease in metal flow and this leads to a decrease in slag entrapment (see Chap. 11, Sect. 1.11). The use of EMBr has been reported to reduce (i) the number of inclusions [295, 297, 298, 301, 303–305] (ii) pinholes [298, 301, 304, 305] (iii) mould-level variations [298, 301] and (iv) the size of the slag rim [306].

Electromagnetic casting (EMC-AC high frequency)

In EMC, an AC current is passed through a solenoid coil which surrounds the mould. The resulting Lorenz force creates a horizontal, circulating secondary field which produces an inward-acting force [294, 307]. This pinch force pushes the melt away from the wall and results in reduced horizontal heat transfer. This, in turn, causes delayed solidification of the shell and results in a short meniscus; both of these effects result in a decrease in capture of inclusions by the shell (see Chap. 11–Sect. 1.11). However, in EMC there is a tendency for unstable metal flow in the meniscus region. Pulsative EMC (P-EMC) was introduced to control the metal flow induced by the magnetic field [307]. The application of P-EMC was found to significantly decrease the number of inclusions (N_{incl}) the depth of the the oscillation marks (d_{OM}) and the number of longitudinal cracks (N_{Lcr}) [307, 294].

9.7.4 Transport Through Slag/Metal Interface

It has been suggested that inclusions need a rest period before they can penetrate the metal/slag interface but it was observed that larger particles, with greater momentum, were capable of piercing the interface directly [292].

9.7.5 Dissolution of Inclusions

One of the principal tasks of the mould slag is to dissolve inclusions. The ability of the slag to dissolve inclusions (usually Al_2O_3) has been reported by several workers [95, 96, 98, 100, 102, 185, 186, 308–310].

9.7.5.1 Factors Affecting Dissolution

(i) *Saturated concentration of inclusions*

The driving force for the dissolution of inclusions, (e.g. Al_2O_3) in the slag is the Al_2O_3 concentration difference (C_{sat}–C), i.e. the difference between saturated and actual concentrations of Al_2O_3 in the slag. There is a wide variation in the reported values for C_{sat} for different inclusions (e.g. Al_2O_3: *ca.* 40% and *ca.* 50% at 1400 and 1500 °C, respectively, [311, 312]; TiO_2 *ca* 10% [153] ZrO_2 ca. 2% [313] and TiN *ca* 0.5% [76].

9.7.5.2 Inclusion Blockages

There is a tendency for undissolved particles to accumulate and to agglomerate, especially where (C_{sat}–C) is low (e.g. TiC, TiN, ZrO_2); the agglomeration (or clustering) occurs through collisions of particles [314]. Collisions increase with increasing metal flow velocities. The agglomerates tend to congregate in the mouth of the infiltration channel and close down the space for slag infiltration and hinder the flow of liquid slag into the mould/strand channel, resulting in a decrease in lubrication. For instance, powder consumption tends to be low when casting Ti-stabilised stainless steels due to the accumulation of TiN, Ti(CN) or *perovskite* ($CaTiO_3$) accretions in the infiltration channel. It has been suggested that the amount of $CaTiO_3$ can be minimised by restricting the amount of CaO, i.e. use a mould slag with low basicity [133]. In extreme cases, the accumulations could cause sticker breakouts. The presence of TiC and TiN particles also results in an increase in viscosity [196, 197]. A mathematical model for the clustering of particles has been reported [315].

9.7.5.3 Chemical Composition of the Mould Slag

The dissolution of inclusions is also affected by the mould slag composition; the dissolution of Al_2O_3 in mould slag is promoted by:

- Additions of those oxides which increase the basicity index (Eq. 9.51), e.g. BaO [100].
- Low Al_2O_3 concentrations (C) in the mould slag, i.e. high (C_{sat}–C) [316, 317].

- High fluoride, Li_2O and Na_2O contents in the mould slag [98, 182, 186, 317].
- B_2O_3 was found to have little effect on the slag's ability to dissolve Al_2O_3 [102].
- Low slag viscosity, the rate of dissolution increasing with decreasing viscosity [318, 319].

$$BI = \{1.53\%CaO + \%MgO + \%CaF_2 + 1.94\%Na_2O + 3.55\%Li_2O\} / (1.4\%SiO_2 + 0.1\%Al_2O_3) \quad (9.51)$$

9.7.5.4 Kinetics of Inclusion Dissolution

The kinetics of inclusion dissolution has attracted attention in recent years with the introduction of high-Al steel casting where large amounts of Al_2O_3 are formed; these must be dissolved quickly to avoid casting problems.

The kinetics of Al_2O_3 dissolution have been studied by several workers [308, 311, 312, 318–320]. In these experiments an Alumina disc was rotated in molten mould fluxes for a known time and the rate of Al_2O_3 dissolution was determined by re-weighing the disc. Confocal laser microscopy [321, 322] and thermodynamic calculations [131] have also been used to determine the rate of Al_2O_3 dissolution in mould slag. Values for the diffusion constant (D) of (2–8) and (4–17) $\times$ 10^{-8} m^2s^{-1} were recorded at 1400 and 1500 °C, respectively [311]. Alumina dissolution is promoted by increasing mass flow rates and agitation. The mass flow rate (J) in the rotating disc experiment is given by the following equation where ν = kinematic viscosity ($=\eta/\rho$) of the liquid and ω = angular velocity (in s^{-1}).

$$J = 0.62D^{2/3}\nu^{-1/6}(C_{sat} - C) \cdot \omega^{0.5} \quad (9.52)$$

A mathematical model has been developed to calculate the dissolution time for alumina particles [311]. However, it has been reported that Marangoni flows enhance the effective diffusion coefficients [323, 324]. Metal/ slag reactions are ionic and the dissolution of an Al_2O_3 particle at the steel/slag interface causes local changes in the charge on the slag anions (Eq. 9.53) and gives rise to local, circulatory (Marangoni) flows which enhance mass flow rates and the dissolution of inclusions [323, 324].

$$4\underline{Al} + 3Si^{4+} = 3\underline{Si} + 4Al^{3+} \quad (9.53)$$

Scheller [323, 324] studied TiO_2 transfer from metal to slag and found $D_{eff} \approx 20D$ where D_{eff} is the effective diffusion coefficient. It is customary to use the dimensionless Sherwood number (Sh) to represent mass transfer (in this case, $\Delta TiO_2/Ti$). It was also noted that the ratio ($\Delta TiO_2/Ti$) increased with increasing pool depth (d_{pool}) for pool depths of 0–6 mm but remained constant for $d_{pool} > 6$ mm [323, 324]. Consequently, Scheller proposed that the modified Bodenstein

No. (B_o*) replace the Sherwood No. (Sh) [323, 324] for the description of mass transfer. The Bodenstein No. (B_o*) is given by B_o* = (total mass transferred)/(total mass flux density from diffusion of various species).

$$Bo^* = \left(d_{pool} \cdot m_{sl}\right)/D_i F_{\rho sl})\left(\Delta C_{\Sigma i}/\Delta C_D\right) \tag{9.54}$$

where m is the mass flux density, C = concentration, $F_{\rho sl}$ = inertia force for slag, the subscript, i = species, e.g. Ti or Al and Σ = total (i.e. by convection + diffusion). Values of *log Bo** were correlated with *log (Ra. Re^2/We)* where *Ra* = Rayleigh No (represents convective forces) *Re* = Reynolds No. (=balance of inertial and viscous forces) and *We* = Weber No. [=balance of inertial and interfacial (or surface) forces].

The dissolution process of Al_2O_3, MgO and $MgAl_2O_4$ inclusions in F-free mould slag has also been explored using Confocal laser microscopy; it was found to be diffusion-controlled [321]; typical dissolution times of *ca.* 200 s were found for 150 μm inclusions. A similar conclusion was also found in rotating disc studies in CA-type mould slags (used to cast high-Al steels) where the rate-determining step for the dissolution of Al_2O_3 was found to be the diffusion of Al_2O_3 in the slag boundary layer; the intermediate phase, CA_2 was formed during the dissolution [319]. The rates of dissolution of MgO, Al_2O_3 and MgO · Al_2O_3 in slag were found to be *ca.* 4 times higher than that of ZrO_2 [313].

9.8 Other Tests Used on Mould Powders

A variety of other laboratory tests are used to characterise both the behaviour of mould powders and the performance of the powder bed [232, 325–328]. Some of these methods have also been covered in the individual chapters, e.g. melting rate tests and methods for measuring pool depths are described in Chap. 4, Sects. 4.4 and 4.7, respectively.

9.8.1 Bulk Density

The thermal insulation provided by the powder bed is essential to prevent solid, steel skulls from forming on the surface of the metal. Thermal insulation increases with increasing bulk density of the casting powder and with increased powder bed depth.

Bulk density is usually measured by pouring the casting powder into a beaker of known volume (*V*) and then determining the mass of the powder added [i.e. ΔW = (mass of filled crucible)—(mass of empty crucible)]. The bulk density is calculated from ($\Delta W/V$). The bulk density tends to be in the following hierarchy (powder > extruded > granulated) as can be seen from Fig. 9.49a [232] since smaller particles (i.e. in powders) tend to pack more compactly.

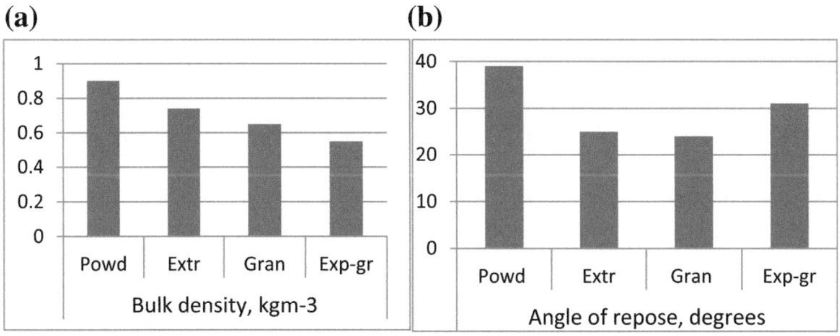

Fig. 9.49 Representations of **a** bulk density and **b** angle of repose as a function of powder type; *Powd* Powder; *Extr* Extruded granules; *Gran* Granules; *Exp-gr* Expanding granules (permission granted; ISS/AIST, re-drawn after [232])

9.8.2 *Flowability*

Spherical granules flow easily and when the metal flow turbulence causes a standing wave to be formed (Fig. 9.50) the spherical granules can flow downhill from the peak and reduce the thermal insulation in the area around the peak. Flowability is particularly important in ingot casting to minimise the formation of "*open eyes*".

A measure of the flowability can derived using the *angle of repose* test [325, 326] in which a known volume of material is allowed to flow into a pile formed on a flat plate and the angle of inclination is measured. Powders with a greater flowability have a smaller angle of repose.

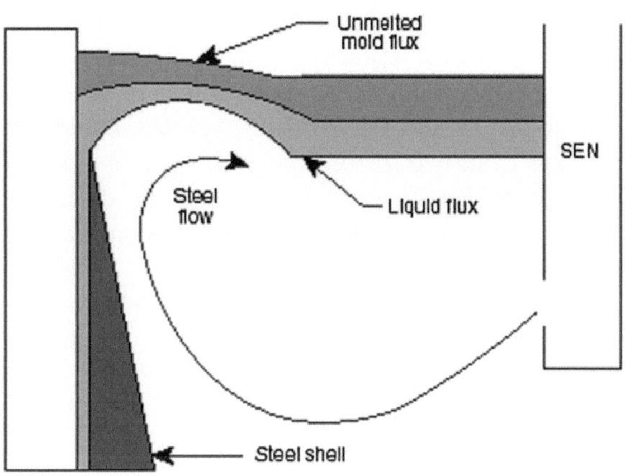

Fig. 9.50 Schematic drawings showing (a) the formation of a standing wave resulting from turbulent metal flow (permission granted, ISS/AIST, [232])

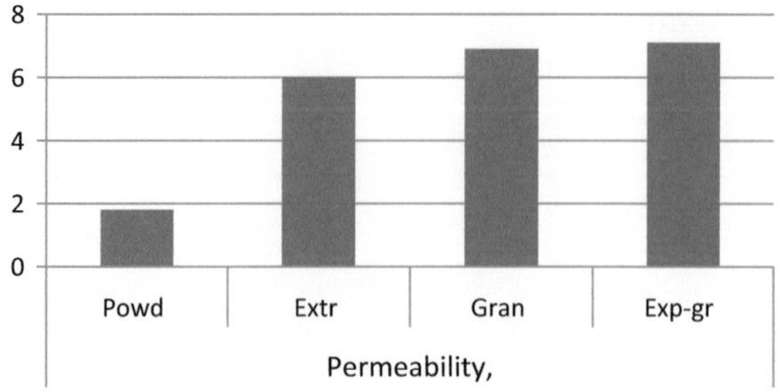

Fig. 9.51 Permeability Index for different powder types, *Powd* Powder; *Extr* Extruded granules; *Gran* Granules; *Exp-gr* Expanding granules (Permission granted, ISS/AIST, re-drawn after [232])

9.8.3 Permeability Index

Vertical heat transfer through the bed contains contributions from gaseous heat transfer (i.e. convection of Ar, CO, CO_2, N_2) passing up through the bed. The permeability test is designed to provide a measure of gaseous conduction through the sintered and powder layers of the bed [232, 237]. The permeability index is determined by measuring the time taken for 1 litre of air to pass through similar volumes of powders. The air flow is regulated by a semi-permeable membrane positioned in front of the sample. It can be seen from Fig. 9.51 that spheroids and granules have much greater permeability than powders because the voids between particles are much greater than those for the larger granules.

9.8.4 Thermal Insulation

The thermal insulation provided by the powder bed is important in controlling the depth of the molten pool (See Sect. 4.2.1). There is a test for determining the insulating properties of various casting powders. The heat flux passing through a known depth of powder is measured in this test (Fig. 9.52a) [232]. The sample is placed on the heat source (e.g. a hot SiC plate) which ensures that the heat transfer is unidirectional, as it is in the powder bed. Typical results are shown in Fig. 9.53. The temperature gradient ($\Delta T/\Delta d$) in the bed is calculated by monitoring temperature readings from two thermocouples immersed at known distances (d) in the powder bed (Eq. 9.55). The heat flux can also be determined using a water-cooled plate (Fig. 9.52b) by measuring the temperature increase (ΔT) for a known mass flow rate of water (m_{H_2O}/dT) (Eq. 9.56) [329, 330]. Another way of measuring the vertical heat flux is to measure the sample surface with infrared detection sensors. Spitzer et al. [329, 330] used the data generated with their apparatus to develop a mathematical model to predict the thicknesses of the various layers in the powder bed.

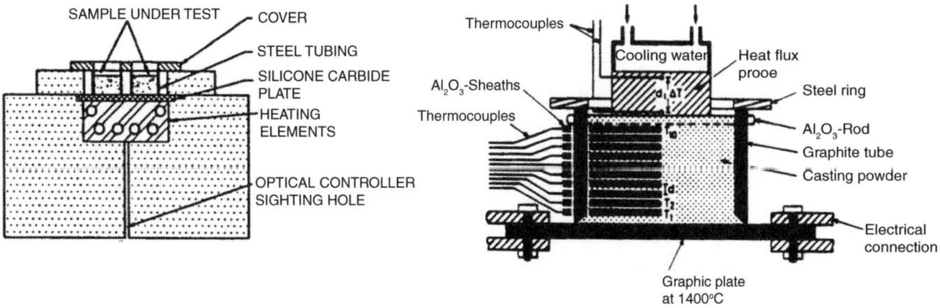

Fig. 9.52 Schematic diagrams showing apparatus used to determine the thermal insulation provided by mould powders **a** due to Diehl [232] and **b** Spitzer [329, 330] (permission granted, (**a**) ISS/AIST [232] (**b**) Verlag Stahleisen [329])

Fig. 9.53 Effective thermal conductivity (k_{eff}) of various types of powders; *Powd* Powder; *Extr* Extruded granule; *Gran* Spherodised granule; *Exp-gr* Spherodised granule containing an expanding agent like starch (permission granted, ISS/AIST, re-drawn after [232])

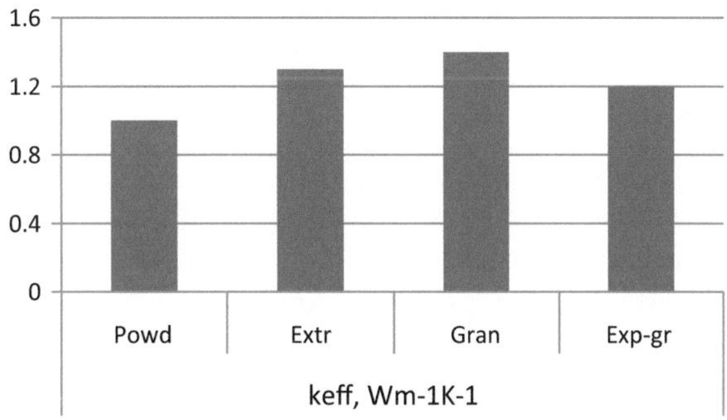

$$\Delta T/\Delta d = (T_1 - T_2)/(d_1 - d_2) = q/k_{bed}. \quad (9.55)$$

$$q = (m_{H_2O}/dT)(T_{exit} - T_{entrance}) \quad (9.56)$$

It can be seen from the effective thermal conductivity values given in Fig. 9.53 that:

(i) Thermal conductivities of powders are significantly lower than those of granules and spheroids.
(ii) The values obtained in these insulation tests [232] are *ca.*10× higher than the values obtained using the Transient Hot Wire (THW) method on powders [61, 62, 229] shown in Fig. 9.34.

This is probably due to the lower contributions from gaseous convection in the more densely packed (lower permeability) powders than in granules (see Sect. 9.3.5, Powder bed). It also indicates that gaseous convection makes

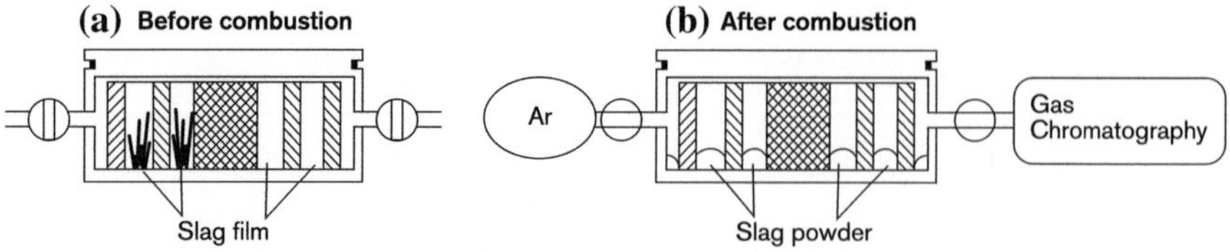

Fig. 9.54 Schematic drawings showing the apparatus used to determine the hydrogen content of slag (Reproduced with permission from EDP Sciences [116])

significant contributions to k_{eff} in these insulation tests (where there is an appreciable temperature gradient across the sample and where the specimen is heated from the bottom). These conditions are not dissimilar to those in the mould.

9.8.5 Measurement of Moisture and Hydrogen

If the moisture content of a casting powder is high it can result in the creation of slag films with pores containing hydrogen. These slag films containing pores, reduce the horizontal heat flux significantly and can result in sticker breakouts because the resulting shell is too thin and too weak [116] [117].

Free moisture contents of casting powders are usually measured from the recorded mass loss when a known mass of powder is heated at 110 °C for a set time.

Crystalline moisture is usually measured by the Karl Fischer method.

The amount of hydrogen present in a known mass of slag film has been measured with the apparatus shown in Fig. 9.54 [116]. It should be noted that this method requires 65 g of slag film which is a large amount to retrieve from the mould. However, Japanese workers have used solid H^1 Nuclear Magnetic resonance equipment to measure the hydrogen content [117, 296] which presumably requires much smaller samples.

9.9 Comparison of Properties of Powders Used in Ingot- (IC) and Continuous Casting (CC)

Since ingot casting is mostly carried out by uphill teeming, it has been assumed below that ingot casting is carried out in this manner. It will be seen below that there are some similarities between mould powders used for continuous casting (CC) and ingot casting (IC) but there are also some significant differences between them.

9.9.1 *Differences in Properties of Mould Powders Used in CC and IC*

The principal difference between powders used in CC and IC lies in their powder consumption or powder requirement. In continuous casting, the mould powder is fed continuously into the mould, where it forms a slag pool which infiltrates into the shell/mould channel; this provides the newly formed shell with the necessary lubrication. The powder consumption (Q_s) in CC should be appropriate for the mould dimensions and casting conditions and $Q_t^{slag} = f^* Q_t^{powd}$ has a value of *ca.* 0.48 kg (slag) tonne $^{-1}$ (steel).

In ingot casting, the slag pool is necessary to seal the metal from the gaseous atmosphere and to provide some thermal insulation. Lubrication is not so important in IC because the steel moves in a slow and steady manner. The *powder requirement* (Q_{IC}) is the amount of powder added at the start of casting to ensure the formation of a protective slag pool and to maintain thermal insulation throughout the cast. Consequently, *powder requirement* ($Q_t^{powd}{}_{IC}$) is somewhat different to *powder consumption* ($Q_t^{powd}{}_{CC}$) in CC. The powder consumption (Q_t^{powd}) is the weight of powder and assuming f^* = 0.75, typical values of $Q_t^{slag}{}_{IC}$ are ca. 3 times those in CC. This is because in IC, the powder requirement, $Q_t^{powd}{}_{IC}$, should ensure that that no "*open eyes*" are formed and the steel meniscus does not freeze. It should be noted that the $Q_t^{powd}{}_{IC}$ value given above is a mean value, whereas the value for fly-ash powders is 15% lower since little molten slag is formed but is 15% higher for low C powders where more liquid slag is formed.

The formation of a slag film is an important function in CC slags. A slag film is probably formed in ingot casting but is restricted because of the high viscosity (especially for fly-ash slags).

Thus in CC, powder consumption should be preferably high enough to provide good lubrication. In IC, powder requirements are already high, so efforts are made to ensure it is not excessive. Powder consumption/requirement values increase with reductions in T_{liq}, viscosity and casting speed.

Although the principal difference between CC and IC lies in the powder consumption/requirement, there are other differences in CC and IC; these are summarised below.

9.9.1.1 Thermal Insulation and Vertical Heat Flux (q_{vert})

The thermal insulation supplied by the powder bed is important to both IC and CC since it controls the depth of the liquid pool. However, it is probably more important in IC since the powder is added to the metal surface in the early moments of casting and must provide thermal insulation throughout the cast, whereas the powder is fed continuously in CC. Thus, thermal insulation is more critical in IC for casts of <10 tonnes and greater care must be taken with flowability; melting rate, etc.

9.9.1.2 Slag/Pool Composition

Slag/metal reactions (e.g. $3FeO_{sl} + 2\underline{Al} = Al_2O_{3\ sl} + 3\underline{Fe}$) change the composition of the slag pool. In CC the slag is continuously replenished, so compositional changes tend to be moderate. In IC the powder is added as a batch at the start; consequently, changes in slag compositions tend to be much larger. The loss of FeO will cause a large increase in q_R but this is offset by a decrease in convection due to the increase in viscosity resulting from Al_2O_3 pick-up.

9.9.1.3 Steel Movement

In IC there is a gradual, upward movement of the steel whereas in CC the steel is withdrawn downwards rapidly and oscillation of the strand is imposed.

9.9.1.4 Lubrication

This is very important in CC (especially in high-speed thin slab casting) but appears to be less important in IC, presumably, because the shell is moving at a much lower speed and there is no oscillation of the mould.

9.9.1.5 Mould Geometry

Slabs are usually continuously cast and slabs have a considerably smaller ratio for R^* (=surface area/volume) of the mould compared with billets and consequently, require considerably more lubrication than billets.

9.9.1.6 Heat Extraction

Cast iron moulds are used in IC, whereas water-cooled copper moulds are used in CC; consequently, the cooling rates and horizontal heat fluxes are considerably larger in CC than in IC (Note k_{Cu}/k_{Fe} = ca. 10).

9.9.1.7 Formation of Slag Film

In CC the formation of a slag film is essential to the control of the heat extraction rate from the shell (q_{hor}) and the process is aided by the oscillation of the mould. Once a slag pool is formed in IC some slag infiltration should occur through the effect of gravity; the flow rate will be dependent on the depth of the slag pool and the fluidity of the slag (i.e. $1/\eta$) the magnitude of q_{hor} does not appear to be an important issue in IC.

9.9.1.8 Sticking of Shell to Mould

The sticking of the shell to the mould is a serious issue for CC. It is tackled by (i) oscillating the mould and (ii) creating a slag film with a liquid layer next to the shell. In general, the sticking of the shell to the mould does not appear to be an issue in IC but can occur in big-end-up moulds where it results in vertical cracks.

9.9.2 Tasks Carried Out by Powders Used in Continuous- and Ingot Casting

The powders used in CC and IC must carry out specific tasks, these are summarised in Table 9.17; It can be seen from Table 9.17 that some tasks are equally important in IC and CC but other tasks are more important in CC than in IC (e.g. control of lubrication and horizontal heat transfer are paramount in CC, whereas, the prime tasks in IC are to provide cover against meniscus freezing and steel oxidation).

9.9.3 Properties and Characteristics of Powders Used in Continuous and Ingot Casting

The properties and casting parameters for mould powders used in continuous casting are compared against those for two different types of powders used in ingot casting in Table 9.18. The latter are (i) *fly-ash*-based powder and (ii) high basicity mould fluxes for casting Al-killed steels (which contain little FeO).

9.9.4 Conclusions from Comparison of CC and IC Mould Powders

1. The powder consumption/requirement is the property showing largest differences with $Q_t^{IC} \approx 3Q_t^{CC}$ due to Q_t^{IC} being applied as a batch which must provide cover through the entire cast whereas Q_t^{CC} is fed continuously.
2. Since powder consumption, Q_t^{IC}, is already high, IC powders tend to be designed to minimise excessive consumption (e.g. high viscosity, low melting rate, etc.).
3. In CC powders, the prime functions are associated with the formation of the slag film, i.e. provision of lubrication and heat extraction from the shell which involves optimisation of properties (e.g. viscosity, T_{br} and f_{crys}) according to the casting conditions and steel grade being cast; lubrication and heat transfer (q_{hor}) do not appear to be important in ingot casting.

Table 9.17 Summary of the various tasks to be carried out by powders for CC and IC

Task	Continuous casting (CC)	Ingot casting (IC)
Seal from atmosphere-Slag pool	*Important:pool formed at an early stage* High (C/S) leads to low O potential Deep pool helps consumption Deep pool minimises C pick-up by steel	*Important: pool formed fairly early in cast* High FeO, low(C/S) leads to high O potential Low η slag leads to "*open eye*" and low d_{pool} Deep pool minimises C pick- up by steel
Thermal insulation (top)	*Important in forming slag pool*: Exothermic reactions reduce T gradients Deeper powder bed reduces $q_{vert.}$ Low viscosity tends to increased convection EMBr reduces $q_{vert.}$ by ca. 30%	*Very Important: $q_{vert}\downarrow$* IR absorption by FeO reduces radiation High pool viscosity reduces convection Exothermic reactions reduce T gradients: High packing density helps
Dissolve inclusions	*Important*: Aided by high (C/S); %Na_2O; % CaF_2; low viscosity	*Important*: Retarded by high viscosity and low (C/S) High FeO promotes Al_2O_3, SiO_2 formation
Lubricate shell	*Very Important: friction$\downarrow$ as $d_l\uparrow$* Control via $d_l\uparrow$ as $T_{br}\downarrow$ Glassy phases in slag film form *scl* at $T > T_g$ Consumption increases with decreasing η Q_s increases as parameter, R^* increases	*Seems less important*: Glasses form *scl* above T_g- move under stress High viscosity lowers consumption Q_{IC} needs to be low to last entire cast
Control heat extraction (q_{hor})	*Very Important*: T_{br} controls d_s: q_R reduced by increasing f_{crys} and D_{crys}	*Seems less important* Lower q_{hor} and cooling rates q_R reduced by FeO content
Sticking of shell/mould	*Very Important*: Formation of liquid layer in slag Avoid blockages to slag flow, e.g. broken rim	*Seems less important*—possibly due to lower shell speeds

Powder requirement in IC is based on total amount of powder added to maintain cover of the steel surface for entire cast and is different to Q_s, Q_t etc. in CC

4. Thermal insulation and the formation of a slag pool seem to be important in both processes but are probably more important in IC because of the need to maintain cover through the entire cast to avoid "*open eye*" formation and the resulting re-oxidation of the steel surface.

Table 9.18 Comparison of important casting parameters and physical properties between (i) mould powders for continuous casting (CC) and (ii) two types of mould powder used in Ingot casting (IC), i.e. fly-ash-based and high basicity powders; { } = Typical values

Property	CC Mould powder/slag	IC Fly-ash powder	IC high (C/S) powder
Slag pool depth	Promoted by low T_{liq}, deep bed, exothermic powders -provide good consumption Deep pool-low Cpick-up -fewer inclusions. Low free carbon content	Seals steel from air Deep pool-low C-pick-up Promotes consumption	Seals steel from air Deep pool-low C-pick-up Promotes consumption Low free carbon promotes slag melting
Powder consumption	Q_s = Appropriate for mould dimensions (R*), $Vc; f, s$, - Affected by melting rate {$Q_t^{slag} = 0.48$ kg tonne^{-1}}	$Q_{t\ IC}$ should be low- aided by high viscosity High C%, low D_C -to keep melting rate low. Very little melting. {Q_t^{powd} = 1.5–2. kg tonne^{-1}}	$Q_{t\ IC}$ should be low- not helped by low viscosity High C%, low D_C -to keep melting rate low {$Q_t^{slag} = 2$ kg tonne^{-1}}
Viscosity, dPas	η = appropriate for V_c; R* High η slag entrapment reduced {η_{1300} = 0.5–2}	High η minimises $Q_{t\ IC}$ -slag entrapment reduced -reduces convection, -slows Al_2O_3 dissolution -{η_{1300} = 300–400}	Low η minimises $Q_{t\ IC}$ -promotes slag entrapment -increases convection, improves Al_2O_3 dissolution {η_{1300} = 3–7}
T_{liq}, T_{br}	T_{br}-appropriate-steel grade T_{br}-controls q_{hor} *via* d_s T_{br}-controls lubrication *via* d_l T_{liq} {1150 °C}	High T_{liq}-shallow pool High T_{liq}-*reduces* $Q_{t\ IC}$ T_{liq} {1150 °C}	
Dissolution of inclusions	- Promoted by high (C/S); % Na_2O; %CaF_2; low η; low γ_{msl}: high (C_{sat}- C_O)	Retarded as cast proceeds - lower (C_{sat}–C_o)-higher η-high Al_2O_3 by slag/metal reaction Slag has high O potential	Better than fly-ash-aided by high (C/S) lower η–No FeO-Less Al_2O_3 Slag-Low O potential
Interfacial tension (γ_{msl})	Sensitive to S%; slag FeO% Low γ_{msl} → Slag entrapment	Low γ_{msl} due to FeO % →Slag entrapment -Aids Al_2O_3 dissolution →Small Ripple marks	γ_{msl} Sensitive to S content-γ_{msl} lower than fly-ash
Thermal insulation	Promoted-deep powder bed-Exothermic powders high packing density	Good thermal insulation High FeO−low q_{vert}. High η-low convection Good powder structure	Moderate insulation due to- no IR absorption, higher q_R Higher convection-low η Poorer powder structure

5. The high FeO content of fly ash is both a blessing and a curse since its absorbs large amounts of radiated heat during solidification but is a continual source of O for Al_2O_3 formation.

9.10 Summary

It is obvious that the performance of a mould slag relies heavily on its physical properties. The lubrication the liquid slag supplied to the shell is directly related to the fluidity (i.e. reciprocal viscosity) and the horizontal heat transfer is largely determined by the solidification temperature (T_{br}) the fraction of crystalline phase (f_{crys}) and the thermal conductivity of the slag film. However, other properties can affect defects, *e.g.* slag/metal interfacial tension affects slag entrapment and scale formation on the steel. Thus, virtually all the thermophysical properties are important. However, some slag properties can be measured accurately (e.g. density can usually be measured to $\pm 2\%$) whereas others (e.g. viscosity) are subject to much greater uncertainty ($\pm 25\%$). Thermal conductivity is particularly vulnerable to uncertainty because (i) it is difficult to quantify the individual contributions from both convection and radiation conduction in the measurements and (ii) there is currently an unresolved dispute over which of two sets of thermal conductivity values should be used for the slag for $T > 700$ °C. The situation is further complicated by the fact that the radiation conductivity is dependent on specimen thickness and there is some indications that the initial glassy slag film is *optically thin*. This poses the questions for simulation experiments, e.g. is the sample of similar thickness to the slag film formed in the mould and how representative is it of the property value for the slag film?

It is surprising that the property database for mould slags is far from complete. For instance, there are no C_p data for mould slags for $T > 750$ °C and there are few density data available. The lack of property data is particularly severe for the specialist mould slags (e.g. F-free or CA-type slags).

Mathematical models (of heat and fluid flow) are now capable of predicting product defects but these models require property data for the mould slag and the steel, as an input. There are a large number of mould slags in use and it is impractical to measure all the required properties since performing measurements is both time consuming and expensive. Thus, it is important to develop models which calculate the required property data from chemical composition, since this is usually available for both mould slag and steel. Some models are available to calculate properties and some of these are referenced in this chapter. However, the uncertainty in the predicted values will always be greater than the uncertainties for the experimental property values. Another problem affecting the quality of property predictions is related to mould powders changing composition with time in the experiment, e.g. CO_2, some F and Na_2O are evolved and Al_2O_3 is picked up during a casting. In order to improve property- composition models, it is necessary to have post-measurement compositions of the slag available; unfortunately, these data are rarely available.

References

1. BO Mysen, *Structure and properties of silicate melts*, (Elsevier, Amsterdam, 1988).
2. BO Mysen, Earth Science Reviews, **27**, 281, (1990).
3. Slag Atlas, (Stahl u Eisen,VDEh, Dusseldorf,1995).
4. BN Roy, A Navrotsky, J Amer. Ceram. Soc., **67**, 606, (1985).
5. T Higo, S Sukenaga, K Kanehashi, H Shibata, T Osugi, ISIJ Intl., **54**, 2039, (2014).
6. JL Li, QF Shu, KC Chou, Ironmaking and Steelmakng, **42**, 154, (2015).
7. TS Kim, JH Park, ISIJ Intl., **54**, 2031, (2014).
8. BT Poe, PF McMillan, J Amer. Ceram. Soc., **77**, 1832, (1994).
9. PF McMillan, WT Petuskey, B Cote, D Massiot, C Landron, JP Coutres, J Non-Cryst Solids, **195**, 261, (1996).
10. TJF Stebbins, Q Zeng, J Non-Cryst. Solids, **262**, 1, (2000).
11. Y Sasaki, M Iguchi, M Hino, ISIJ Intl., **47**, 346, (2007).
12. Y Sasaki, M Iguchi, M Hino, ISIJ Intl., **47**, 638, and 1370 (2007).
13. T Asado, Y Yamada, K Ito, ISIJ Intl., **48**, 120, (2008).
14. JH Park, ISIJ Intl., **47**, 1368, (2007).
15. S Hayakawa, A Nakao, C Ohtsuki, A Osaka, S Matsumoto, Y Miura, J Mater. Res., **13**, 729, (1998).
16. M Hayashi, N Nabeshima, H Fukuyama, K Nagata, ISIJ Intl., **42**, 352, (2002).
17. VS Esaulov, GF Konovalov, SI Popel, VI Sokolov, Steel USSR, **6**, 311, (1976).
18. M Susa, S Kubota, M Hayashi, KC Mills, Ironmaking and Steelmaking, **28**, 390, (2001).
19. L Baak, A Olander, Acta Chem. Scand., **9**, 1350, (1955).
20. PM Bills: J Iron Steel Inst., **201**, 133, (1963).
21. A van Bernst, C Delaunois, Verres Defract, **20**, 435 (1966).
22. T Nakamura, K Morinaga, T Yanagase, Nippon Kinzoku Gakkaishi, **41**, 1300, (1977).
23. A Shankah, PhD thesis, *Studies of high- alumina blast-furnace slags*, KTH, Stockholm (2007).
24. S Yagi, K Mizoguchi, Y Suginahara, Bull. Kyushu Inst. Technol. Sci. Technol., **40**, 33, (1980).
25. SF Zhang, X Zhang, H Peng, LY Wen, GB Qui, ML Hu, CG Bai, ISIJ Intl., **54**, 734, (2014).
26. JF Li, Q Shu, KC Chou, Canad. Met. Quart., **54**, 85, (2015).
27. T Mukongo, C Pistorius, A Garbers-Craig, Ironmaking Steelmaking, **31**,135, (2004).
28. AB Fox, KC Mills, D Lever, C Bezerra, C Valadares, I Unamono, J Laraudogoitia, J Gisby, ISIJ Intl., **45**, 1051, (2005).
29. H. Huang, J. L. Liao, K. Zheng, H. H. Hu, F. M. Wang, Z. T. Zhang, Ironmaking and Steelmaking, **41**, 67, (2014).
30. Z Wang, QF Shu, XM Hou, KC Chou, Ironmaking and Steelmaking, **39**, 210, (2012).
31. QF Shu, Z Wang, J L.Klug, K Chou, P R. Scheller, Steel Res. Intl., **84**, 1138, (2013).
32. G Urbain, *Steel Research*, **58**, 111, (1987).
33. KC Mills, L Yuan, Z Li: Estimation of the thermo-physical properties of slags Part 1 Viscosity to be published.
34. T Iida, H Sakai, Y Kita, K Murakami, High Temp. Mater. Processes, **19**, 153, (2000).
35. KC Mills, ISIJ Intl., **33**, 148, (1993).
36. KC Mills, S Sridhar, Ironmaking and Steelmaking, **26**, 262, (1999).

37. VK Gupta, SP Sinha, B Raj, Steel India, **21**, 22 (1998) and **17**, 74, (1994).
38. J Duffy, MD Ingram, J Non-Cryst. Solids, **21**, 373, (1976).
39. T Nakamura, Y Ueda, JM Toguri, J Jap. Inst. Metals, **50** (5), 456, (1986).
40. GH Zhang, KC Chou, G Xue, KC Mills, Met. Mater. Trans. B, **43B**, 64, (2012).
41. QF Shu, KJ Hu, BJ Yan, JY Zhang, KC Chou, Ironmaking and Steelmaking, **37**, 387, (2010).
42. Q Shu, KC Chou, Ironmaking and Steelmaking, **40**, 571, (2013).
43. GH Zhang, KC Chou, Met. Mater. Trans. B, **41B**, 131, (2010)..
44. H Gaye, J Welfinger, Proc. 2nd Intl. Symp. Metall. Slags and glasses, Lake Tahoe, NV ed. by HA Fine and DR Gaskell (TMS-AIME, Warrendale, PA, 1984) p. 357.
45. L Zhang, S Sun, S Jahanshai, Met. Trans. B, **29B**, 177, (1998).
46. L Zhang, S Sun, S Jahanshai: Met. Trans B, **29B**, 187, (1998).
47. GH Zhang, KC Chou, KC Mills, ISIJ Intl. **52**, 355, (2012).
48. GH Zhang, KC Chou, J Min. Metall. **B 48**, 1/ 10, (2012).
49. GH Zhang, KC Chou, **40**, 376, (2013).
50. KC Mills, L. Yuan; Z Li, GH Zhang, KC Chou, Proc. 9th Intl. Conf. Molten slags, fluxes and salts (Molten 12*),* Beijing, China,2012 (Chinese Met. Soc., 2012)
51. KC Mills, L. Yuan, Z Li, GH Zhang, KC Chou, High Temp. Mater. Processes, **31**, 301, (2012).
52. A Fluegel, Electrical resistivity of silicate glass melts based on SciGlass Database. http//glassproperties.com. 2007 p. 21.
53. M.Susa, KC Mills, MJ Richardson, R Taylor, D Stewart, Ironmaking and Steelmaking, **21**, 279, (1994).
54. S Ozawa, M Susa, Ironmaking and Steelmaking, **32**, 487, (2005).
55. S Ozawa, M Susa, T Goto, R Kojima, KC Mills, ISIJ Intl., **46**, 413, (2006).
56. M Susa, A Kushimoto, R Endo, Y Kobayashi, ISIJ Intl., **51**, 1587, (2011).
57. T Watanabe, M Hayashi, K Nagata, Paper presented at ISIJ –VDEh-Jernkontoret, Joint Symposium, Osaka, Japan, April, 2013, p. 192.
58. K Tsutsumi, R Nakamura, J Ohtake, T Nagasaka, M Hino, ASIA Steel Intl. Conf., 2000, (ASIA Steel' 2000), The Chinese Society for Metals, Beijing, Vol. C, (2000), p. 366
59. JW Cho, H Shibata, T Emi, M Suzuki, ISIJ Intl., **38**, 268, (1998).
60. K Nishioka, T Maeda, M Shimizu, ISIJ Intl., **46**, 427, (2006).
61. P. Andersson, C Eggertson, Ironmaking and Steelmaking, **42**, 456, (2015) and 465.
62. P Andersson, *Thermal conductivity of powders used in continuous casting of slabs and billets*, in press.
63. M Hanao, M Kawamoto, ISIJ Intl., **48**, 180 (2008) see also Tetsu-to Hagane **92** (1), 13, (2006).
64. M Hanao, M Kawamoto, A Yamanaka, ISIJ Intl., **52**, 1310, (2012).
65. MD Seo, CB Shi, JW Cho, SY Kim: Proc. 8th Europ. Conf. Continuous Casting, Graz, Austria, 2014 (Austrian Met. Mater. Soc., 2014).
66. GH Wen, S.Sridhar, P Tang, X Qi, YQ Liu, ISIJ Intl., **47**, 1117, (2007).
67. H Nakada, K Nagata, ISIJ Intl., **46**, 441, (2006).
68. N Takahira, N Hanao, Y Tsukaguchi, ISIJ Intl., **53**, 818, (2013).
69. M Hanao, Y Tsukaguchi, M Kawamoto, Proc. 4th Intl. Congress Science and Technol 2008, Gifu, Japan, 2008, (ISIJ, Tokyo, 2008) p. 694.
70. P Grieveson, S Bagha, N Machingawuta, K Liddell, KC Mills, Ironmaking and Steelmaking, **15**, 181, (1988).
71. T Watanabe, H Fukuyama, K Nagata, ISIJ Intl., **42**, 489, (2002).
72. M Hayashi, T Watanabe, H Nakada, K Nagata, ISIJ Intl., **46**, 1805, (2006).
73. K. Sorimachi, Fachber. Huttenpraxis Metallweitver., **20** (4), 244, (1982).
74. M Kawamoto, T Kanazaka, S Hiraki, S Kumakura, Proc. 5th Intl. Conf. Molten slags, fluxes and salts, Sydney (1997) (ISS, Warrendale, PA, 1997) p. 777.
75. B Tarrant, G Brooks: Iron and Steelmaker, **2003** (5), 52, (2003).

76. T. Mukongo, MEng Thesis, *Effect of titanium pick up on mould flux viscosity in continuous casting of Ti-stabilised steel.* Dept Mater. Sci. & Metall. Eng., Univ. Pretoria, 2003.
77. P Rocabois, JN Pontoire, J Lehmann, H Gaye, J Non- Cryst. Solids, **283**, 98, (2001).
78. S Riaz, Ironmaking and Steelmaking, **39**, 409, (2012).
79. Y Wang, Yu Wang, L Fan, B Xie, XP Liang, X Zeng, ZW Wei, Proc. 9th Intl. Conf Molten slags, fluxes and salts (Molten 2012), Beijing, China, 2012 (Chinese Met. Soc., 2012).
80. S. Sankararanarynan, Proc. 75th Steelmaking Conf., 1992, (ISS, Warrendale, PA, 1992) p 607.
81. JL Klug, R Hagermann, NC Heck, JFC Viela, PR Scheller, Proc. 5th Intl. Congress Science and Technol. Steelmaking, Dresden, (2012) http//dx.doi.org/10.4322/tmm.2012.005.
82. C Righi, R Carli, V Ghilardi, Proc. 6th Intl. Conf. Metallurgical slags and fluxes, Stockholm, Sweden, paper 102 (2000).
83. Z Li, R Thackray, KC Mills, Proc. 7th Intl. Conf. Molten slags, fluxes and salts, Capetown, South Africa, 2004 (SAIMM, Johannesburg, 2004) p. 813.
84. M Hanao M Kawamoto, M Hara, T Murakami, H Kikuchi, K Hanazaki, Tetsu-to-Hagane **88**, 23, (2002).
85. MV Fonseca, ODC Afrange, A Lavinas, AA Ramos, CA Valadares, Proc. 5th Intl. Conf. Molten slags, fluxes and salts, Sydney, 1997 (ISS, Warrendale, PA, 1997) p 851.
86. H Fukuyama, T Watanabe, H Takahashi, H Tabata, T Oshima, K Nagata, *Metallurgical and Materials Processing: Principles and Technologies; vol 1, Materials processing fundamentals and new technologies,* ed. by F Kongoli, K Ingoki, C Yamauchi and HY Sohn (TMS (2003) p. 937.
87. R Kinoshita, CAMP- ISIJ, **20**, 781, (2007).
88. C Orrling, AW Cramb, A Tilliander, Y Kashiwaya, Iron and Steelmaker, **27** Part 1, 51–63, (2000).
89. S Lachmann S., Scheller P. R., Proc. 6th Europ. Conf. on Continuous Casting, Riccone, Italy, 2008.
90. Y Harada, K Kusada, S Sukenaga, H Yamamura, Y Ueshima, T Mizoguchi, N Saito, K Nakashima, ISIJ Intl., **54**, 2071, (2014).
91. WL Wang, K Blazek, AW Cramb, Met. Mater. Trans. B, **39B**, 66, (2008).
92. J Rudnizki, R Shepherd, E Balichev, S Karrasch, F Kruger, Proc 8th Europ. Conf. Continuous casting, Graz, Austria, 2014 (Austrian Met. Mater. Soc., 2014).
93. PS Kharlashin. Azovetal, **1999** (4), 34, (1999).
94. H Heimbach, K Schulz, J Markardt, HJ Ehrenberg, Stahl u Eisen, **117**, 105 (1997).
95. Y Lu, GD Zhang: Mater. Sci. Forum, **676/677**, 877 (2011).
96. H Abratis, F Hofer, M Junemann, H Sardmann, H Stoffel, Stahl u Eisen, **116**, 85, (1996).
97. SY Choi, DH Lee, DW Shin, JW Cho, JM Park, J Non-Cryst. Solids, **157** (8), 15, (2004).
98. Y Lu, GD Zhang, X Yu, Appl. Mech and Mater., **71/78**, 2899, (2011).
99. QC Wei, YQ Ding, KD Peng, J Chonqing Univ., **4**, 110, (1995).
100. Y Lu, X Fang, GD Zhang, Adv. Mater. Res., **287/290**, 805, (2011).
101. AB Fox, K Mills, D Lever, M C Bezerra, CA Valadares, Unamono, J Laraudogoitia, J Gisby: Proc. 36th Steelmaking Seminar, Vittoria, ES, Brazil, May (2005) available on line at http://www.carboox.com/pdf/paper2005sfs.pdf.
102. Y Lu, GD Zhang, MF Jiang, Advanced Mater. Res., **233/235**, 805, (2011).
103. LY Chen, GH Wen, CL Yang, F Mei, CY Shi, P Tang, Ironmaking and Steelmaking, **42**, 105, (2015).
104. Q Wang, YJ Lu, SP He, KC Mills, Z Li, Ironmaking and Steelmaking, **38**, 297, (2011).
105. X. J. Fu, G. H. Wen, P. Tang, Q. Liu, Z. Y. Zhou, Ironmaking and Steelmaking, **41**,342, (2014).
106. CB Shi, MD Seo, JW Cho, SH Kim, Proc. 8th Europ. Conf. Continuous Casting, Graz, Austria, 2014, (Austrian Met. Mater. Soc., 2014).
107. Z Li, KC Mills, MC Bezerra, Proc. XXXV Semin. De Fusao Refino e Solidifacao Metals, Salvador, Brazil (2004) p. 281.

108. H Sakai, T Kawashima, T Shiomi, K Watanabe,T Iida: Proc. 5th Intl. Conf. Molten slags, fluxes and salts, Sydney (1997) (ISS, Warrendale, PA, 1997) p. 787.
109. H Hashimoto, T Watanabe, K Nagata, CAMP-ISIJ, **17**, 849, (2004).
110. M Susa, A Kushimoto, H Toyota, M Hayashi, R Endo, Y Kobayashi, ISIJ Intl., **49**, 1722, (2009).
111. P R Scheller, Proc. 6th Intl. Conf on Slags, fluxes and salts, Stockholm, 2000, Session 6, Filename 405 pdf.
112. YG Moldana, FA Acosta, AH Castillejos, BG Thomas, Iron and Steel Technology, 2013, (July),1, (2013).
113. P.V. Riboud, Y. Roux: *Fundamental study of the behaviour of casting powders,* Report EUR 9560 (Europ. Comm. Sci. and Tech. Publ., Luxembourg, 1985).
114. PV Riboud, C Gatellier, H Gaye, JN Pontoire, P Rocabois, ISIJ Intl., **36**, 522, (1996).
115. K Tsutsumi, T Nagasaka, M Hino, ISIJ Intl., **39**, 1150, (1999).
116. JN Pontoire, Revue de Metallurgie, CIT, **93**, 1237, (1996) see also Proc. 3rd Europ. Conf. Continuous Casting,Madrid, Spain (1998) p. 973.
117. Y Uemiya, T Mizoguchi,T Kajitani, Proc. 9th Intl. Conf Molten slags, fluxes and salts (Molten 12) Beijing (2012) (Chinese Met. Soc, 2012).
118. L Hering, HP Heller, HW Fenske, Stahl u Eisen, **17**, 61, (1992).
119. Y Meng, BG Thomas, Proc. ISS Tech., Indianapolis, (2003) see also Met.. Mater. Trans. **34B**, 685 and 707, (2003).
120. TJ Billany, AS Normanton, KC Mills, P Grieveson, Ironmaking and Steelmaking, **18**, 403, (1991).
121. R J O'Malley, J Neal, Proc. Intl. Conf. New Developments in Metallurgical Process Technol., Dusseldorf, METEC Congress (1999) p. 73.
122. MCC Bezerra, ODC Afrange, AC Valadares, Proc. Mills Symp., London (2002). (Inst of Materials, London, 2002) p. 293.
123. S Riaz, KC Mills, K Nagata, AS Normanton,. Proc. Mills Symp., London (2002) (IOM, London, 2002) p. 465.
124. L Courtney, S Nuortie-Perkkio, C Valadares, MJ Richardson, KC Mills: Ironmaking and Steelmaking, **28** (5), 412, (2001).
125. KC Mills L Courtney, AB Fox, B Harris, Z Idoyaga, MJ Richardson, Thermochica Acta, **391**, 175, (2002).
126. H Lidefelt, Swedish Inst. Metals Research Report JK D45; IM 1755, (1983) Stockholm.
127. SH Chang, Proc. Conf. Continuous casting of steel in Developing countries, Beijing (1993) p. 832.
128. SP He, X Long, JF Xu, T Wu, Q Wang, Ironmaking and Steelmaking, **39**, 593, (2012).
129. A, Lopez, Z Idoyaga, F Plazaola, S Riaz, KC Mills, Proc. 4th Europ. Conf. Continuous casting, Birmingham, UK, 2002 (IOM Commun., London, 2002) vol 1, p. 360.
130. Q Wang, Y Wang, Jh Chi, B Xie, YM He, B Zhu, WM Chen, J Chongqing Univ –Eng. Ed, **5**, 65, (2004).
131. IH Jung, MA van Ende, Proc. 8th Europ. Conf. Continuous casting. Graz, Austria, 2014 (Austrian Met. Mater. Soc., 2014).
132. KC Mills, A Olusanya, RF Brooks, R Morrell, S Bagha, Ironmaking and Steelmaking, **15**, 257, (1988) see also KC Mills, *Funamental study of the behaviour of casting powders*. ECSC Final Report 7210 CA 131/810 (1983) (Europ. Comm. Sci. and Tech Publ. Luxembourg).
133. T Kishi, H Tsubio, H Takeuchi, T Nakano, M Yamamiya, T Ando, Nippon Steel Tech. Report, **34**, 11, (1987) see also T Kishi, Y Nagano T Nagano, Trans. ISIJ, **25**, B302, (1985).
134. C Orrling, Y Fan, N Phinichka, S Sridhar, AW Cramb, JOM, **51** (7)(1999). http://www.tms.org/pubs/journals/jom/9907/orrling/orrling-9907.html.
135. German Standard DIN 51740.
136. LJ Zhou, WL Wang, R Liu, BG Thomas, Met. Mater. Trans. B, **44B**, 1284, (2013).
137. KC Mills, PV Riboud, K Schwerdtfeger, Report on ECSC Contract 7210 CA/131/311/810 "*Fundamental study of the behaviour of casting powders*" (1983).

138. P Artelt, Stahl u Eisen, **97** (23), 1171, (1977).
139. JA Moore, RJ Phillips, TR Gibbs, Proc. 74th Steelmaking Conf., Washington, DC,1991, (ISS, Warrendale, PA, 1991) p 615.
140. V Ghilardi, R Carli, GL Faravelli, Proc. 3rd Europ. Conf. Continuous casting, Madrid, 1998, p 971.
141. M Hanao, M Kawamoto, T Watanabe, ISIJ Intl., **44**, 827, (2004).
142. J Chen, J Xie, F He, E Wan, Z Xu, C Yang, G Liu, Ironmaking and Steelmaking, **42**, 126, (2015).
143. KC Mills, S Karagadde, L Yuan, PD Lee, F Shahbazian, ISIJ Intl., **56**, 264, (2016)
144. KC Mills, L Yuan, Z Li, RT Jones, J South Afr. Inst. Min. Metall., **111**, 649, (2011).
145. S Riaz, Private communication Tata Steel Teesside Research Laboratories, Middlesbrough, UK. (2008).
146. S Feldbauer, I Jimbo, A Sharan, K Shimizu, W King, J Stepanek, J Harman, AW Cramb, Proc. 78 th Steelmaking Conf., 1995, (ISS, Warrendale, PA, 1995) p. 655.
147. JW Kim, J Choi, OD Kwon, IR Lee YK Shin, JS Park, Proc. 4th Intl. Conf. Molten slags and fluxes, Sendai (1992) (ISIJ, Tokyo, 1992) p. 468.
148. Z Elahipanah, MSc Thesis "*Thermo-physical properties of mould flux slags for the continuous casting of steel.*" KTH, Stockholm, (2012) available on line.
149. M Persson, M Gornerup, S Seetharaman, ISIJ Intl., **47**, 1533, (2007).
150. I Marschall, N Kohl, H Harmuth, GM Xia, Proc. of 9th Intl. Conf. Molten slags, fluxes and salts (Molten 12), Beijing, China, 2012 (Chinese Met. Soc., 2012).
151. V Kircher, I Marschall, N Kolbl, H Harmuth, Proc. 8th Europ. Conf. Continuous Casting, Graz, Austria, 2014 (Austrian Met. Mater. Soc., 2014).
152. RJ Phillips, WF Salem, WR Emling, HD Baker, Proc. 73rd Steelmaking Conf. (1990) (ISS, Warrendale, PA, 1990) p 247.
153. PW Johnstone, G Brooks, Proc. 5th Intl. Conf. Molten slags, fluxes and salts, Sydney (1997) (ISS, Warrendale, PA, 1997) p. 845.
154. D Wang, M Jiang, CJ Liu, PY Shi, J North Eastern Univ. (Natural Sci.) **26** (11), 1082, (2005).
155. P Tang, GH Wen, ZB Wang, SB Fan, ZB Zhang, SC Li, H Li, Proc. ISS Tech Conf. (2003) (ISS, Warrendale, PA, 2003) p. 567.
156. S Sridhar, KC Mills; ODC Afrange; HP Lorz, R Carli, Ironmaking and Steelmaking, **27**, 238, (2000).
157. KC Mills, AB Fox, R P Thackray, Z Li, Proc. 7th Intl. Conf Molten slags, fluxes and salts, Capetown, (2004) (SAIMM, Johannesburg, 2004) p. 713.
158. JV Dubrawski, JM Camplin, J Thermal Analysis, **40**, 329, (1993).
159. KC Mills, AB Fox, ISIJ Intl., **43**, 1479, (2003).
160. JD Mackenzie, *Physico-chemical measurements at high temperatures* ed. by JOM Bockris, White and JD Mackenzie, (Butterworths, London 1959) Chapter 15, p. 313.
161. LJ Wu, P Heller, XG Bi, PR Scheller, Proc. 9th Intl. Conf. Molten slags, fluxes and salts (Molten 12) Beijing, China, 2012 (Chinese Met. Soc., 2012).
162. T Iida, T Tanaka, Proc. 4th Intl. Conf. Metallurgical slags and fluxes, Sendai (1992). (ISIJ, Tokyo, 1992) p. 444.
163. M Tokuda, K Nakajima, T Yamamoto, H Nakajima, K Kawaguchi, The Sumitomo Search, **44**, 368, (1990).
164. SJ Roach, H Henein, Canad. Met. Quart., **42**, 175, (2003).
165. AP Froba, A Leipertz, Proc. (Abstracts) of 16th Europ. Conf. on Thermophysical properties, London, (2002), p. 59.
166. N Kawasaki, K Watanabe, Y Nagasaka, High Temp.-High Press., **30**, 91, (1998).
167. Y Kobayashi, Y Nagasaka, Proc. (Abstracts) 16th Europ. Conf. on Thermophysical properties, London, Sept. (2002) p. 103.
168. KC Mills, M Halali, HP Lorz, A Kinder, R Pomfret, BA Walker, Proc. 5th Intl. Conf. Metallurgical slags and fluxes, Sydney, Australia, 1997 (ISS, Warrendale, PA.) p. 535.
169. Z Zhang, G Wen, P Tang, S Sridhar, ISIJ Intl., **4**8, 739, (2008).

170. X Yu, GH Wen, P Tang, H Wang, Ironmaking and Steelmaking, **36**, 623, (2009).
171. JL Li, QF Shu, YA Liu, KC Chou, Ironmaking and Steelmaking, **41**, 732, (2014).
172. Z Wang, Q Shu, KC Chou, Steel Research, **84**, 768, (2013).
173. S Yoshimura, CAMP-ISIJ, **20**, 837, (2007).
174. K Okazawa: CAMP-ISIJ, **16**, 125, (2003).
175. S Riaz, Ironmaking and Steelmaking, **39**, 409 (2012).
176. W McCauley, D Apelian, Canad. Met. Quart., **20**, 247, (1981).
177. PV Riboud, M Olette,H Leclerc, W Pollak, Proc. 61st NOH- BOS Conf., Chicago,1978, (ISS, Warrendale, PA, 1978) p. 411.
178. VL Zhuk, SV Timofeeva,VN Moshenskii, Steel USSR **21** (6), 255, (1991).
179. M Lanyi, CF Rosa. Metall. Trans., **12B**, 387, (1981).
180. HY Chang, TF Lee, YC Ko, ISS Trans., **4**, 27 (1984).
181. HY Chang, TF Lee, T Ejima, Trans. ISIJ, **27**, 797, (1987).
182. M Kawamoto, K Nakajima, T Kanazawa, K Nakai, Tetsu-to- Hagane, **80**, (3), 219, (1994).
183. A, Lopez, Z Idoyaga, F Plazaola, S Riaz, KC Mills, Proc. 4th Europ. Conf. Continuous casting, Birmingham, UK, 2002 (IOM Commun.,London 2002), vol 1 p. 360.
184. JH Park, DJ Min, HS Song, YD Lee,. Proc. 85th Steelmaking Conf., Nashville, TN, 2002 (ISS, Warrendale, PA, 2002) p. 775.
185. T Lu, X Fang, G Zhang, Proc. Conf. *Mater. For Renewable Energy and Environment* (2011)- p. 869 (2011).
186. Y Lu, X Fang, XF Yu, Advanced Mater. Res., **455**, 805, (2012).
187. F Shahbazian, S Du, KC Mills, S Seetharaman, Ironmaking Steelmaking, **26**, 193, (1999).
188. F Shahbazian, S Du, S Seetharaman, ISIJ Intl., **42**, 155, (2002).
189. WL Wang, K Blazek, AW Cramb, Met. Mater. Trans. B, **39B**, 66, (2008).
190. JL Klug, DR Silva, SL Freitas, MMSM Pereira, NC Heck, ACF Viela, D Jung, Steel Res. Intl., **83**, 791, (2012).
191. C Zhang, DX Cai, Proc. 8th Europ, Conf. Continuous Casting, Graz, Austria, 2014 (Austrian Soc. Metall. Materials, 2014).
192. K Tsutsumi, K Watanabe, J Kubota, S Hatori, Y Miki, T Suzuki, T Omoto, Proc. 7th Europ. Cont. Casting Conf., Dusseldorf, 2011, (VDEh, Dusseldorf, 2011) Session pp 1 / 4.
193. K Watanabe, K Tsutsumi, M Suzuki, H Fujita, S Hatori T Suzuki, T Omoto, ISIJ Intl.**54**, 868, (2014).
194. Q Liu, GH Wen, JZ Li, XJ Fu, P Tang, W Li, Ironmaking Steelmaking, **41**, 292 (2014).
195. A Kusano, N Sato, M Okimori, S Fukunaga,K Nishihara, M Sato, Y Minigawa, Proc. 74th Steelmaking Conf. (1991) (ISS, Warrendale, PA, 1991) p. 111.
196. YL Chen, GH Zhang, KC Chou, Met. Mater. Trans. B, **46B**, 155, (2015).
197. YL Zhen, GH Zhang, KC Chou, KC Mills, Canad Met. Quart., **54**, 340, (2015).
198. JA Kromhout: PhD Thesis "*Mould powders for the high speed continuous casting of steel*". Univ of Delft, NL, (2011) p. 165.
199. M Kawamoto, CAMP-ISIJ, **6**, 1176, (1993).
200. PV Riboud, Y Roux, LD Lucas, H Gaye: Fachber. Huttenprax. Metallweiterverarb., **19**, 859, (1981).
201. M Hanao, M Kawamoto, T Tanaka, M Nakamoto, ISIJ Intl., **46**, 346, (2006).
202. S Du, J Bygden, S Seetharaman, Met. Trans. B, **25B**, 519 and 589. (1994).
203. M Hayashi, S Seetharaman, CAMP- ISIJ, **16**, 860, (2003).
204. KC Mills, L Yuan, Z Li, GH Zhang, High Temp.- High Press., **42**, 232, (2013).
205. QF Shu, Steel Research Intl., **80**, 107, (2009).
206. M. Susa, K Nagata, KC Mills: Ironmaking and Steelmaking, **20**, 312, (1993).
207. R Gardon, Proc. 2nd Intl. Conf. Thermal Conductivity, Ottawa (1962) p. 167 and J. Amer. Ceram. Soc. **44**, 305, (1961).
208. H Ohta, H Shibata, T Emi, Y Waseda, Tetsu-to Hagane, **61** (4), 350, (1997).
209. R Taylor, KC Mills, Ironmaking and Steelmaking, **15**, 187, (1988).
210. Y Waseda, M Masuda, K Watanabe, H Shibata, H Ohta, K Nakajima, High Temp. Mater. Processes, **13**, 267, (1994).

211. H Shibata, JW Cho, T Emi, M Suzuki, Proc.5th Intl. Conf. Molten slags, fluxes and salts, Sydney, (1997) (ISS, Warrendale, PA., 1997) p. 771.
212. H Ohta, M Masuda, K Watanabe, K Nakajima, H Shibata,Y Waseda, Tetsu-to- Hagane, **80**, 463, (1994).
213. H Ohta, H Shibata, H Hasegawa, Y Kawakri, Y Shinoki, S Kitamura, Y Waseda, Proc. 9th Intl. Conf. Molten slags, fluxes and salts (Molten 12), Beijing, China, 2012 (Chinese Met. Soc., 2012).
214. H Ohta, Y Shiraishi, Proc. 2nd Intl. Conf. Metallurgical slags and fluxes, Lake Tahoe, NV, USA, 1984, ed. HA Fine and DR Gaskell, (Met. Soc. AIME, Warrendale, PA, 1984) p 863.
215. M Kishimoto, M Maeda, K Mori, Y Kawai, Proc. 2nd Intl. Conf. Metallurgical slags and fluxes, Lake Tahoe, NV, USA, ed. HA Fine and DR Gaskell, (Met. Soc., AIME, Warrendale, PA, 1984) p. 891.
216. K Nagata, KS Goto, Proc. 2nd Intl. Conf. Metallurgical slags and fluxes, Lake Tahoe, NV, (1984), ed. HA Fine and DR Gaskell, (TMS-AIME, Warrendale, PA, 1984) p. 875.
217. M Hayashi, H Matsui, Unpublished thermal conductivity measurements, Tokyo Inst. Tech., Tokyo March, 2015.
218. JJ Macho, G Hecko, B Golomowski, M Frazee, Proc. Conf. on Thinner Slab Casting, Hamilton, ONT, Canada (McMaster Univ, Hamilton, 2005) p. 131.
219. KS Goto, HW Gudenau, K Nagata, KH Lindner, Stahl u Eisen, **105**, 1387, (1985).
220. S Bagha, N Machingawuta, P Grieveson, Proc. 3rd Intl. Conf. Molten slags and fluxes; Glasgow, 1988, (Inst. Of Metals, London, 1988) p 235.
221. JS Powell Unpublished data from NPL Teddington, cited in Slag Atlas [3] p 598.
222. M Hayashi, AA Riad, S Seetharaman, ISIJ Intl., **44** (5), 691, (2004).
223. M Gornerup, M Hayashi, CA Dacker, S Seetharaman, Proc. 7th Intl. Conf. Molten slags, fluxes and salts, Capetown, South Africa, 2004 (SAIMM, Johannesburg, 2004) p. 745.
224. BJ Monaghan, RF Brooks, Ironmaking and Steelmaking, **29**, 115, (2002).
225. H Ohta, K Nakajima, M Masuda, Y Waseda, Proc 4th Intl Conf Molten slags and fluxes, Sendai, 1992 (ISIJ, Tokyo, 1992) p. 421.
226. Y Kim, K Morita, ISIJ Intl., **53**, 2077, (2013).
227. KC Mills, M Susa, V Ludlow, Proc. 13th Proc. Tech. Conf., Nashville, TN, 1995 (ISS, Warrendale, PA, 1995) p. 157.
228. T El Gammal, U Schoneberg, Stahl u Eisen, **112** (1), 45, (1992).
229. F Neumann, J Neal, MA Pedroza, AH Castillejos, FA Acosra, Proc 79th Steelmaking Conf. (1996) (ISS, Warrendale, PA., 1996) p 249.
230. SG Bratchikov, GF Konvalov, MV Frolov, Teplotekh. Protsessov Vplavski Stahli Splavov, **1977** (5), 130, (1977).
231. M Ichihashi, M Kawamoto, Y Tsukaguchi, N Nishida, T Kanazawa, S Hiraki, Tetsu to Hagane, **81**, (10), 1132, (1995).
232. S Diehl, JA Moore, RJ Phillips, Proc. 78th Steelmaking Conf., Nashville, TN, (1995) (ISS, Warrendale, PA., 1995) p. 351.
233. S Ozawa, R Endo, M Susa, Tetsu-to Hagane, **93**, 8, (2007).
234. H Hasegawa; T Kowasari, Y Shiroki, H Shibata, H Ohta, Y Waseda, Met. Mater. Trans. B, **43 B**, 1411, (2012) see also H Ohta H Shibata: J Manufact. Sci and Production, **13**, 115 (2013).
235. A.Fluegel: http//glassproperties.com/thermal –conductivity/ (2011).
236. K Tsutsumi, KWatanabe, M Suzuki, M Nakada, T Shiomi, Proc.7th Intl. Conf. Molten slags, fluxes and salts, Capetown, South Africa, 2004 (SAIMM, Johannesburg, 2004) p. 803.
237. H Nakato, S Takeuchi, T Fujii, T Nozaki, M Washio, Proc 74th Steelmaking Conf., Washington, DC (1991) (ISS, Warrendale, PA, 1991) p. 639.
238. Yongsug Chung, PhD Thesis, "*Surface and interfacial phenomena in iron-alloys*" Carnegie-Mellon Univ., Pittsburgh, PA (1999).
239. T Tanaka, M Nakamoto, J Lee, Proc. Conf. Metal Separation Technology, Copper Mountain, CO. (2004) ed. by R E Aune and M Kekkonen (Helsinki Univ. Technol, 2004). p. 135.

240. Y Chung, AW Cramb, Met. Mater. Trans. B, **31B**, 957, (2000).
241. KC Mills, BJ Keene, RF Brooks, A Shirali, Phil. Trans. Roy. Soc, London, A **356**, 911, (1998).
242. H Gaye, D Lucas, M Olette, PV Riboud, Canad. Met. Quart., 23, 179, (1984).
243. PV Riboud, LD Lucas,: Canad. Met. Quart., **20**, 199, (1981).
244. A Jakobsson, M Nasu, H Manguiro, KC Mills, S Seethraman, Proc. Marangoni and interfacial phenomena in Materials Processing, Royal Soc. London, ed. by ED Hondros, M McLean and KC Mills, (IOM, London, 1998) p. 181.
245. AW Cramb, Proc.5th Intl. Conf. Molten slags, fluxes and salts, Sydney, 1997, (ISS Warrendale, PA, 1997) p. 35.
246. M Ferrari, L Liggieri, F Ravena, C Amodio, A Passerone R Miller, J Colloid Interface Sci., **186**, (1997) 40/45 and 46/52.
247. M Wegener, L Muhmood, S. Sun, AV Deev: Met. Mater. Trans B,**46B**, 318, (2015).
248. VI Yavoisky, YM Nechkin, IV Zinkovsky, VG Padalka, Sb. Mosk Inst Stali Splavov, **74**, 87, (1973).
249. AM Yakushev, VM Romashin, NV Ivanova, Steel USSR, **15**, 425, (1985)..
250. J Krusina, T Mylsevic, Kovove Mater., **16** (4), 510, (1978).
251. J Elfsberg, T Matsushita, Steel Res. Intl., **82** (4) p. 404.
252. A Kusano, N Sato, M Okimori, S Fukunaga, K Nishihara, M Sato, Y Minigawa. Proc. 74th Steelmaking Conf. (1991) (ISS, Warrendale, PA, 1991) p. 111
253. Y Cheng, Y Wang, D Li, H Miao, Continuous Casting, **2008** (4). 42,
254. T Duberstein HP Heller, P R Scheller: *Determination of the thermophysical properties of molten slags by maximum bubble pressure*. www.pyrometallurgy.co.za/MoltenSlags2012/W180pdf.
255. GF Konovalov, Izv. VUZ Chern Met., **1974** (2), 15, (1974).
256. R Hagemann, HP Heller. S Lachmann, S Seetharaman, PR Scheller, Ironmaking and Steelmaking, **39**, 508, (2014).
257. KC Mills, BJ Keene, Intl. Materials Rev., 22, 1, (1987) and Proc. Symp. Mineral matter and ash in coal, Philadelphia, PA (1984) ACS Monograph Series **301** (1986) p. 197
258. M Nakamoto, A Kiyose, T Tanaka, L Holappa, M Hamalainen, ISIJ Intl., **47**,38, (2007).
259. M Hanao, T Tanaka, M Kawamoto, K Takatani, ISIJ Intl., **47**, 935, (2007).
260. M Nakamoto, T Tanaka, L Holappa, M Hamalainen, ISIJ Intl., **47**, 211, (2007).
261. KC Mills, S Karagadde, L Yuan, PD Lee, F Shahbazian, ISIJ Intl., **56**, 264, (2016).
262. R Olivares, MP Brungs, H Liang, Met. Trans. B, **22B**, 305, (1991).
263. T Matsushita, T Ishikawa, PF Paradis, K Mukai, S Seetharaman, ISIJ Intl., **46**, 606, (2006).
264. P Quested, RF Brooks, AP Day, KC Mills, Proc. Conf. on Fluid flow phenomena in Materials Processing, San Diego, CA, 1999.
265. Y Bottinga, D Weill, Amer. J Sci., **272**, 438, (1972).
266. M Persson, T Matsushita, J Zhang, S Seetharaman, Steel Res. Intl., **78**, 102, (2007).
267. M Persson, J Zhang, S Seetharaman, Steel Res. Intl., **78**, 290, (2007).
268. A Fluegel, J Amer. Glass Technol., **90** (8), 2622, (2007).
269. A Fluegel, Thermal expansion calculation on silicate glasses at 210 °C based on systematic analysis of global databases., http//glassproperties.com.
270. A Fluegel, DA Earl, AK Varshneya, TP Seward III, Phys. and Chem. Glasses, Europ. J Glass Sci. and Technol., Part B, **49** (5), 245, (2008).
271. I K Priven: Glass Technol., **45** (6), 244, (2004).
272. JF Stebbins, ISE Carmichael, IK Moret, Contrib. Mineralogy and Petrology, **86**, 131, (1984).
273. DA Ditmars, in *Compendium of Thermophysical property measurement methods; Volume 1: Survey of Measurement techniques*, ed. by KD Maglic, A Cezairliyan and VE Peletsky, (Plenum Press, New York, 1984) Chapter 13, p 527.
274. MJ Richardson, in *Compendium of Thermophysical property measurement methods; Volume 1 Survey of Measurement techniques* ed. by KD Maglic, A Cezairliyan and VE Peletsky, (Plenum Press, New York, 1984) Chapter 17, p. 669.
275. Factsage: website www.factsage.com: for use see for example - reference [131]

276. MTDATA; website www.ntech.npl.co.uk- see J Robinson, AT Chapman, Pn Quested, BJ Monaghan, J Gisby and KC Mills: Proc. 2nd Intl. Conf. Science and Technol. of Steelmaking, Swansea, 2001 (1OM, London, 2001) p. 149.
277. Thermocalc: website www, thermocalc.com
278. SH Firoz, R Endo, M Susa, Ironmaking Steelmaking, **34**, 437, (2007).
279. JF Holzhauser, KH Spitzer, K Schwerdtfeger, Final Rept. ECSC Contract 7210 CA/ 137 (1988)(Europ. Comm. Sci. Publ., Luxemborg, 1988).
280. J Diao, B Xie, JP Xiao, Ironmaking and Steelmaking, **36**, 610, (2009).
281. J Chen, J Xie, F He, E Wan, Z Xu, C Yang, G Liu, Ironmaking and Steelmaking, **42**, 126, (2015).
282. A Yamauchi, K Sorimachi, T Sakuraya, T Fujii, ISIJ Intl., **33**, 140, (1993). See also Tetsu-to Hagane, **79** (2), 167, (1993).
283. Y Seko: BSc Thesis Tokyo Inst Technol (2006).
284. KC Mills: ISIJ Intl. **56**, 1, (2016).
285. M Kawamoto, T Watanabe, T Ikeda, Tetsu-to Hagane, **81**(12), 1132, (1995).
286. Y. Meng, B.G. Thomas, Proc. ISS Tech. Conf. (2003), Indianapolis, see also Met. Mater. Trans., **34B**, 685 and 707 (2003).
287. T Matsushita, T Watanabe, M Hayashi, K Mukai, S Seetharaman; Intl. Mater. Rev., **56**, 287, (2011).
288. H Ehrenburg, Proc. 3rd Europ. Conf. Continuous casting, Madrid, 1998 (UNESID, Madrid, 1998) p. 645.
289. G. Skoczylas, Proc. 79th Steelmaking Conf., 1996, (ISS, Warrendale, PA,1996) p. 269.
290. A Kiyose, K Miyazawa, W Yamada, K Watanabe, H Takahashi, ISIJ Intl., **36**, S 155, (1996).
291. D Gotthelf, P Andrezjewski, E Julius, H Haubrich, Proc. 3rd Europ. Conf. Continuous Casting, Madrid (1998) (UNESID, Madrid, 1998) vol 2; p. 825.
292. H Sharan, AW Cramb, Met. Mater. Trans. B, **26B**, 318, (1995) see also., ISS Trans., **15**, 95 (1995).
293. Era Kapilsharami, V Sahajwalla, S Seetharaman, Ironmaking and Steelmaking, **31**,509, (2004).
294. T Matsumiya, ISIJ Intl., **46**, 1800, (2006).
295. K Miyazawa, Sci. Technol. Adv. Mater., **2**, 59, (2001).
296. H Yamamura, T Kajitani, J Nakashima, M Yamasaki, S Mineta, Nippon Steel Technical Report, **104**, 54, (2013).
297. E Favre, S Kunstreich, MC Nove, D Rotelec, W Courths, E Korte, Proc 3rd Intl Conf. Continuous Casting, Madrid, 1998 ((UNESID, Madrid,1998) p. 595.
298. H Take, H Osanai, J Hasunuma, T Yamamoto, H Bada, H Tozawa, Proc. Conf. on Quality of ordinary steel in Iron- and steel-making, Bangkok, (1994) Session 3, Paper 1.
299. P Gardin, JM Galpin, MC Regnier, JP Radot, IEEE Trans. On Magnetics, **31**, 2088, (1995).
300. D W van der Plas, C Platvoet, B Diesesme, JP Radot, JM Galpin: Proc. 2nd Europ. Conf. Continuous Casting, Dussseldorf, 1994 (Stahl u Eisen, Dusseldorf, 1994), p. 109.
301. M Washio, M Sugizawa, S Moriwaki, K Kariyaa, S Idogawa, S Takeuchi, Revue de Metallurgie, CIT, **90** (April), 507, (1993).
302. R Koldewijn Unpublished Corus Internal Rept (2007) cited in KC Mills, J Kromhout, A Hamoen, R Boom, Proc. of Admet Conf., Dnipetrovsk, Ukraine, May 2007(Natl. Metall. Acad. Ukr., Dnipropetrovsk, 2007), vol 2 p. 174.
303. SG Kollberg, HR Hackl, PJ Hanley, Iron and Steel Engineer, **73** (7), 24, (1996).
304. E Takeuchi, JOM, **1995** (May), 42, (1995).
305. E Takeuchi, H Harada, H Tanaka, T Ishii, T Toh, M Zeze, M Hojo, K Shigematsu, Nippon Steel Tech. Report, **61**, 29, (1994).
306. J Kromhout, RS Schimmel, Proc. 8th Europ. Conf. Continuous casting, Graz, Austria, June, 2014. (Austrian Met. Mater. Soc., 2014).
307. M Tani, T Toh, K Umetsu, K Tanaka, M Zeze, K Tsunenari, K Hayashi, S Fukunaga: Nippon Steel Tech. Report, **104**, 62, (2013).
308. P Rocabois, J Lehmann, C Gatellier, JP Teres, Ironmaking and Steelmaking, **30**, 95, (2003).

309. PV Riboud M Olette, J Leclerc, W Pollak, Proc. 61st Conf NOH- BOS Conf., Chicago, IL, 1978, (ISS, Warrendale, PA, 1978) p. 411.
310. T Emi, Proc. 74th Steelmaking Conf., Washington, DC, 1991, (ISS Warrendale, PA, 1991) p. 623.
311. M Bruhl, J Sardemamm, F Oeters. Proc. Scanjet VII part II 7th Intl Conf. Refining Processes (1996) p. 171.
312. X Yu, RJ Pomfret, KS Coley, Met. Trans B, **28B**, 275, (1997).
313. AB Fox, ME Valdez, J Gisby, RC Attwood, PD Lee,S Sridhar, ISIJ Intl., **44**, 836, (2004).
314. P Misra, S Sridhar, AW Cramb, Met. Mater. Trans. B, **32B**, 963, (2001).
315. H Lei, Y Zhao, DQ Geng, ISIJ Intl., **54**, 1629, (2014).
316. K Koyama, K Nagano, Y Nagano, T Nakano, Nippon Steel Tech. Report, **34**, 41, (1987).
317. T Araki *et al.* Proc. 5th Japan-USSR Joint Symp. On physical chemistry of metallurgical processes. Publ. ISIJ Special Report No **22**, 263, (1975).
318. YQ Liu, LJ Wang, KC Chou, ISIJ Intl., **54**, 728, (2014).
319. JL Li, QF Shu, YA Liu, KC Chou, Ironmaking and Steelmaking, **41**, 732, (2014).
320. VS Esaulov, GF Konovalov, SI Popel, Izv VUZ Chern Met., **1976** (8), 36, (1976).
321. M Valdez, K Prapacom, AW Cramb, S Sridhar,. Steel Research, **72**, 291, (2000).
322. L Teng: KTH Report "*Inclusion absorption capacity of mould flux slag for Ingot casting*" (IPTINGOT) Dept. Materials Science and Eng. KTH Stockholm (2013).
323. P Scheller, Proc. Mills Symp., London (2002) (IOM, London, 2002) p. 477.
324. P Scheller, Proc. 3rd Europ. Conf. Cont. Casting, Madrid,1998 (UNESID, Madrid,1998) p. 797.
325. RE Fash, WF Salem, DC Evans, Proc. 74th Steelmaking Conf., Washington, DC, 1991. (ISS, Warrendale, PA, 1991).
326. R Branion, DA Dukelow, GD Lawson, J Schade, M Schmidt, HT Tsai, Proc. 78th Steelmaknig Conf. (1995) (ISS, Warrendale, PA,1995) p. 647..
327. MJ Frazee, Proc. 78th Steelmaking Conf. (1995) (ISS, Warrendale, PA, 1995) p. 639.
328. NT Mills, BN Bhat, Iron and Steelmaker, **1978** (Oct), 18, (1978).
329. KH Spitzer, *et al.*, Stahl u Eisen, **108**, 441 (1988).
330. KH Spitzer, K Schwerdtfeger: "*Investigation of the isolating properties and melting behaviour of mould powders for continuous casting*". Tech. Univ. Clausthal Report **7249** (Dec 1989).

Chapter 10
Selection of Mould Fluxes and Special Mould Fluxes for Continuous Casting

Abstract The various factors influencing the selection of mould powder compositions are discussed in this chapter. Mould powders contain two types of component (i) the minerals which form the slag film and (ii) Carbon which controls the melting rate. Both are vital to the successful performance of the mould powder. The composition of the mould powder is determined by the mould dimensions, the casting conditions and the steel grade being cast. Empirical rules have been developed to provide the required values of the powder consumption (Q_s^{req}), the viscosity (η^{req}) and the break temperature (T_{br}^{req}) for satisfactory casting (i.e. free from longitudinal cracking and sticker break-outs). More than 85% of mould powders conform to these empirical rules. However, other factors can also influence the powder selection process. These include

(i) Increasing the slag viscosity or the interfacial tension to reduce slag entrapment.
(ii) Increasing the interfacial tension to minimise scaling of the steel surface.
(iii) Minimising carbon pick-up (especially for LC and ULC steels).
(iv) Providing both support in the mould and optimal heat transfer when casting round billets.
(v) Providing reasonable heat transfer in mould configurations with corners acting as heat sinks.
(vi) Handling the large amounts of Al_2O_3 formed when casting high-Al steels whilst maintaining good lubrication throughout the casting.
(vii) Handling Ti-stabilised steels which form either TiN or Ti(CN) in the slag pool (which has low solubility in mould slag) or $CaO{\cdot}TiO_2$ (which has a high T_{liq}).
(viii) Reducing the fluoride content of slag to minimise environmental and health concerns.

All of these problems are discussed in detail and possible solutions are outlined.

The original version of this chapter was revised: For detailed information please see correction. The correction to this chapter is available at https://doi.org/10.1007/978-3-319-53616-3_12

Within this book minor changes to some of the figure permissions and the readability of the chapters have been made and several figures have been replaced with new versions.

K.C. Mills and C.-Å. Däcker, *The Casting Powders Book*,
DOI 10.1007/978-3-319-53616-3_10

Symbols, Abbreviations and Units

D_C	Diameter of carbon particles (m)
f	Frequency (Hz or cpm)
f^*	fraction of powder forming slag
Q_s	Powder consumption (kgm^{-2})
q	Heat flux density (Wm^{-2})
R^*	(Surface area/Volume)—mould (m^{-1})
s	stroke (m)
T	Temperature (°C)
T_{br}	Break or Solidification temperature
t	Thickness of mould (m) or time (s)
V_c	Casting speed (m min^{-1})
w	Width of mould (m)
η	Slag viscosity (dPas)
ρ	Density (kg m^{-3})
EMBr	Electromagnetic braking
IR	Infrared radiation
LC	Low-carbon steel
MC	Medium-carbon steel
SEN	Submerged entry nozzle
ULC	Ultra-low-carbon steel

In chemical formulae

A	Al_2O_3
C	CaO
F	FeO
Fl	CaF_2
M	MgO
S	SiO_2
T	TiO_2

10.1 Introduction

It is obviously important to select the best mould powder for the given casting conditions in order to obtain problem-free casting and the minimum of defects in the steel product. Mould powder components can be divided into (i) the mineral components (oxides and fluorides) and (ii) carbon particles. The mineral components are the most important since they form the mould slag and, subsequently, the slag film which largely determines the performance of the mould powder. The carbon particles play a minor role, i.e. in controlling the melting rate of the powder;

however, it is absolutely essential that the melting rate matches the demand for liquid slag. Consequently, the selection of the carbon content is also important.

It is shown below that selection of the mineral components is dependent upon the casting conditions and the steel grade being cast. Much of our present knowledge is based on "ad hoc" research where compositional changes were made and the results observed. Mould powders giving a good performance were obtained in this way and it was only later that the logic underlying the selection of mould powder composition was discovered. Certain empirical rules were deduced for the selection of mould powders. However, with our expanding knowledge of how mould slags work, there is evidence that scientific knowledge is being used to develop efficient mould powders (such as in the development of slags to cast both high-Al steels and round billets which are described below).

Using the empirical rules developed to predict slag properties required for the given casting conditions and steel grade, it was found that 85–90% of the experimental values were consistent with the predicted slag properties. The other 10–15% usually does not fit because they have been modified to carry out further duties, e.g. reduce either slag entrapment or SEN erosion rates. It has always been the case that mould powders are expected to be "*flexible*" or "*forgiving*", i.e. will help to alleviate problems caused by the casting conditions imposed. These special cases are identified below along with the special slags developed for certain conditions.

10.2 Selection of Mineral Compositions of Mould Powder for Given Casting Conditions

The molten mould slag must perform the following functions satisfactorily:

- It must form a liquid slag pool which seals off the molten steel from the gaseous atmosphere and thereby prevents oxidation and nitration of the steel.
- The powder bed must provide thermal insulation to prevent freezing of the steel meniscus (i.e. *skull formation*).
- The liquid slag pool should absorb inclusions and gas bubbles from the steel.
- The *liquid layer* of the slag film formed between the shell and the mould must be sufficiently thick to provide good lubrication of the newly formed shell.
- The *solid layer* of the slag film must reduce the horizontal heat flux to a suitable level for the steel grade being cast.

Although all of these five tasks are important, the selection of mould powder compositions is mainly determined by the latter two conditions. It should always be remembered that mould powders are selected for the effect that they have on the newly formed shell. The principal factors affecting the selection of mould powder compositions and properties are

- The mould geometries selected for the casting.
- The casting conditions used in the casting.
- The type of steel being cast.

10.2.1 *Effect of Mould Geometry on Mould Powder Selection*

The mould geometry affects the level of lubrication required by the newly formed shell.

It is customary to classify the mould dimensions into four main classes

- Billets.
- Blooms.
- Slabs.
- Thin slabs.

In addition, there are (i) round billets (ii) beam blanks and (iii) hollow billets (for pipe production) [1] which all have their own specific demands.

The liquid friction forces (F_l) acting on a specific face of the shell tend to increase with increasing distance from the corner (Fig. 10.1) [1–3]. Thus the frictional forces acting at the centre-line of slabs tend to increase with increasing slab width (w) [2, 3]. The powder consumption, Q_s (in kgm^{-2}) is frequently used as a measure of the lubrication supplied to the shell. Since the frictional forces increase with increasing width, it follows that more lubrication must be supplied as the width increases. Consequently, the *required* powder consumption, Q_s^{req} increases with

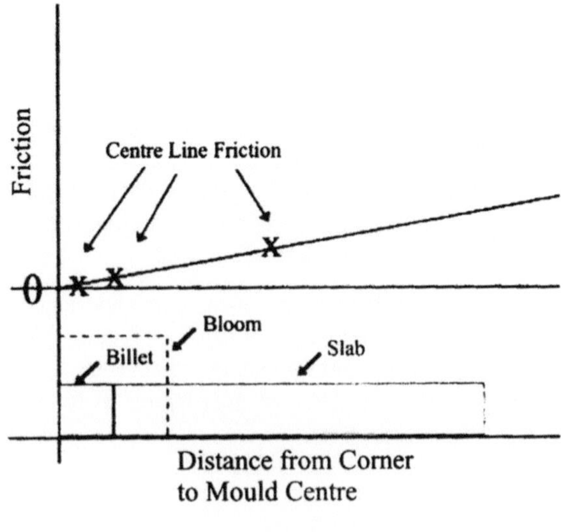

Fig. 10.1 Diagram showing centre-line friction increasing with increasing distance from the corner; X = friction at *centre-line* for 3 mould geometries; *solid line* billet; *dotted line* bloom and *solid line* slab; (*courtesy* A.B. Fox [3])

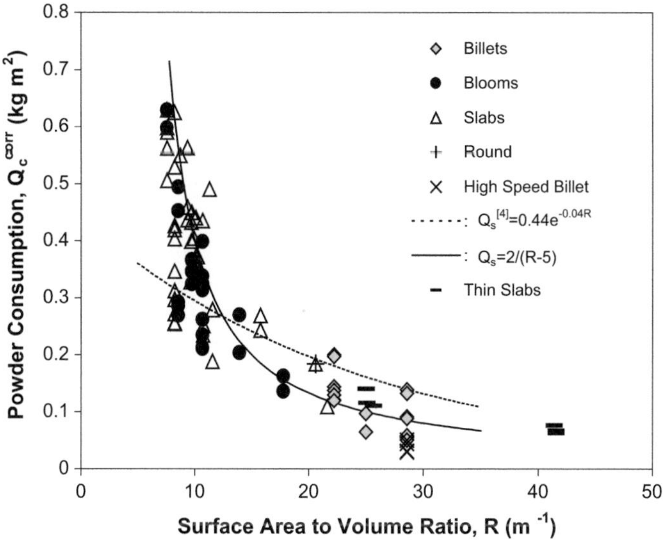

Fig. 10.2 Required powder consumption, Q_s^{req}, as a function of the parameter, $R*$ for different mould geometries; ◇ = billets; o = blooms; Δ = slabs; ▬ = thin slabs; + = Rounds; *X* high-speed billets; *dotted line* equation, $Q_s = 0.44\ \exp(-0.04R*)$ [6] *solid line* ($Q_s = 2/(R* - 5)$) [4, 5]. Typical ranges of $R*$ values; slabs (8–15) blooms (12–16) billets (20–30) thin slabs (25–40); (permission granted, UNESID [5])

increasing width (w) of the wide face. Furthermore, Q_s^{req} values are in the hierarchy slabs > blooms > billets as can be seen in Fig. 10.2 [4, 5].

Although relationships have been derived for Q_s^{req} as a function of the distance from the centre-line and corner ($d_{cent\text{-}corner}$) [3], it is customary to use the parameter, $R*$ rather than ($d_{cent\text{-}corner}$) [6]. The parameter, $R*$ represents the ratio {surface area/volume} of the mould and is calculated by Eq. 10.1 where w and t are the width and thickness of the mould. For round billets, $R* = 2/r$, where r is the radius of the mould.

$$R^* = 2(w + t)/\,(w.t) \tag{10.1}$$

The mould powder contains both carbon [which is oxidised to CO (g) or CO_2 (g)] and carbonates (which form CO_2 (g) during the heating process in the powder bed). Consequently, not all the powder forms slag and it must be remembered that it is the slag which provides the lubrication of the shell. Consequently, it is necessary to calculate the fraction of the powder forming slag ($f*$) using Eq. 10.2 where $\%C_{free}$, $\%C_{total}$ are free and total C contents and LOI is the loss on ignition. Thus, it is necessary to differentiate between the *powder* and the *slag* which are denoted by superscripts ^{powd} and ^{slag}, respectively

$$f^* = 0.01\,\{100 - \%C_{\text{free}} - (44/12)\,(\%C_{\text{total}} - \%C_{\text{free}})\} = (100 - \%\text{LOI})/100 \tag{10.2}$$

Powder consumption can be expressed in various ways, the parameter Q_t is the powder consumption measured on plant with units of kg (powder) tonne (steel)$^{-1}$ and represents the amount of mould powder used per tonne of steel. Another powder consumption term is the melting rate, Q_{MR} (in kgs^{-1} or kg min^{-1}) which is a measure of the rate of slag supply. It is essential that the *actual* melting rate should match the *required* powder consumption, Q_{MR}^{req}. These various terms (Q_t, Q_s and Q_{MR}) can be calculated by Eqs. 10.3–10.4, respectively where ρ is the density and d_l is the thickness of the molten slag layer.

$$Q_s^{slag} = f^* 7.6 \cdot Q_t / R^* = \rho d_l \tag{10.3}$$

$$Q_{MR}^{slag}\,(\text{kgs}^{-1}) = 2\,(w+t) \cdot Q_s \,.\, V_c / 60 \tag{10.4}$$

It can be seen from Fig. 10.2 that Q_s^{slag} is a function of the parameter R^* and that wide slabs (with R^* = *ca.*10) require considerably more powder consumption (i.e. lubrication) than billets (with R^* = *ca.*30). The relationship between Q_s^{slag} and R^* is shown in Fig. 10.2 and can be represented by Eq. 10.5 [4, 5]; it is also compared with an earlier relationship reported by Neumann et al. [6] (i.e. Q_s^{slag} = 0.44 exp ($-0.04R^*$) in this figure.

$$Q_s^{slag} = 2/(R^* - 5) \tag{10.5}$$

Thus it is possible to calculate the *required* powder consumption (Q_s^{slag}) for the given mould dimensions.

10.2.2 Effect of Casting Conditions on Mould Powder Selection

The powder consumption, Q_s^{slag}, is also affected, by factors other than the mould dimensions, principally by the casting speed and the viscosity (η) and, to a lesser extent, by the frequency (f) stroke (s) and the break temperature (T_{br}).

Empirical relationships have been reported for the required powder consumption, Q_s^{req} as functions of casting speed and viscosity, such as those shown in Eqs. 10.6–10.8 [3, 7–9].

Wolf [7]:

$$Q_{s\,req}^{powd} = 0.7/\eta^{0.5} \cdot V_c \tag{10.6}$$

Ogibayashi [8]:

$$Q_{\mathrm{s\,req}}^{\mathrm{powd}} = 0.6/\eta \cdot V_{\mathrm{c}} \tag{10.7}$$

Subsequently, Eq. 10.5 was modified to provide a better fit with plant data [3, 9].

$$Q_{\mathrm{s\,req}}^{\mathrm{slag}} = 0.55\,/\eta^{0.5} \cdot V_{\mathrm{c}} \tag{10.8}$$

The effect of other casting variables on the powder consumption has been reported by a number of investigators; the reported relationships vary considerably on the effect of the variable on the powder consumption [3]. A typical equation involving *f*, *s* and T_{br} is shown in Eq. 10.6 [10].

$$Q_{\mathrm{s}} = \frac{k\beta}{T_{\mathrm{sol}}^{1.6}\eta^{0.5}} \frac{s^{0.4}}{V_{\mathrm{c}}} \cos^{-1}\left[\frac{1000V_{\mathrm{c}}}{2\pi fs}\right] \tag{10.9}$$

The effects of *f*, *s* and T_{br} tend to be relatively small and thus empirical rules involving only V_{c} and η work reasonably well. Thus, the *required powder consumption* ($Q_{\mathrm{s\,req}}^{\mathrm{slag}}$) can be expressed in terms of a *required viscosity*, η^{req} (at 1300 °C) for the given casting speed, etc. using Eqs. 10.6–10.8. It should be noted that mathematical models of the slag infiltration into the shell/mould channel indicate Q_{s} increases with increasing frequency whereas plant trials indicate the reverse relationship. It has been suggested that increasing frequency increases the number of *periods of confused flow* accompanying the changes in mould direction (little slag infiltration occurs in these periods) and thus, increased frequency lowers powder consumption [11]. It should also be noted that the amount of slag required per tonne of steel is reasonably constant ($f^*Q_{\mathrm{t}} = 0.48$ kg tonne^{-1} [3]) for most casting but is not valid for high-viscosity, billet slags and cases where slag infiltration is restricted (e.g. casting of Ti-stabilised stainless steels).

10.2.3 Effect of Steel Grade on Mould Powder Selection

The type of steel being cast also affects the selection of the mould powder composition. When molten steels with C contents in the range 0.06–0.17% (denoted MC steels) solidify, they form a δ-Fe shell, which, subsequently, undergoes a *peritectic* phase transition to γ-Fe (*austenite*) (Fig. 10.3).

$$\mathrm{Fe\ (liq)} + \delta - \ \mathrm{Fe} \rightarrow \gamma - \mathrm{Fe} \tag{10.10}$$

The transition ($L + \delta \rightarrow \gamma$) is accompanied by a volume change of (0.4–0.6%) due to the tighter packing in the *austenite* (*cf. ferrite*) and there is also a 4% mismatch between the thermal shrinkage coefficients of δ-Fe and γ-Fe [12]. These changes give rise to wrinkled shells with considerable variations in shell thickness.

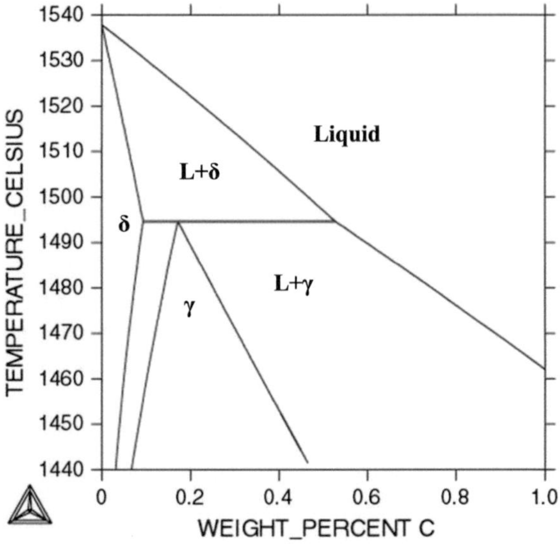

Fig. 10.3 The Fe–C phase diagram for the *peritectic* range (0.08 < wt%C < 0.55) [13]; drawn by Swerea KIMAB

The mismatch in shrinkage coefficient, coupled with an uneven shell, creates an accumulation of stresses which are relieved by longitudinal cracking. The strategy used to combat longitudinal cracking involves the creation of a thin, uniform shell and this is achieved using a thick, crystalline, slag film. A thick slag film is obtained by producing a mould slag with a high solidification (or break) temperature (T_{sol} or T_{br}) and this is achieved by increasing the slag basicity (*C/S*) up to 1.3, if necessary. Increased basicity also promotes further crystallisation in the slag film.

High-carbon (HC) shells tend to be weak and thus thin shells are prone to sticker break-outs and hot tears. Consequently, thick shells are needed to provide the necessary strength. This is achieved by promoting high heat transfer between the shell and mould by use of a glassy slag film with a low T_{br}; i.e. the reverse of the practice used for MC steels prone to longitudinal cracking. Slags used to cast HC steels tend to have a low basicity.

Consequently, when selecting a mould powder it is necessary to differentiate between MC, HC and other steels on the basis of steel composition.

10.2.4 Routines to Differentiate Between Steel Grades

Several routines have been explored to differentiate those steels undergoing the *peritectic* reaction [14–16]. In plain carbon steels the *peritectic* range lies between 0.06 and 0.17%.

Wolf [17, 18] introduced the term "*ferrite potential*" as a measure of the amount of peritectic reaction occurring in the steel. Although the ferrite potential is affected by the carbon content, it is also affected by other alloying elements; some of these elements stabilise the *ferrite* (Cr, W, Mo, Al and Si) whilst others stabilise the *austenite* (Ni, Mn, Co, N and Cu). For *low-alloy steels,* the ferrite potential (*FP*) can be calculated by Eq. 10.11 where *CP* is the carbon potential which is defined in Eq. 10.12.

Low alloy

$$\mathrm{FP} = 2.5(0.5 - \mathrm{CP}) \tag{10.11}$$

$$\begin{aligned}\mathrm{CP} = {} & (\mathrm{wt\%C}) + 0.04(\%\mathrm{Mn}) + 0.1(\%\mathrm{Ni}) + 0.7(\%\mathrm{N}) - 0.14(\%\mathrm{Si}) \\ & - 0.04(\%\mathrm{Cr}) - 0.1(\%\mathrm{Mo}) - 0.24(\%\mathrm{Ti})\end{aligned} \tag{10.12}$$

For *stainless steels* the *FP* is given by Eq. 10.13 where "*Ni*" and "*Cr*" are defined in Eqs. 10.14 and 10.15, respectively, [17, 18].

Stainless steel

$$FP = 5.26\left(0.74 - \text{“}Ni\text{”}/\text{“}Cr\text{”}\right) \tag{10.13}$$

$$\text{“}Ni\text{”} = (\mathrm{wt\%Ni}) + 0.31(\%\mathrm{Mn}) + 22(\%\mathrm{C}) + 17.5(\%\mathrm{N}) \tag{10.14}$$

$$\text{“}Cr\text{”} = (\mathrm{wt\%Cr}) + 1.5\,(\%\mathrm{Si}) + 1.65\,(\%\mathrm{Mo}) + 2(\%\mathrm{Nb}) + 3(\%\mathrm{Ti}) \tag{10.15}$$

The peritectic reaction is at a maximum at *FP* = 1.0 and the range, *FP* = (0.8–1.05) corresponds to the crack-sensitive range (Fig. 10.4).

Other investigators have reported coefficients which differ from those given in Eq. 10.12, these are summarised in [14–16] (Table 10.1).

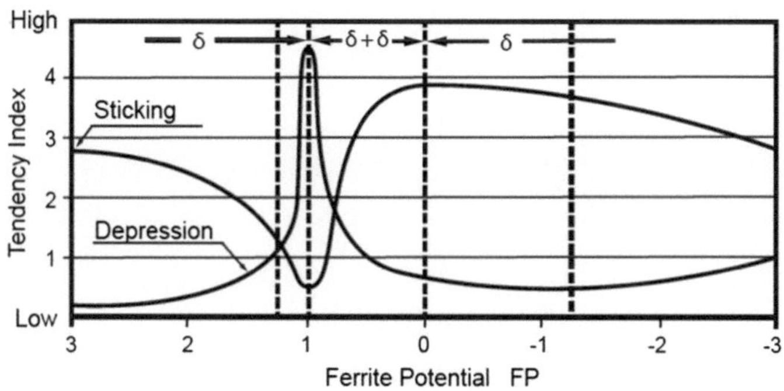

Fig. 10.4 The tendencies of steels towards longitudinal cracking (depression) and sticker break-outs as functions of the ferrite potential [17, 18] (permission granted, VDEh, Verlag Stahleisen, [18]; re-drawn by Swerea KIMAB)

Table 10.1 Coefficients for selected equations to calculate Carbon potentials (after [14, 18]); Al_t = total Al

Ref.	Mn	Si	N	P	S	V	Ti	Cu	Cr	Ni	Mo	Others
Wolf	0.01	−0.1			−0.7				−0.04	0.04	−0.1	
Trico	0.02	0.009	0.05	0.008	0.17	0.009	0.007	0.007	0.003	0.02	−0.007	Al_t 0.05 Nb 0.04 Sn 0.0006 B 1.32 Ca −0.24
BSSTC	0.043	−0.14	1.06	0.029	0.11	−0.13	−0.024	0.037	−0.083	0.1	−0.003	
SMS	0.14	−0.037		−0.04	0.222			0.003		0.023	−0.004	
Howe	0.04	−0.14	0.7				−0.24		−0.04	0.1	−0.1	

10.2.5 Plots of T_{br} as a Function of Slag Viscosity

Figure 10.5 shows a plot of T_{br} as a function of slag viscosity for a number of mould slags used to cast MC and HC steels along with low and intermediate C steels (LC and IC) [5, 9]. It can be seen that slags used for MC, crack-sensitive steels fall on the upper curve, those used with HC steels on the lower curve and those for LC and IC steels tend to fall between the two curves. Such plots represent *inverse lubrication* (viscosity) as a function of *inverse heat transfer* (T_{br}) and thus, indicate the optimum properties for casting. It should be noted that the heat transfer is also dependent upon the fraction of crystalline phase (f_{crys}) in the slag films. This type of plot works because slag basicity affects both T_{br} and f_{crys} in the same way (i.e. the heat flux decreases).

The data shown in Fig. 10.5 were expressed as three curves (Fig. 10.6) and Eqs. (10.16–10.18); these can be used to calculate the required break temperature (T_{br}^{req}) from the required slag viscosity (η^{req}) for three different steel grades.

Crack-sensitive

$$CP = 0.06 - 0.18 \quad T_{br}(^{\circ}C) = 1157 + 60 \ln \eta^{req} \tag{10.16}$$

Sticker-sensitive

$$CP > 0.4 \quad T_{br}(^{\circ}C) = 1051 + 76.4 \ln \eta^{req} \tag{10.17}$$

Others (intermediate)

$$CP\, 0 - 0.06 \text{ and } 0.18 - 0.4 \quad T_{br}(^{\circ}C) = 1103 + 68.5 \ln \eta^{req} \tag{10.18}$$

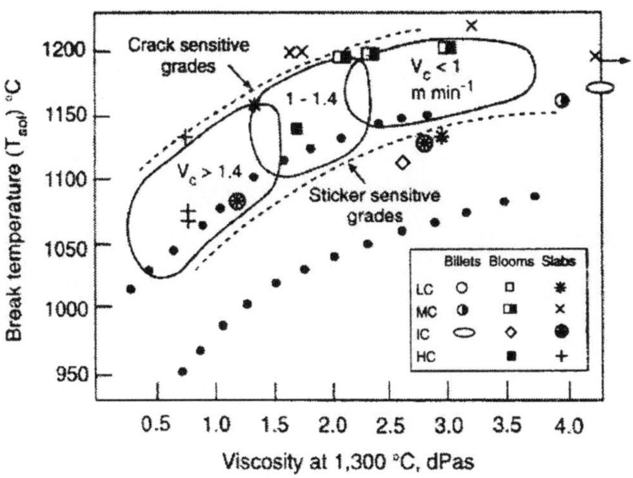

Fig. 10.5 The break temperature (T_{br}) as a function of slag viscosity for slags used in the casting of slabs, blooms and billets; (permission granted, UNESID [5])

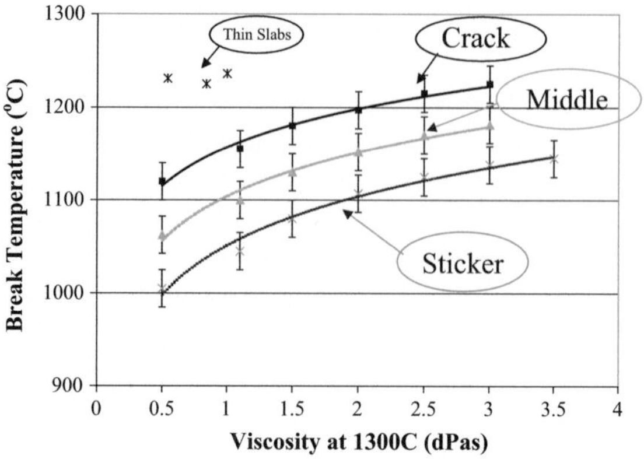

Fig. 10.6 The required break temperature (T_{br}^{req}) as a function of required viscosity (η^{req}) for different steel grades; *upper curve* crack-sensitive (MC) steels; *lowest curve* sticker-sensitive (HC) steels; *middle curve* other steels; (courtesy of A.B. Fox [3])

Thus, the required powder consumption can be expressed in terms of the required values for Q_s, viscosity and break temperature. As mentioned above, it was found that 85–90% of the plant data conformed with the calculated values of Q_s^{req}, η^{req} and T_{br}^{req} [9].

10.2.6 Other Casting Conditions Affecting Powder Consumption

There are reasons why the above empirical rules do not apply to 10–15% of the mould powders. These exceptions fall into two classes

(i) Where slag compositions are deliberately altered in order to combat other problems, e.g. by increasing viscosity (thereby reducing Q_s) to tackle high levels of slag entrapment or SEN erosion.
(ii) Where other conditions (e.g. metal and Ar flows) affect the powder consumption, etc.

The first class of exceptions is discussed in Sect. 10.3. With regard to the second group, there are a number of conditions which affect powder consumption, these are

Steels containing Titanium

TThe Ti in these steels reacts with nitrogen and carbon in the bed to form Ti(CN) particles which have a low solubility limit in the slag and thus remain largely as Ti (CN) particles in the slag pool. These particles agglomerate and block the entrance to the mould/shell channel (see Fig. 2.16) and thereby reduce the powder consumption [19] and also increase the slag viscosity; thus, Q_s values tend to be low when casting these steels.

Metal flow

Electromagnetic braking (EMBr) is frequently used to deal with the effects of turbulent metal flow in the mould. One consequence of applying EMBr is that it

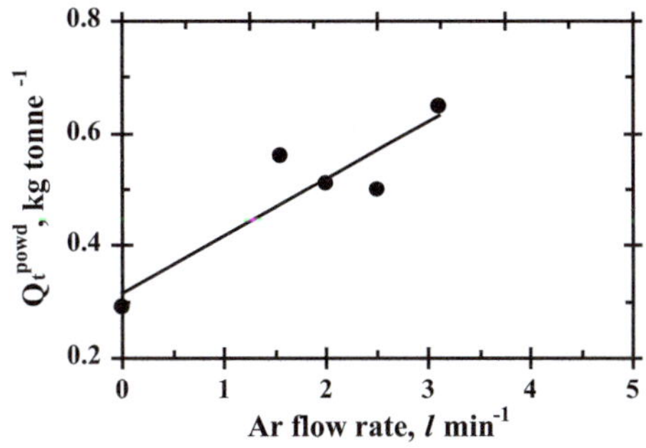

Fig. 10.7 Powder consumption, Q_t (in kg tonne^{-1}) as a function of Ar flow rate (courtesy of A.B. Fox [21])

reduces the vertical heat transfer [20] which leads to shallow slag pool depths and low powder consumption. There are a number of possible remedies for dealing with these conditions, e.g. (i) using a deeper powder bed (ii) incorporating exothermic agents into the mould powder (iii) reducing the C% in the powder.

Argon flow

It can be seen from Fig. 10.7 that the powder consumption, Q_t is affected by the argon flow rate [21]. This is thought to be due to the increased vertical heat transfer resulting from the gaseous convection which promotes a deeper pool. However, at higher flow rates (>5 litre min^{-1}) the Ar flow causes a switch in the direction of the metal flow (double roll to single roll) and higher vertical heat flux values [22, 23].

10.3 Selection of Carbon Components of Mould Powders

As mentioned above, the principal function of the carbon particles is to control the melting rate. Carbon is non-wetting to molten slag and thus hinders the agglomeration of molten slag globules and the formation of a molten slag pool (Fig. 10.8). However, the carbon particles are gradually oxidised by both O_2 (g) and CO_2 (g) present in the bed and eventually, the liquid slag forms a molten slag pool.

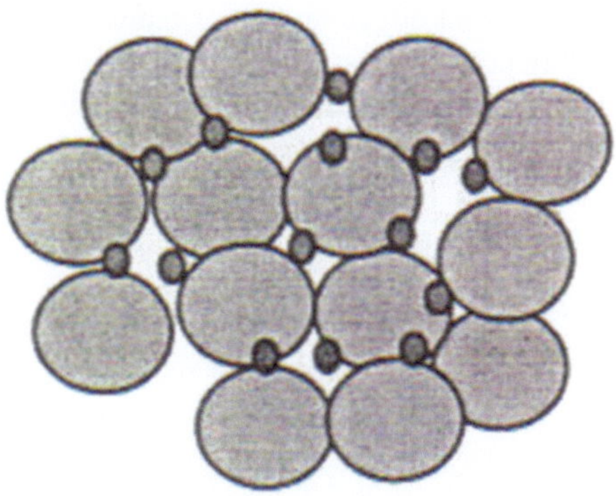

Fig. 10.8 Schematic diagram showing non-wetting carbon particles (•) preventing the agglomeration of liquid slag globules (O)

The melting rate (Q_{MR}) must match the required powder consumption (Q_s^{req}). Several factors affect the melting rate with Q_{MR} increasing (Q_{MR} ↑) as (i) %C decreases (ii) the mean size of Carbon particles increases (D_C ↑) (iii) the bulk density of the powder decreases (ρ_{bulk} ↓) and (iv) as the carbonate content increases (%CO_3^{2-} ↑) [24]. The melting rate is usually controlled by the carbon content and, to a lesser extent, by the size of the carbon particles. Different types of carbon (e.g. lamp black or coke breeze tend to have small and large diameters, respectively) and powder manufacturers obtain the appropriate mean particle size by blending the various forms of carbon.

Since the carbon is dependent upon both C% and D_C and manufacturers rarely provide detailed data for carbon size distribution, thus it is difficult to derive empirical rules in a similar way to that for the mineral content. Plots correlating melting rate (Q_{MR}) with just one carbon parameter (either C_{free}% or C_{total}%) were found to show significant scatter [3].

However, care must be taken when reducing the C% since carbon is a fuel (exothermic) and if the reduction of C% is too stringent the steel meniscus can freeze and form a *skull*. Carbon-free powders have been developed and are available commercially [25–28].

10.4 Mould Powder Selection for Special Conditions

It was reported above that 85–90% of the mould powders in an extensive database of plant data fitted the empirical rules used to cast described above [9], the 10–15% of powders which do not conform with these empirical rules arise because the mould powder is used to alleviate special conditions arising in the casting. These special conditions are described below.

Reducing slag entrapment

The drive for higher production rates has resulted in increases in casting speed, High casting speeds, in turn, have led to high metal flow velocities and turbulent flow in the mould. Slag entrapment is principally caused by these turbulent flows. There are several mechanisms which are responsible for slag entrapment [29–32].

The most effective way of reducing entrapment is by reducing the effect of the turbulence, for instance, by introducing EMBr or by adjusting SEN immersion depths to minimise the turbulence. However, some workers have chosen to change the mould slag properties to minimise slag entrapment. Slag entrapment can be reduced by increasing (i) the slag viscosity and (ii) the interfacial tension (γ_{msl}) between the metal and slag [29, 30].

Viscosity

A number of workers have reduced slag entrapment using mould powders with higher viscosities. This practice does reduce slag entrapment but also reduces

Fig. 10.9 Viscosity as a function of shear rate for Newtonian (O; *dashed line*) and non-Newtonian slags (●; *solid line*) (permission granted, ISIJ [34])

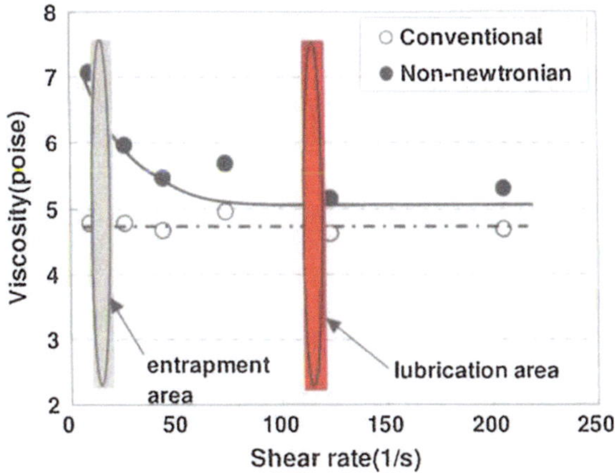

powder consumption and consequently, should be used with caution. The slag viscosity can be raised by (i) increasing SiO_2 and Al_2O_3 content or by (ii) reducing the network-breaking oxides (e.g. CaO) or the flux content (e.g. Na_2O).

Tsutsumi et al. [33, 34] used an interesting approach involving non-Newtonian slags. The viscosity of a non-Newtonian slag decreases at high shear rate (Fig. 10.9). These workers argued that the shear rate was low at meniscus level where slag entrapment occurs but was much higher in the region where slag infiltration occurs. Thus, viscosity is high (7 dPas) at meniscus level but lower in the infiltration region (5 dPas). Non-Newtonian slags contain about 0.2% Si_3N_4 and form some Si–N bonds which are readily broken at high shear rates, thereby, causing a decrease in viscosity.

Interfacial tension

High interfacial tension between the metal and slag tends provides some resistance to slag entrapment. The interfacial tension (γ_{msl}) can be calculated by Eq. 10.19 where φ is an interaction coefficient, which frequently adopts a value of *ca.* 0.5.

$$\gamma_{msl} = \gamma_m + \gamma_{sl} - 2\varphi(\gamma_m \cdot \gamma_{sl})^{0.5}. \tag{10.19}$$

The principal way of increasing the surface tension of the slag is to reduce the content of the surface-active components of the slag (B_2O_3; K_2O; Na_2O; CaF_2). However, the surface tension of metal (γ_m) is the dominant term in Eq. 10.19 (γ_m = *ca* 4 γ_{sl}). Consequently, the most effective way of increasing γ_m is to reduce the S content of the steel but the reduction of B_2O_3, K_2O, Na_2O and CaF_2 in the slag will also help.

10.4.1 Thin-Slab Casting

The parameter, R^* given in Eq. 10.1 has a high value for thin slabs (e.g. 30–40) which suggests that the required powder consumption (Q_s^{req}) is low like that for casting billets. However, the frictional forces acting on the centre regions of the wide faces of the thin slabs are almost certainly higher than the friction forces exerted on billets and consequently, require more lubrication (i.e. powder consumption) than that calculated from Eq. 10.5. In thin-slab casting, frequently the depth of the slag pool is low and the powder consumption is well below the calculated Q_s^{req} value (Fig. 10.10) [35]). In thin-slab casting, speeds are frequently very high in order to maintain high production rates, Electromagnetic braking (EMBr) is frequently used to minimise metal flow turbulence in the mould. The use of EMBr results in a decrease in vertical heat flux [20] which causes a reduction in slag pool depth; so frequently, slag pool depths are very low in thin-slab casting. Consequently, the mould flux may be modified to (i) increase powder consumption and (ii) increase the depth of the slag pool.

The powder consumption can be increased by decreasing the slag viscosity; but in practice, the slag viscosities are already very low (because of the high casting speeds used).

The depth of the slag pool can be increased by (i) improving the thermal insulation of the mould and (ii) increasing the melting rate of the powder. The thermal insulation is improved by using (i) deeper powder beds and (ii) exothermic agents (e.g. Ca/Si or Fe/Si) in the powder.

The melting rate can be increased by (i) reducing the $C\%$ in the powder $(\%C \downarrow)$ (ii) increasing the average particle size of the carbon particles ($D_C \uparrow$) (iii) increasing the carbonate content ($\%CO_2 \uparrow$) and (iv) reducing the bulk density of the powder ($\rho_{bulk} \downarrow$) [24]. The C content tends to be the dominant factor and a reduction in $C\%$ does result in a deeper pool. However, some workers have reduced both $C\%$ and

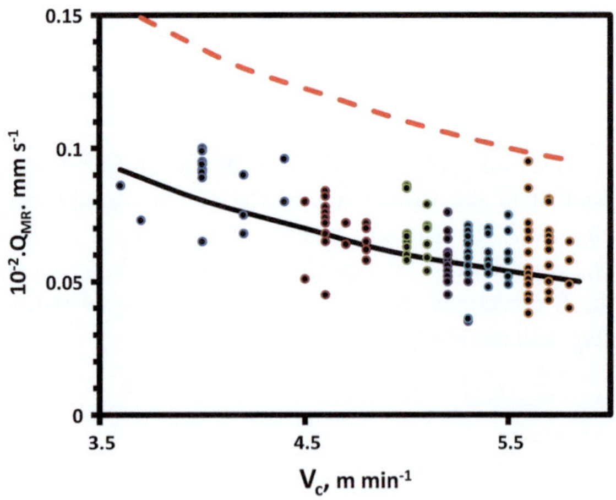

Fig. 10.10 Powder consumption as a function of casting speed; •, *solid line* measurements on thin slab caster; *dashed line* values predicted b equation $Q_{s\ req}^{slag} = 0.55/\eta^{0.5} \cdot V_c$ (permission granted, Tata Steel, re-drawn from [35]); Tata Steel point out that the results were obtained with an experimental powder, thus, the results shown may not be typical for a regular thin-slab powder

D_C, simultaneously; the latter tends to work against any improvement obtained with the reduction in $\%C$. Furthermore, care must be taken to avoid reducing the carbon content too low since Carbon is an exothermic agent and freezing of the steel surface ("*skull formation*") can occur if the $C\%$ is too low. In summary, a deeper pool can be obtained by (i) using a deeper powder bed (ii) reducing the $C\%$ in the powder.

10.4.2 Round Billets

It has been pointed out that when casting round billets, it is necessary to provide good support to the shell all around the mould [36]. It is also necessary to control the heat extraction rate in order to minimise longitudinal cracking; this is especially so when casting *peritectic*, MC steels. This is usually achieved by using a powder with a (*C/S*) ratio >1 to create a slag film containing crystalline phases. The crystallisation causes a reduction in heat flux by (i) scattering of IR radiation and (ii) the formation of an interfacial resistance ($R_{Cu/sl}$) and (iii) porosity in the slag film resulting from the shrinkage accompanying crystallisation. The interfacial resistance at the copper/slag interface is often referred to as "*surface roughness*" or as a "gas *gap*". The *surface roughness* between mould and slag results in poor support for the round billet. Consequently, a different type of mould powder was developed which produced a glassy slag film (i.e. containing no crystalline phases) with no surface roughness [36]. The heat flux was reduced by adding transition metal oxides (e.g. MnO) to the powder to absorb the IR radiation [36]. This powder was reported to be successful in casting round billets [36]. Hanao et al. [37] reported that a slag forming *melilite* (a solid solution ranging from *akermanite* (C_2MS_2) to *gehlenite* C_2AS) had proved successful in casting round billets; *melilite* slags maintain a high fraction of glass in the slag film and tend to have higher viscosities higher than those of conventional slags.

10.4.3 Mould Powder Selection for Moulds with Large "Corner" Regions

Some mould geometries contain extensive corner regions, such as moulds used for casting beam blanks [38] or where the "*football*" meets the regular section in funnel —thin-slab moulds. These corner regions promote high heat transfer rates. Thus the resulting shell is much thicker in these regions and shells exhibit considerable variation in shell thickness around these "*corner*" regions. The variations in shell thickness lead to stresses during the solidification process, especially when casting *peritectic*, MC steels which, in turn, lead to longitudinal cracking in the corner regions.

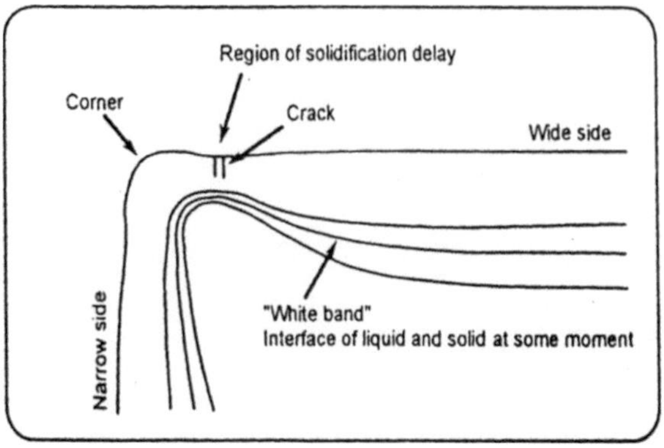

Fig. 10.11 Schematic diagram showing the thinning of the shell in the corner regions caused by strong metal flows (Permission granted, UNESID [39])

Most of the usual remedial actions do not involve mould powder selection (e.g. reducing the sharpness of the corners, coatings, variable water-cooling for different regions, etc.). However, the formation of a thick, crystalline slag film minimises the variations in shell thickness and result in less longitudinal cracking. This can be achieved by using a powder with a high break temperature which is usually achieved by increasing the basicity of the mould powder.

A similar approach can be used (i.e. to increase T_{br} of the slag) to combat longitudinal corner cracking caused by thinning of the shell in the corner due to a high metal flow velocity (Fig. 10.11) [39]; such conditions frequently occur after reducing the width dimension of the mould with a concomitant increase in casting speed to maintain throughput. The use of a powder with higher T_{br} would alleviate the severity of this problem.

10.4.4 Casting High-Al (Trip, Twip) Steel Grades

Steels containing high Al (1% or more) have excellent mechanical properties combining high strength with good ductility. However, these steels have proved difficult to cast with conventional mould powders because of the huge amounts of Al_2O_3 (>30%) produced by reaction shown in Eq. 10.20 (see Sect. 6.2.12).

$$2\underline{Al} + 3SiO_{2\,slag} = 3\underline{Si} + Al_2O_{3\,slag} \quad (10.20)$$

The Al_2O_3 pick-up results in increased viscosity and liquidus temperatures (T_{liq}) and causes increased crystallisation of the slag film. Alumina particles remain undissolved in the slag and collide to form agglomerates which can

- Block off liquid slag infiltration into shell/mould channel and
- Promote the formation of large slag rims, which are crystalline and tend to crack; cracked slag rims can result in a blockage to slag infiltration in the shell/mould channel and cause depressions.

Both of these events can result in sticker break-outs. The cracked slag rims also cause depressions and false sticker-break-out alarms.

Similar reactions to Eq. 10.19 occur between Al and other slag components like FeO, MnO, B_2O_3, etc. Conventional mould powders can be used to cast steels with up to 0.6% Al but a different family of casting powders is needed to cast steels with >1% Al. Three types of mould powder are used to cast these steels (with 1% Al), (i) low basicity slags (ii) high basicity powders (including those forming *melilite*) and (iii) calcium aluminate (CA) type powders (see Sect. 6.2.12). Most of the recent work has focused on casting powders which consist of (calcium aluminates + fluxing agents) [40–43]. The (%CaO/%Al_2O_3 (or C/A) ratio) is usually kept around 1 to hold T_{liq} down and SiO_2 contents are kept around 5–10% to minimise Al_2O_3 formation via Eq. 10.20. Recent work has shown that FeO, MnO and B_2O_3 are not very satisfactory fluxing agents since they (like SiO_2) are reduced and form Al_2O_3 [43]. This leaves CaF_2, Na_2O, K_2O and Li_2O as the prime candidates to flux the calcium aluminate slag; more work is still needed to derive satisfactory mixture of fluxes to cast these steels.

Another factor which is important in the selection of powders for casting these steels is that they should create a deep slag pool to dissolve as much Al_2O_3 as possible. Deep pool formation is promoted by mould powders containing (i) exothermic agents (ii) smaller granule size and (iii) low-carbonate content.

High Al, Mn steels have low T_{liq} values and the vertical heat flux is insufficient to melt the mould powder. Consequently, liquid slag feeding [44] will have to be developed to satisfactorily cast these steels.

10.4.5 Fluoride-Free Powders

There has been a drive to reduce the use of fluorides in slags used in all steelmaking operations. Fluorides pose a health hazard to plant personnel, a threat to the environment and cause plant corrosion since they (i) emit HF(g) by reaction with moisture and (ii) fluorides are leached from slag wastes. The addition of fluorides results in

- A decrease in both liquidus temperature and viscosity of the slag.
- The formation of crystalline *cuspidine* ($3CaO{\cdot}2SiO_2{\cdot}CaF_2$) in the slag film which controls the heat transfer between the shell and the mould.

The slag composition can readily be adjusted to accommodate the replacement of CaF_2 to provide similar T_{br} and viscosity values (e.g. by using alternative fluxes like Na_2O) but the replacement of both CaF_2 and *cuspidine* for control of the horizontal heat flux has proved more difficult. Several crystalline compounds have been suggested as a substitute for *cuspidine*, e.g. *perovskite* ($CaTiO_3$) [42, 45, 46] *melilite* [37] and for *peritectic* MC steels. $Na_2O_2{\cdot}CaO{\cdot}3SiO_2$ [47]. However, it has been claimed the TiO_2 additions to the mould powder result in the formation of Ti (CN) particles which subsequently, result in sticker break-outs [48]; this allegation has been disputed [46].

The current situation is that satisfactory F-free mould powders have been developed to replace F-containing (conventional) mould powders for casting billets [49–52]. The challenge is to develop a satisfactory F-free powder for slab casting, especially for MC *peritectic* steels. Another approach is to reduce the amount of CaF_2 added and rely on other crystalline phases to augment the scattering of IR radiation provided by the remaining *cuspidine* [46].

10.4.6 Reducing SEN Erosion Rates

SEN erosion rates are affected by several factors such as metal and slag flow velocities (ii) dissolution of the refractory by the mould slag and (iii) the dissolution of any carbon in the refractory by the steel. There are three mould slag-related factors

- The slag viscosity (with SEN erosion rates decreasing as the viscosity increases) since viscous resistance reduces the flow velocity of the slag phase [53].
- The concentration (C) difference ($C_{sat} - C_0$) where subscripts $_{sat}$ and $_0$ represent, respectively, the saturated limit and the actual concentrations of the refractory (e.g. ZrO_2) which affect the dissolution rate.
- The CaF_2 content of the slag (since ZrO_2 undergoes a high-temperature phase transition and CaO, MgO or Y_2O_3 are used to stabilise the ZrO_2) due to the attack of CaF_2 on the stabilising oxide [54].

Thus the mould slag can be adjusted in the following ways to reduce SEN erosion rates:

- By increasing the slag viscosity (via increases in SiO_2, Al_2O_3 or decreasing Na_2O, CaO, etc.) but this will have the effect of decreasing the powder consumption [53].
- By adding to 2% ZrO_2 to the powder since C_{sat} occurs at *ca.* 2% ZrO_2 in mould slags and this reduces the driving force for dissolution ($C_{sat} - C_0$) to zero.
- By reducing the CaF_2 content since it reduces the attack on the oxide stabilising the ZrO_2 [49, 53–55].

10.4.7 Minimising Carbon Pick-up

Carbon pick-up by the steel is a significant problem when casting ULC and LC steels. The mould powder is the main source of the carbon pick-up. The carbon is added to control the melting rate. In order to minimise C-pick-up powder manufacturers have developed powders where carbon has been replaced by other minerals [25–28]. The molten mould slag is non-wetting to the carbon and thus carbon particles prevent the slag globules from agglomerating and retard the formation of a molten slag pool (Fig. 10.8). The carbon particles are continuously oxidised by any O_2 (g) or CO_2 (g) present in the powder bed. Smaller carbon particles have higher ratios of (surface area/mass) than large particles. Thus smaller particles provide more non-wetting separation but oxidise more rapidly. The fact that the melting rate decreases with decreasing mean carbon particle diameter (D_C) indicated the non-wetting separation is the more important of these two processes.

Thus, any substitute for carbon must (i) be non-wetting to molten slag globules and (ii) react with O_2 and CO_2 in the powder bed. Various nitrides (e.g. BN, Si_3N_4) have been used to replace carbon and some success has been reported [25–28] but these powders have not been widely adopted because they tend to increase the size of slag rims.

The best way of reducing C-pick-up by the steel is to operate with a deep slag pool to keep the shell far away from both the C particles floating in the slag pool and the slag rim. This entails operating with (i) deep powder bed (ii) exothermic agents in the powder and (iii) a lower C% in powder to promote melting rate.

10.4.8 Minimising Scale Formation

Some mould powders, which are perfectly satisfactory in every other way, have been rejected because they promote the formation of scales on the surface of the steel. The retention of the mould slag (on the steel surface) in the secondary cooling is key since it provides the necessary SiO_2 to form low-melting $FeO{\cdot}SiO_2$ which penetrates into the grain boundaries [56–67]. The retention of slag is promoted by a low interfacial tension between steel and slag (γ_{msl}) [68]. The principal components causing low interfacial tension are K_2O, Na_2O and Li_2O but it might be expected that B_2O_3 would also promote low γ_{msl}. Thus, where scaling is a problem, the powder should be reformulated to provide similar Q_s^{req}, η^{req} and T_{br}^{req} values but with lower contents of Na_2O, etc.

10.5 Summary

Mould powders contain two types of material (i) mineral components which form a liquid which provides the slag film and (ii) the carbon particles which regulate the melting rate of the powder. Both are vital to the success of the casting process.

Satisfactory mould slags (i.e. mineral compositions) can be derived for the specified mould dimensions, casting conditions and steel grade using empirical rules based on plant data where no problems or defects were experienced [4, 5]. These empirical rules provide *the required values* of powder consumption (Q_s^{req}); viscosity (η^{req}) and break temperature (T_{br}^{req}) for these conditions. It has been estimated that 85% of the powders in the database conformed with the required values calculated using these empirical rules [3]. The 15% with property values which did not conform with those derived with the empirical rules did so because

- The viscosity of the mould slag has been deliberately increased to minimise slag entrapment or SEN erosion rates; this is particularly prevalent in billet casting where the required powder consumption is relatively low.
- For powders used in casting steels containing Ti, the latter reacts with the mould powder to form Ti(CN) which has a low solubility in the slag and forms agglomerated, Ti(CN) particles; these block the entrance to the shell/mould channel [19] and thereby reduce powder consumption to a low level [3].
- The required break temperature and f_{crys} are deliberately increased to minimise variations in shell thickness (which lead to longitudinal corner cracking) which can arise in the casting of beam blanks and where there are sharp corners (e.g. in funnel moulds) and also where the metal flow causes thinning of the shell in the corner [39].
- Certain in-mould conditions affect powder consumption, e.g. Q_s increases with increasing Ar flow rate and tends to decrease when EMBr is applied (because of the reduction in vertical heat transfer).

The carbon, in addition to the mineral content, of the mould powder must also be carefully controlled. This involves selection of both the carbon content and the mean carbon particle size and these must be selected to ensure that the melting rate of the powder matches the demand for liquid slag (i.e. Q_s^{req}). Some workers have decreased the carbon content of the powder in order to either increase the depth of the molten slag pool or to reduce C-pick-up by the steel. However, care must be taken with the reduction of C% since carbon is a fuel (exothermic) and if the reduction of C% is too stringent the steel meniscus can freeze and form a skull. Carbon-free powders have been developed and are available commercially. The best way of minimising C-pick-up is operate with a deep slag pool.

The elimination or reduction of F-emissions has led to the development of F-free powders in which *cuspidine* (formed in the slag film) is fully or partially replaced by other crystalline phases. This has proved a difficult task in the case of slab casting but some progress has been made. These F-free powders result in lower SEN erosion rates.

High-Al steels (with >0.6% Al) have proved difficult to cast with conventional mould powders. Consequently, a new family of slags based on calcium aluminates has been developed to maintain Al_2O_3 at a manageable level. Although these have proved promising, further developments are still required to maintain some glassy phase (to provide lubrication and minimise fracture of crystalline, slag rims) [41]. High-Al steels have low liquidus temperatures and liquid mould slag feeding will be necessary to cast steels with higher Al and Mn contents.

References

1. H Harada, E Anzai, E Takeuchi, Canad. Met. Quart., **39**, (3), 307, (2000).
2. S Ogibayashi, T Mizoguchi,T Kajatani, Intl. Workshop on *Thermophys. Data for the Development of Mathematical models of solidification*. Gifu City, Japan (1995).

3. AB Fox, PhD Thesis, "*Mould fluxes- their properties and performance*" Imperial College, London, (2003).
4. KC Mills, S Sridhar, AS Normanton, ST Mallaband, Proc. Brimacombe Conf., Vancouver, (2000) p. 781.
5. S Sridhar, KC Mills, V Ludlow, ST Mallaband, Proc. 3rd Europ. Conf. Continuous Casting, Madrid, 1998 (UNESID, Madrid, 1998) p. 807.
6. F Neumann, J Neal, MA Pedroza, AH Castiliejos E, FA Acosta G, Proc. 79th. Steelmaking Conf., 1996 (ISS, Warrendale, PA) p. 249.
7. M Wolf, Proc. 2nd Europ. Conf. Continuous Casting, Dusseldorf, 1994, (VDEh, Dusseldorf, 1994) p. 78.
8. S Ogibayashi, Y Yamaguchi, T Mukai, T Takahashi, Y Mimura, K Koyama, Y Nagano, T Nakano, Nippon Steel Technical Report, **34**, 1, (1987).
9. KC Mills, AB Fox, ISIJ Intl., **43**, 1479, (2003).
10. K Tsutsumi, S Murakami, S Nishioka, Tetsu-to-Hagane, **84**, 617, (1998).
11. PE Ramirez-Lopez, KC Mills, PD Lee, B Santillana, Met. Mater. Trans., B, **43B**, 109, (2012).
12. SN Singh, KE Blazek, J Metals, **26** (10), 17, (1974).
13. Thermocalc see www.thermocalc.com.
14. MB Santillana, PhD Thesis, *Thermo-mechanical properties and cracking during solidification of thin slab cast steel.* Univ. Delft, NL (2013) Chapter 3.
15. J Miettinen, AA Howe, Ironmaking and Steelmaking, **27**, 212, (2000).
16. KE Blazek, O Lanzi, P Gano, D Kellogg, Proc. AISTech., 2007, (AIST, Warrendale, PA, 2007) PR-351–141 - 2007.
17. M Wolf, W Kurz, Met. Mater. Trans., **12**, 85, (1981).
18. M Wolf, Proc. METEC Congress'94;. 2nd Europ. Conf. Continuous Casting, Dusseldorf, 1994, (VDEh, Dusseldorf, 1994) vol. 1, pp 78.
19. T Mukongo, C Pistorius, A Garbers-Craig, Ironmaking and Steelmaking, **31**, 135, (2004).
20. R Koldewijn, Unpublished Corus Internal Rept. (2007) cited in KC Mills, J Kromhout, A Hamoen, R Boom, Proc. Admet Conf., Dnipropetrovsk, 2007, (Natl. Metall. Acad. Ukr., Dnipropetrovsk, 2007) vol. 2, p. 174.
21. T Mallaband (Metallurgica, UK) Unpublished results cited by AB Fox, ref. [3].
22. PE Ramirez- Lopez, PN Jalali, J Bjorkvall, U Sjostrom, C Nilsson, ISIJ Intl., **54**, 342, (2014) and Proc. 8th Europ. Conf. Continuous Casting, Graz, Austria, 2014 (ASMET, Vienna, 2014).
23. E van Vliet, DW van der Plas, SP Carless, A A Kamperman, AE Westendorp, Proc. 7th Europ. Conf. Cont. Casting, Dusseldorf, 2011, (VDEh, Dusseldorf, 2011) Session 4.
24. M Kawamoto, K Nakajima, T Kanazawa, K Nakai, ISIJ Intl., **34**, 593, (1994).
25. C Lefebvre, JP Radot, JN Pontoire, Y Roux, Revue de Metallurgie, CIT, **94**, 489, (1997).
26. B. Debiesme, J Radot, D Coulombet, C Lefebvre, Y Roux, C Demarval, US Patent No 5 876, 482 (1999) and US Patent No 6, 328, 781 (2001).
27. S Terada, S Kaneo, T Ishikawa, Y Yoshida, Proc. 74th Steelmaking Conf., Washington, DC, 1991, (ISS, Warrendale, PA, 1991) p. 635.
28. J Macho, G Hecko, B Golomowski, M Frazee, McMaster Iron & Steelmaking Symp. "*Thinner slab casting*", 2005, (McMaster Univ., Hamilton, Ont., 2005) p. 131.
29. LC Hibbeler, BG Thomas, Proc. AISTech 2013, (AIST, Warrendale, PA, 2013) p. 1215.
30. S Feldbauer, AW Cramb, Proc. 79th Steelmaking Conf., 1996 (ISS, Warrendale, PA, 1996) p. 595.
31. K Tsutsumi, K Watanabe, M Suzuki, M Nakada, T Shiomi, Proc. 7th Intl. Conf. Molten slags, fluxes and salts, Cape Town, 2004, (SAIMM, Johannesburg, 2004) p. 803.
32. J Yoshida,T Ohmi, M Iguchi, ISIJ Intl., **45**, 1160, (2005).
33. K Tsutsumi, K Watanabe, J Kubota, S Hatori, Y Miki, T Suzuki, T Omoto, Proc. 7th Europ. Cont. Casting Conf., Dusseldorf, 2011, (VDEH, Dusseldorf, 2011) Session p. 1.
34. K Watanabe, K Tsutsumi, M Suzuki, H Fujita, S Hatori, T Omoto, ISIJ Intl., **54**, 865, (2014).
35. J Kromhout, PhD thesis, *Mould powders for high speed continuous casting of steel.* Univ. of Delft, NL (2011), p. 99.

36. M Kawamoto, Paper presented at International Workshop on "*Thermo-physical data for the development of mathematical models of solidification*" Gifu City, Japan, (1995).
37. M Hanao, Y Tsukaguchi, M Kawamoto, Proc. 4th Intl. Congress Science and Technol., Gifu, 2008, (ISIJ, Tokyo, 2008) p. 694.
38. R Nishimachi, Y Ogura, SEAISI Quarterly, **25**, (4), 44, (1996).
39. H Tai, M Morashita, T Miyake, Proc. 3rd Europ. Conf. Continuous Casting, Madrid, 1998, (UNESID, Madrid, 1998) p. 447.
40. K Blazek, HB Yin, G Skoczylas, M McClymonds, M Frazee, Proc. EEEC-METEC Dusseldorf, 2011 (VDEh, Dusseldorf, 2011).
41. JW Cho, KE Blazek, MJ Frazee, HB Yin, JE Park, SW Moon, ISIJ Intl., **53**, 62, (2013).
42. GH Wen, S Sridhar, P Tang, X Qi, YQ Liu, ISIJ Intl., **47**, 1117, (2007).
43. Q Liu, GH Wen, JZ Li, XJ Fu, P Tang, W Li, Ironmaking and Steelmaking, **41**, 292, (2014).
44. JK Park, JW Cho, KH Moon, SH Lee, KH Kim, HS Jeong, Proc. 7th Intl. Conf. Clean Steel, Balatonfured, Hungary, 2007, (Hung. Min. Metall. Soc., Budapest 2007), p. 264.
45. H Nakada, K Nagata, ISIJ Intl., **46**, 441, (2006).
46. LY Chen, GH Wen, CL Yang, F Mei, CY Shi, P Tang, Ironmaking and Steelmaking, **42**, 105, (2015).
47. N Takahira, N Hanao, Y Tsukaguchi, ISIJ Intl., **53**, 818, (2013).
48. Q Wang, YJ Lu, SP He, KC Mills, Z Li, Ironmaking and Steelmaking, **38**, 297, (2011).
49. MC Bezerra, CA Valadares, IP Rocha, JP Bulota, MC Carboni, IL Scripnic, CR Santos, K Mills, D Lever, Proc. 36th Steelmaking Seminar, Porto Allegre, RS- Brazil, 2005, (ABM, Sao Paulo, 2005). p. 190 available on line at www.carboox.com/pdf/fluxante_2007.pdf.
50. AB Fox, KC Mills, D Lever, C Bezerra, C Valadares, I Unamono, J Laraudogoitia, J Gisby, ISIJ Intl., **45**, 1051, (2005).
51. AB Fox, K Mills, D Lever, M C Bezerra, CA Valadares, I Unamono, J Laraudogoitia, J Gisby, Proc. 36th Steelmaking Seminar, Vittoria ES- Brazil, 2005, (ABM, Sao Paulo, 2005) p. 222, available on line at www.carboox.com/pdf/paper2005sfs.pdf,
52. Z Li, K Mills, MC-Bezerra, Proc. 35th Seminar de Fusao Refino e Solidificao Metals, Salvador, Brazil, 2004 (ABM, Sao Paulo, 2004) p. 281.
53. Y Nakamura, T Ando, K Kurata, M Ikeda, Trans. ISIJ, **26**, 1052, (1986).
54. DW Bruce, NS Hunter, Proc. 2nd Intl. Conf. Continuous Casting, Dusseldorf, 1994, (VDEh, Dusseldorf, 1994) p. 156.
55. Q.Wang, SP He, KC Mills, Proc. 4th Intl. Conf. Cont. Casting Steel in Developing Countries, Beijing, China (2008). (Chinese Soc. Metals, Beijing, 2008), p. 715.
56. MM Wolf, Ironmaker & Steelmaker, **2000**, (Jan), 22, (2000).
57. MM Wolf, Ironmaker & Steelmaker, **2000**, (Feb.), 65, (2000).
58. MM Wolf, Ironmaker & Steelmaker, **2000**, (March), 69, (2000).
59. MM Wolf, Ironmaker & Steelmaker, **2000**, (April), 58, (2000).
60. MM Wolf, Ironmaker & Steelmaker, **2000**, (May), 78, (2000).
61. MM Wolf, Ironmaker & Steelmaker, **2000**, (June), 22, (2000).
62. MM Wolf, Ironmaker & Steelmaker, **2000**, (July), 63, (2000).
63. MM Wolf, Ironmaker & Steelmaker, **2000**, (Aug), 75, (2000).
64. MM Wolf, Ironmaker & Steelmaker, **2000**, (Sept), 90, (2000).
65. MM Wolf, Ironmaker & Steelmaker, **2000**, (Oct), 114, (2000).
66. MM Wolf, Ironmaker & Steelmaker, **2000**, (Nov), 67, (2000).
67. MM Wolf, Ironmaker & Steelmaker, **2000**, (Dec), 45, (2000).
68. H Nakato, S Takeuchi, T Fujii, T Nozaki, N Washio, Proc. 74th Steelmaking Conf., 1991, (ISS, Warrendale, PA, 1991) p. 639.

Chapter 11
Using Mould Fluxes to Minimise Defects and Process Problems

Abstract The quality of the cast product is affected by any defects present. The causes of each individual defect are discussed, along with the factors affecting their formation and proposed treatment to minimise the severity of the defect. The following defects and process problems are covered in this chapter: (i) longitudinal cracks, (ii) longitudinal corner cracks, (iii) sticker breakouts, (iv) oscillation marks, (v) transverse cracks, (vi) star cracks, (vii) depressions, (viii) overflows, (ix) entrapment of slag, gas and inclusions, (x) formation of scales on the surface of the steel, (xi) Carbon pick-up by the steel, (xii) SEN erosion and (xiii) fluoride emissions. These defects are discussed from the viewpoint of how the mould powder can help to alleviate the problem.

Symbols, Abbreviations and Units

d	Thickness or depth (m)
f	Frequency (Hz or cpm)
h	Height or depth (m)
L	Length or pitch (m)
Q	Powder consumption (kg (slag) m^{-2})
q	Heat flux (W m^{-2})
s	Stroke length (m)
T	Temperature (°C)
T_{br}	Break temperature (°C)
t	Thickness of mould (m) or time (s)
V_c	Casting speed (m min^{-1})
V_m	Velocity of mould (m s^{-1})
w	Width of mould (m)
C/S = %CaO/%SiO_2	Basicity
EMBr	Electro-magnetic braking
EMC	Electromagnetic casting

The original version of this chapter was revised: For detailed information please see correction. The correction to this chapter is available at https://doi.org/10.1007/978-3-319-53616-3_12

Within this book minor changes to some of the figure permissions and the readability of the chapters have been made and several figures have been replaced with new versions.

K.C. Mills and C.-Å. Däcker, *The Casting Powders Book*,
DOI 10.1007/978-3-319-53616-3_11

EMS	Electro-magnetic stirring
HC	High carbon steel
LC	Low carbon steel
LCAK	Low carbon Al-killed steel
MC	Medium carbon steel
MTM	Mould thermal monitoring
OM's	Oscillation marks
ULC	Ultra-low carbon steel

11.1 Introduction

Continuous casting is a highly successful process. Nevertheless, it is affected by both process problems and defects which affect the quality of the cast product. In this chapter, data concerning these various problems and defects are examined and analysed to determine the underlying causes of the problem and the mechanisms involved. From this knowledge, ways of dealing with the various problems will be proposed. Frequently, the underlying problem originates from changes in the casting conditions (e.g. changes in casting speed), but the mould slag is expected to minimise the impact of the problem. So, one focus of this chapter will be to highlight what changes in the mould fluxes will help to alleviate the problem. The various problems and defects are discussed individually below.

11.2 Longitudinal Cracking

Longitudinal cracking occurs on the surface and the sub-surface of the cast steel product. The cracking is a major problem since cracks are usually removed by grinding (or scarfing) which is time-consuming and leads to a loss in yield.

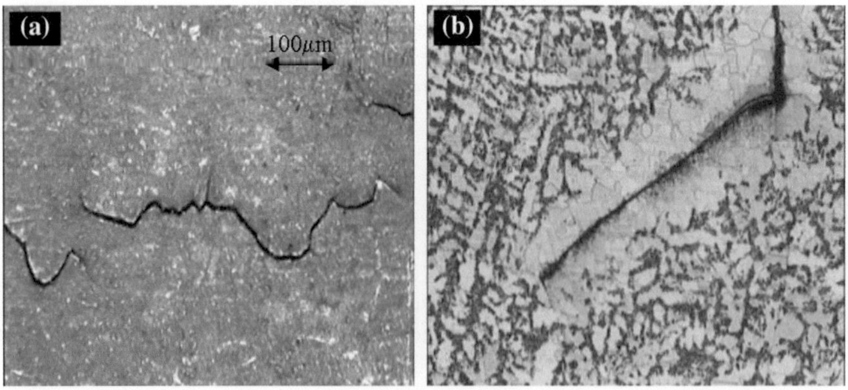

Fig. 11.1 Photographs showing **a** Gross crack **b** Small sub-surface cracks (permission granted, ISS/AIST, [2])

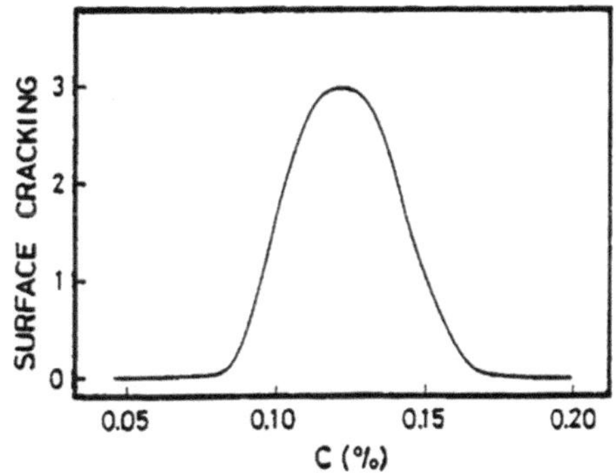

Fig. 11.2 Schematic diagram showing incidence of surface cracks as a function of wt%C in steel (permission granted, ISIJ, [3])

Longitudinal cracks can be classified into two types (Fig. 11.1) [1, 2]:

(i) *Gross cracks* (up to 400 mm long) usually accompanied by longitudinal depressions; this form of crack is usually associated with severe casting problems (such as poor mould level control).
(ii) *Sub-surface, shallow cracks* found when casting MC, *peritectic*, steel grades; these cracks (typically, 30–40 mm) are much smaller than the gross cracks.

There are several factors affecting the formation of longitudinal cracks such as

- The type of steel.
- The rate of heat extraction from the shell (i.e. horizontal heat flux, q_{hor}) in the meniscus region.
- The lubrication and powder consumption supplied to the shell.
- The metal flow and the use of electromagnetic appliances to reduce turbulence.

11.2.1 Type of Steel

Longitudinal cracking is particularly prevalent in *peritectic*, MC steels (C% = 0.085–0.145) as shown in Fig. 11.2 [3]. On solidification, the steel forms δ-Fe (*ferrite*) which at lower temperatures, transforms to γ-Fe (*austenite*). *Austenite* shells are stronger but less ductile than *ferrite* shells (Fig. 11.3) [4].

The transition ($L + \delta \rightarrow \gamma$) is accompanied by a volume change of (0.4–0.6%) due to the tighter packing in the *austenite,* and there is also a 4% mismatch between the thermal shrinkage coefficients of δ-Fe and γ-Fe. This mismatch results in stresses. These stresses are much larger in thicker regions of the shell than in thinner regions and result in a hoop stress around the mould. This hoop stress is relieved by longitudinal cracking.

Wolf [5] introduced a parameter (the *ferrite potential*, FP) to describe the amount of *peritectic* reaction taking place in the steel. In this system, the maximum amount of peritectic reaction occurs at FP = 1.0. The calculation of ferrite potential for various steels requires knowledge of other parameters, namely, the carbon equivalent for peritectic reaction (CP) (Eq. 11.1), and for stainless steels the "*Ni*" and "*Cr*" terms (given in Eqs. 11.2 and 11.3, respectively) where all contents of elements in steel are in mass%.

Fig. 11.3 Ultimate tensile strength at 1400 °C at high strain rate as function of %C in the steel (permission granted, Verlag Stahleisen, re-drawn after [4])

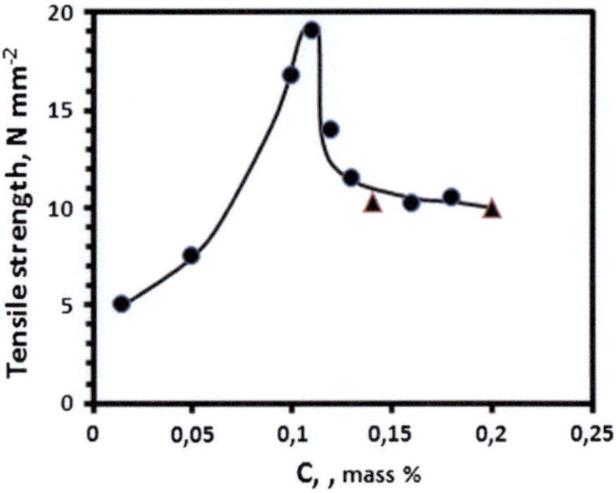

$$CP = \%C + 0.04\,Mn + 0.10\,Ni + 0.7\,N - 0.14Si - 0.04\,Cr - 0.1\,Mo - 0.24Ti \quad (11.1)$$

$$''Ni'' = \%Ni + 0.31\,Mn + 22\,C + 17.5\,N \quad (11.2)$$

$$''Cr'' = \%Cr + 1.5\,Si + 1.65\,Mo + 2\,Nb + 3\,Ti \quad (11.3)$$

The ferrite potential can then be calculated by Eq. 11.4 or 11.5.

Low alloy steel:

$$FP = 2.5\,(0.5 - CP) \quad (11.4)$$

Stainless steel:

$$FP = 5.26\{0.74 - (''Ni''/''Cr'')\} \quad (11.5)$$

The characteristics of the shells formed for steels with various ferrite potentials are summarised in Table 11.1 and are depicted in Fig. 11.4.

According to Fig. 11.4, the region with largest risk of depressions lies between FP = 0.8 and 1.1.

Table 11.1 Characteristics of shells formed for different bands of ferrite potential (FP)

FP	Phase	Properties of shell
<0.0	*Austenite*	Strong and more ductile with increasing FP
0.0–1.0	Initially forms *ferrite*, then transforms to *austenite*	Irregular, strong shell
>1.0	*ferrite*	Weak, ductile shell

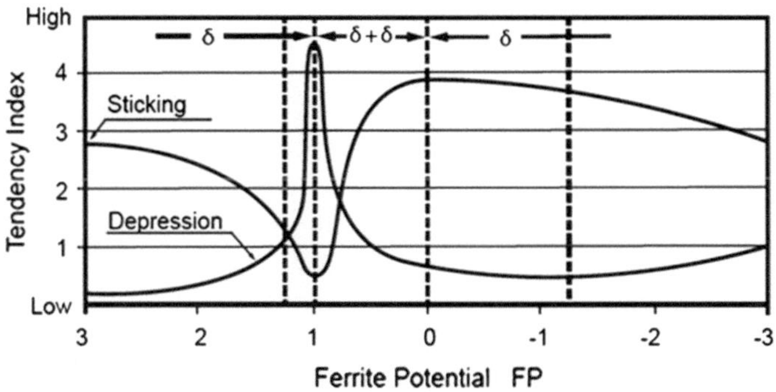

Fig. 11.4 Schematic diagram showing tendencies to form longitudinal cracks and sticker breakouts as functions of ferrite potential (FP). It can be seen that longitudinal cracking is low in the δ-Fe and γ-Fe regions but high in two-phase region (permission granted, VDEh, Verlag Stahleisen [5]; re-drawn Swerea KIMAB)

11.2.2 Heat Flux

The longitudinal cracks are thought to be initiated in the meniscus region of the shell. Longitudinal cracking appears to be related to

(i) The magnitude of the heat extraction rate from the shell (i.e. horizontal heat flux, q_{hor}).
(ii) Local variations in heat flux (e.g. where a hot spot is formed in the mould) which results in a shell of variable thickness.

These are studied separately below. The strategy usually adopted is to minimise longitudinal cracking by producing a shell which is both as thin and as uniform as possible.

11.2.2.1 Horizontal Heat Flux

The relation between longitudinal cracking and horizontal hear flux (q_{hor}) has been studied by a number of investigators [6–17].

Hiraki et al. [6] established that longitudinal cracking occurred when the horizontal heat flux in the meniscus region exceeded a certain critical value (q_{crit}) and that the level of cracking increased sharply with further increases in heat flux above $q_{\mathrm{crit.}}$ Inspection of Fig. 11.5 shows that q_{crit} has values of 1.5 M W m^{-2} for ULC (hypo-peritectic) steels and 2.5 M W m^{-2} for LC steels. Thus, a much lower heat flux is required to avoid longitudinal cracking in a ULC, hypo-peritectic steel (FP > 1.0) than for a LC steel (1.0 < FP < 1.2). It can be seen in Fig. 11.5 that longitudinal cracking increases sharply when q_{crit} is exceeded. Similar findings have been reported by other workers [7, 9, 11, 14].

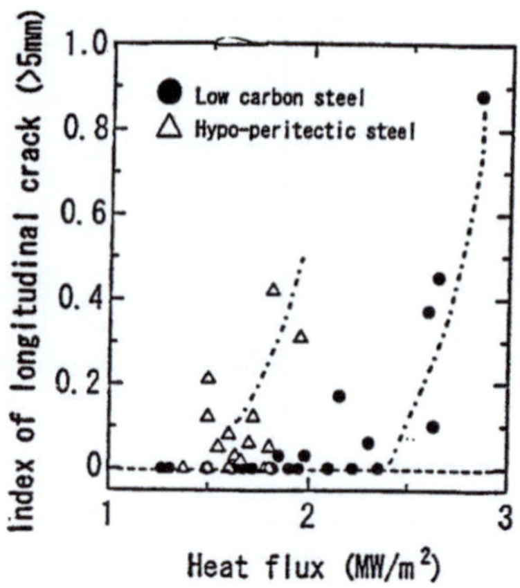

Fig. 11.5 The incidence of longitudinal surface cracks as a function of horizontal heat flux for LC (●) and hypo-peritectic steel (Δ); index of longitudinal cracks = Σ(surface cracks >5 mm/cast length); q_{crit} occurs at 1.5 and 2.5 M W m^{-2} for hypo-peritectic and LC steels, respectively (permission granted, ISS/AIST, [6])

The following methods have been used to reduce horizontal heat flux:

- Decreasing the water flow rate [18–20].
- Using Meniscus-free casting (in which a refractory material is placed inside the mould in the meniscus region [21–23].
- Machining grooves in the copper mould around the meniscus level to form gas gaps and create a thicker slag film [24–31]; grooved moulds provide a 10% decrease in horizontal heat flux [25, 31].
- Coating the mould with a material with a lower conductivity than copper [20, 32–37].
- By using a mould powder which provides a thick, crystalline, slag film [38], the thickness of solid slag film is related to the break temperature (Fig. 11.6a, b) and a high-crystalline fraction (f_{crys}) provides (a) an interfacial resistance ($R_{Cu/sl}$, or gas gap) caused by the shrinkage accompanying crystallisation of the slag film and (b) the reflection of the radiant energy emitted by the shell by crystalline phases in the slag film.

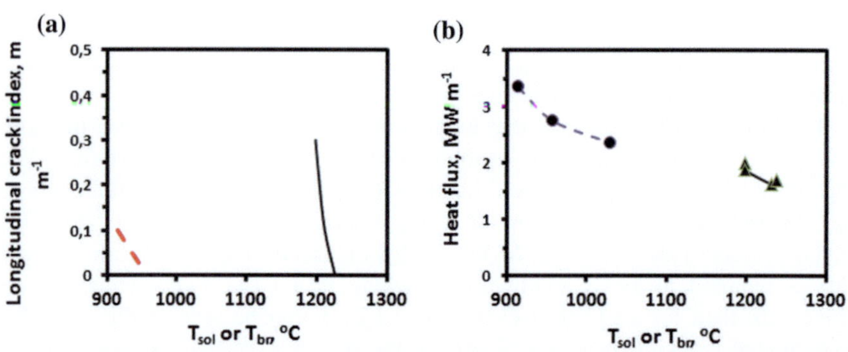

Fig. 11.6 Diagrams showing the effect of break (or solidification) temperature, T_{br}, of mould slags on **a** longitudinal cracking and **b** horizontal heat flux, q_{hor}, when casting hypo-, *peritectic* steel; **a** ▬ and **b** ▲low-C steels, **a** *solid line* and **b** ●, respectively (permission granted, ISS/AIST [6], re-drawn)

- By adding transition metal oxides (e.g. MnO, FeO, NiO,) to the casting powder to create a slag film which absorbs (and hence reduces) the radiant energy emitted by the shell [39, 40].
- The use of non-sinusoidal oscillation has been reported to decrease the heat flux by 10% [41] but this claim has been disputed.

It has been reported that longitudinal cracking increases as negative strip time (t_n) increases [42]. The effect of slag basicity (C/S) and negative strip time on longitudinal cracks for a micro-alloyed, peritectic, medium-carbon steel was explored in a study covering 27 heats at SSAB, Oxelösund [43]. It can be seen from Fig. 11.7 that longitudinal cracking (i) decreases with increasing basicity (since heat flux, q_{hor}, tends to decrease with increasing basicity) but (ii) exhibits a maximum at ca. $t_n = 0.16$. It has been suggested that the quadratic relationship in Fig. 11.7 is related to the fact that longitudinal cracks tend to initiate in the base of oscillation marks and occur more frequently as the depth (d_{OM}) increases. Values of d_{OM} tend to decrease with lower t_n and q_{hor} (i.e. with high basicity). However, at higher t_n values the higher slag infiltration tends to lead to provide more even cooling [43].

11.2.2.2 Local Variations in Heat Flux

Solidification (and shell growth) occurs mostly in the period of the oscillation cycle when the mould is descending, and there is little steel growth in the period when the

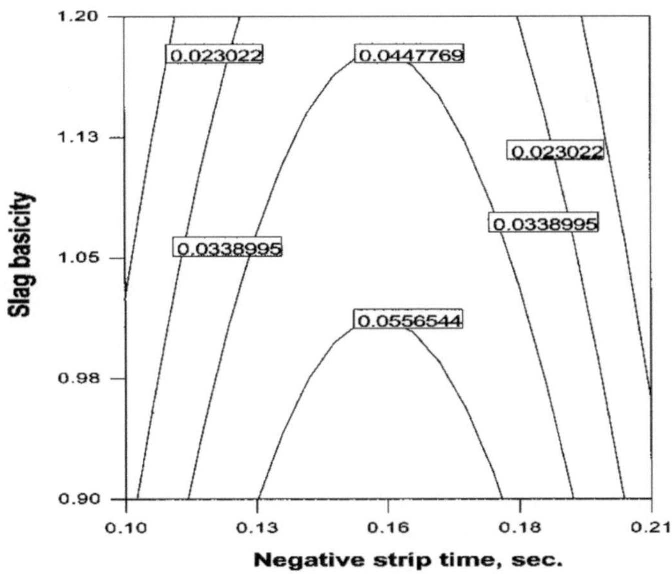

Fig. 11.7 Longitudinal crack length (m/m) as function of slag basicity and negative strip time (Reproduced with permission from EDP Sciences [43])

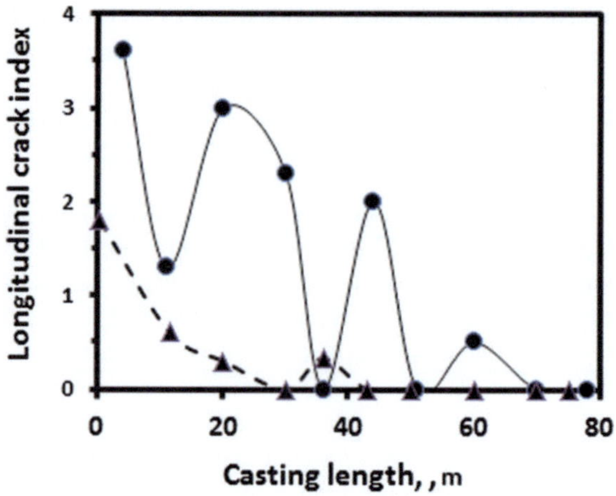

Fig. 11.8 Longitudinal crack index as a function of cast length of the strand; note cracking levels are reduced as the cast proceeds and by using powders containing exothermic agents (-— ▲ -) *cf.* no exothermic agents (—●—); drawn using data from [44]

mould is ascending. This leads to a shell of varying thickness ("*sausage-like*"). Thus, there are variations in shell thickness arising naturally from the mould oscillation. However, further local variations in shell thickness can arise if a "*cold spot or a hot spot*" develops in the mould. Such a *cold (or hot) spot* will create even greater differences in shell thickness. The accumulations of stress (due to the mismatch in shrinkage coefficients of δ- and γ-Fe phases) are much greater in the thick regions of the shell than in the thinner regions and result in an increased risk of longitudinal cracking.

"*Cold spots*" can be created (i) when high-conductivity, SENs are too close to the mould (this is especially important in thin slab casting), (ii) by overcooled corners (see Sect. 3.4.5). Hot spots can be formed where the flow of metal impinges on the shell, which causes melt-back and thus a thin region in the shell.

Longitudinal cracking tends to be prevalent at the beginning of the cast; this can be seen in Fig. 11.8 where the incidence of longitudinal cracking is high in the first metres of the strand but gradually decreases as the thermal conditions and the slag film stabilise [44]. It can also be seen that the use of casting powders containing exothermic agents help to stabilise the thermal conditions and thus reduce longitudinal cracking [44]. This may be due to the exothermic agents reducing, sequentially, the vertical heat flux, the length of the shell and the depth of the oscillation mark.

11.2.3 Lubrication and Powder Consumption

Ogibayashi et al. [45] pointed out that frictional forces acting on the shell increase as the distance from the corner increases (Fig. 11.9) i.e. in the centre of the broad face; it is for this reason that longitudinal cracks tend to occur in the centre of the broad face. Consequently, it is necessary to provide good lubrication in the centre of the strand where the shell is thinnest. They attributed longitudinal cracking to (i) poor lubrication and (ii) to irregular shell growth at low casting speeds [45]. Although, higher casting speeds lead to a thinner shell and a more even shell, friction (and longitudinal cracking) tend to increase overall with increasing V_c due to the lower powder consumption and a lower critical friction coefficient.

11.2.4 Metal Flow, Use of EMBr, EMC and EMS

Longitudinal cracking has been correlated with the amplitude of the waves running along the steel surface (Fig. 11.10a). These waves are formed by the backwash resulting from the impact of the metal flow on the mould wall. The velocity of the metal flow (V_{metal}) is linearly related to the casting speed (Fig. 11.10b).

Various electromagnetic applications are used in continuous casting (see Sect. 5.8). *Elecromagnetic stirring* (EMS) is used to homogenise both the steel composition and the temperature of the steel in the mould. EMS has been reported to reduce both longitudinal cracking and entrapped inclusion content (Figs. 11.10a [52] and Fig. 11.11, respectively). Presumably, reduced longitudinal cracking arose from the minimisation of temperature differences and thus, differences in shell thickness [45, 46]. *Electromagnetic braking* (EMBr) is used to reduce (i) the metal level variations (i.e. waves) and (ii) both V_{metal} (Fig. 11.10b, c) and the penetration of the metal flow into the mould. EMBr has been reported to reduce longitudinal cracking [46–49].

In *Electromagnetic casting* (EMC), the applied field creates a "*pinch force*" which pushes the shell back and allows more slag infiltration, thereby forming a thicker slag film. This delays solidification and produces a short and thin shell which results in shallow oscillation marks and decreased longitudinal cracking [50].

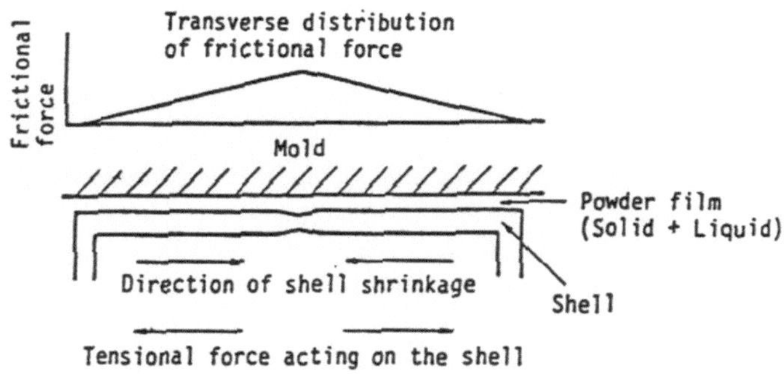

Fig. 11.9 Schematic diagrams showing (*upper*) the frictional forces across the mould and (*lower*) thickness of the steel shell across the mould. (Permission granted private communication, NSSM Corp. and authors [45])

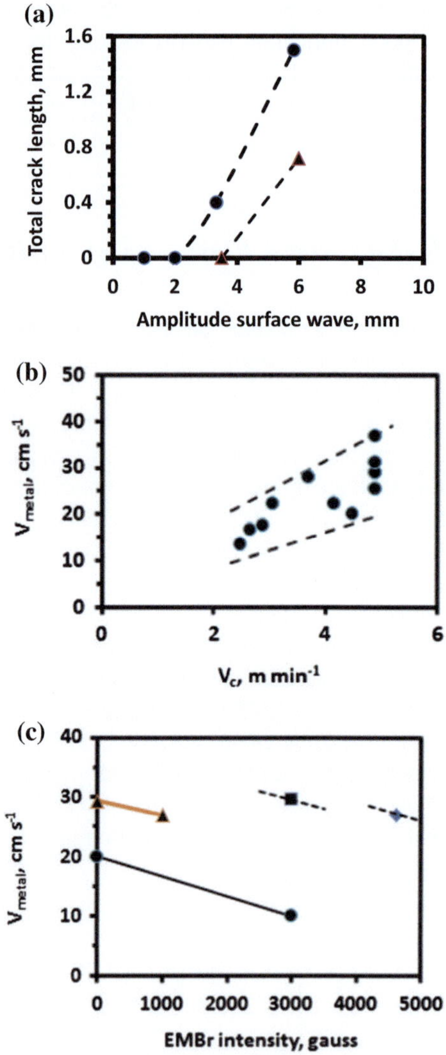

Fig. 11.10 Diagrams showing longitudinal crack length as function of **a** wave amplitude (determined by plate-dip tests) *upper curve* = without EMBr and *lower curve* = with EMBr **b** metal flow velocity as functions of casting speed and **c** metal flow velocity as a function of EMBr intensity; V_c (m min^{-1}), ● = 2; ▲ = 3; ■ = 3.5; ♦ = 5 (permission granted, ISIJ, re-drawn after [52])

Increased levels of longitudinal cracking were found to occur when asymmetric flow patterns developed as a result of SEN clogging [53]; presumably, this is due to either the metal flow creating hot spots where the metal flow impacts or to the creation of large surface waves (Fig. 11.10a).

11.2.5 Causes and Mechanisms

The initial solidification of the shell in the meniscus region of the mould is not a smooth process. Heat extraction is affected by the oscillation cycle with rapid shell

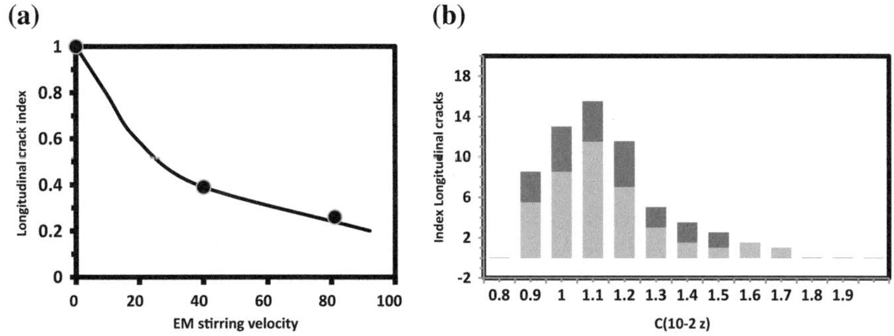

Fig. 11.11 Diagrams showing Index of longitudinal cracking reduced **a** with EMS stirring velocity (permission granted, NSSM Corp. [46]) and **b** by using EMBr (with EMBR = ▪; without EMBr = ▪) showing longitudinal cracks prevalent in *peritectic* range and the reductions obtained with EMBR (permission granted, NSSM Corp.; re-drawn after [46])

growth occurring when the mould is descending (i.e. in negative strip periods) and very little solidification occurring during the ascent of the shell (i.e. in positive strip periods). The maximum solidification (and heat flux) occurring in the period between late i_n and early t_p. These cyclical periods of growth of the shell during the oscillation cycle tend to result in a corrugated shell [53, 55].

Longitudinal cracking results from a sequence of events. It was mentioned above that there is a 4% difference in the thermal shrinkage coefficients of the δ-Fe and γ-Fe phases (Fig. 11.12) which results in stresses [56]. The accumulation of stress is much greater in the thick regions of the shell than in the thinner regions. These local

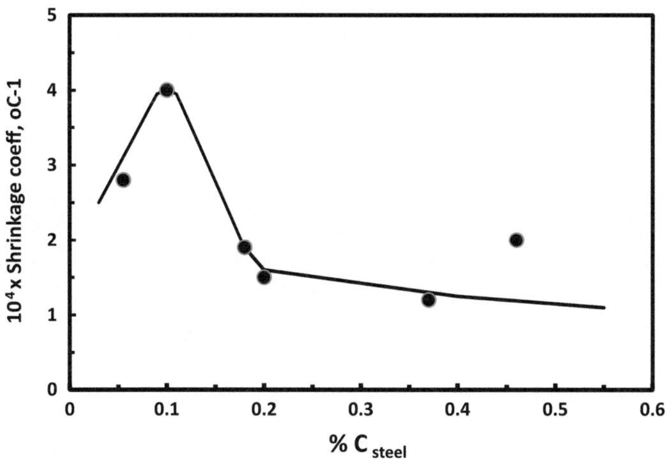

Fig. 11.12 Average shrinkage coefficient as a function of carbon content of the steel; note enhanced shrinkage coefficients in the *peritectic* range, C% = 0.06–0.15 (permission granted, re-drawn after [56])

stress gradients result in hoop stresses which are subsequently relieved by longitudinal cracking. Thus, large variations in shell thickness lead to enhanced longitudinal cracking. Bouts of longitudinal cracking can be detected from variability in the heat flux.

11.2.6 Ways of Dealing with Longitudinal Cracking

Longitudinal cracking is principally caused by local differences in shell thickness, so, remedial treatments are based on providing a thinner, more uniform shell. This can be achieved in the following ways.

11.2.6.1 By Decreasing the Horizontal Heat Flux in the Meniscus Region ("Mild Cooling")

This can be achieved by

- Modifying the casting powder so that it forms a thick, crystalline slag film with a high T_{br} [6, 38], this is usually obtained by increasing the (%CaO/%SiO_2) ratio to ca. 1.3 and having a sufficient amount of Fluorine for the formation of *cuspidine*, or by making additions of 2–3% ZrO_2 (which acts as a nucleant for crystallisation).
- Using a glassy mould slag with high IR absorption ability with a high content of FeO, MnO or NiO [39] could be a solution for special cases such as round-billet casting. In this case, the solid slag film must provide a good support for the strand and the creation of a gas gap (resulting from slag crystallisation) must be avoided.
- Using mould coatings with lower thermal conductivities than copper, preferably Nickel [37].
- Machining grooves in the meniscus region of the mould [24–27].
- Reducing the water flow rate but this has only a small effect [18–20].
- Using non-sinusoidal mould oscillation [41].
- It is thought that some longitudinal cracks are initiated at the base of oscillation marks [57]; consequently, taking steps to reduce d_{OM} should also be beneficial in reducing longitudinal cracking, these include (i) reducing the *vertical* heat flux and the length of the shell and (ii) by reducing t_n and stroke (and any other parameters which decrease t_n, such as increases in *Vc* and *f*).

11.2.6.2 By Reducing the Variations in Shell Thickness

The variations in shell thickness can be achieved by

- Proper design of the SEN to ensure an optimum and even steel flow in the mould.
- Using EMBr and EMS to smooth out any temperature gradients.

- Avoiding asymmetric flows arising from SEN clogging.
- Ensuring sufficient melting of the mould powders
- Enhanced levels of longitudinal cracking tend to occur in the early moments of the cast (Fig. 11.8), but using casting powders containing exothermic agents helps to establish steady state conditions quicker and thereby, reduce longitudinal cracking.

11.2.6.3 Monitoring to Avoid Longitudinal Cracking

Readings from moulds fitted with thermocouples, for example, breakout detection systems, can be used to detect longitudinal cracking. Presumably, the thermocouples detect the presence of a "*buckled shell*" on the heat flux which would be expected to result in (i) variability of the heat flux (i.e. temperature response) and (ii) a decrease in heat flux. It has been reported that periods of longitudinal cracking can be detected via

- Temperature variability (Fig. 11.13a) [1, 2, 58].
- A sudden decrease in temperature which successfully detected 80% of gross cracks and 40% of sub-surface cracks (Fig. 11.13b) [52].

11.2.6.4 Improving Powder Consumption

It has been suggested that longitudinal cracking can arise from insufficient lubrication and the use of an incorrect (low) taper [42, 45].

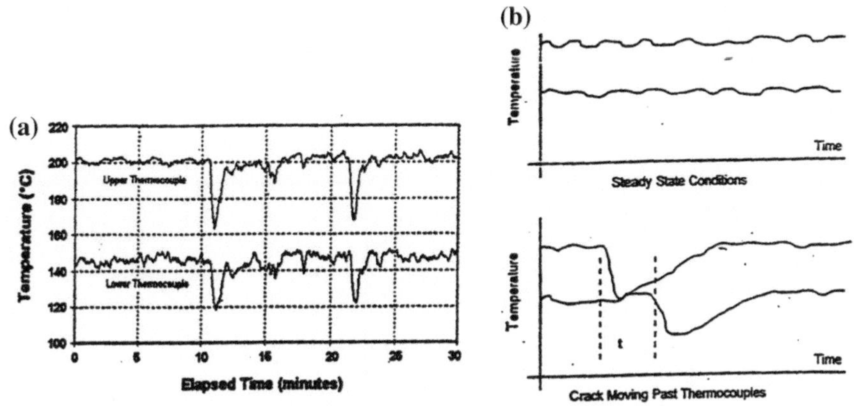

Fig. 11.13 Mould thermal monitoring (MTM) of temperature showing **a** variability in temperature associated with longitudinal cracking [58] and **b** conventional trace (*upper*) and drop in temperature associated with longitudinal cracking (*lower*) [52] (permission granted, **a** 10 M/Taylor and Francis, [58] **b** EPD Sci. [52]

Another way of successfully avoiding longitudinal, midway, cracks for slab casting is by increasing the superheat [15]. This leads to delayed solidification of the shell and leads to improved powder consumption.

11.3 Longitudinal Corner Cracking

11.3.1 Published Information

The causes of longitudinal corner cracking are similar to those responsible for longitudinal (midway) cracking. Narzt et al. [12] reported that the incidence of longitudinal corner cracking could be reduced by lowering the total heat flux (Fig. 11.14) i.e. by reducing horizontal heat transfer by using a casting powder with a high basicity which produces a thick, crystalline slag film.

Significant decreases in the incidence of longitudinal corner cracking have been recorded after applying EMBr to reduce both the metal flow velocity and surface instabilities [13].

It has also been reported that longitudinal corner cracking increases with (i) increasing S content of the steel [59] and (ii) increasing negative strip time, t_n [42] (i.e. similar relations to those for longitudinal cracking).

11.3.2 Causes, Mechanisms

The causes of longitudinal corner cracking are similar to those for longitudinal cracking, namely, sharp differences in shell thickness, which, for certain steel grades (especially medium-carbon steels) lead to stresses resulting from the mismatch of thermal shrinkage coefficients of the δ-Fe and γ-Fe. These stresses result in strains which are relieved by cracking. There are two different types of longitudinal corner cracking:

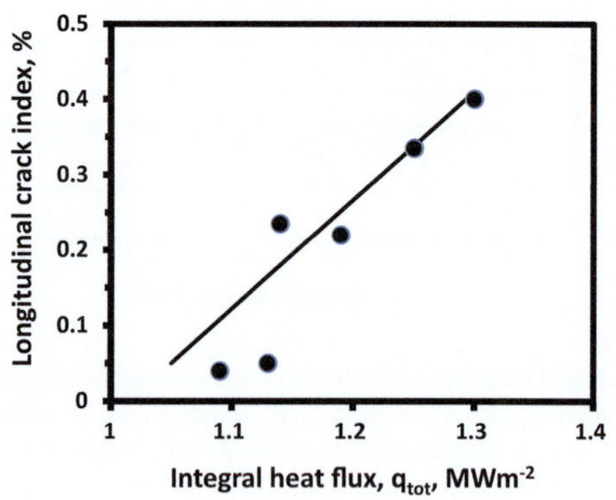

Fig. 11.14 Incidence of longitudinal corner cracking as a function of total (*horizontal*) heat flux (permission granted, UNESID, re-drawn from [12])

- The type caused by overcooled corners which is prevalent in billet casting and some other mould geometries [60]; a low taper leads to an increase in longitudinal cracking [42] due to the formation of an air gap and hence a shell of variable thickness.
- The type caused by "*thinning of the shell in the corner*" resulting from the local impact of the metal flow [61–63]; this type is prevalent in slab casting but can also be a result of improper narrow-side geometry (i.e. corner radius [60] and taper [42, 60, 64].

11.3.2.1 Longitudinal Corner Cracking Arising from Overcooled Corners

It can be seen in Fig. 11.15 that the heat in the corner regions has a much greater area to flow into than that in the centre. Consequently, the heat flux is much higher in the corners than in the centre of the mould. The high heat flux results in a thick shell in the corner, and during the subsequent contraction, slag normally flows in to fill the gap. However, in some cases, excessive contraction can lead to the formation of a gas gap which results in a sharp decrease in heat flux. Large variations in shell thickness lead to longitudinal corner cracking. The distance between corner and centre is much smaller in billets and leads to large, thermal gradients in the horizontal direction. These gradients lead to sharp changes in shell thickness and are particularly large in the corner region which results, sequentially, in stresses and then cracking.

"Football" moulds

There is very little room for the SEN in thin slab casting moulds and the "*football mould*" is one possible solution to this problem (Fig. 11.16). However, it can be seen that four "*corners*" are created around the SEN. There is enhanced heat flow through these corners which creates a thicker shell in these regions. Longitudinal corner cracking tends to occur in these regions.

Beam blank casting

Beam blanks can be cast in the desired shape. However, these moulds contain "*corners*" which can lead to enhanced cooling and a thicker shell in these regions.

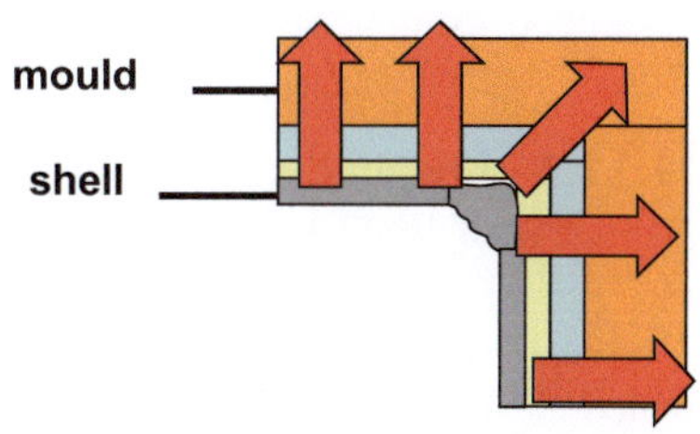

Fig. 11.15 Schematic diagram showing the path for heat extraction in the mould; note that there is a much greater area for heat removal in the corners than in the central regions of the mould

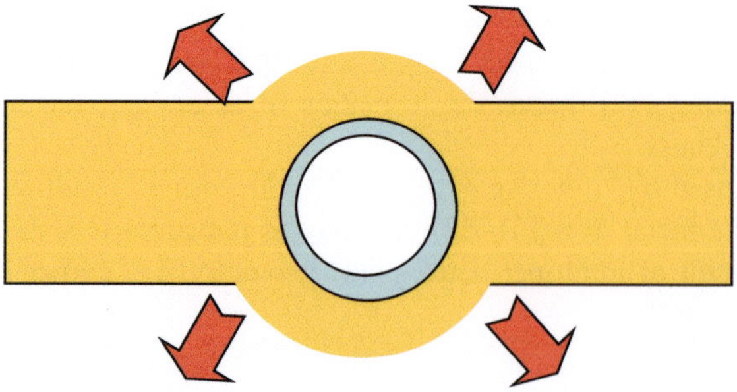

Fig. 11.16 Schematic drawing showing "*football mould*" used in thin slab casting and the creation of four "corners" with enhanced heat flow (see *arrows*) which are potential sites for longitudinal corner cracking; the SEN is shown in *blue*

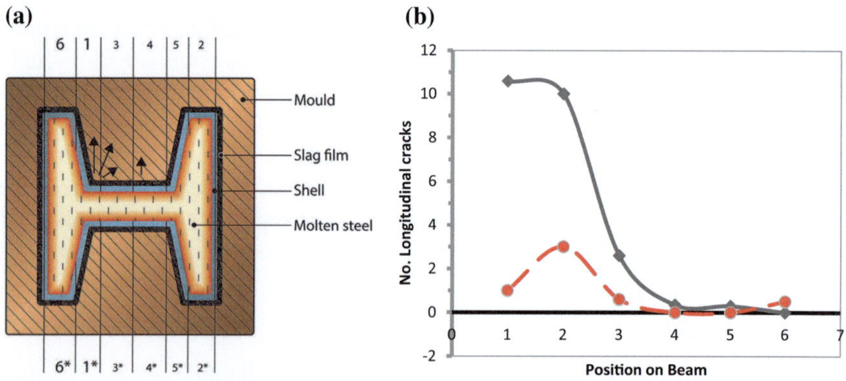

Fig. 11.17 **a** Schematic diagram showing the mould for casting H beams showing mould (*brown*) molten steel (*yellow*) solidified steel shell (*blue*) and slag film (*black*) and *black arrows* show heat flow paths (courtesy of Daniel Guerra). **b** Longitudinal cracking at different positions in H beam; —◆—sites 1, 2 etc. ● - -, 1*, 2* etc. (drawn from data reported by [65]); Note longitudinal cracking is prevalent at internal corners where heat flux density is high

The consequent variation in shell thickness can lead to longitudinal corner cracking. It has been reported [65] that the cracking

- was worse for the internal corners (B and E in Fig. 11.17).
- was worse when using mould slags with low T_{br} (or T_{sol}).
- tended to increase with increasing sulphur content of the steel.

11.3.2.2 Longitudinal Corner Cracking Resulting from Thinning of Shell by the Metal Flow

A typical example of this type of longitudinal cracking is shown in Fig. 11.18a) [61, 62] which clearly shows a "*white band*" resulting from negative segregation. The white

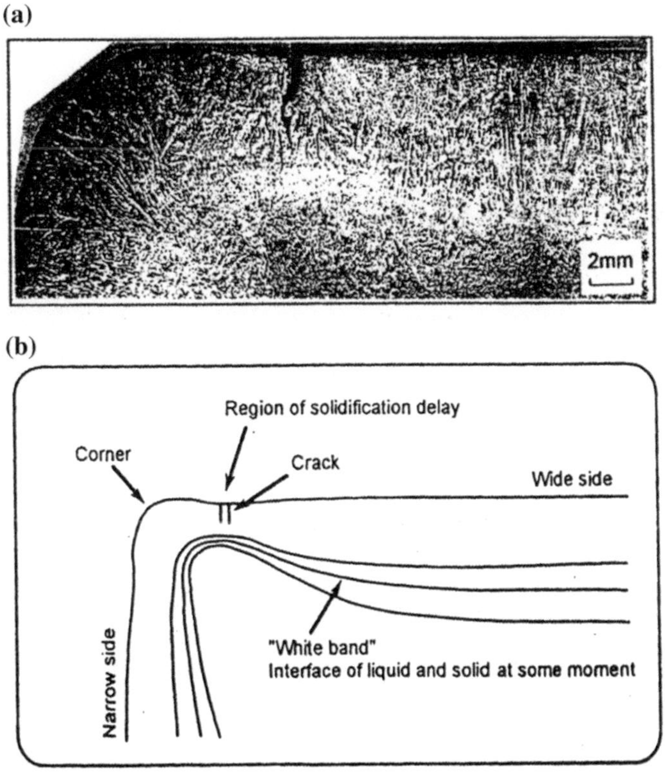

Fig. 11.18 **a** Photograph showing occurrence of "*white band*" and **b** drawing outlining shell thickness and white band and location of corner cracks (permission granted, UNESID, [61])

band reveals the location of the instantaneous solid/liquid interface (Fig. 11.18b) and it can be seen that the shell is thinner in the vicinity (on both sides) of the corner. This was attributed to direct impingement of the molten steel flow on the shell which causes melt-back and thinning of the shell and thus, a shell of variable thickness.

These casting conditions can occur when the mould dimensions are changed (i.e. the slab width is reduced and the casting speed increased in order to maintain the steel throughput at its current level). Both of these changes (in mould width and casting speed) result in increased metal flow velocity, which causes thinning of the shell in the impingement region and a shell of variable thickness. Significant decreases in the incidences of longitudinal corner cracking have been recorded after applying EMBr to reduce the metal flow velocity [13].

11.3.3 Ways of Dealing with Longitudinal Corner Cracking

The normal strategy to deal with this problem is to reduce the overall heat flux and create a more uniform, thinner shell. Obviously, the use of casting powders which produce a high-melting (i.e. high T_{br}) thick, crystalline slag film will be beneficial in most cases.

11.3.3.1 Longitudinal Corner Cracking Due to Overcooled Corners

Longitudinal corner cracking of this type can possibly be reduced by

- Replacing sharp corners with smooth chamfered corners [64, 66].
- Reduce water flow rates in the corner regions (*cf.* those for themed-face) [18, 19] but it has been reported that reduced water flow rates have little effect on the total heat flux [20], shutting off the water flow in the corners has been found to reduce longitudinal corner cracking [19].
- Apply coatings to the corner region of the mould to reduce heat transfer.

11.3.3.2 Longitudinal Corner Cracking Due to Shell Thinning

In this case, it is beneficial to reduce the melt-back of the shell due to the impact of the metal flow. This can be carried out by

- Redesigning the SEN port so as to give "more diffusive flow" into the corner regions of the mould [61, 62].
- Reducing the superheat.
- Reducing the casting speed.

11.4 Sticker Breakouts

Sticker breakouts are a major problem and result in hot metal streaming from the mould. They pose a safety hazard to the operating staff and result in loss of production; they have been estimated to cost between 0.1 and 1 million $ per incident. There are probably several causes for sticker breakouts but all involve the formation of a thin shell at some location and a loss of lubrication at that site. The solidified shell retrieved from the mould has the following distinctive characteristics:

- The "*stuck shell*" fans out from the *sticking point*, marked as d) in Fig. 11.19 and in detail in Fig. 11.20.
- The regular oscillation marks disappear and (in the stuck shell) are replaced by ripple marks with a characteristic V-shape fanning out from the breakout point (Fig. 11.19b).
- The shell is at its thinnest at the sticking point with the stuck shell increasing in thickness as we travel upwards (Fig. 11.19a).

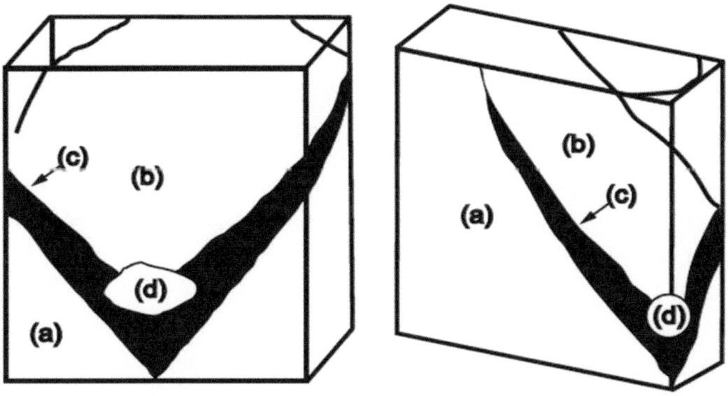

(i) (ii)

Fig. 11.19 Schematic drawings of the appearance of the strand after sticker breakout where the breakout occurs in (i) the broad face and (ii) the corner; **a** denotes regions of regular oscillation marks, **b** denotes ripple marks on stuck shell, **c** denotes the line of constrained shell and **d** denotes the sticking (or breakout) point (or position) (courtesy of Swerea KIMAB)

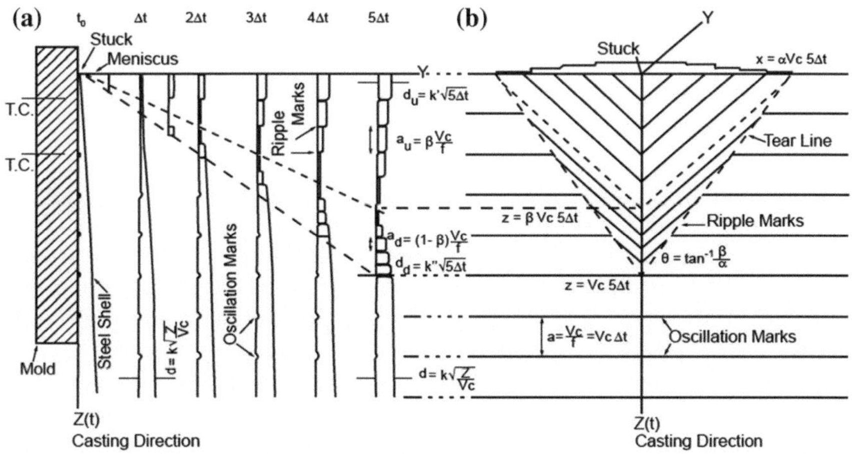

Fig. 11.20 Schematic diagrams showing (a) the shell is thinnest at the sticking point and (b) the appearance of the stuck region of the shell (permission granted, ISS/AIST [67])

11.4.1 Factors Affecting Sticker Breakouts

There is probably more than one mechanism involved in the creation of sticker breakouts, but all are thought to be linked to the formation of a shell which is too thin and too weak. A number of factors play a part in the creation of a sticker breakout, these are discussed below.

11.4.1.1 Steel Grades

Sticker breakouts are prevalent whilst casting high-Carbon (C > 0.4%) steel grades. The shells formed when casting these steels have poor mechanical strength due to micro-segregation between the austenite grains. Wolf [5] proposed that sticker breakouts were prevalent when the ferrite potential of the steel (FP) was less than 0.6 (where FP is given by Eq. 11.4 and CP is the carbon potential of the steel which can be calculated by Eq. 11.1).

11.4.1.2 Mould Dimensions and in Mould Conditions

A statistical analysis of plant data indicated that sticker breakouts tended to increase with (i) increasing mould width, w [68], (ii) decreasing mould length, L_{mould} [68, 69], (iii) increased casting speed (leading to a thinner shell) and (iv) increased mould level variations [68]. The effects of mould width and length are consistent with the formation of a thinner shell. Mould level variations will cause (i) carbon pick-up by the steel in contact with slag rim and the floating carbon particles (see Sect. 11.4.1.5 below) and (ii) the slag rim to fracture (see Sect. 11.4.1.3 below).

11.4.1.3 Steelmaking Conditions Creating Large Amounts of Al_2O_3 or TiN or ZrO_2

Sticker breakouts are thought to occur when an agglomerate gets trapped in the shell/ mould gap and blocks off the flow of liquid slag in this region. These agglomerates can arise from (i) the fracture of slag rim (ii) collisions of solid particles which lie undissolved in the liquid slag and (iii) liberated accretions ripped from the tundish stopper and SEN ports [70]. It might be expected that large mould level variations would also aid the fracture of slag films [68, 71].

There is a considerable amount of anecdotal evidence that sticker breakouts tend to occur when the steelmaking conditions are creating more Al_2O_3 than usual. It has been suggested that turbulent collisions of the Al_2O_3 particles causes the particles to agglomerate. Some Al_2O_3 accretions can also accumulate on the SEN and stopper rod refractories. However, Al_2O_3 pick-up by the slag is known to increase the crystallinity of slag films and rims [72] and which tend to fracture easily [73, 74]. Furthermore, large amounts of Al_2O_3 formation also lead to the formation of large, slag rims [74]. Thus, when large amounts of Al_2O_3 are formed the blockages to the liquid slag supply may come from either agglomerated Al_2O_3 particles or from fractured slag rims. In the casting of high-Al steels, the Al_2O_3 pick-up (>30% for 1%Al steels) results in frequent sticker breakouts.

Steels containing Ti (e.g. ULC and Ti-stabilised stainless steels) have also been reported to be prone to sticker breakouts [75]. Large amounts of TiN or Ti(CN) tend to form and these have low, saturated solubilities in mould slag (C_{sat} = ca. 0.5%) and these undissolved particles form agglomerates through collisions [76–78].

A high incidence of breakouts has been reported when using F-free mould slags containing TiO_2 [75].

It has also been reported for Ti-stabilised stainless steels, that the Ti in the steel reacts with SiO_2 in the mould slag to form *perovskite* ($CaO{\cdot}TiO_2$) [75, 79] which has a high-melting point. Consequently, it precipitates into the molten slag where it (i) increases the viscosity of the slag and (ii) agglomerates and blocks off the supply of molten slag to the shell. It is necessary to keep the basicity (C/S) of the slag < 1 to avoid *perovskite* precipitation [79].

ZrO_2 also has a low, saturated solubility in mould slag (C_{sat} = ca. 2%) [80]. Slag films attached to the stuck portion of the shell were found to contain high-ZrO_2 contents (up to 17%) [81]. Consequently, ZrO_2 probably behaves like TiN and forms agglomerated particles which shut off the flow of liquid slag in local sections of the mould. However, ZrO_2 will (i) increase T_{br} for the mould slag by 20–40 °C [80] and (ii) particles will act as nucleants for the precipitation of *cuspidine* and other crystalline phases [81]. These will result in a thicker and more crystalline slag film slag film, respectively, and thus will reduce the horizontal heat flux and shell thickness and thereby increase the incidences of sticker breakout.

11.4.1.4 Loss of Lubrication and Frictional Forces

The local loss of lubrication caused by a blockage to the liquid slag flow will increase the frictional forces acting on the shell. The shear stress and ferrostatic forces tend to increase as the shell moves further down the mould [67]. When the conditions result in a flaw in the solidified shell, breakout will occur when the shear stresses exceed the yield stress. The frictional forces calculated for these conditions were found to be very similar to those recorded under normal conditions [82]. A mathematical model has been developed to calculate the frictional forces, the shear stress and the high temperature, strength of the shell through the oscillation cycle [83, 84]. Tsuneoka et al. [67] developed a steady state, heat transfer model to simulate the proposed mechanism for sticker breakouts and obtained good agreement between predicted and measured temperature transitions.

Loss of lubrication can also occur when the melting rate of the mould powder fails to meet the required powder consumption. This can occur when large standing waves are formed and the granules run downhill from the peak leaving the peak with little mould slag. This can cause high friction and, in the worst case, breakouts; this is exemplified in Fig. 11.21.

11.4.1.5 Carbon Pick-up by Liquid Steel Near the Sticking Point

A metallographic examination of the stuck shell in the mould showed that the shell near the sticking point had a carburised structure, and there were some molten metal droplets with a carburised structure and an enhanced C content (ca. 4%) [85]. These droplets also contained cavities thought to have been formed by CO (g) during

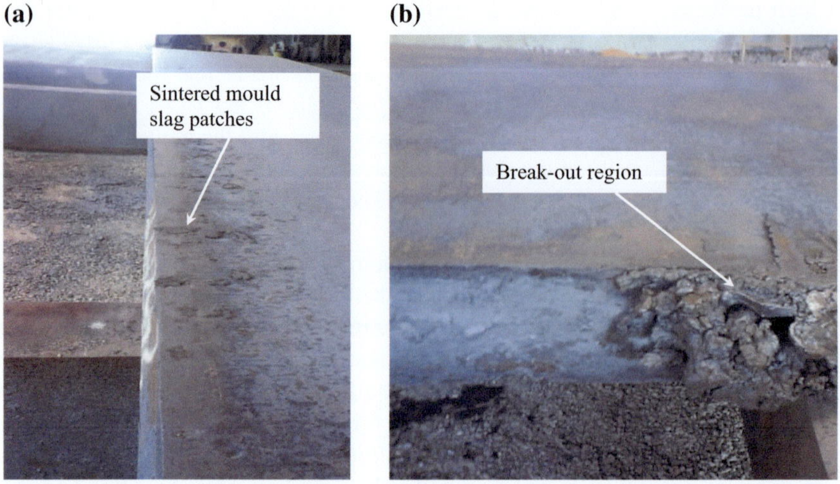

Fig. 11.21 Photographs showing **a** breakout slab with a region without mould slag (only sintered slag) **b** the breakout region (courtesy C-Å Däcker)

solidification. Thus, it was concluded that the carburised structures that there was local Carbon pick-up by the steel [85]. There are several sources of carbon:

- Unmelted casting powder in the bed [85] especially when there are large mould level variations (see also Fig. 11.21).
- The carbon particles floating near the top of the slag pool.
- Slag rims undergoing large mould level variations and fractured slag rims.
- Refractories (e.g. SEN, stopper rods) [70].

Pick-up of Carbon by the steel will reduce its melting point and steel will tend not to solidify when the mould is descending (in negative strip time) which results in a flaw in the shell.

11.4.1.6 Low Heat Transfer Across the Slag Film

Slag films producing a very thin solidified shell can also lead to sticker breakouts. The overly thin shells can result from several events and lead to sticker breakouts:

- Excess crystallisation in the slag film.
- Pick-up of ZrO_2 by the slag film which (i) increases T_{br} and d_s (ii) nucleates *cuspidine* and other crystalline phases; both of these events decrease q_{hor} and reduce d_{shell} [80, 81].
- Hydrogen in steel which results in a porous slag film with low thermal conductivity [86–88].
- Cyclical changes in casting speed which affect the thickness of the solid slag film [89].

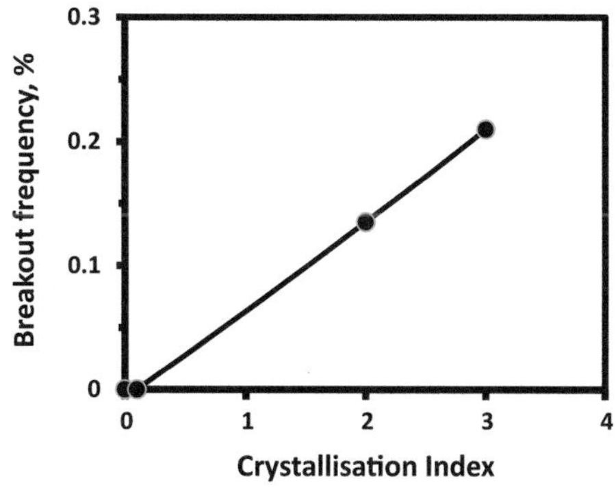

Fig. 11.22 The effect of crystallisation index on the incidence of sticker breakouts (permission granted, ISIJ, re-drawn from [90])

These various cases are described individually below.

Crystallisation in Slag Film

It can be seen from Fig. 11.22 that the incidence of sticker breakouts was found to increase when using casting powders with a high-crystallisation index (which were determined by metallographic examination [90]). The horizontal heat flux decreases with increasing crystallisation and thus the resulting steel shell will be both thinner and weaker [90, 91]. Thus, it would appear that incidences of sticker breakout can be reduced by using mould powders forming glassy slag films which lead to thicker, stronger steel shells. This is especially important when casting low carbon steel with an initial solidification of *δ-ferrite* which is very soft.

ZrO_2 Pick-up in Slag Film

The pick-up of up to 17% ZrO_2 in slag films attached to the stuck portion of the shell [81] has been discussed above (Sect. 11.3.1.3). It was pointed out that ZrO_2 additions will reduce the horizontal heat flux and thus, result in a thin shell. The most probable sources of ZrO_2 are the refractories used for the SEN and stopper rods [80]. The pick-up of ZrO_2 provides some measure of support for the mechanism involving the ripping off of accretions attached to the refractories by the metal flow [70].

Hydrogen in Slag Film

Incidences of sticker breakout have been correlated with high concentrations of hydrogen in the slag film [86–88]. Slag films with high-hydrogen contents contain many gas pores; the porous slag film has a low thermal conductivity. Consequently, the horizontal heat flux is low and a very thin shell is formed which is prone to

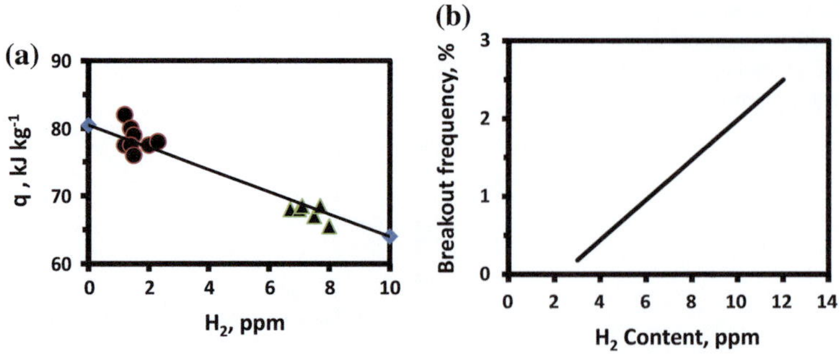

Fig. 11.23 **a** Heat extraction from the mould as a function of H_2 content of steel showing the effect of degassing (● = de-gassed; ▲ = not de-gassed), and **b** frequency of sticker breakout as a function of hydrogen concentrations in the slag film (permission granted, EDP Sci., re-drawn from [86])

sticker breakout. The effect of H_2 concentration in the steel can be seen in Fig. 11.23 where degassing of molten steel results, sequentially, in lower H_2 contents, lower porosity in slag film, higher horizontal heat flux and thicker shells [86] The incidence of sticker breakouts was found to increase linearly with the measured H_2 content in the slag film [86].

The principal sources of hydrogen in the metal are (i) moisture in the mould powders and refractories and (ii) water/steam leaks in the steel processing. It has been proposed [87, 88] that moisture in the mould powder results in the formation of hydroxyl ions (OH^-) formed by the reaction:

$$H_2O(g) + (O^{2-})_{slag} = 2(OH^-)_{slag} \tag{11.6}$$

These hydroxyl ions then react with Al or Si in the metal (denoted $\underline{Al}$)

$$2(\underline{Al}) + 6(OH^-)_{slag} = 6[\underline{H}] + 4(O^{2-})_{slag} \tag{11.7}$$

$$(\underline{Si}) + 4(OH^-)_{slag} = 4[\underline{H}] + 2(O^{2-})_{slag} \tag{11.8}$$

Ueshima et al. [87] pointed out that sticker breakouts can occur in de-gassed Si-killed steels, whereas, they rarely occur in de-gassed, Al-killed steels. The slag film in contact with the Si-killed de-gassed steel was found to contain micro-pores (Fig. 11.24b) whereas the slag film in contact with the de-gassed, Al-killed steel was found to contain little porosity (Fig. 11.24a) [87, 88]. In Al-killed, de-gassed steels, the strong reducing power of Al results in the consumption of the hydroxyl ions and there is no accumulation of OH^- ions [87, 88]. In contrast, in Si-killed, de-gassed steels the reductive power of Si is not sufficient to consume enough of the hydroxyl ions and there is an accumulation of OH^- ions in the slag. In the latter case, when mould slag infiltrates into the mould/strand channel, the slag is in contact with the cold shell and thus becomes supersaturated with OH^- ions which, subsequently, form H_2O (g) in the slag film by the reaction:

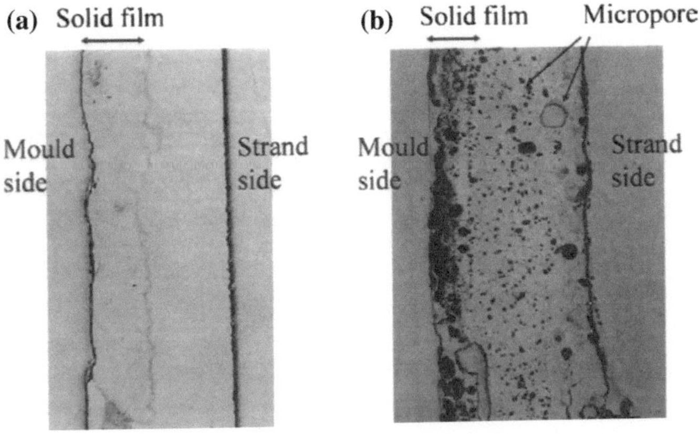

Fig. 11.24 Cross sections of slag film taken from below meniscus in the mould **a** with Al-killed, de-gassed steel and **b** Si-killed de-gassed steel (permission granted, ISIJ [88])

$$2(OH^-)_{slag} = H_2O(g) + (O^{2-})_{slag}. \tag{11.9}$$

It has been reported that sticker breakouts occur when the steel contains >7 ppm hydrogen [87, 88, 92, 93]. The partial pressure, $p_{H2,}$ can exceed 1 atm. during steel solidification and then passes into the molten slag where it forms a porous slag film.

Cyclical Changes in Casting Speed

Sticker breakouts occur when the total heat flux is low since this leads to a thin shell which is too weak to withstand the ferrostatic pressure. Sticker breakouts can occur as a result of cycling the casting speed. The following changes all occur when the casting speed is increased ($V_c \uparrow$):

(i) The mould temperature (T_{mould}) increases (Fig. 11.25a, b).
(ii) The horizontal heat flux density (q_{hor} in W m^{-2}) increases but the total heat flux q_{tot} (in J) decreases because of the reduced residence time (see Sect. 3.3).
(iii) A thinner shell is produced ($d_{shell} \downarrow$) (see Sect. 3.3).
(iv) The increased q_{hor} results in partial melt-back of the solid slag film (i.e. $d_s \downarrow$).

This sequence of events can be summarised as ($V_c \uparrow \rightarrow T_{mould} \uparrow \rightarrow q_{hor} \uparrow \rightarrow d_s \downarrow$) but ($q_{tot} \downarrow \rightarrow d_{shell} \downarrow$).

Inspection of Fig. 11.25a, b indicates (i) d_{shell} is thin when T_{mould} is high and (ii) that the changes to the slag film (d_s) and hence (d_{shell}), are not instantaneous but take time to adjust to the changes in $q_{hor.}$ The danger of sticker breakout is greatest when (i) d_{shell} is thin (i.e. when T_{mould} is high in Fig. 11.25) and (ii) when d_s is thicker than the equilibrium (or steady state) value (i.e. $d_s^t > d_s^{eqm}$ or $T_{mould}^t > T_{mould}^{eqm}$. The periods of the cycle where there is a danger of sticker breakout in Fig. 11.25a, b are marked by A.

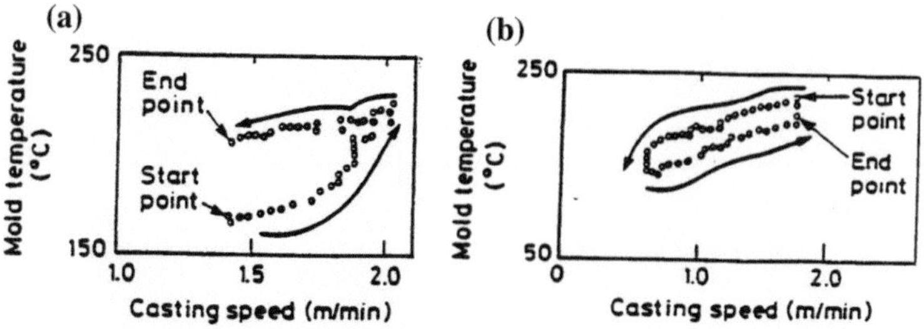

Fig. 11.25 Mould temperatures (related to heat flux) as functions of casting speed for a cycle where **a** casting speed is first increased then decreased and **b** where V_c is first decreased and then increased (permission granted, NSSM Corp. [89])

11.4.2 Causes, Mechanisms

11.4.2.1 Formation of a Pseudo-Meniscus

It has been proposed that the pseudo-meniscus is formed following the occurrence of a "*constraint*" [67]. It was also proposed that this pseudo-meniscus separated from the sound shell during periods of negative strip (i.e. when the strand is descending faster than the mould) but the shell is "*healed*" in periods of positive strip. The shear stress increases as the pseudo-meniscus moves down the mould. Breakout finally occurs when the shear stress exceeds the yield stress of the steel. This mechanism of rupture and repair is depicted in Fig. 11.26. An analysis of the various forces acting on the shell showed that the shell does not rupture under normal conditions but when a constraint occurs, the rupture will propagate downwards once the shear stresses exceed the yield stress. A 3-dim., non-steady state, heat transfer model was made to simulate the proposed mechanism; this obtained good agreement between predicted and measured temperature transitions [67].

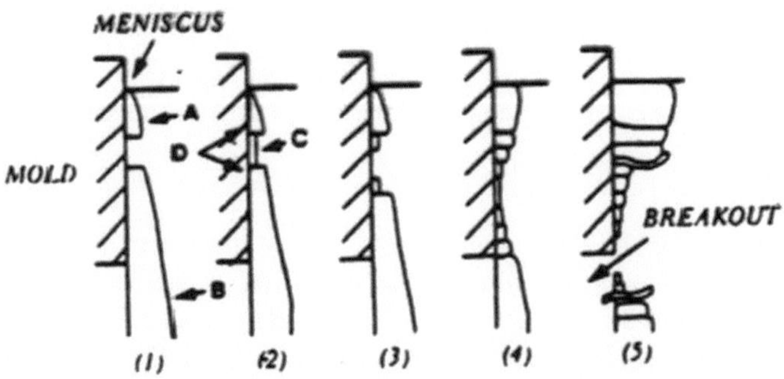

Fig. 11.26 Schematic drawing showing the mechanism for the evolution of sticker breakouts **A** = Stuck shell: **B** = normal shell before constraint: **C** = newly solidified shell: **D** = ripple marks. Note the drawing does not include a slag film (permission granted, ISS/AIST, [67])

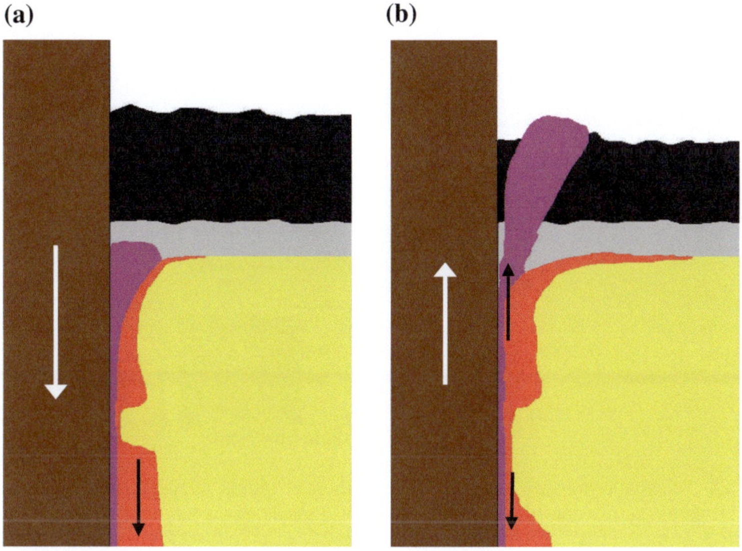

Fig. 11.27 Schematic description of the start of a breakout, **a** *downward* mould movement **b** *upward* mould movement pushing up the slag rim (*purple*); (courtesy of C-Å Däcker)

In some cases, the step-by-step withdrawal of the damaged steel shell can be observed in the mould and a breakout can be avoided by employing a sharp reduction in the casting speed. If the mould powder used is designed for soft cooling (i.e. by using a high-basicity powder), a slag rim will be created in the meniscus region. When the friction is too high the upper part of the steel shell will get stuck to the mould and the shell is ripped off when the lower part follows the strand (Fig. 11.27a). The upper, stuck part will oscillate with the mould but cannot move freely. The ripped area will be filled with steel that freezes to form a thin shell.

The effect is that the ripped shell is pushed up when the mould is ascending, which, in turn, pushes the slag rim into the powder layer (Fig. 11.27b). In this way, sticker breakouts can be avoided by observant operators, especially in slab casting, where the signs are more obvious.

11.4.3 Ways of Dealing with Sticker Breakout

11.4.3.1 Use Sticker Detection Systems

It is customary to minimise the number of sticker breakouts by using a sticker detection system [1, 2, 52, 58]. These systems consist of an instrumented mould containing several rows of thermocouples). During potential breakouts, the mould

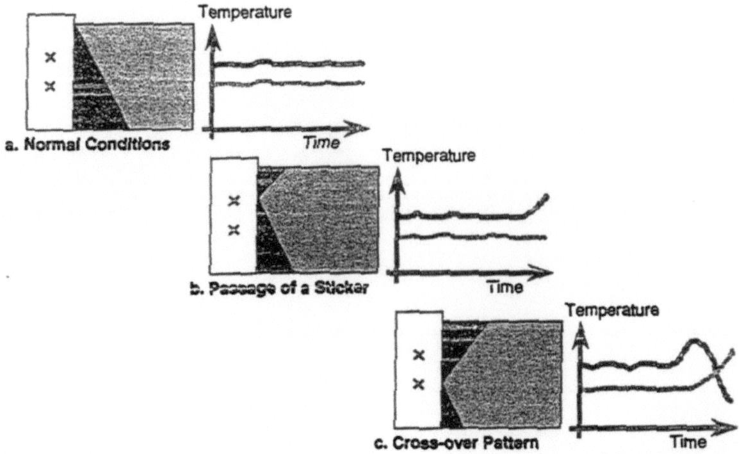

Fig. 11.28 Schematic drawing showing the evolution of mould temperature readings during a sticker breakout (permission granted, IOM/Taylor & Francis, www.tandfonline.com [58])

temperatures show the characteristic "*crossover*" in readings, (as shown in Fig. 11.28) associated with the formation of a pseudo-meniscus [58, 94]. This behaviour triggers an alarm allowing the plant personnel to take remedial action. The major problem, to date, has been the large number of false alarms and considerable effort has been devoted to reducing the incidence of false alarms. Success rates, where 80% of the alarms are genuine breakouts, have been claimed [58, 94]. Monitoring of the mould friction has also been proposed for the detection of sticker breakouts [95].

When a breakout alarm is sounded, remedial action must be taken. The first action is to reduce the casting speed which allows the shell to thicken and strengthen. However, in thin slab casting involving casting speeds of >4 m min^{-1}, there is very little time to take the remedial action [96]. Furthermore, the reduction in casting speed requires additional responses, i.e.

- The reduction of water flow rates in the secondary cooling zone;
- Adjustments to the mould level control since if the metal inflow is not reduced, an overflow is developed which is followed by a period of poor mould level control [96].

POSCO has installed a sticker-recovery system which is outlined in Fig. 11.29 which has been estimated to save 6 million $ p.a. [96].

Siemens VAI has developed a breakout prevention system called SIMETAL Mould Expert. The system works with two or three rows of thermocouple and sophisticated software which determines when there is an obvious risk of break out and seeks to avoid false alarms. When the system has identified a breakout situation the system automatically reduces the casting speed to a low level. After a healing period, the casting speed is manually increased, step-by-step back, to normal production.

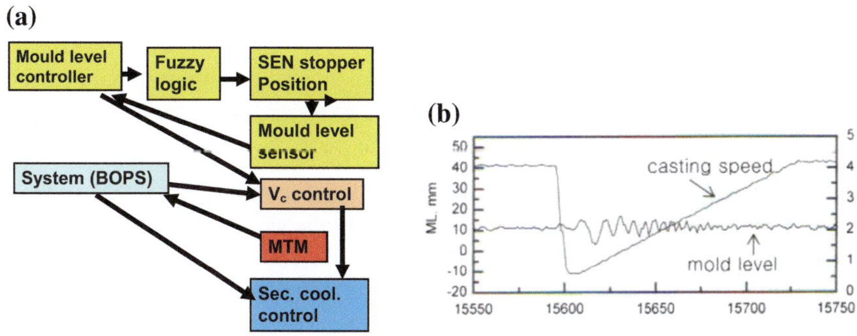

Fig. 11.29 Schematic diagrams showing **a** the steps taken in the sticker-recovery procedure and **b** traces of casting speed and mould level during a period of sticker recovery using the sticker-recovery procedure; *BOPS* Breakout Prediction System *MTM* Mould Thermal monitoring (permission granted, ISIJ [96])

Another advantage, of using a breakout system based on instrumented moulds, is that it can be used as a detailed tool for evaluation of plant trials with, for example, new mould powders.

11.4.3.2 Use Casting Powders Which Help Form a Thicker, Stronger Shell

(i) Use mould powders forming a thin glassy slag film with a low T_{br}; this results in a thicker (and hence) stronger shell. This is especially important for steels which solidify to form a soft, *ferritic*, low-strength, shell.
(ii) Avoid pick-up of ZrO_2 by the slag which results in a high T_{br}, a low heat flux and a thin shell.
(iii) Degas the liquid steel to reduce the hydrogen content since pores of H_2 and H_2O in the slag film impart a low thermal conductivity which leads to a thin shell.
(iv) Minimise H_2 levels by (a) storing mould fluxes in dry conditions (b) pre-heating the casting powder to reduce moisture contents (iii) inspect steel-making process for possible water/steam leaks
(v) Surprisingly, Ueshima et al. [87] reported that the incidence of sticker breakouts associated with high H_2 contents could be reduced by using a mould powder with a very high basicity (%CaO/%SiO_2) = 1.8 (this would be expected to produce a crystalline slag film with high T_{br}, resulting in a thin shell) and the powder would be vulnerable to water absorption.

11.4.3.3 Minimise the Numbers of TiN and ZrO_2 Particles in the Steel

TiN and ZrO_2 have low solubilities in molten mould slag so they remain largely undissolved and tend to agglomerate in the metal phase. Consequently, it is possible that sticker breakouts in Ti-containing steels can be reduced by the following:

(i) Maintaining a deep slag pool which will dissolve more of the TiN and will keep TiN particles away from the carbon (in the sinter bed or floating at the top of the slag pool).
(ii) Minimising the TiO_2 content in the casting powder (TiO_2 has been used to produce crystalline phases in F-free fluxes as an alternative to *cuspidine*) [75] and where TiO_2 is used in the powder, it is necessary to have a basicity of <1 to prevent *perovskite* formation [79].
(iii) Use a tundish two-layer cover with a melting slag on top of the steel followed by an insulation powder. The molten slag layer will help to reduce the pick-up of nitrogen from the atmosphere.

The same steps can be taken with ZrO_2 particles (some mould fluxes contain 2% ZrO_2 to minimise SEN erosion).

11.5 Oscillation Marks (OM's)

11.5.1 Characteristics of Oscillation Marks

Oscillation marks are regular, transverse indentations formed on the steel surface (Fig. 11.30). Oscillation marks are considered to be defects. Although oscillation marks do not impose much threat in themselves, cracks tend to form in the base of the mark due to micro-segregation. The severity of these cracks tends to increase as the depth of the mark (d_{OM}) increases. The segregation occurs as low-melting metal is squeezed out during negative strip time (when the mould is descending faster than the shell) [97, 98].

Oscillation marks are a natural occurrence and are related to the oscillation characteristics. They have also been found on the surface of steel cast with rapeseed oil. However, these marks were found to be shallower than those observed when using mould fluxes. However, marks at regular spacing can also be observed in static (ingot casting); these are known as "*ripple marks*" and are probably formed by an overflow of the newly formed shell by molten steel (similar to the overflow mechanism described below) or as folding marks when the shell formation is slower due to higher casting speed [99, 100]. Ripple marks have also been identified for static casting in the predictions of a mathematical model which coupled the fluid flow, heat transfer and shell solidification in the mould [101].

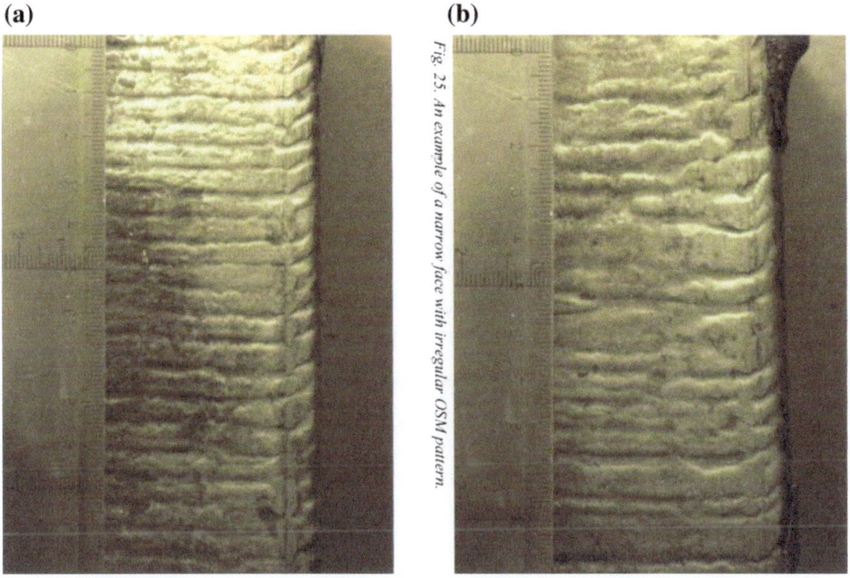

Fig. 11.30 Photographs of oscillation marks on narrow face of stainless steel slabs showing **a** regular pattern of marks and **b** an irregular pattern of marks (courtesy of C-Å Däcker [102])

The pitch between oscillation marks (L_{OM}) is also determined by the oscillation characteristics where the theoretical value for the pitch is given in Eq. 11.10 where s = stroke and f = frequency.

$$L_{OM} = V_c/f \tag{11.10}$$

It can be seen from Fig. 11.30 that oscillation marks vary in depth and pitch. In some cases, these oscillation marks have a regular pattern (Fig. 11.30a), whereas, in other cases the OM pattern is more irregular Fig. 11.30b [102]. These irregular marks occurred more commonly when casting *peritectic* steel grades where the newly formed shell is both buckled and strong. The variations in depth and pitch reflect the fluctuations in flow occurring in the mould; when all the factors come together, a deep OM is obtained but when they cancel one another out, a shallow OM is formed [103].

The interior of the shell also exhibits oscillation marks (Fig. 11.31) [103–105]. Thus, the internal thickness mirrors the external thickness. The variation in shell thickness is due to the variations in shell growth during the oscillation cycle (i.e. rapid growth occurring between (t_n^{mid} and t_p^{early}) and the limited, or zero, growth between (t_p^{mid} and t_n^{early}).

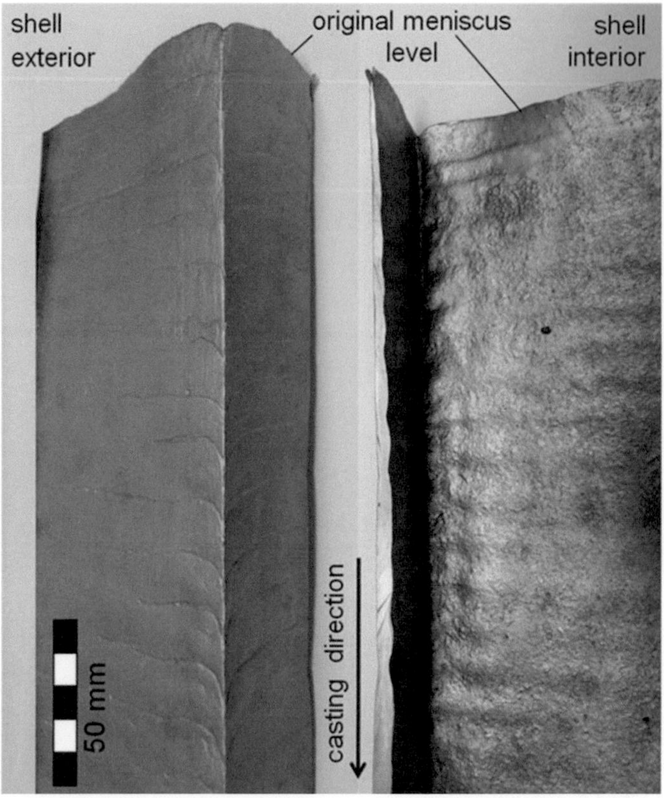

Fig. 11.31 Breakout shell showing internal depressions matching external depressions (permission granted, AIST [55])

11.5.2 Mould Oscillation

The mould is oscillated to prevent the steel shell from sticking to the mould. The oscillation characteristics are usually defined in terms of the *positive* (t_p) and *negative strip times* (t_n) [106] as shown in Eq. 11.11. The negative strip time is the time in the oscillation cycle when the mould is descending faster than the mould ($V_m > V_c$) and is calculated by Eq. 11.12 where f = frequency and s = stroke length. The positive strip time is the remainder of the oscillation cycle, i.e. $t_p = (t_{cycle} - t_n)$. In terms of the position of the mould and slag rim, the mould is at its highest position in late t_p and its lowest position in early t_p. Thus, the mould is descending throughout t_n and is ascending for the major part of $t_{p:}$

$$t_p + t_n = t_{cycle} \tag{11.11}$$

$$t_n = \left(\frac{60}{\pi f}\right) \cdot \arccos\left(\frac{V_c}{\pi s f}\right). \tag{11.12}$$

It can be seen from Eq. 11.12 that there are two terms involving the frequency: the first term tends to be dominant and the second term tends to partially cancel the effect of the first term. Thus, t_n increases as f decreases as V_c decreases and s increases ($t_n \uparrow$ as $f\downarrow$, $s \uparrow$, $V_c\downarrow$).

11.5.3 Factors Affecting Depth of OM's (D_{OM})

The factors affecting the depth of oscillation marks (d_{OM}) have been studied by a number of investigators using plant observations, hot physical models and mathematical modelling studies. The various factors, reported to affect the depth of oscillation marks, can be divided into the following categories.

11.5.3.1 Oscillation and Casting Variables

Several of casting variables have been shown to affect the depth of the oscillation mark. There is general agreement on the effect of some parameters affecting d_{OM} but disagreement in the published literature over the effect of other variables.

(i) There is general agreement that the depth of the oscillation mark (d_{OM}) decreases as the *negative strip time* decreases ($d_{OM} \downarrow$as $t_n \downarrow$) [43, 107–111, 113, 114] (see Fig. 11.32a).

(ii) It has been reported that the effect of frequency, stroke and casting speed on d_{OM} can be predicted through their effect on t_n, (as given in Eq. 11.12), i.e. $d_{OM} \downarrow$ as $t_n \downarrow$as f, $\downarrow s\downarrow$, $Vc \uparrow$) [55].

(iii) There is also general agreement on the relationship between d_{OM} and *stroke length*, *s*, where the relation ($d_{OM} \downarrow$ as $s\downarrow$) has been proposed in [106–109, 113].

(iv) However, there is disagreement on the reported relations between d_{OM} and *casting speed* where ($d_{OM} \downarrow$ as $V_c \uparrow$) is supported by [55, 73, 106, 113–117, 119] and ($d_{OM} \downarrow$ *as* $V_c \downarrow$) by [106, 111, 118].

(v) Similarly, there are differences in the reported relationships between d_{OM} and *frequency*, *f*, with ($d_{OM} \downarrow$ as, $f \downarrow$) supported by [55, 73] and ($d_{OM} \downarrow$ as, $f \uparrow$) by [46, 107, 108, 113, 114, 119, 120, 121]; however, it has been reported that the $d_{OM} - f$ relation goes through a maximum at ca. 100 cpm [111] and this may be due to f having opposing effects on the two terms in Eq. 11.12. Values of d_{OM} have been reported to decrease as the parameter ($s/f^{0.5}$) decreased [108].

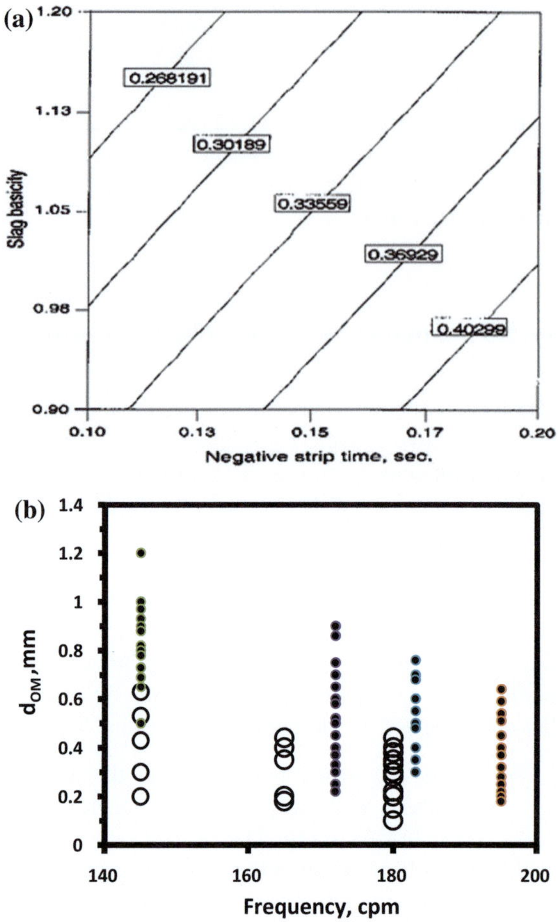

Fig. 11.32 Depth of oscillation mark as a functions of **a** negative strip time and basicity and **b** frequency; ● = with Ar blow; O = without Ar blow (permission granted, **a** EPD Sci. [43] and **b** ISIJ [108], re-drawn)

(vi) The depth of an oscillation mark has been reported to decrease with decreasing *superheat*, ΔT, (i.e. $d_{OM} \downarrow$ as $\Delta T \downarrow$) [108, 118]; this a surprising result since increased superheat would be expected to delay shell solidification (see Sect. 11.5.3.3 below) and hence, lead to an increase in the separation distance between the slag rim and the shell tip which would lead to shallower oscillation marks (i.e. $d_{OM} \downarrow$ as $\Delta T\uparrow$) [55].

(vii) It has been suggested that the *separation distance* between the slag rim and the shell tip ($d_{\text{rim to tip}}$) is an important variable affecting d_{OM} (with $d_{OM} \downarrow$ *as* $d_{\text{rim to tip}} \uparrow$) [55]; note a short stroke would increase $d_{\text{rim to tip}}$ and thus decrease d_{OM}.

(viii) A regression analysis of plant data from 27 trials indicated that d_{OM} decreased when the *basicity* (C/S) of the mould powder increased ($d_{OM} \uparrow$ as (C/S)↑) [43] (Fig. 11.32a) and the relation, d_{OM} (mm) = 0.49 + 1.11 t_n − 0.30 *(C/S)* [43]; this was attributed to an increase in both T_{br} and slag film thickness with increasing (C/S) which resulted in a shorter shell and delayed solidification (see Sect. 11.5.3.3) [43].

(ix) The use of *non-sinusoidal oscillation* has been reported to decrease d_{OM} [51, 108]; this is consistent with a decrease in negative strip time ($t_n \downarrow$).

(x) Values of d_{OM} were greater when *Ar flow* was used (*cf.* when no Ar flow was applied in the SEN) (Fig. 11.32b); this is consistent with the Ar flow boosting the vertical heat flux and thereby, increasing the length of the shell (see Sect. 11.5.3.2). Ar flow has been reported to increase powder consumption (see Fig. 2.15) and thus this finding is consistent with d_{OM} increasing with increasing powder consumption.

(xi) Oscillation mark depths have been found to increase with *mould level variations* [105, 118]; both d_{OM} and the pitch, $L_{OM,}$ decreased for low positions of the (slag/steel interface) in the mould (Fig. 11.33) [105].

(xii) Reductions (of 50%) in d_{OM} and in the variability of d_{OM} have been reported when using *Electromagnetic devices* fitted to the mould (see Sect. 5.8); the effects of using EMS [50] EMC [50, 51] and EMBr [46] on d_{OM} are shown in Fig. 11.34a, b and c, respectively. This was attributed to a shorter shell and delayed solidification resulting from (a) the homogenisation of the steel temperature with EMS (b) the "*pinch force*" generated in EMC and (c) the increased meniscus temperature and the reduced q_{vert} in EMBr.

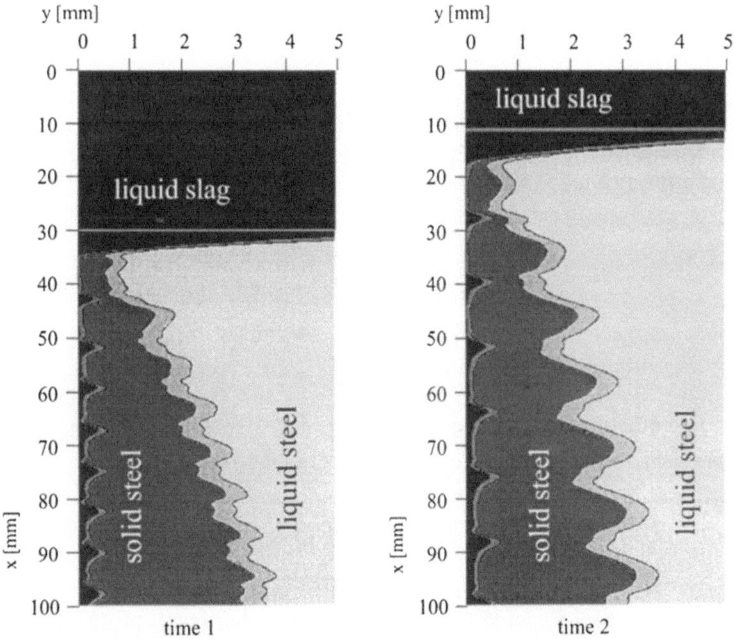

Fig. 11.33 Schematic diagrams showing the predicted effect of mould level change on depth and pitch of oscillation marks with d_{OM} and L_{OM} increasing as metal level increases (permission granted, Eurotherm and H Steinruck [105])

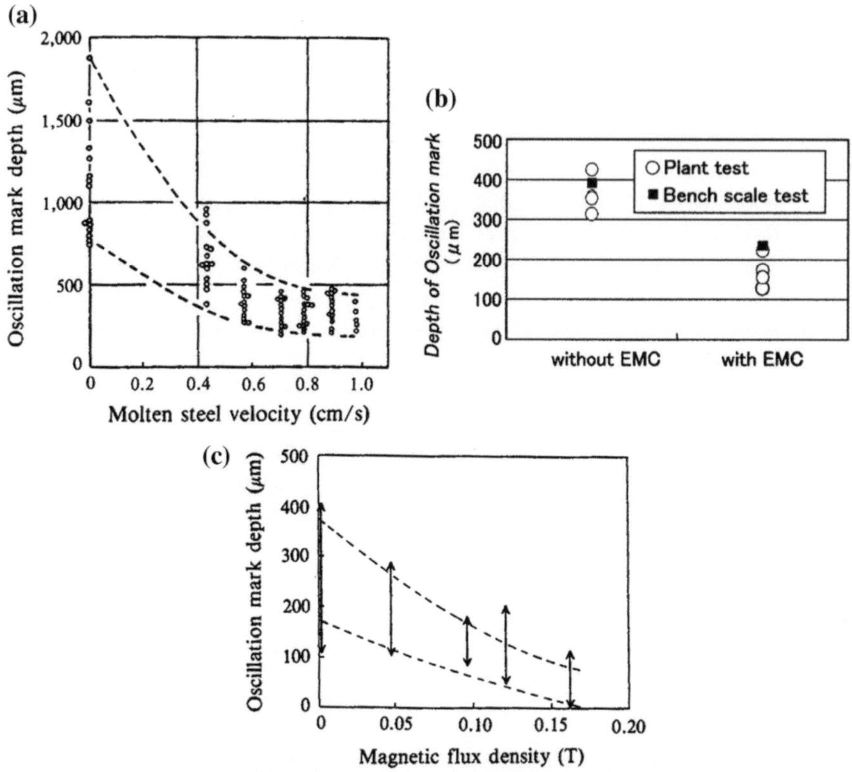

Fig. 11.34 The depth of the oscillation mark as functions of **a** the molten steel velocity in EMS [46], **b** the application of EMC [51] and **c** the electromagnetic flux density applied in EMBr [46] (permission granted NSSM Corp.; **a** [46], **b** [51], **c** [46])

11.5.3.2 Reducing the Length of the Solidified Meniscus

It has been suggested that the depth of the oscillation mark can be reduced by reducing the length of the solidified shell [122]; it can be seen that d_{OM} would be smaller in the case of Fig. 11.35a than for Fig. 11.35b. The length of the solidified shell can be reduced by reducing either (i) the vertical heat flux (q_{vert}, by increasing the depth of the powder bed, or incorporating exothermic agents) (ii) reducing the horizontal heat flux (q_{hor}, e.g. by using EMBr, EMC or creating a slag film with high T_{br} and f_{crys}). A shorter, solidified meniscus will also reduce inclusion and bubble entrapment [122]. Reduced Ar flow rates will tend to decrease oscillation mark depth [108] (Fig. 11.32b) by reducing q_{vert} (since gaseous convection increases vertical heat transfer and hence, increases the length of the shell).

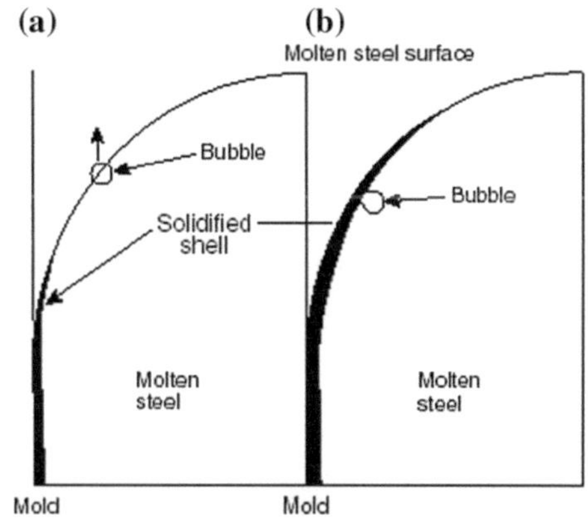

Fig. 11.35 Schematic drawings showing the formation of **a** a short solidified shell and **b** a long solidified shell (permission granted, ISS/AIST [122])

11.5.3.3 Delaying Solidification Further Down the Mould

It was pointed out above, that the distance between slag rim and steel tip ($d_{rim/tip}$) is one of the key factors affecting the depth of oscillation marks with $d_{OM} \downarrow$ as $d_{rim/tip} \uparrow$. Thus reducing the length of the solidified shell can be seen as one way of increasing $d_{rim/tip}$ and thereby reducing d_{OM}. Another way of increasing $d_{rim/tip}$ is to initiate solidification further down the mould. This achieved in ways which are similar to those used to reduce the length of the shell.

Increasing the Superheat or the Steel Meniscus Temperature

It has been reported that ($d_{OM} \downarrow$ as $\Delta T \uparrow$) from the results obtained in a sensitivity study (Table 11.2) [55] but the reverse relation ($d_{OM} \downarrow$ as $\Delta T \downarrow$) has been reported in two studies [108, 118]. The latter finding could be attributed to the effect of powder consumption on d_{OM} ($d_{OM} \uparrow$as $Q_s \uparrow$) However, the application of EMBr has been reported to increase the steel meniscus temperature by ca. 10 °C [47, 123] and also to cause a reduction in d_{OM} which provides support for ($d_{OM} \downarrow$ as $\Delta T \uparrow$). Applying EMBr also reduces q_{vert} [124] and leads to a decreased slag pool depth. The confusion over the effect of superheat probably arises because of the effect of these competing effects of Q_s and q_{vert}.

Reducing Horizontal Heat Flux (q_{hor})

The horizontal heat flux in the meniscus region can be reduced by using (a) a mould powder with high basicity (C/S) and T_{br} (Figs. 11.6 and 11.7) (b) Meniscus-Free casting (in which a refractory feeder-head is added to the top of the mould) to reduce the horizontal heat transfer in the meniscus region [21–23] or (c) Electromagnetic casting (EMC) where the *pinch-effect* reduces q_{hor} in the meniscus region [50, 125].

Table 11.2 Summary of the change needed in the various factors to bring about a decrease in the depth of the oscillation mark (d_{OM}) (Reproduced with permission, from [55])

	V_c	ΔT	t_n	t_p	f	s	γ	η	T_{br}
d_{OM} ↓ as	↑	↑	↓	↑	↑	↓	↑	↓	↓

Delayed solidification has the benefit of taking the shell away from the turbulent, metal/ slag interface which results in less slag, inclusion and bubble entrapment.

11.5.3.4 Changing Mould Slag Properties

It has been reported that mould slag properties can affect the depth of the oscillation marks.

Viscosity (η)

Most information supports the view that viscosity affects d_{OM} through its effect on powder consumption (Q_s); since Q_s decreases with increasing viscosity. Thus, d_{OM} would be expected to decrease as η increases (d_{OM} ↓ as η ↑) [110, 115, 127]. Additional support for this relation comes from the slag flow mechanism (Sect. 11.4.4.3) [55] where increased viscosity resists the downward flow velocity in the slag pool and reduces the depth of the "*dimple*". However, the reverse relation (i.e. d_{OM} ↓ as η ↓) has been reported in Table 11.2 [55] and in [106, 126]. This confusion may arise because d_{OM} is dependent upon both q_{hor} and Q_s and the following relations apply (d_{OM} ↓ as Q_s ↓ → η ↑ → (*C/S*) ↓) and (d_{OM} ↓ as q_{hor} ↓ → (*C/S*) ↑ → η ↓), respectively; thus, the viscosity has opposing effects on these two factors affecting d_{OM}.

Interfacial Tension ($\gamma_{m/Sl}$)

It has been proposed that d_{OM} decreases as interfacial tension increases (d_{OM} ↓ as $\gamma_{m/sl}$ ↑) since a high $\gamma_{m/sl}$ results in a smaller radius of curvature and hence a smaller distance between metal and mould [53, 101]. It has also been proposed that $\gamma_{m/sl}$ influences the pitch of the oscillation marks (L_{OM}) through its effect on the radius of curvature [128]; it was proposed that L_{OM} was greater for the flow mechanism than for the overflow mechanism [128] (see Sect. 11.5.4).

Break Temperature (T_{br})

It can be seen from Fig. 11.32a) that d_{OM} decreases with increasing slag basicity (C/S), ($d_{OM} = 0.49 + 1.11t_n - 0.30$(C/S)). Since the break temperature and f_{crys}

tend to increase with increasing basicity, it would be expected that d_{OM} would decrease with increasing T_{br} (d_{OM} ↓as (C/S) ↑as T_{br} ↑). It might be argued that a higher T_{br} leads to a lower powder consumption and hence a decreased d_{OM}. Conversely, a sensitivity study using a mathematical model found the reverse relation with T_{br}, i.e. d_{OM} decreased as T_{br} decreased (d_{OM} ↓ as T_{br} ↓) as shown in Table 11.2.

11.5.3.5 Powder Consumption

Values of d_{OM} have been reported to decrease with decreasing powder consumption (d_{OM} ↓as Q ↓) [110, 115, 127, 129]. It has also been reported that oscillation marks became indistinct and irregular when the slag pool becomes very shallow [130]. Jenkins et al. [131] noted that oscillation marks disappeared when slag infiltration was cut off. This behaviour could be interpreted as evidence that d_{OM} is affected by the slag flow (i.e. powder consumption, Q). Jenkins [131] also observed that there was a pressure increase as the rim approached the slag/steel interface.

Itoyama [108] considered that the shell was deformed by the positive pressure exerted by the liquid slag during negative strip time. They modified an experimental caster so that, in addition to the vertical sinusoidal or non-sinusoidal oscillation, a horizontal oscillation could be simultaneously applied to the narrow faces. It was found that horizontal oscillation was beneficial in reducing d_{OM} especially using mode B (i.e. when the distance between the faces was widened following the mould descent and reduced at the beginning of the descent).

11.5.3.6 Increasing the Strength of the Solidified Shell

Fukuda et al. [127] reported that a stronger shell would result in shallower oscillation mark (Fig. 11.36). This was attributed to the formation of a thicker shell (by increasing horizontal heat transfer). However, a shell with a finer grain size might

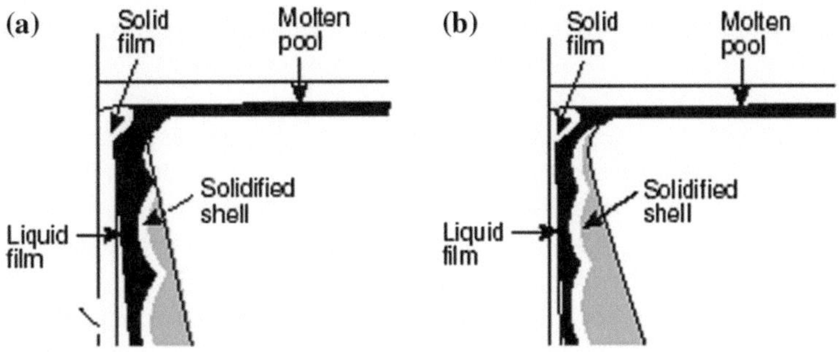

Fig. 11.36 Schematic diagrams showing that **a** a thin shell results in deeper oscillation marks and **b** a thicker shell leads to shallower oscillation marks (permission granted, ISS/AIST, [127])

be expected to give the same effect. This finding is in conflict with the view that d_{OM} can be reduced by decreasing q_{hor}.

11.5.3.7 Effect of Metal Flow on d_{OM}

Eung et al. [132] reported that the depths of hooks (d_{hook}) could be reduced by improving the metal flow in the mould; this was achieved by modification of the SEN port design. There is a linear relation between d_{OM} and d_{hook} [113].

11.5.4 Causes, Mechanisms

Two mechanisms have been proposed for the formation of oscillation marks; these are known as the "*Hook or Overflow*" type and the "*Depression or Folded*" type [104, 133–138]. Recently, another mechanism has been proposed based on the predicted flows of a mathematical model (Slag flow mechanism) [53, 55].

The curved meniscus (between steel and molten slag) solidifies to form a hooked shell tip (Fig. 11.57a). The ferrostatic pressure subjects this shell to deformation. The type of oscillation mark formed (Overflow or Folded) is determined by the ability of the shell to resist the deformation and this, in turn, is largely determined by the thickness of the shell.

11.5.4.1 Hook or Overflow OM Mechanism

Hooks are formed when there is insufficient heat supplied to the meniscus region, especially (i) at lower casting speeds and (ii) in regions near the narrow faces, where temperatures are at their lowest and (iii) for steels with low liquidus temperatures (e.g. *ferritic* stainless steels and high-Al steels). If the shell is strong enough to avoid deformation, the molten steel overflows the steel tip and a new meniscus is formed which solidifies to give a new oscillation mark (Fig. 11.37a). In some cases, the overflowing liquid remelts the meniscus tip (Fig. 11.37b) and the hook, or nail, disappears but in other cases the hook, or nail, survives (Fig. 11.37a). The critical difference lies in the thickness of the shell. In the first case, the shell is strong enough to avoid deformation, causing the steel meniscus to overflow the tip (i.e. overflow mechanism shown in Fig. 11.38a; such a case is shown in Fig. 11.39a [139]. Several workers have discussed the mechanisms responsible for the formation of hooks [55, 140–142]. The mean, hook depth (d_{hook}) and thickness (t_{hook}) were found to decrease as (i) $d_{hook}\downarrow$ as $t_n\downarrow$ as $V_c\uparrow$ as $f\uparrow$ and (ii) $t_{hook}\downarrow$ as $V_c\uparrow$as as $f\uparrow$and (iii) the length of hook, $l_{hook}\downarrow$ as $\Delta T\uparrow$ [140]. Thus, it can be concluded that d_{hook} and l_{hook} are linearly related to d_{OM} and that d_{hook} and l_{hook} are linked [140]. The number of hooks has been reported to increase with increasing stroke length [50, 116, 118, 140, 141].

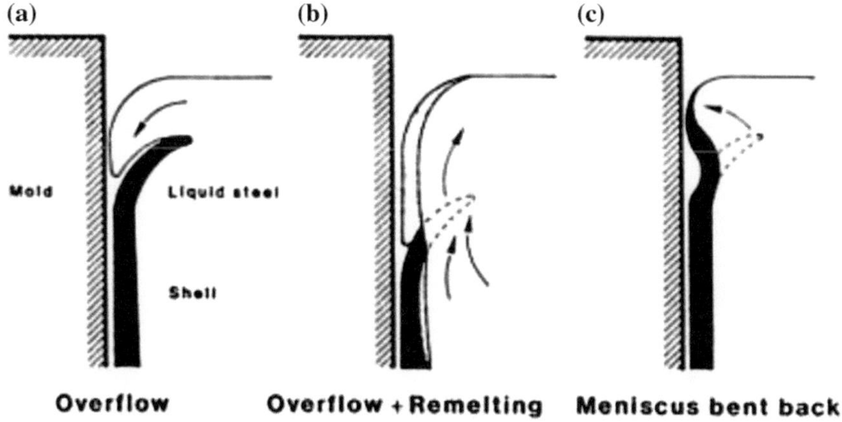

Fig. 11.37 Schematic diagram showing types of oscillation mark formed by **a** Overflow mechanism with the formation of nail or hook **b** Overflow mechanism with remelting of the hook and **c** Folding mechanism where the shell is bent back during positive strip time (permission granted, ISS/AIST [138])

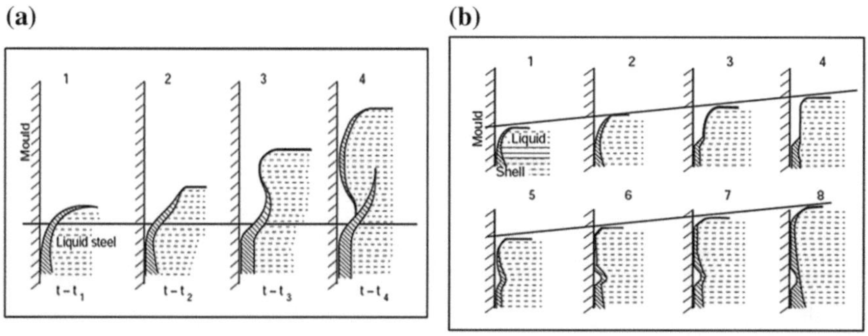

Fig. 11.38 Schematic diagrams showing the evolution of oscillation marks during positive strip time by **a** Overflow mechanism and **b** Folding mechanism (permission granted, IOM/Taylor and Francis [136])

11.5.4.2 Depression or Folded OM Mechanism

If the shell is unable to resist the deformation, it is folded back during positive strip time as shown in Figs. 11.37c and 11.38b. The pitch of oscillation marks (L_{OM}) has been reported to vary according to whether it is an overflow or folded type [128] with $L_{OM}^{mean} < L_{OM}^{folded}$ and $L_{OM}^{mean} > L_{OM}^{overflow}$. An example of a folded oscillation mark is shown in Fig. 11.39b.

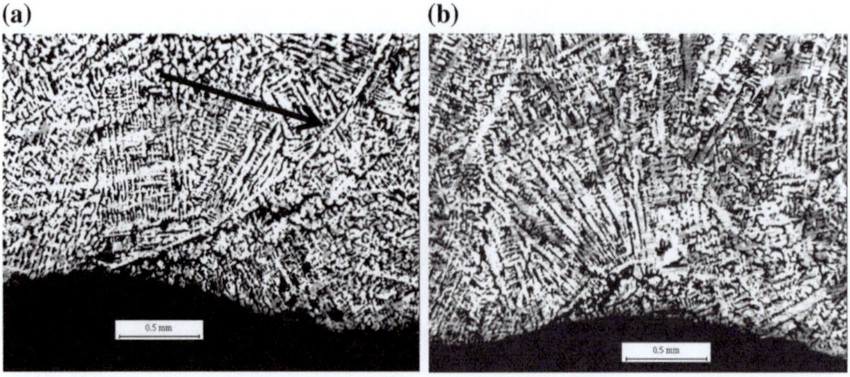

Fig. 11.39 Micrographs showing **a** surface structure with a hook (marked with an *arrow*) in base of OM in a *peritectic*, carbon steel [139]; casting direction *right* to *left* and **b** surface structure for a folding mark at the base of an OM (permission granted, Europ. Comm. Sci.Tech. Publ. [139])

11.5.4.3 Slag Flow Mechanism for OMs

This mechanism was proposed as a result of the analysis of the predicted fluid and heat flows and shell solidification in the mould [53, 55, 142]. It was observed that there were changes in the direction of the flow in the slag pool at different parts of the oscillation cycle. This is particularly important in the region of the mouth of the channel between shell and mould. These changes can be seen in Fig. 11.40 where the following stages occur:

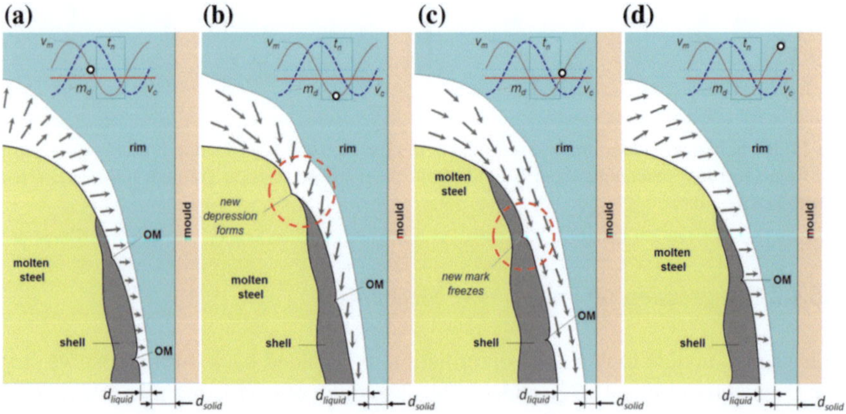

Fig. 11.40 Schematic diagram showing the direction of slag flow at different parts of the oscillation cycle **a** midway through, t_p **b** halfway through t_n **c** at lowest position of mould in early t_p **d** at highest position of mould in late t_p (permission granted, ISS/AIST, [55])

(i) When the mould is ascending, the flow in the slag pool is hot (*tropical*) and radially outward and there is little, or zero, shell growth at this time (Fig. 11.40a).

(ii) When the mould and slag rim reach their highest position (in t_p^{late}) there is a *tide change* and, subsequently, the slag flow becomes downward and since this flow originates in the cooler parts of slag pool it is cold, or *arctic*, flow (Fig. 11.40b, c); this cold convective flow causes increases in both the heat flux and shell growth which increase gradually as the mould descends further during t_n (Fig. 11.40a, d).

(iii) The mould descends throughout t_n and reaches its lowest point in early t_p and during the period (t_n^{late} to t_p^{early}) the pressure of the cold downward flow in the slag pool causes the formation of a "*dimple*" in the liquid adjacent to the shell tip; this dimple solidifies rapidly because of the cold (*arctic*) convective flow and the dimple is the initial oscillation mark which deepens during further solidification (Fig. 11.40b, c).

Thus, in summary, oscillation marks are formed because of (i) tide changes in the direction of flow in the slag pool and the gradual development of a cold, convective flow when the mould is descending and (ii) the pressure associated with the cold downward flow creates a "*dimple*" in the liquid adjacent to the shell tip and the dimple rapidly solidifies to form the initial mark.

It was concluded that (i) the distance between the slag rim and shell tip ($d_{rim/tip}$) is an important factor (since $d_{OM} \uparrow$ as $d_{rim/tip} \downarrow$) and (ii) the effect of casting parameters can be predicted from their effect on t_n [55]. Table 11.2 was derived from modelling predictions. The mechanism proposed for the formation of hooks is shown in Fig. 11.41 [55].

11.5.5 Ways of Dealing with Deep OMs

The depths of oscillation marks can be reduced by decreasing (i) negative strip time, t_n (ii) the vertical heat flux, q_{vert} (iii) the horizontal heat flux, q_{hor} and (iv) powder consumption, Q_s; these are discussed below. However, some variables affect these parameters in different (and sometimes, in opposing) ways. An example of this is that high, slag viscosity causes (i) ($\eta \uparrow \rightarrow Q_s \downarrow \rightarrow d_{OM} \downarrow$) and (ii) involves low basicity (C/S) slags giving ($\eta \uparrow \rightarrow$ *(C/S*$\downarrow$*)* $\rightarrow q_{hor} \uparrow$- $\rightarrow d_{OM} \uparrow$). A second example is increasing superheat, ΔT, which delays solidification ($d_{OM} \downarrow$) but reduces slag viscosity ($\eta \downarrow \rightarrow Q_s \uparrow \rightarrow d_{OM} \uparrow$).

11.5.5.1 Reduce Negative Strip Time (t_n)

It can be seen from Eq. 11.12 above that the negative strip time (t_n) can be reduced by decreasing the stroke length ($s \downarrow$) or by increasing the casting speed ($V_c \uparrow$).

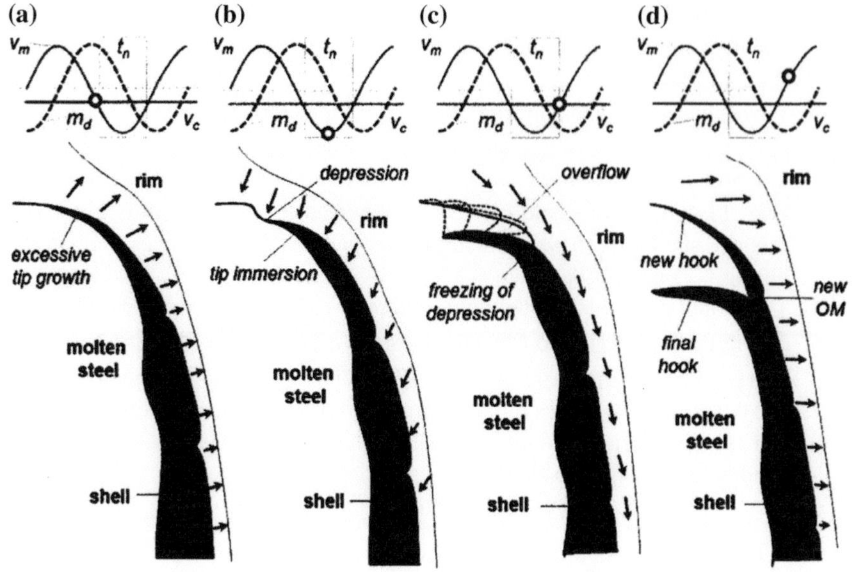

Fig. 11.41 Schematic diagrams depicting the evolution of hook formation during the oscillation cycle **a** at t_n^{start}, **b** t_n^{mid}, **c** at t_n^{late} (near lowest point of mould) and **d** at t_p^{mid} (permission granted, ISS/AIST, [55])

The effect of frequency on t_n is more complicated since Eq. 11.12 contains two terms involving the frequency, with the first term being the dominant term and the second term tending to cancel out the effect of the first term. Overall, t_n tends to decrease as frequency increases ($t_n \downarrow$ as $f \uparrow$).

11.5.5.2 Increase the Distance Between Base of the Slag Rim and Steel Tip ($d_{rim/tip}$)

The argument here is that if $d_{rim/tip}$ can be increased, it will reduce the depth of the *dimple* formed next to the shell tip. This could be achieved in the following ways:

(i) Decreasing the stroke length ($s \downarrow$).
(ii) Increasing mould slag viscosity ($\eta\uparrow$) resists the downward, flow velocity and hence decreases the depth of the dimple and thus, reduces d_{OM} but, as explained above, glassy, low basicity slags also cause increases in both q_{hor} and d_{OM}.
(iii) Delaying solidification to a position further down the mould and which should also help to reduce inclusion levels; this can be achieved by

- increasing superheat or meniscus temperature.
- using Meniscus-free casting which uses a refractory to reduce the horizontal heat flux in the meniscus region of the mould [21–23].
- Using EMC which reduces the horizontal heat flux in the meniscus region.

11.5.5.3 Reduce the Length of the Shell Tip

As stated above, this too may be regarded as increasing the distance between slag rim and shell tip ($d_{\mathrm{rim/tip}}$). Reducing the length of the shell tip should also help to decrease slag, inclusion and bubble capture; this can be achieved by reducing the vertical heat transfer by

(i) Using a deeper powder bed.
(ii) Reducing the thermal gradient in the powder bed by using casting powders containing exothermic agents [13].
(iii) Using Electromagnetic braking [125] to slow down the metal flow and decrease the metal penetration into the mould; EMBr has been reported to cause a 30% decrease in vertical heat transfer [124].
(iv) Reducing the argon flow rate [108].
(v) Increasing slag viscosity will reduce convective heat transfer in the slag pool but will also reduce powder consumption.

11.5.5.4 Adjusting Mould Slag Properties

There is some uncertainty about the effects of changing mould slag properties, but it has been reported that d_{OM} can be decreased by increasing the interfacial tension (γ_{msl}); this can best be achieved by lowering the S content of the steel but reducing the Na_2O, K_2O and Li_2O contents of the casting powder should also be helpful. The result presented in Fig. 11.32a [43] indicated that d_{OM} decreased with increasing basicity; however, increased basicity would also be expected to promote the sequence ($(C/S)\uparrow \rightarrow \eta\downarrow \rightarrow Q_s\uparrow \rightarrow d_{\mathrm{OM}}\uparrow$) The results in Fig. 11.32a suggest that the delayed solidification resulting from the decrease in q_{hor} (and increase in $d_{\mathrm{rim/tip}}$) is greater than the opposing effect of increased powder consumption).

11.5.5.5 Use of Mathematical Models Covering OM Formation

Mathematical models have reached the stage of development where they can provide an insight into the processes and mechanisms occurring in the mould. Furthermore, it is shown above that there are sometimes opposing effects taking place which affect d_{OM}. Thus, mathematical models may prove the best way of predicting the overall effect. Mathematical models covering oscillation mark formation have been reported by several groups [53, 55, 106, 113, 119, 140, 142–144].

11.6 Transverse and Corner Cracking

Transverse and corner cracks occur on both the surface (Fig. 11.42) [139] and in the interior of steels (Fig. 11.43), [139, 145]. Transverse cracks result from

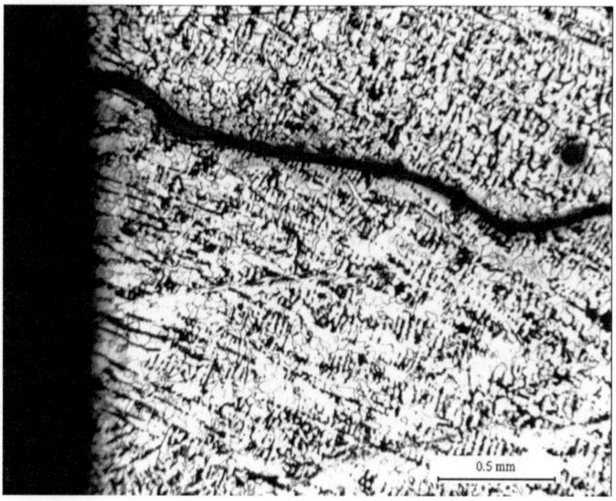

Fig. 11.42 Micrograph showing a transverse crack on the surface of a micro-alloyed, low carbon steel (permission granted, Europ. Comm. Sci. Tech. Publ. [139])

- Cracks formed in the base of deep oscillation marks as a result of stress concentrations arising from micro-segregation [97], (Fig. 11.42[139]).
- Surface strains occur during bending in a low ductility, temperature region for crack sensitive steel grades (especially, *peritectic*, micro-alloyed, low carbon steels); these regions occur, in the upper part of the strand for a straight mould machine and, alternatively, at the straightening unit at the end of the strand. The crack occurs at the site (surface or interior) which is experiencing the ductility trough (Fig. 11.42 or 11.43, respectively).
- Improper taper in the mould.
- Improper process parameters and the use of an unsuitable mould powder.

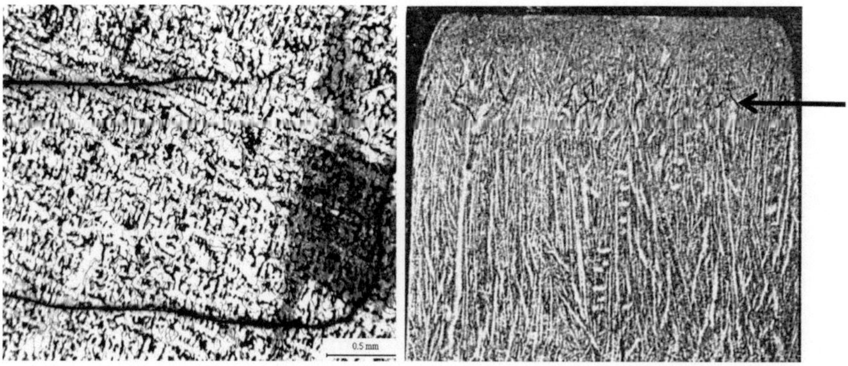

Fig. 11.43 Micrographs showing transverse cracks in the interior of a micro-alloyed peritectic, carbon steel (permission granted, Europ. Comm. Sci. Tech. Publ. [139])

11.6.1 Factors Affecting Transverse Cracking

One form of transverse cracking is associated with the bending and straightening operation in continuous casting where the surface of the steel experiences a tensile stress. There is loss of ductility in the steel between 900 and 700 °C and a "*ductility trough*" is encountered in this temperature range (Figs. 11.44 and 11.44). Transverse cracking occurs when the bending and straightening processes coincides with the ductility trough. The usual strategy adopted is to prevent the steel from falling into this temperature range during the straightening process. Further details on this problem can be obtained from the review carried out by Wolf [146].

Transverse cracking on the upper surface of slabs has proved a serious problem when casting micro-alloyed, low carbon, *peritectic* steels (e.g. pipe line steels X70) [154]. The problem has been attributed to "abnormal grain growth of the austenitic grains, in combination with precipitation of AlN and NbC, in the grain boundaries". During straightening and bending of the strand, these grains are subjected to a very slow deformation speed which causes creep and, subsequently, transverse cracking. This finding (Fig. 11.45) has been corroborated in a number hot ductility tests using a Gleeble machine [145, 155].

It can be seen from Fig. 11.44 that the ductility drops dramatically at low strain rates; this explains the problems encountered at the straightening unit where the deformation rate is low. It also shows that the variations in cooling on the strand surface (resulting from contact with the support rolls and water cooling) tend to

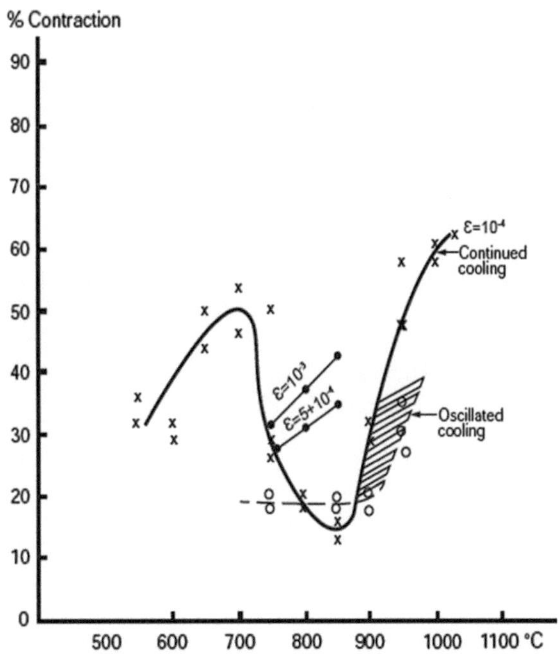

Fig. 11.44 Schematic drawing showing ductility of micro-alloyed, peritectic steel as function of temperature, strain rate and cooling curves (Reproduced with permission from Jernkontoret [145])

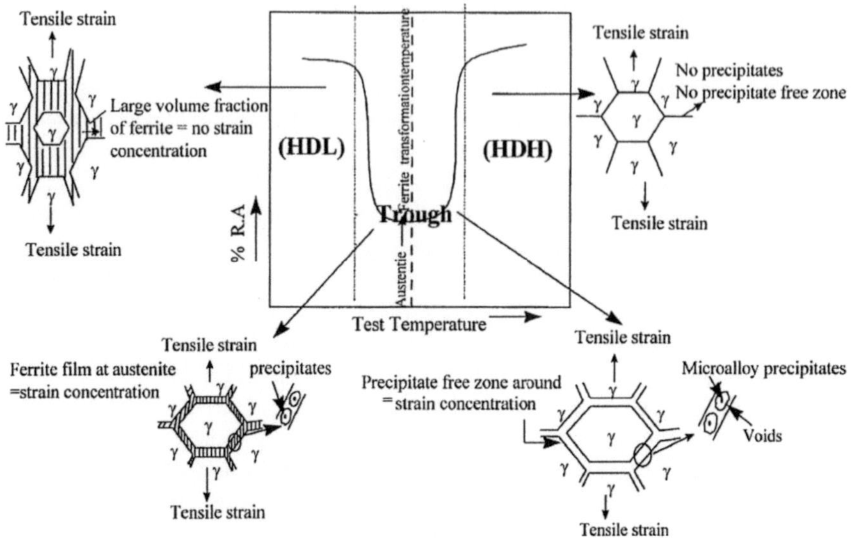

Fig. 11.45 Schematic drawing showing the micro mechanism of hot ductility in micro-alloyed steels (Reproduced with permission from ISS/AIST, after [147])

expand the low-ductility area. The origin of the low-ductility trough is explained in more detail in Fig. 11.45 [147].

For the explanation, the ductility behaviour has been divided into four regions:

(1) **HDH** (High Ductility, High temperature): in this region the *austenite* grains have a high ductility since no precipitation takes place.
(2) **Trough in *austenite***: Precipitation of micro alloying elements (e.g. Nb, V, Al, N) takes place (mainly in the grain boundaries) where the slow tensile strain results in creep and the formation of voids; this causes a marked reduction in ductility. The effect is accentuated by abnormal grain growth of the *austenite* grains.
(3) **Trough in *ferrite***: The pro-eutectoid formation starts in the grain boundaries as a very thin film with micro alloy precipitates. The *ferrite* is softer than the *austenite*, so the strains concentrate here and this also becomes a low-ductility region.
(4) **HDL** (High Ductility, Low temperature). At lower temperatures the f*errite* fraction increases leading to a strong increase in ductility.

The abnormal grain growth of *austenite* grains can be seen in Fig. 11.46a and the micro-alloyed precipitates can be seen in the FEG-SEM image (Fig. 11.46b).

Micro-segregation leads to the surface enrichment of residual (or tramp) elements (Cu, Sn, Sb and As) and also leads to brittleness. Face brittleness (or hot shortness) is associated with the presence of Cu-rich phases in the surface region and the application of stresses. This arises because the solubility limit of Cu in *austenite* is <0.2%. If the surface concentration of Cu exceeds this value, it will

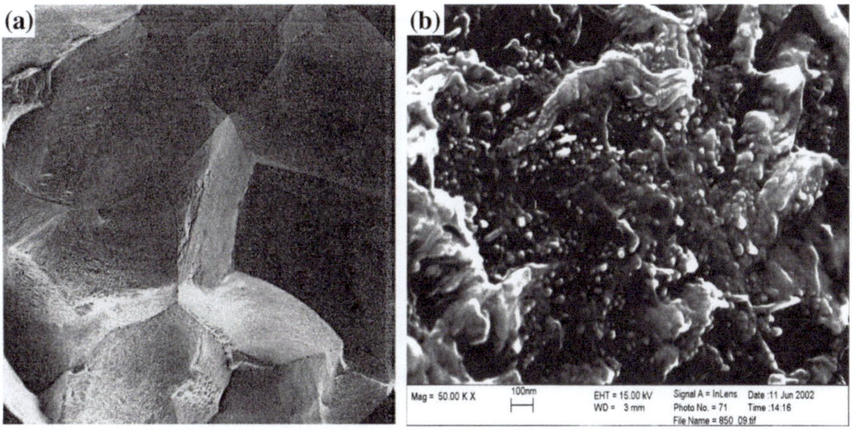

Fig. 11.46 Micrographs of a surface fracture of tensile test samples **a** Surface from fracture, $x = 26$, **b** SEM image of fracture surface, $x = 50.000$ showing precipitates (permission granted, Swerea KIMAB [148])

form a Cu-rich liquid, which is preferentially oxidised and the resulting liquid penetrates into the grain boundaries. Other elements, such as Sn and Sb, also promote the formation of the oxidised liquid phase. Nickel additions are usually made to alleviate this problem, since Ni increases the solubility limit of Cu in *austenite* and favours the occlusion of Cu from the scale [149]. The Copper equivalent (Cu_{eq}) is used as a measure of the propensity to form these liquid phases which penetrate the grain boundaries (Cu_{eq} = %Cu + n(%Sn + %Sb)- %Ni) where n has a value between 6 and 8). The transverse cracking was found to increase with increasing copper equivalent (Fig. 11.47) [149, 150].

The effects of mould taper and casting speed on transverse corner cracks (cf. transverse facial cracks) due to excessive mould taper (on narrow face) have been investigated by Ogibayashi [45]. The results, shown in Fig. 11.48, show the minimum powder consumption, Q_s^{min} (or liquid slag film thickness, d_l^{min}) required to avoid cracking at two casting speeds, V_c = 1.8 and 2.2 m min^{-1}. It can be seen that in their study:

Fig. 11.47 Schematic drawing showing the Transverse crack length as a function of the copper equivalent (Drawn from data due to [149])

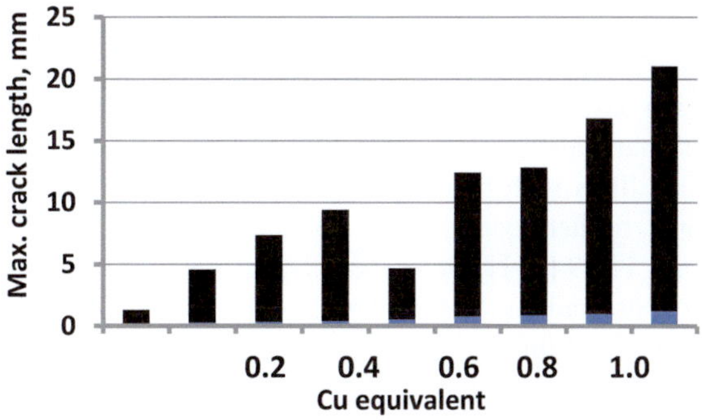

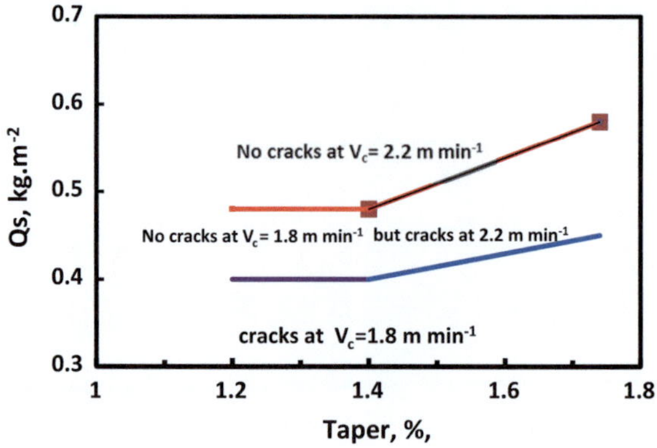

Fig. 11.48 The minimum powder consumption to avoid transverse cracking (Q_s^{min}) as a function of mould taper on narrow face; cracks are formed if powder consumption lies below curve (—— for V_c = 1.8 and (—■— = 2.2 m min^{-1}) but no cracks where consumption above curve (Reproduced with permission granted as private communication from the authors of Nippon Steel and Sumitomo Metal Corp. [45])

- Q_s^{min} is reasonably constant (at both speeds) for tapers of <1.4% but increases steadily for tapers >1.4%.
- Q_s^{min} increases with increasing casting speed; thus, higher casting speeds will require a higher powder consumption to avoid cracking (Note that Q_s^{min} increases with increasing V_c and the situation is further complicated by the fact that Q_s tends to decrease with increasing V_c (e.g. $Q_s^{slag} = 0.55/\eta^{0.5}V_c$).

A statistical evaluation of 36 heats involving transverse and corner cracks in micro-alloyed, low carbon steel showed that cracking decreased with decreasing negative strip time (due to its effect on d_{OM}) [43]. It was also found that it was important to use a mould with corner shrinkage compensation and a mould powder with high basicity (to reduce q_{hor}). The transverse crack length ($TcrL$ = crack length (m/m)) was found to show linear relations (Eq. 11.13) with (i) t_n (ii) the parameter, c^* = (narrow side concavity + taper) and (iii) basicity (C/S) [43]. It was also reported that transverse cracking decreased with increasing slag viscosity [43].

$$TcrL = 26.8 + 12.4 \cdot t_n - 1.6 \cdot c * -5.2 \cdot (C/S) \qquad (11.13)$$

The reduction in q_{hor} was obtained using "*soft cooling*" mould slags (i.e. with high basicity, T_{br} and f_{crys}) [43]. However, high-T_{br} values result in a thin liquid slag film and a slag film with high f_{crys} but make the slag film brittle and fragile. Thus, slag films with high basicity create high frictional forces (which act on the shell) and, furthermore, tend to fracture easily. Such behaviour can be clearly seen in Fig. 11.49 [151] where the sharp increase in mould temperature corresponds to the fracture and loss of the slag film which is gradually replaced (over ca. 8 min) with a new slag film.

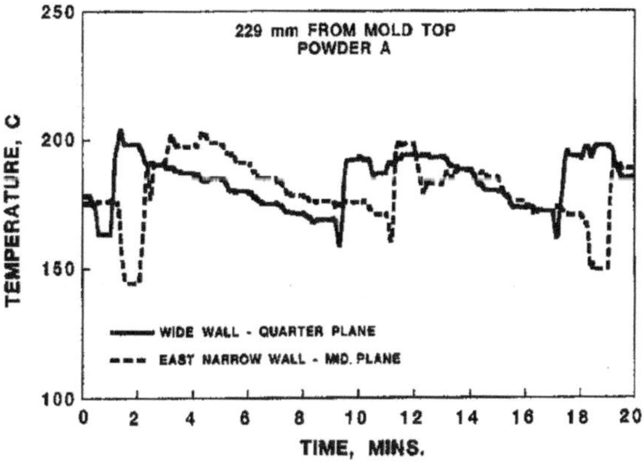

Fig. 11.49 Mould temperature (229 mm from mould top) as a function of time when casting with a mould powder with C/S = 1.62; *solid line* = wide face, 1/4 plane; *dashed line* = narrow face midway; the "step decreases then increases" in mould temperature correspond to the influx of molten slag and fracture of slag film and the formation of a new slag film (permission granted, ISS/AIST, [151])

11.6.2 Ways of Dealing with Transverse and Corner Cracking

11.6.2.1 Bending at the Low-Ductility Trough for Crack Sensitive Steel Grades

This problem is particularly severe when casting of micro-alloyed, low carbon steels and requires significant oxygen scarfing and grinding to remove the surface cracks. The problem has been minimised by the development of new casting machines which seek to avoid the low-ductility trough. This can be done in two ways:

(i) *Straightening at temperatures above the low-ductility trough.* This is achieved using proper secondary cooling and a high casting speed, since it is necessary to maintain a liquid core to attain a reasonably-high temperature [145]. However, the resultant thin shell is vulnerable to bulging between the rolls and to internal cracking; this has led machine builders to provide better roll support (using smaller rolls).

(ii) *Straightening at temperatures below the low-ductility trough.* This procedure is used for thick-slab casting. For this case, there is no need for a close distance between the rolls because the strand is fully solidified and has a strong surface.

11.6.2.2 Improper Taper in the Mould

The taper should match the shrinkage of the shell as closely as possible. An excessive taper results, sequentially, in overcooled corners, high-mould friction and corner cracks [64]. An insufficient-taper results in loss of contact between mould and shell with a consequent reheating of the shell (the "rebound effect").

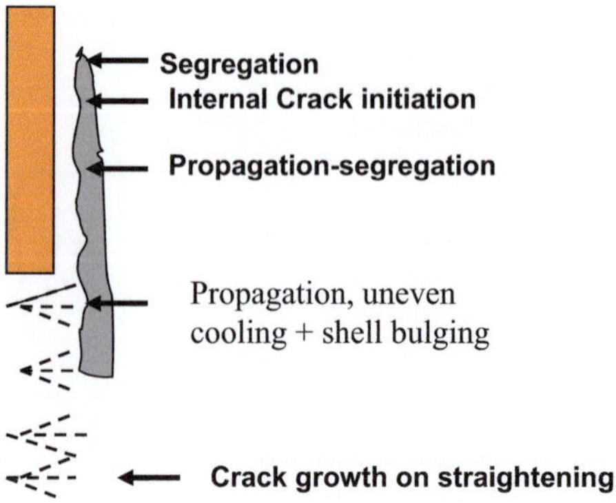

Fig. 11.50 Schematic diagram showing the sequence of events leading to transverse cracking in secondary cooling zone

This causes abnormal grain growth and an extra tension in the surface when the shell and enters the secondary cooling zone (Fig. 11.50). A parabolic taper is required for the meniscus region but the optimal taper is dependent upon casting speed, mould length and steel grade [64]; a mathematical model to predict the optimal taper has been reported [64]. However, in practice, casters tend to handle a large number of steel grades and use a variety of casting speeds.

The loss of contact in the corners is a problem which can be solved by using a mould design which compensates for the corner shrinkage [43].

11.6.2.3 Optimisation of Process Parameters and Use of a Suitable Mould Powder

A number of actions must be taken to minimise the risk for transverse and corner cracking:

1. Reduce the depth of the oscillation marks by
   - Decreasing stroke length, negative strip time (t_n) or increasing frequency.
   - Increasing the rim to shell tip distance by decreasing both stroke length and the vertical heat transfer (with a thick bed of mould powder) and the horizontal heat flux by using a "*soft cooling*" mould powder with a high-break temperature.
2. Increase the strand temperature to avoid the low-ductility range of the steel (ca. 820 °C) in the straightening region (i.e. reduce heat lost by the strand) by
   - Use a casting speed that is high enough to ensure that the surface temperature is above the critical temperature range.
   - Use a mould flux with high values of T_{br} and f_{crys}.
   - Reduce the velocity of the water-sprays, or by shutting off the sprays to the corner of the strand [152].
3. Minimise the concentration of Cu, Sn and Sb in the steel or add Nickel to reduce the copper equivalent.

11.6.2.4 Use of Electromagnetic Braking

It has been reported that EMBr reduces the number of transverse, corner cracks; this was ascribed to the reduction of the surface waves [153]. It was concluded that the strength of the applied magnetic field has to be adjusted according to the steel grade being used. [153]. However, the beneficial effect of EMBr may arise from its reduction of d_{OM} due to delayed solidification resulting from the 10 °C increase in meniscus temperature.

11.7 Star Cracking

Star (or Spongy) cracking tends to occur in the lower half of the mould. A photograph of a star crack is shown in Fig. 11.51 [156].

11.7.1 Factors Affecting Star Cracking

Star cracking was, in the past, usually associated with the presence of copper droplets on the strand [157] due to contact of the shell with the mould. This resulted

Fig. 11.51 Photograph showing a star crack (permission granted, Taylor & Francis. [156])

in the copper melting to form a droplet which, subsequently, penetrated into the grain boundaries where it oxidised. The usual procedure adopted to avoid star and spongy cracking was to coat the mould with high-melting metals such as Ni, Cr, Mo, etc. However, Billany et al. [158] observed that star and spongy cracking could occur in Ni-coated moulds. It was also observed that cracking levels (Fig. 11.52):

- did not correlate with overall heat flux values.
- but did correlate with periods where there were large variations in heat flux (Fig. 11.52).

It was, subsequently, proposed that star cracking resulted from heat flux variations caused by a lack of lubrication (i.e. the liquid slag film is too shallow) which results in stresses which cause the fracture of the solid slag film which requires time to repair the (see Fig. 11.49 [151]). O'Malley et al. [159] also observed cracking of the slag film in the bottom half of the mould; this was accompanied by periods of heat flux variability until the slag film was fully restored. Crystalline slag films (with high basicity) are prone to fracture.

Meng and Thomas [73] developed a model to account for the fracture of the slag film in the lower half of the mould, from which they concluded:

- If powder consumption decreases to a critical value (Q_{crit}, i.e. d_l decreases) then an axial stress is produced which can exceed the fracture strength of the slag film; this, subsequently, results in film fracture and movement of the slag film.
- The critical powder consumption value (Q_{crit}) was determined from the viscosity –T curve.
- A thick, crystalline slag film will result in a hotter shell and a thicker liquid film, i.e. it will provide more lubrication.

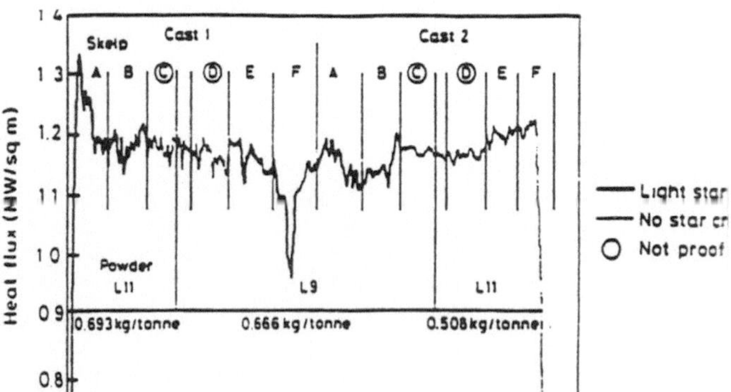

Fig. 11.52 Heat flux variations during casting using two casting powders, L11 (cast 1 periods *A–C* and cast 2 periods *D–F*) and L9 (cast 1 *D–F* and cast 2*A–C*); star cracking occurred with powder L9 and was accompanied by heat flux variability (permission granted, Taylor & Francis [158])

- Glassy slag films tend to fracture near the mould exit but any advantage is offset by the fact that (Q_{crit}) for a glass is higher than that for crystalline slag film.
- A crystalline, slag film will fracture higher up the mould because of the high-break temperature.

11.7.2 Ways of Dealing with Star Cracking

- Use a mould powder which provides a liquid slag film (of sufficient thickness to lubricate the shell) throughout the mould. However, this is not a straightforward task. We can select a slag with a lower T_{br} which would be expected to increase the liquid slag film thickness (d_l). However, the resulting, thinner, solid slag film will lead to a high-heat flux and more rapid solidification of the slag. The latter would tend to cancel any gains obtained with a thicker liquid film.
- Reducing the overall heat loss in the mould should help to reduce star cracking; this can be achieved by using a low-conductivity, coating on the mould or by applying EMBr to reduce heat losses.
- Possibly the most effective treatment is to increase the casting speed since this will reduce the amount of heat lost in the mould, although it would lead to a high-heat flux density, the shorter residence time will more than offset this effect.
- Reduced water flow rates in the mould would also be beneficial.
- Increased superheat could also be beneficial to combat star cracking.

11.8 Depressions

Under certain casting conditions, depressions are formed on the surface of the strand. Typical examples are shown in Fig. 11.53a, b.

11.8.1 Longitudinal Depressions

11.8.1.1 Factors Affecting the Formation of Longitudinal Depressions

There have been two in-depth studies on the formation of longitudinal depressions [129, 131]. These depressions had the following dimensions (50–500 mm long) (1–2 mm wide) × (0.3–1 mm deep) [129] and 10 mm long × 2 mm deep, respectively [131]. The depressions were found to occur with and without entrapped scums [129]. The following observations were made:

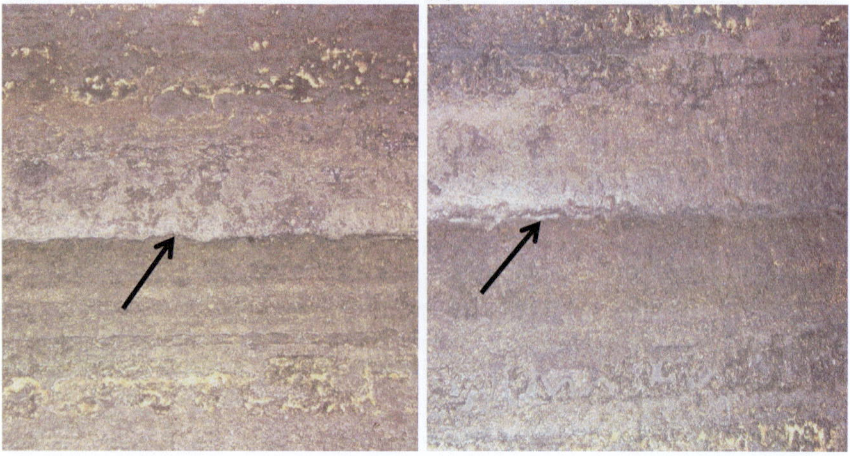

Fig. 11.53 Photographs of Longitudinal depressions (indicated by *arrows*) with cracks on a stainless bloom (steel 411), **a** depression with a crack, **b** depression with a crack penetrated with liquid steel (courtesy of C-Å Däcker, Swerea KIMAB)

- The depressions corresponded to periods of mould level instability [129, 131] (Fig. 11.54a)
- The depressions tended to occur at regular intervals (0.5–1 m apart) in regions where no oscillation marks could be seen (Fig. 11.54b) [131].
- Depressions occurred when the mould level was at its peak (Fig. 11.54b) [131]; slag infiltration in the meniscus region will be blocked, if fluctuations in mould level exceed (slag pool depth + mould level amplitude) [129].
- Depressions corresponded with low values of the horizontal heat flux [131].
- Cracks were frequently associated with the depressions.
- Metallographic studies of the steel showed that depressions corresponded to lower cooling rates than the norm [131].
- It was concluded that depressions were caused by irregular, and insufficient, slag infiltration [129].
- It was suggested that the mould level fluctuations were caused by the erosion of the refractory used as the stopper rod in the tundish [129].
- Depressions occurred when the SEN immersion (d_{SEN}) depth was too deep; these workers highlighted the importance of optimising the SEN immersion depth (d_{SEN}^{opt}) since, if $d_{SEN} > d_{SEN}^{opt}$ this leads to depressions and if $d_{SEN} < d_{SEN}^{opt}$ this will give turbulent metal flow and slag entrapment. Kim et al. [129] proposed that the mould level turbulence index (I_t) was helpful in determining d_{SEN}^{opt} (Eq. 11.14 where w and t are the mould dimensions and R_c is the casting speed in tonne min^{-1}. As I_t increases d_{SEN} should be increased.

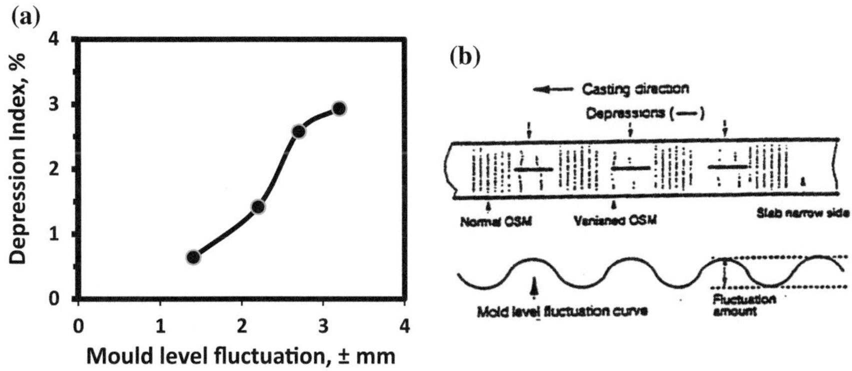

Fig. 11.54 **a** Depression index as a function of mould level fluctuations when casting 304 stainless steel at V_c = 0.8–1.0 m min^{-1} (re-drawn after [129]) and **b** the appearance of the slab; depressions (—) occur when the oscillation marks (|) disappear and also show that these periods occur when mould level is at its highest point [131] (permission granted, ISS/AIST [129, 131])

$$I_t = 0.33(w/t)R_c V_c \qquad (11.14)$$

- Kim et al. [129] concluded that as depressions appeared to be associated with insufficient slag infiltration. They replaced mould powder A, with powders B and C (B provided higher powder consumption through an increase in the (CaO/SiO_2) ratio and the Na_2O content); Powder B was adopted since Powder C led to deeper oscillation marks (Fig. 11.55) possibly because of the high-powder consumption.

It should be pointed out that if insufficient slag infiltration is responsible for depressions, this could arise because the carbon levels in the powder are too high. This would cause

- A reduction in the melting rate and powder consumption.
- Incorporation of excess carbon into the slag rim.

11.8.1.2 Causes, Mechanisms

There seem to be general agreement that longitudinal depressions are associated with

- Large fluctuations in mould level.
- Insufficient slag infiltration at some periods.

Jenkins [131] proposed the following mechanism:

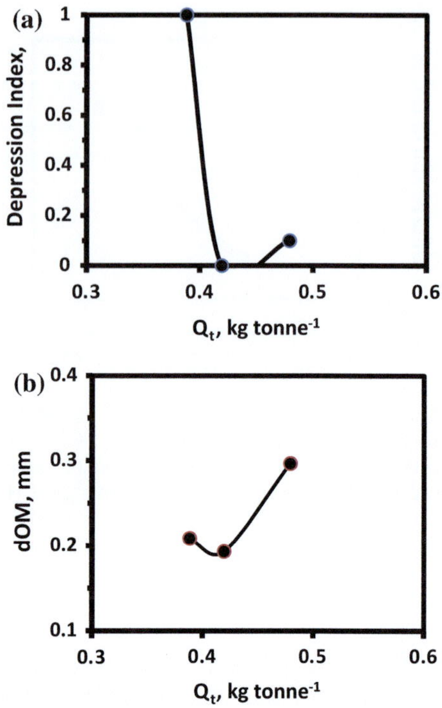

Fig. 11.55 Effect of mould powder consumption (Q_t) on **a** depression index and **b** depth of oscillation mark, using three mould powders (permission granted, ISS/AIST re-drawn from [129]

(1) A rising mould level causes deeper oscillation marks (Fig. 11.56a) (in line with Aigner's findings [105] see Fig. 11.33)
(2) Further rises in mould level result in narrowing of the rim/shell gap (Fig. 11.56b) and the formation of "*glaciation marks*".
(3) Even further rises in mould level cause freezing of the shell, above the slag rim, resulting in blocking off of slag infiltration (Fig. 11.56c).
(4) Then the shell captures the slag rim and drags it down the mould as the mould level drops (Fig. 11.56d); there is little, or no, slag infiltration during this period and glaciated regions (i.e. with no OMs) are formed.
(5) As the mould level continues to drop, a new slag film is formed and normal oscillation marks reappear on the strand (Fig. 11.56e).

Jenkins [131] was uncertain as to whether the captured rim was held in place during its descent in the mould or whether it escaped leaving an imprint on the shell filled with gas and slag. In both cases, the depression would result in lower horizontal heat transfer to the mould and this is consistent with the observations.

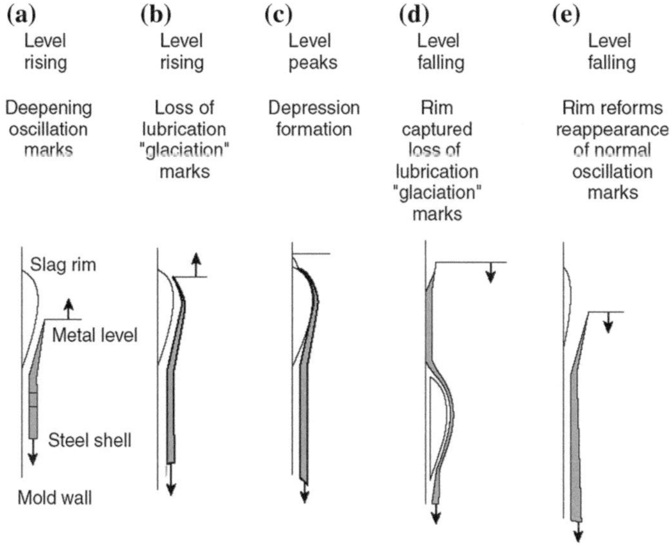

Fig. 11.56 Proposed mechanism for the formation of longitudinal depressions (permission granted, ISS/AIST [131])

11.8.1.3 Ways of Dealing with Longitudinal Depressions

1. *Mould level variations*—the action taken will depend upon what is causing the mould level fluctuations;
   - If it is due to strand bulging, the use of a mould powder with a lower T_{br} which produces a lower crystalline fraction in the slag film would help (but maybe at the expense of more longitudinal cracking)
   - Kim et al. [129] attributed mould level variations to erosion in the tundish stopper rod and decreased the incidence of depressions by improving the corrosion resistance of the stopper rod.
   - Jenkins [131] obtained better mould level control by "*detuning*" the controller.
2. Improving slag infiltration (i.e. powder consumption) by reducing the slag viscosity but this may cause deeper oscillation marks.
3. In some cases, the carbon content of the powder may be too high resulting in a low-melting rate and a low powder consumption value.
4. "*Soft cooling*" with a high basicity mould powder has also proven to be effective.

11.8.2 Transverse Depressions

Jenkins [160] observed transverse depressions (up to 10 mm deep) were formed in the first few metres of the cast when casting billets of high-C steel (Fig. 11.57).

11.8.2.1 Published Information

Jenkins made the following observations concerning the casts exhibiting transverse depressions:

- Depressions tended to be prevalent in the first few metres of the cast.
- In these casts, the mould temperatures tended to be low for up to 60 min of casting.
- Depressions correlated with mould level fluctuations and tended to occur when the mould level was rising [160].
- Depressions did not correlate with the movement of the strand.
- Depressions were assumed to be associated with the movement of slag rim/film.

11.8.2.2 Causes, Mechanisms

Since it was associated with mould level variations, it was assumed that the transverse depression mechanism was similar to that for longitudinal depressions and involved the capture of the slag rim [160] (see Fig. 11.58). During large mould level changes, the rising shell envelopes the slag rim and cuts off slag infiltration; it then captures the slag rim and drags it down the mould. The captured slag rim is a good thermal insulator and local mould temperatures drop suddenly as it passes down the mould.

Fig. 11.57 Photograph showing transverse depressions formed in HC steel billets (permission granted, ISS/AIST, [160])

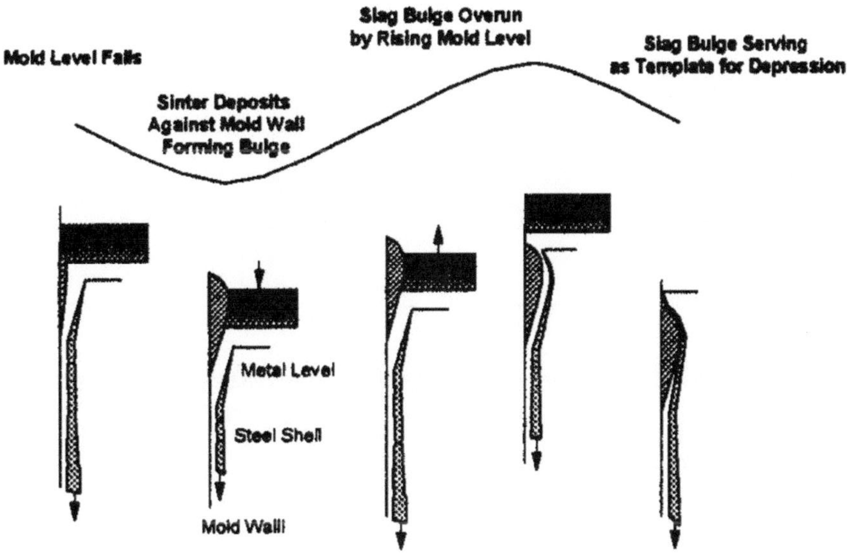

Fig. 11.58 Proposed mechanism for the formation of transverse depressions, where mould level changes allow the rising shell to capture the slag rim (permission granted, ISS/AIST, [160])

11.8.2.3 Ways of Dealing with Transverse Depressions

Jenkins [160] found that the incidence of transverse depressions could be significantly reduced by allowing an additional 30 s before adding the operating powder; this allowed the starter powder to form a glassy slag film and restricted the size of the rim (or *rope*) formed.

It must be emphasised that this type of problem tends to be rare today because of the development of more sophisticated, mould level control systems.

11.8.3 Off-Corner Depressions

Off-corner depressions or "*gutters*" are depicted in Fig. 11.59a. They have been observed in the casting of stainless steel and plain carbon (0.12%C) steels [161]. The depressions run in the longitudinal direction and may contain sub-surface cracks.

11.8.3.1 Published Information

Thomas et al. [161] carried out an investigation of off-corner depressions which included finite-element modelling studies.

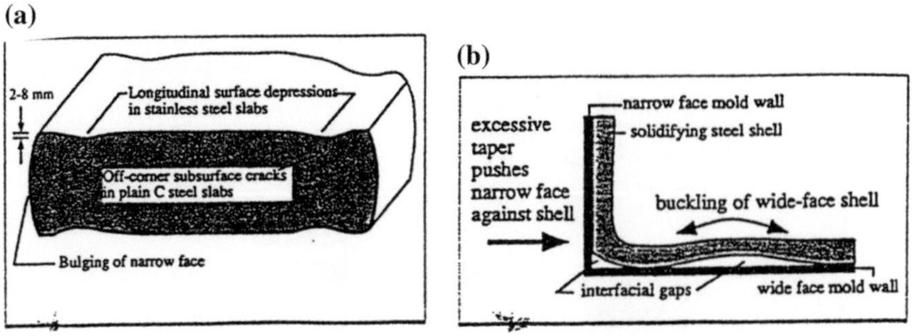

Fig. 11.59 Schematic drawings showing **a** the appearance of off-corner depressions [161] and **b** how the thin shell is buckled by excessive mould taper (permission granted, ISS/AIST, [161])

11.8.3.2 Causes, Mechanisms

Thomas et al. [161] concluded that the depression was caused by:

- A hot, thin shell in the corner region.
- Cyclic bending of the shell due to bulging between the rolls, resulting in the bending of the thin, off-corner region of the shell.
- Improper mould geometry, no corner-shrinkage, compensation of the narrow face and no compensation for parabolic solidification.

Thomas et al. [161] suggested that the hot, thin shell could be formed by inadequate liquid slag feeding. However, it is also possible that the thin hot shell could be caused by the nature of the metal flow, as shown in Fig. 11.18 in Sect. 11.3.2.2. An excessive taper will then bring the narrow face close to the shell and which will then cause buckling of the shell in the corner region of the wide face (Fig. 11.59b).

Li and Thomas [64] pointed out that the shell in the corners of billets can be bent out by ferrostatic pressure which induces stresses perpendicular to the direction of dendritic growth; this leads to sub-surface cracks located 10–15 mm from the corner. These stresses increase with increasing casting speed and can lead to hot tears in extreme cases. Producing cold corners can help to minimise this problem.

11.8.3.3 Ways of Dealing with Off-Corner Depressions

Thomas et al. [161] suggested that this problem could be treated in the following ways:

- Minimise the formation of a hot, thin shell in the corner region by
  - Improving the metal flow pattern.
  - Reducing the superheat.

- Ensuring uniform spray intensities on shell perimeter.
- Increasing powder consumption.

- Ensuring taper on the narrow face is correct for the steel being cast.
- Reducing bulging in both wide faces by maintaining roll alignment.
- Compensating for corner shrinkage by better design of the narrow face mould [43].

11.9 Overflows

11.9.1 Factors Affecting Overflows

When overflows (or C-type effects) occur, the overflowing steel forms a protruding, double-skin (typically, with a thickness of 0.3 mm); the overflow has a higher carbon content than that of the strand [162]. These C-type effects have a "*tounge-like*" appearance. It can be seen from Fig. 11.60 that the overflow corresponds to a period where the oscillation marks disappear.

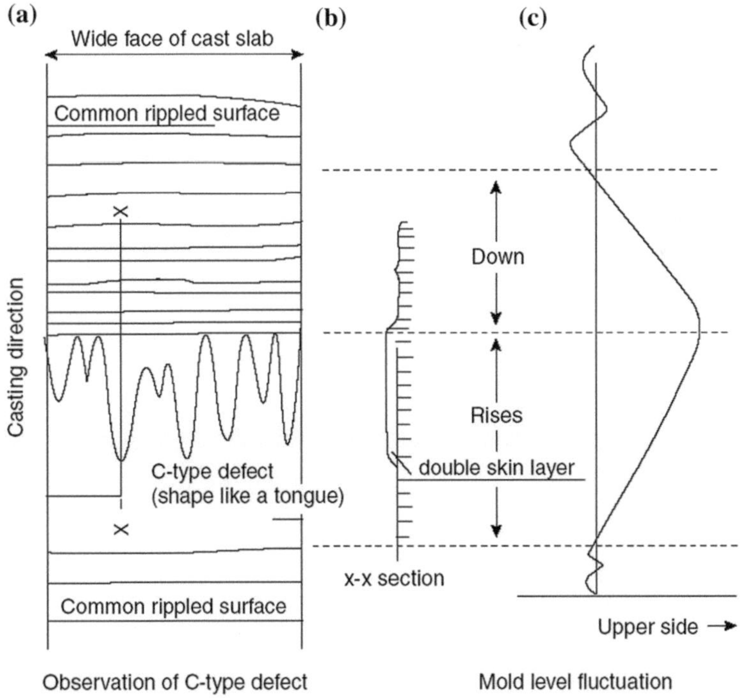

Fig. 11.60 Schematic diagrams of C-type effects showing **a** general appearance **b** steel surface profile of overflow for X–X section in (**a**) and **c** mould level variations corresponding to Fig. 11.60 (**a**) (permission granted, ISS/AIST, [162])

11.9.2 Causes, Mechanisms

These overflows result from sudden changes in mould level. If the slag pool has insufficient depth, the sudden rise in mould level results in the following sequence of events [162]:

- It brings molten steel into contact with the carbon-rich, sinter layer; this results in carbon pick-up by the steel and the formation of a low-melting steel.
- The molten steel cuts off the flow of molten slag into the mould/steel channel in a similar way to that shown in 11.58d, e; this results in the disappearance of oscillation marks.
- When the mould level drops, the higher C, low-melting steel flows over the previously formed shell producing a second skin of steel.

11.9.3 Ways of Dealing with C-Type Effects

11.9.3.1 Increase the Depth of the Slag Pool

The slag pool depth can be increased by

- Reducing the vertical heat flux by increasing the bed thickness, or incorporating exothermic agents into the casting powder.
- Increasing the melting rate; this can be achieved reducing the carbon content of the mould powder.

11.9.3.2 Reduce Carbon Content of the Mould Powder

Reducing the carbon content of the mould powder results in

- A reduction of the carbon pick-up by the steel.
- An increase in melting rate of the mould powder which, in turn, will assist the formation of a deeper slag pool.

11.10 Slag, Gas Entrapment and Sliver Formation

Entrapment (of slag, or gas) is also referred to as, *entrainment, engulfment* and *emulsification*. The entrapment of slag, oxides and gas bubbles remains a serious problem, especially as casting speeds are raised to improve productivity. Mould slag, non-metallic inclusions (e.g. Al_2O_3) and gas bubbles get trapped by the newly

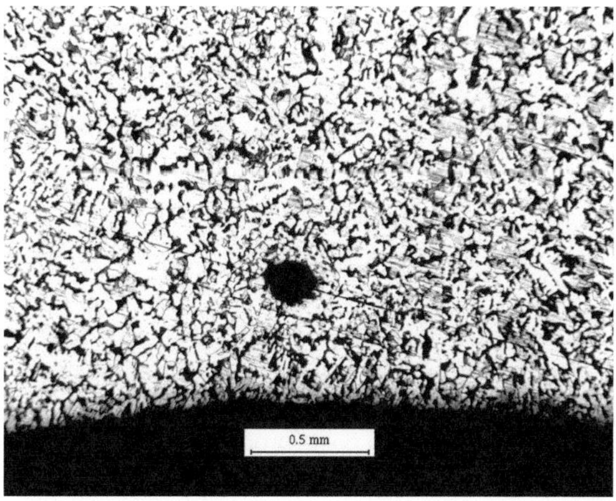

Fig. 11.61 Micrograph showing slab surface with porosity due to argon injection (permission granted, Europ. Comm. Sci. Tech. Publ. [139])

formed shell in the meniscus region. An example of gas entrapment is given in Fig. 11.61. These trapped inclusions and gas pores reduce the mechanical strength of the cast steel. In the case of inclusions, their thermal expansion coefficients are much lower than those of the steel matrix. Thus, when the steel is heated, stresses build up around the inclusion, which result in local cracks, with consequent loss in mechanical strength of the steel. Slag entrapment has become much more important in recent years with the use of increased casting speeds to meet productivity demands. The velocity and turbulence of the metal flow in the mould is a key factor in all three forms of entrapment. Entrapment can occur by a variety of mechanisms, [163, 164]; these various mechanisms are discussed first and then the slag, gas and inclusion entrapment are discussed, individually, in Sects. 11.10.1, 11.10.2 and 11.11.3, respectively.

Since entrapment is so closely tied to metal flow in the mould, it is not surprising that much of our knowledge of the mechanisms of entrapment have been derived from water-modelling and mathematical-modelling studies. These room temperature, physical models use the Froude and Reynolds numbers to represent the metal flow pattern and meniscus level instabilities and the Weber number to represent the effects of interfacial tension [163, 164]. However, the similarity criteria are not well satisfied (especially in the case of the Weber No.) in these water models [163, 164]; thus, some of the information obtained tends not to be quantitative.

11.10.1 Metal Flow Conditions Leading to Entrapment

The metal flow leaving the SEN collides against the mould wall; this creates a reverse flow across the metal surface. The reverse flow is accompanied by surface waves, vortex formation and turbulence. The high, interface velocities shear the slag into filaments which, subsequently, detach and are released into the metal stream. Slag entrapment increases with increasing meniscus fluctuations [163].

Entrapment can occur by a variety of mechanisms. Hibbeler and Thomas [163, 164] have published an excellent review of both the various mechanisms in play and the conditions promoting individual mechanisms [164, 165]. Nine different mechanisms are identified [163, 164] and summarised below but for more details the reader should consult the original articles.

11.10.1.1 Mould Level Fluctuations

This mechanism is shown in Fig. 11.62 [163, 164]. It occurs when there is a sudden drop in mould level and the mould movement, causes shearing of the slag, (in contact with the shell tip) resulting in "*necking and detachment*" of the slag filament formed [163]. It has been proposed that mould level variations associated with mould oscillation are unlikely to trigger entrapment but larger level fluctuations, due to transient changes in metal flow pattern, are more likely to cause of entrapment by this mechanism [163].

Mould level fluctuations (and hence, entrainment) have been reported to increase [163] with

(i) Increasing casting speed (V_c ↑) [165, 166]; however, it is also affected by increases in the Ar flow rate [165, 167, 168].
(iii) Increasing bore of SEN (D_{SEN} ↑) [167].
(iv) Increasing SEN port angle [165, 167].
(v) Decreasing SEN immersion depth (h_{immers}↓) [165].
(vi) Decreasing mould width (w↓) [163, 164].
(vii) Increasing Ar flow rate (V_{Ar} ↑) [165, 167, 168] and high, SEN/Ar, contact angle (θ) with increasing θ (i.e. decreasing wettability) leading to increased fluctuations [168] but increased casting speed reduces the effect of the Ar flow rate [165].

Meniscus level fluctuations can be suppressed by optimising the above factors for the selected casting speed and mould width. It should also be noted that meniscus

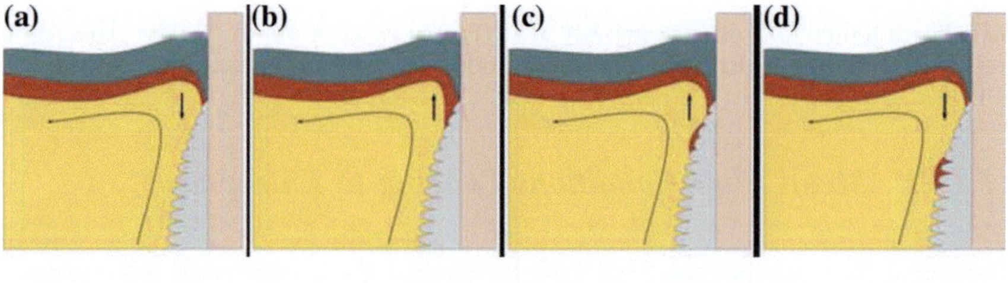

Fig. 11.62 Schematic drawings showing slag entrapment caused by mould level fluctuations; **a** contact of metal with slag pool during metal level descent, **b** mould level rise and *necking* of a slag filament during ascent, **c** detachment of slag droplet and entrainment during ascent and **d** entrapment in the shell dendrites during the subsequent descent (permission granted, ISS/AIST, [164])

fluctuation behaviour is different in thin slab casters to that experienced in thick-slab casting, [163, 164]. Meniscus level variations due to bulging can be reduced by promoting heat transfer (q_{hor}) in the mould to create a thicker shell. The magnitude of surface waves can also be reduced using EMBr [13, 153] (Sect. 5.8.4).

The meniscus fluctuations exhibit, at least, two frequencies, i.e. ca. 0.2 and 2 Hz [169–172]. The 2 Hz variations are associated with the mould oscillation, and the 0.2 Hz frequency has been variously ascribed to bulging [169] and to fluctuations at the nozzle ports [171]. Direct observation of the meniscus showed that the 2 Hz fluctuations have an amplitude of ca. 70% of that for oscillation but tended to increase with increasing casting speed [170].

The number of defects tends to increase with increasing metal flow velocity (V_{metal}), consequently, equations have been derived to characterise the strength of the metal flow up the narrow face. The "*F' value*" is a measure of this flow. It is calculated from Eq. 11.15 [165] where Q_{metal} = steel throughput, V_{coll} = velocity of metal at point of collision on the narrow face and θ_{coll} and h_{coll} are defined in Fig. 11.63 [163, 164].

$$F' = \rho Q_{metal} \cdot V_{coll}(1 - \sin\theta_{coll})/4/h_{coll}. \tag{11.15}$$

The F' factor has also been linked (Eq. 11.16) to both the amplitude of the fluctuations (A_{fluct}) and the velocity of the metal flow along the steel surface, V_{surf}, (which is associated with several types of entrapment)

$$A_{fluct}(\mathrm{mm}) = 3F' = 35\,V_{surf}\left(\mathrm{ms}^{-1}\right). \tag{11.16}$$

Thus, using the critical F' limits to avoid defects (3–5 N m^{-1}) given above, it can be shown that the *critical* surface velocity $\left(V_{surf}^{crit}\right)$ corresponds to 0.2–0.4 m s^{-1} [166]. However, critical surface velocities for other slag entrainment mechanisms tend to vary depending upon the mechanism involved.

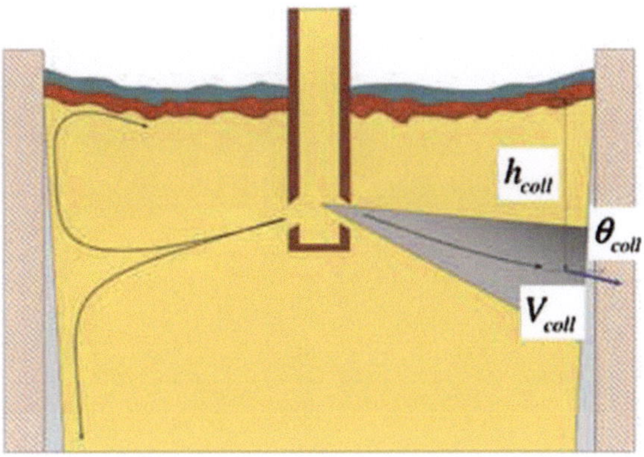

Fig. 11.63 Schematic drawing showing definitions of terms used in calculating the F' factor (permission granted, ISS/AIST, [163])

11.10.1.2 Meniscus Freezing and Hook Formation

Hooks are formed when there is insufficient heat supplied to the meniscus region, especially, at lower casting speeds and in regions near the narrow faces where temperatures are at their lowest. Such conditions also pertain when there is no Ar flow or when casting low-melting steels (e.g. *ferritic* stainless steels). The hooks trap slag, inclusions (e.g. Al_2O_3) and gas bubbles (Fig. 11.64a–c). Furthermore, any overflow of the frozen meniscus by the molten metal can cause slag to be trapped against the hook.

The formation of hooks occurs when the SEN immersion depth (h_{immers}) is too shallow since this creates a downward, *single roll*, flow which promotes low vertical heat transfer and a danger of meniscus freezing. It has also been reported that when the Ar flow velocity exceeds a critical value (5 L min^{-1}) a *double roll* flow system is transformed to a *single roll flow* [173, 174].

Hook formation can be minimised by

(i) Reducing the number of hooks by increasing the heat supplied to the meniscus region; this can be done by increasing (a) superheat, ($\Delta T \uparrow$) (b) redesign of SEN port angles (c) flow velocity when casting at low casting speed (V_{surf} = > 0.2 m min^{-1}) [166] and by decreasing the depth of immersion [163, 164].

(ii) Decreasing the length of the hooks can be achieved by decreasing the vertical heat flux in the bed by (a) increasing the depth of the powder bed (b) using powders containing exothermic agents [13] and (c) increasing q_{vert} (via gas convection) by increasing Ar gas flow (but must be <5 L min^{-1})

(iii) Using electromagnetic devices since (a) EMS homogenises the meniscus temperature and minimises freezing near the cooler, narrow face and (b) EMBr provides a 10 °C increase in meniscus temperature which reduces hook formation (however, the reduction of the metal flow velocity reduces heat losses (q_{vert}) which leads to a shorter meniscus shell (hook)).

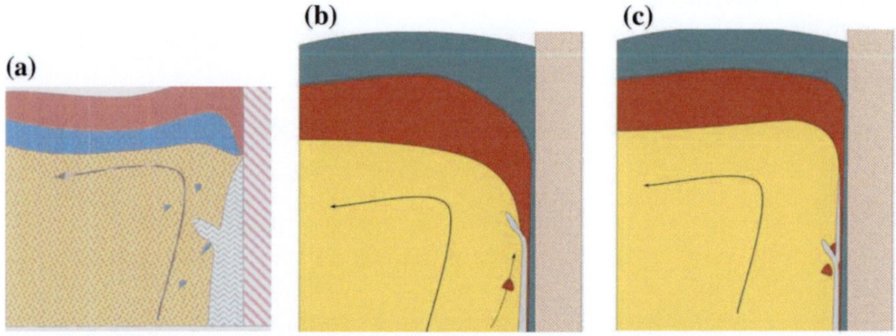

Fig. 11.64 Schematic representation of **a** entrapment by hooks [163], **b** a particle rising towards a hook and **c** its subsequent capture by the hook (permission granted, ISS/AIST **a** [163], **b**, **c** [164])

Slag entrapment by hook formation can be reduced by

(a) Reducing both the number and size of the hooks (see (i) and (ii) above).
(b) Using EMS and EMBr (as in (iii) above).
(c) Using mould slags with high viscosity (but this will reduce the powder consumption) and high interfacial tension (best achieved by reducing the S content of the steel).

11.10.1.3 Argon Bubble Interactions

Argon bubbling is used to reduce nozzle clogging and is also beneficial in reducing the hook formation by increasing the vertical flux via gas convection. However, it can lead to both slag and gas entrapment since slag tends to coat the gas bubbles. Entrapment can occur by the two mechanisms shown in Figs. 11.65 and 11.66. The mechanism shown in Fig. 11.65 involves the formation of a foam in the slag pool. The interfacial forces favour the covering of the SEN with slag and this results in slag crawling down the SEN carrying the foam with it (Fig. 11.65a). When it reaches the point where the foam and the metal stream come in contact (Fig. 11.65b) gas and slag are released into the metal stream. This mechanism tends to be prevalent when the bubbles are small with low inertia; small bubbles are formed when the steel throughput is high.

Slag and gas entrapment by this mechanism increases with

(a) decreasing bubble diameters which have low inertia (promoted by increased steel throughput),
(b) increasing slab width,
(c) increasing argon injection rates,
(d) increasing slag viscosity, but this may be a consequence of high SiO_2 producing high η and low γ_{msl} which aids crawling,

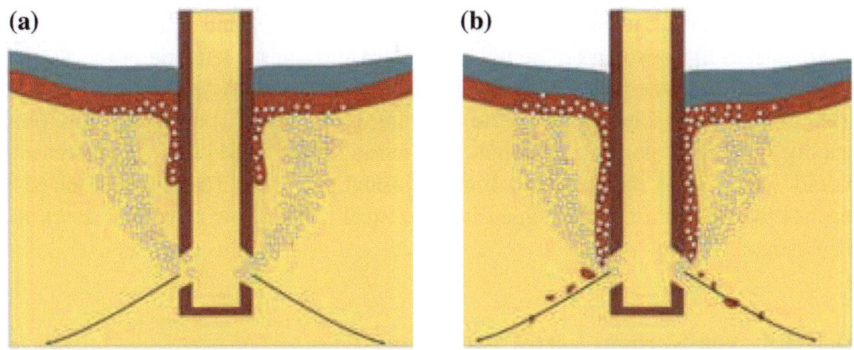

Fig. 11.65 Schematic diagrams of slag foam being carried down the SEN by slag crawling down the SEN to **a** a sub-critical depth (no entrainment) and **b** a critical depth where entrainment occurs (permission granted, ISS/AIST, [163, 164])

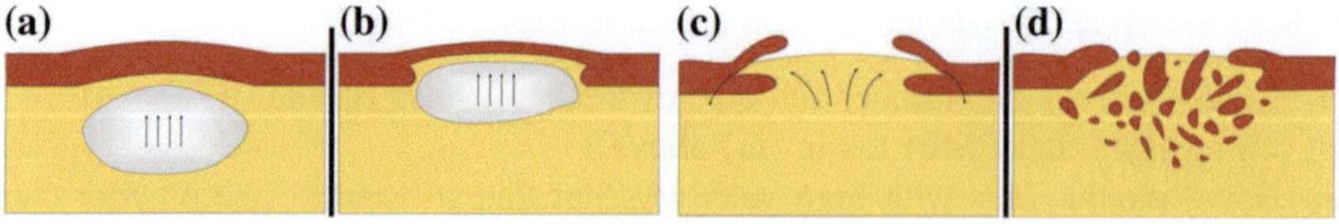

Fig. 11.66 Schematic drawings showing **a** gas bubble rising, **b** bubble covered with steel, pushing into slag pool, **c** bubble penetrating the slag and powder layers and **d** draining of metal carrying slag droplets (permission granted, ISS/AIST, [164])

(e) decreasing slag density, and
(f) decreasing interfacial tension (slag/ metal).

The *critical* gas injection rate (i.e. above which entrapment occurs) decreases with (i) increasing casting speed, (ii) decreasing slag layer viscosity, (iii) decreasing nozzle port angles, and (iv) decreasing wettability between the slag and SEN material [168].

Larger bubbles have enough inertia to penetrate both the slag pool and the powder layer and escape into the atmosphere. A thin layer of steel covers the top surface of the bubble (Fig. 11.66b) and when the bubble collapses and escapes, this steel drains back carrying slag with it (Fig. 11.66d). [175–178].

The only physical property found to affect this mechanism is the interfacial tension (γ_{msl}) with entrapment increasing as γ_{msl} decreases but entrapment has been reported to increase with increasing throughput.

11.10.1.4 Slag Crawling

Interfacial tension forces favour the covering of the SEN with slag; this results in slag crawling down the SEN. However, Hibbeler and Thomas [163, 164] point out slag crawling is aided by the pressure differences arising from asymmetric flow. Asymmetric flow (like that shown in Fig. 11.67a results in both a pressure increase and meniscus elevation (shown on the right side of the SEN and a pressure decrease and meniscus drop shown on the left). Low pressure (on the left side of the SEN in Fig. 11.66a) promotes slag crawling down this left side. Asymmetric flows can arise from several sources, (i) nozzle clogging [179, 180] (ii) turbulent flow (emanating from SEN ports) (iii) stopper rod misalignment [181] and (iv) increasing mould width since this leads to higher throughput [163, 164]. When the slag reaches the SEN port, it is dragged into the steel by the flow travelling down the SEN (Fig. 11.67b).

11.10.1.5 von Karman Vortex Formation

The SEN tends to block part of the metal flow and vortices are formed in the wake of a metal flow. These vortices are referred to as von Karman vortices and they

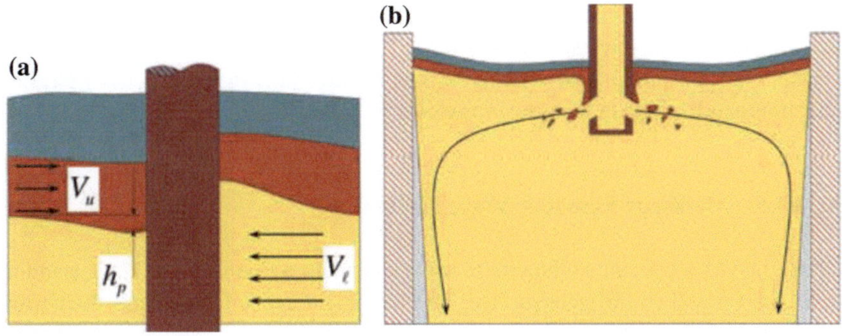

Fig. 11.67 Schematic representation of entrapment by slag crawling **a** high pressure and meniscus elevation shown here on the right side of the SEN and decreased pressure and a meniscus drop on the left side of the SEN, **b** entrapment of slag droplets in the metal stream (permission granted, ISS/AIST, [164])

occur on the low-pressure side of the SEN (on the left of the SEN in Figs. 11.68 and 11.69b) [182–184]. Karman vortices can cause slag entrapment (i) between dendrites of the shell on the wide face near the SEN or (ii) by creating a funnel deep enough to be captured by the metal flow leaving the SEN [182–184]. Asymmetric flow in the mould can also cause entrapment by the latter mechanism. The vortex formation is affected by the SEN geometry and positioning. The *vortex depth* increases with increasing meniscus velocity [185, 186] and the *vortex diameter* increases with increasing SEN misalignment [183, 184], The *frequency of vortex formation* (f_{vort}) increases with (i) increasing casting speed [184–187] (ii) increasing slab width [184, 187, 188] (iii) increasing port angle [183, 184] (iv) increasing SEN misalignment [182, 183] and (v) decreasing SEN immersion depth [184, 187, 188]. The relationship $f_{vort} = \{3/2F'\} - 10$ was derived in water model studies with a mean value of $f_{vort} = 1/2F'$ [164, 167]. The location of the vortex is affected by the type of flow; for a *double roll* system the vortex appears about 30 mm from the

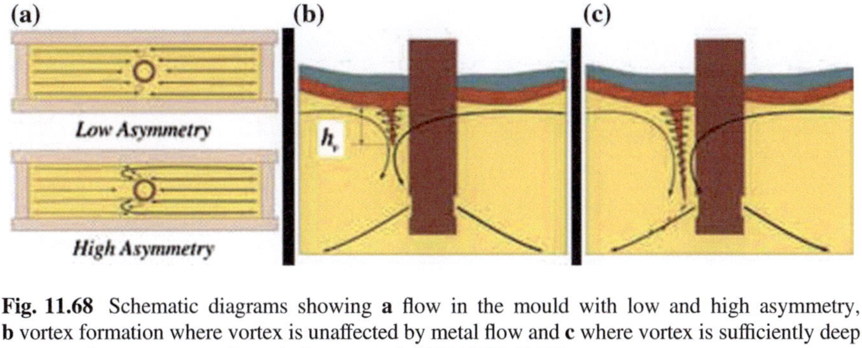

Fig. 11.68 Schematic diagrams showing **a** flow in the mould with low and high asymmetry, **b** vortex formation where vortex is unaffected by metal flow and **c** where vortex is sufficiently deep to cause slag entrapment (permission granted, ISS/AIST, [164])

SEN and moves towards the narrow face [184], whereas, for a *single roll* the vortex appears near the narrow face [188].

Quantitative equations have been derived by several investigators from physical modelling studies; these have been reviewed by Hibbeler and Thomas [163, 164].

11.10.1.6 Meniscus Standing Wave Instability

Surface waves are formed when there is flow beneath a free surface. Thus, standing waves are formed on the steel surface. The standing wave is formed near the narrow face for a *double roll* flow system. If these standing waves become unstable they can turn over, entraining liquid slag in the process. It is customary to determine the critical wave height (h_{wave} = difference between crest and trough of the wave) above which entrainment occurs. The stability criteria, shown in Eq. 11.17, [191], is an example (where λ = wavelength and is defined as distance between outside of the SEN and the narrow face [192]).

$$(h_{wave}/\lambda)_{crit} = 0.21 + 0.14(\rho_{sl}/\rho_m)^2 \quad (11.17)$$

The critical height has been determined in various physical modelling studies in terms of the velocity of the metal leaving the SEN and the diameter of the SEN port (D_{port}) [171, 188, 192]. The various factors affecting the wave height are defined in Fig. 11.69. The derived equations have been critically reviewed by Hibbeler and Thomas [163, 164]. However, the wave height (h_{wave}) was found to increase as (i) casting speed increased (V_c ↑) [193–195], (ii) SEN port diameters increased (D_{port} ↑), (iii) SEN immersion depth decreased (h_{immers} ↓), (iv) SEN port angles were aimed more upward, (v) decreasing mould widths (w ↓) [172, 188, 190–195] and (vi) decreasing mould thickness (t ↓) [196].

Fig. 11.69 Definition of the various casting parameters affecting height of the standing wave (permission granted, ISS/AIST, [164])

11.10.1.7 Shear Layer (Kelvin–Helmholtz) Instability

This movement of the molten steel at the meniscus creates an aerodynamic drag force which causes the liquid slag to move in the same direction as the metal. However, the velocity in the slag phase is much smaller in the slag than that in the metal because the viscosity of the slag is much higher than that of the steel. The velocity difference between the two liquid layers results in shearing of the slag layer and slag filaments are created in the metal. Subsequently, *necking and detachment* of the filament occurs and slag droplets are swept along in the metal stream, as shown in Fig. 11.70. This phenomenon is known as the Kelvin–Helmholtz instability.

Various quantitative equations have been derived for the critical velocity (V_{crit}), i.e. the velocity above which slag entrainment occurs; these equations have been reviewed by Hibbeler and Thomas [163, 164]. An equation involving most the major terms is given in Eq. 11.18 [163, 164, 191]. Thus, the critical velocity increases (and slag entrapment decreases) with (i) increasing interfacial tension (γ_{msl}) (ii) increasing density difference ($\rho_m - \rho_{sl}$) and increasing slag viscosity (η). Other equations indicate the critical velocity increases as the depth of the slag pool increases.

$$V_{crit} = (4g\,(\rho_m - \rho_{sl})\gamma_{msl}\{\eta_m + \eta_{sl})^4/(\eta_m^2 \cdot \rho_{sl} + \eta_{sl}^2 \cdot \rho_m)^2\}^{0.25} \qquad (11.18)$$

11.10.1.8 Upward Flow Impinging on Meniscus

In *double roll* flow patterns, the upward movement of the metal results in filament formation and subsequently, *necking* of the filaments, as shown in Fig. 11.71. The detachment of the slag film can occur by *dragging* of the slag filament [197]

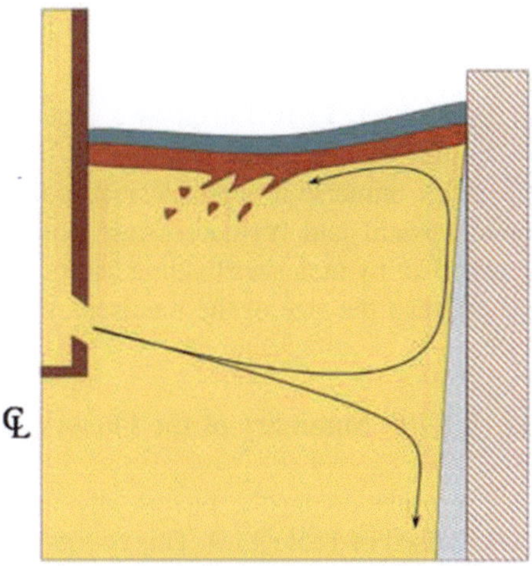

Fig. 11.70 Schematic drawing showing *necking and detachment* of slag filaments by shear layer (Kelvin–Helnhotz) instability; (permission granted, ISS/AIST, [163, 164])

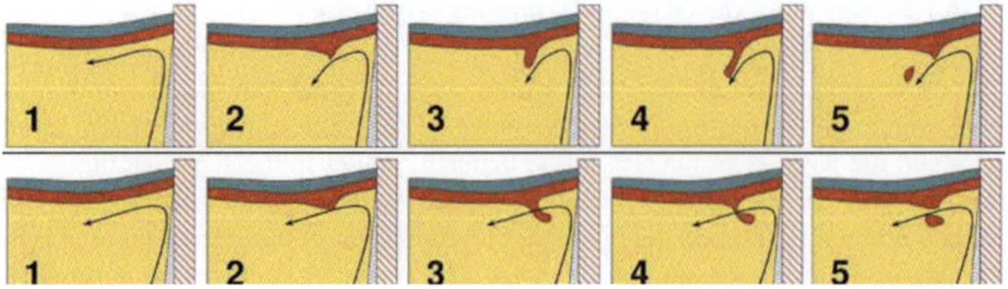

Fig. 11.71 Schematic drawings showing the sequence for slag entrapment by Kelvin–Helmholtz instability (*upper*) for the *dragging* mode and (*lower*) for the *cutting* mode(permission granted, ISS/AIST, [163, 164])

(Fig. 11.71-upper) or by *cutting* of the slag filament [198] (Fig. 11.71 lower). The released slag droplets are, subsequently, trapped by the shell.

A number of quantitative equations for the critical surface velocity (V_{crit}) have been derived from physical modelling studies; these have been reviewed by Hibbeler and Thomas [163, 164]. The critical velocity (V_{crit}) decreases and entrapment increases as (i) viscosity decreases, (ii) the interfacial tension decreases, (iii) the density difference decreases and (iv) the depth of the slag pool increases.

11.10.1.9 Meniscus "Balding"

A rising metal flow (in a *double roll* system) with sufficient momentum can push the slag layer away and can expose the metal to both the powder and to the atmosphere (Fig. 11.72a). This mechanism is known as "*meniscus balding*". The bald patch (or *eye*) appears near the narrow faces and causes reoxidation of the steel and the formation of inclusions (e.g. Al_2O_3 and entrained powder); this process is more severe when the "*eye*" coincides with the trough of the standing wave.

Meniscus balding can also occur from excessive Argon flows which lead, sequentially, to a *single roll* flow system and to bald patches close to SEN (Fig. 11.72b).

From Sect. 11.10.1.6, it was noted that wave height (h_{wave}) increases as (i) casting speed increases (V_c ↑), (ii) SEN port diameters increases (D_{port} ↑), (iii) SEN immersion depth decreases (h_{immers} ↓), (iv) SEN port angles are aimed more upward and (v) decreasing mould widths (w ↓) and thickness (t ↓). Slag entrapment by meniscus balding can be minimised by creating a slag layer which is deeper than the size of the meniscus wave.

11.10.1.10 Summary of the Factors Affecting Entrapment

The information on the factors affecting the different entrapment mechanisms is summarised in Table 11.3. This covers the factors included in the various equations

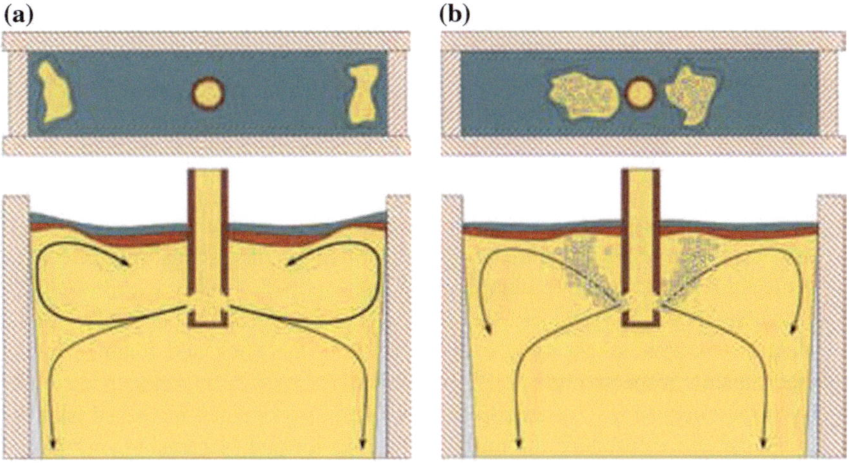

Fig. 11.72 Schematic drawings showing "*meniscus balding*" caused by **a** upward spouts in a *double roll* flow system and **b** excessive Ar flow leading to a *single roll* flow system; the *upper* figures show the location of the bald patches (permission granted, ISS/AIST, [164])

Table 11.3 Summary of the effect that casting variables have on the various slag entrapment mechanisms; E = entrapment, l_{hook} = length of hook, N = number, Q_{metal} = steel throughput, V_{Ar} = Ar flow rate; ΔT = superheat; () = Intuitive deduction, $\uparrow$ = increase, $\downarrow$ = decrease

Mechanism	Factors affecting entrapment
1. Mould level fluctuations	Fluctuations and E $\uparrow$ as (i) $V_c \uparrow$(ii) $D_{SEN} \uparrow$(iii) $h_{immers}\downarrow$ (iv) port angle $\uparrow$ (v) $w\downarrow$ (vi) $V_{Ar} \uparrow^{a}$ (vii) SEN /Ar contact angle $\uparrow$(viii) ($\eta \downarrow$)
2. Hook formation	$N_{hooks} \uparrow$ and E $\uparrow$ as (i) $V_c \downarrow$(ii) $\Delta T \downarrow$(iii) *low* h_{immers} (iv) port angle $\uparrow$ (v) $q_{vert} \downarrow$ $l_{hook} \uparrow$ and E $\uparrow$ as (i) $q_{vert\ bed} \uparrow$
3. Ar bubble interactions (a) Slag foam (b) Bubble collapse	(a) Foam formation and E $\uparrow$ as (i) $D_{bubble} \downarrow$ as$V_c \uparrow$ (ii) $w\uparrow$ (iii) $V_{Ar}\uparrow$ (iv) $\eta\uparrow$ (v) $\gamma_{msl} \downarrow$ (vi) $\rho_{sl} \downarrow$ and (vii) $Q_{metal} \uparrow$(i.e., $\uparrow$ V_c, w, t) (b) E $\uparrow$ as $\gamma_{msl} \downarrow$ ($\eta \downarrow$)
4. Slag crawling	(i) $h_{pen} < h_{immers}$ and $h_{pen} \uparrow$as $\rho_{sl}\uparrow$ $V_{sl}\uparrow$ (ii) ($\eta \downarrow$) (iii) Asymmetric flow $\uparrow$
5. Von Karman vortices	E$\uparrow$ as (i) $\gamma_{msl} \downarrow$ (ii) ($\eta_{sl} \downarrow$) (iii) Asymmetric flow $\uparrow$
6. Standing wave instability	$h_{wave} \uparrow$and E$\uparrow$as $\uparrow$(i) $V_c \uparrow$ (ii) $D_{port} \uparrow$ (iii) $h_{immers} \downarrow$ (iv) $w \downarrow$ (v) $t \downarrow$ (vi) SEN port angle more upward (vii) ($\eta \downarrow$) ($\gamma_{msl} \downarrow$)
7. Shear layer (KH) instability	$V_{crit} \downarrow$ and E$\uparrow$ as (i) $\gamma_{msl} \downarrow$(ii) $\eta_{sl} \downarrow$ (iii) $d_{pool} \downarrow$ (iv) $\Delta\rho\downarrow$
8. Upward flow on meniscus	$V_{crit} \downarrow$ and E$\uparrow$ as (i) $\gamma_{msl} \downarrow$(ii) $\eta_{sl} \downarrow$ (iii) $d_{pool} \uparrow$ (iv) $\rho_{sl}\uparrow$
9. Meniscus balding	E $\uparrow$ as (i) $d_{pool} \downarrow$; (ii) ($\eta \downarrow$) (iii) ($\gamma_{msl} \downarrow$)

[a]Interaction between V_c and V_{Ar}; increasing V_c tends to decrease effect of V_{Ar} and increased V_{Ar} tends to reduce effect of V_c; when $V_{Ar} > 5$ L min^{-1} double roll transforms to single roll

derived in physical modelling studies [163, 164]. It also includes some intuitive deductions, e.g. increased slag viscosity would be expected to resist meniscus balding and standing wave instability but reduce slag pool convection and q_{vert}; these deductions are denoted by parenthesis.

The effect of increases in the various casting variables on the entrapment (denoted E) is given in Table 11.4. It can be seen form this table that

- Increases in the casting variables have contrasting effects on the different mechanisms; only the slag/metal interfacial tension has a consistent effect (i.e. decreasing entrapment with increasing γ_{msl}).
- Increases in casting speed and slag fluidity (i.e. reciprocal viscosity, $1/\eta$) tend to increase entrapment, the only exceptions being the hook and foam formation mechanisms, respectively.
- Even the effect of q_{vert} on entrapment is contradictory since increased heat flux to the surface will reduce the number of hooks formed (decreasing entrapment) but increased heat transfer through the bed will increase the length of the hooks (increasing entrapment).
- The metal flow system formed (Fig. 11.72) is very important in slag entrapment, especially if an asymmetric flow system is established.
- The metal flow system is dependent upon the casting speed, the SEN immersion depth, the Argon flow rate, etc.
- The casting speed (V_c) and Ar flow rate (V_{Ar}) tend to oppose each other, so that increasing V_{Ar} tends to *cushion* the metal flow (and hence, its effect on

Table 11.4 The effect of increases in the various, casting parameters on the different mechanisms of entrapment; E↑ = increase in entrapment; E↓ = decreased entrapment; 1, 2, 3, *etc.* refer to mechanisms listed in Table 11.3; () = by intuitive deduction

Ariable /mechanism	1	2	3a	3b	4	5	6	7	8	9
Casting speed (V_c ↑)	E ↑	E ↓	E ↑		(E ↑)	(E ↑)	E ↑	E ↑	E ↑	(E ↑)
SEN immers depth (h_{immers} ↑)	E ↓	(E ↑)			E ↑		E ↓			
Ar flow rate (V_{Ar} ↑)	E ↑	(E ↓)	E ↑							
Mould width (w↑) Thickness (t ↑)	E ↓		E ↑				E ↓ E ↓			
Superheat (ΔT ↑)		E ↓								
SEN Bore (D_{bore} ↑)	E ↑									
Slag pool depth (d_{pool}↑)								E ↓	E ↑	(E↓)
Slag viscosity (η↑)	E ↓	(E↓)	E ↑	E↓	E↓	E↓	E↓	E ↓	E ↓	E ↓
Interfacial tension (γ_{msl} ↑)			E↓	E↓		E↓	(E↓)	E↓	E↓	(E↓)
Slag density diiff ($\rho_{\mathrm{metal}} - \rho_{\mathrm{sl}}$ ↑)			E↓					E ↑	E ↑	
Vertical heat flux (q_{vert} ↑)		E ↓								

entrapment) and can even cause a change from *double roll* system to *single roll* when $V_{Ar} > 5$ L min^{-1}. Similarly, increased V_c tends to oppose the effects of V_{Ar} on entrapment.

The various flow systems are shown in Fig. 11.73 and the conditions leading to various forms of entrapment are given in Table 11.5.

The equations derived in physical modelling studies may not always provide quantitative values for entrapment by specific mechanisms for the actual caster, because of differences in the similarity criteria for water and steel; this is especially true where the interfacial tension is involved. Nevertheless, the trends summarised in Tables 11.4 and 11.5 are valid for the casting mould. It should be remembered that gas bubbles are coated with liquid slag and are used to remove inclusions from the steel, so mechanisms involving gas bubbles apply to slag, gas and inclusion entrapment.

11.10.2 Slag Entrapment

Published data on slag entrapment have been classified below according to the mechanism involved.

It is apparent from the previous section that entrapment is mainly affected by the metal flow velocity and the flow pattern. However, slag entrapment can be reduced by using a mould slag with a high-interfacial tension (i.e. reducing S content of steel) and high viscosity (but only at the expense of decreased powder consumption).

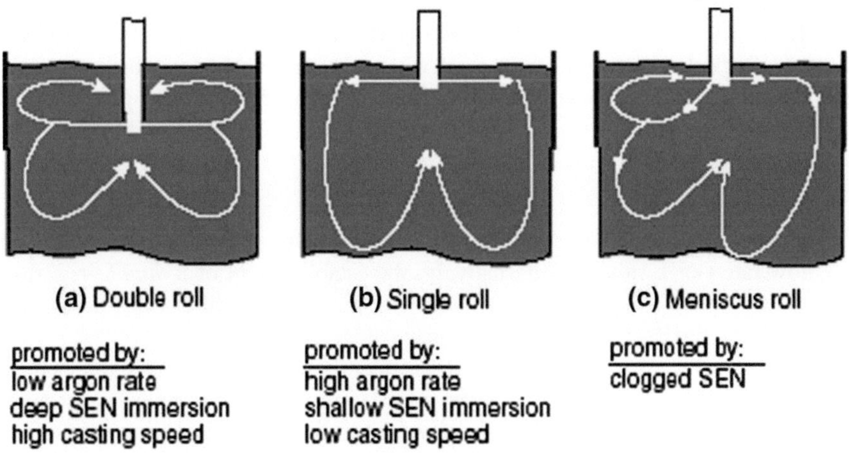

Fig. 11.73 Schematic drawings showing various metal flow patterns **a** Single roll, **b** Double roll and **c** Asymmetric meniscus roll (permission granted, UNESID, [53])

Table 11.5 Operating conditions leading to different modes of entrainment (from data given in [207])

Mode	Conditions	Findings and outcomes
1	Low *Vc*; low SEN immersion depth	Downward recirculation which tore slag particles away
2	Higher *Vc*; higher SEN immersion	Recirculation flow at narrow face causes *necking and detachment* of slag
3	Excessive argon flow	When (i) Ar flow <3 L min^{-1}; decreased entrainment (ii) >3 L min^{-1}; increased entrainment
4	Foam formation around SEN Important at high *Vc*	When foam reaches SEN port it is swept away and results in pencil pipe defects. Lower Ar flow rate

11.10.2.1 Mould Level Fluctuations

The normal mould level variations in continuous casting are unlikely to cause slag entrapment by the mechanism depicted in Fig. 11.62 [163]. However, periodic flow oscillations occur continually and tend to magnify as casting speed is increased [193]. As mentioned in Sect. 11.10.1.1, the velocity of the flow running up the narrow wall can be characterised by the *F′ factor* (defined in Eq. 11.15 and Fig. 11.63) [165]. It has been found in a large number of plant trials, that good surface quality was obtained when the *F′* factor had values between 3 and 5 N/m (grade not stipulated) [165], or between 2 and 3 N/m for LCAK steels [166].

11.10.2.2 Hook Formation and Meniscus Freezing

Hooks and frozen skulls, (formed on the steel surface) can trap slag droplets, gas bubbles and inclusions. They are formed when insufficient heat is supplied to the meniscus region. These conditions arise when casting with (i) low casting speeds, (ii) low superheat, (iii) steels with low-melting temperatures, (iv) poor thermal insulation in the powder bed and (v) when a *single roll* flow system occurs. The problem can be solved by

(i) Reducing the *number of hooks* formed (by improving the heat transfer to the meniscus); this can be achieved by increasing, casting speed, superheat and Ar flow rate (<5 L min^{-1}). Alternatively, the establishment of a *double roll* flow system or by using EMS to smooth out temperature differences across the mould or EMBr to increase the meniscus temperature.
(j) Reducing the *size of the hook* (the underlying principle is explained in Fig. 11.35 [122]), this can be achieved by improving the thermal insulation of the powder bed (by increasing the powder bed thickness, or adding exothermic agents to the mould powder [44]) or by using EMBr or EMC to delay solidification. Note increased slag viscosity will reduce the vertical heat transfer by convection in the slag pool but will reduce powder consumption also.

11.10.2.3 Argon Bubble Interactions

Slag coats the surface of gas bubbles and there are two mechanisms in which slag is released and, subsequently, captured by the shell. These involve (i) slag foam moving down the SEN (Fig. 11.65) and (ii) the penetration and collapse of larger bubbles (Fig. 11.66). Smaller bubbles have less inertia than bigger bubbles so they find it difficult to penetrate the slag pool so tend to create a foam in the slag pool. Bubble diameters decrease with increased steel throughput ($= \rho_m V_c \cdot w \cdot t$) so foam formation is promoted by higher casting speeds and mould dimensions. High values of viscosity and interfacial tension will resist the escape of the bubbles and thereby, promote foam formation.

Larger bubbles have more inertia and can penetrate the liquid slag pool. When the bubble collapses the slag coating forms droplets which drain with the help of the draining metal. High values of slag viscosity and interfacial tension will resist the movement of bubbles.

11.10.2.4 Slag Crawling

The slag crawling mechanism is shown in Fig. 11.67 [164]. Slag crawling is aided by the pressure difference induced across the SEN resulting from the interference of the SEN to the reversing flows (travelling from the narrow faces towards the SEN) [200]. The slag penetration depth (h_{pen}) increases with increases in the *Blockage factor* which is usually taken to be (D_{SEN}/t) where D_{SEN} is the diameter of the SEN and t is the thickness of the mould. The magnitude of the slag penetration depth, h_{pen}, is dependent on the interfacial velocities and densities in the metal and slag phases [200]. The densities of most steels and mould slag do not vary much. Slag entrapment takes place when $h_{pen} > h_{immers}$ and so entrapment increases for low SEN immersion depths. It has been proposed that this mechanism is the most important contributor to the slag entrapment [200].

11.10.2.5 von Karman Vortices

Some workers consider this mechanism to be the most potent source of slag entrapment in continuously cast steel [189, 201]. The slag droplets held in the tube of the vortex escape when the vortex becomes unstable. Entrapped mould slag droplets in the metal can be seen in the specimens of hot physical modelling studies where Karman vortices were induced (Fig. 11.74a) [189, 201]. These Karman vortices have also been identified in the predicted flows at the slag/metal interface using a model coupling heat and fluid flow and shell solidification (Fig. 11.74b) [53].

Hot and cold modelling studies involving Karman vortices showed that the amount of entrapped slag decreased with increasing slag viscosity (Fig. 11.75a) and inceasing interfacial tension (Fig. 11.75b). An empirical relation, given in

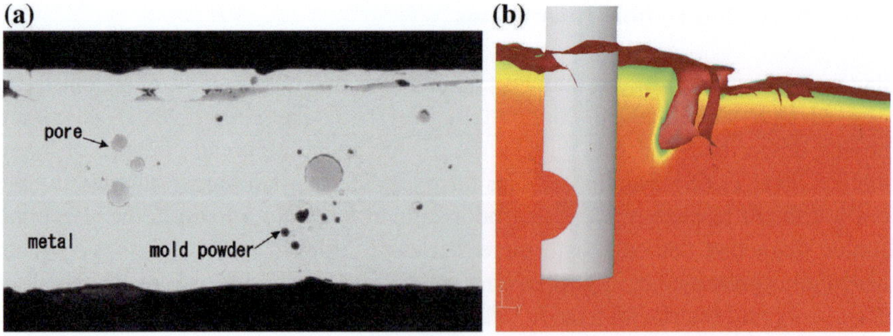

Fig. 11.74 **a** Photograph showing slag entrapped by von Karman votices during hot modelling experiments [189] and **b** Formation of a Karman vortex close to the SEN predicted by a mathematical model (permission granted, **a** SAIMM, [189], **b** ISIJ, [53])

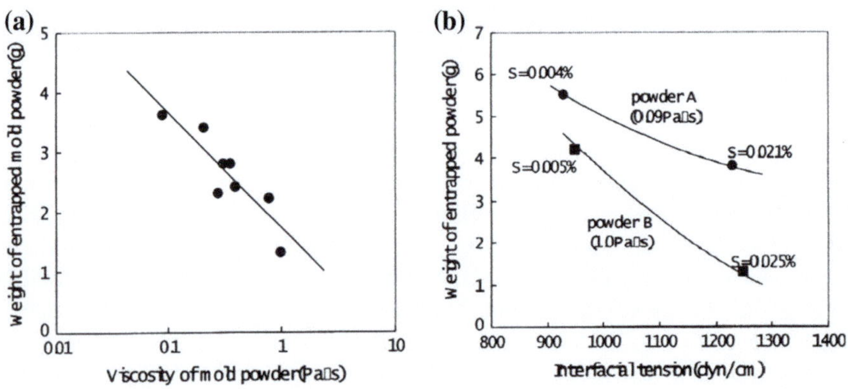

Fig. 11.75 The mass of entrapped casting slag as a function of **a** viscosity in Pas and **b** interfacial tension in m N m^{-1}for 2 powders A, B (permission granted, ISIJ, [201])

Eq. 11.19 was obtained (note the units are viscosity, η_{sl} in Pas, and interfacial tension, γ_{msl}, in m N m^{-1}) [189, 201]. Increased viscosity can be obtained by reducing slag basicity and increased interfacial tension can be best achieved by lowering the S and O contents of the metal but decreasing the B_2O_3, K_2O and CaF_2 contents in the slag would also help.

$$m_{\text{entrap}} = 1.06 \times 10^{-7} (\eta_{\text{sl}})^{-0.255} \cdot (\gamma_{\text{msl}})^{-2.18} \qquad (11.19)$$

11.10.2.6 Standing Wave Instability

Standing waves are formed on the steel meniscus and if they become unstable, slag can be trapped in the overturning wave. The wave instability arises either, because of the turnover of the wave, or by a shearing mechanism. It has been pointed out that the former mechanism is unlikely to be a major cause of slag entrapment since the critical surface velocity (V_{crit}) for this mechanism is higher than that for the shearing mechanism [163, 164]. Surface velocities have been reported to increase in thin slab casting. The wave height (h_{wave}) was found to increase with increases in (i) casting speed (V_c ↑), (ii) SEN port diameters increased (D_{port} ↑) and with decreases in, (iii) SEN immersion depth (h_{immers} ↓), (iv) mould widths (w ↓) and (v) as SEN port angles were aimed more upward (with the jet impingement occurring 25–50 mm lower for every 0.2 m min^{-1} increase in casting speed [187].

11.10.2.7 Shear Layer (Kelvin–Hemholtz) Instability

This is a natural phenomenon which occurs when two liquid layers are travelling at different speeds (ΔV_{msl}). The reversing flow (formed near the narrow-face wall) shears the slag to form a slag filament (depicted in Fig. 11.70). The shear force causes *necking* of the slag filament. Then Kelvin–Helmholtz instabilities cause the slag filament to break up resulting in detachment of the slag droplet. The predictions of a mathematical model of heat and fluid flow (shown in Fig. 11.76 [54, 55, 202]) show cases of the *necking and detachment* mechanism. This mechanism is considered to be a major source of slag entrapment.

Slag entrapment occurs when the velocity difference between slag and metal phases exceeds the critical velocity difference (i.e. when $\Delta V_{msl} > \Delta V_{crit}$). It has been reported that the critical velocity difference (ΔV_{crit}) increases (i.e. entrapment decreases) with increasing interfacial tension, viscosity, density difference ($\rho_{metal} - \rho_{sl}$) and slag pool depth (ΔV_{crit} ↑, as γ_{msl}↑ [203], as η_{sl} ↑ [203], ($\rho_{metal} - \rho_{sl}$↑) [203] and as d_{pool} ↑) [204]. Thus, slag entrapment by this mechanism can be minimised by (i) increasing interfacial tension (by minimising the S content of the steel or the FeO, Na_2O, B_2O_3 contents of the slag), (ii) by using low basicity (C/S) slags with higher viscosity and (iii) by using a deep slag pool; note, ($\rho_{metal} - \rho_{sl}$↑) does not vary much, so has little effect.

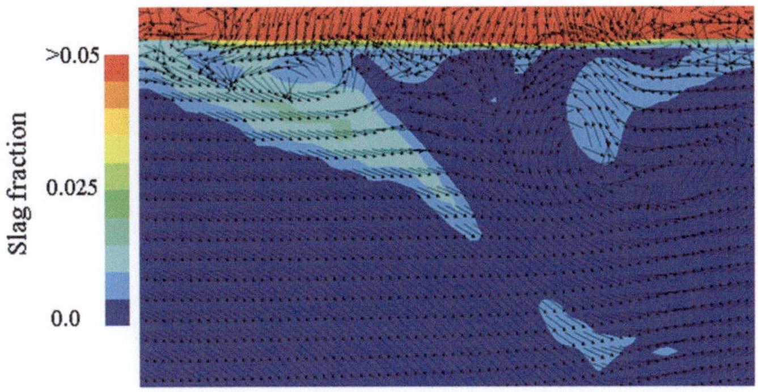

Fig. 11.76 The predictions of a mathematical model showing "*necking and detachment*" of slag (*light blue*) in steel (*dark blue*) (permission granted, P.E. Ramirez-Lopez [202])

Magnetic fields applied in a direction perpendicular to the flow do not affect the interface but a magnetic field applied parallel to the flow does stabilise the interface [205, 206]; its effect is akin to that of increasing the interfacial tension but it does not affect ΔV_{crit}.

11.10.2.8 Upward Flow on Meniscus

With this mechanism, the metal flow impacts against the mould wall and creates an upward flow which in turn, creates a vortex (Fig. 11.77a); the shearing force associated with the vortex produces a slag filament (Fig. 11.77b). Subsequently, the shearing force causes *necking and detachment* of the filament to occur. Since the vortex is produced near the narrow face, the slag droplets are deposited near the edges of the slab (Fig. 11.77c). It can be seen from Fig. 11.71 that there are two flow modes causing necking and detachment, namely, i.e. the *dragging* and *cutting* modes.

Water-modelling studies have indicated that the velocity at the interface (V_{surf}), decreases (and entrapment decreases) when the interfacial tension, slag viscosity and density difference ($\Delta\rho = \rho_m - \rho_{sl}$) all increase (i.e. $V_{surf} \downarrow$ as $\gamma_{msl}\uparrow$ as $\eta\uparrow$ as $\Delta\rho \uparrow$) and when the depth of the slag pool decreases ($d_{pool} \downarrow$) as shown in Fig. 11.78 [198, 207]. In practice, (i) the densities of slags and metals vary little and (ii) the interfacial tension is principally determined by the Sulphur content o the steel ($\gamma_{msl}\uparrow$as S (ppm) $\downarrow$) however, high FeO and MnO concentration in the slag also reduce interfacial tension and (c) slag viscosity is largely determined by the basicity (*C/S*, with $\eta\uparrow$ as *(C/S)* $\downarrow$). Although entrapment can be decreased by reducing the depth of the slag pool, this would have a deleterious effect on powder consumption and C-pick-up by the shell.

A special case of the upward flow mechanism occurs in billet casting where downward pouring is used. For high SEN immersion depths it has been reported that downward flows in the mould result, periodically, in the formation of vortices which, in turn, result in periodic backflows (Fig. 11.80) [208]. These backflows

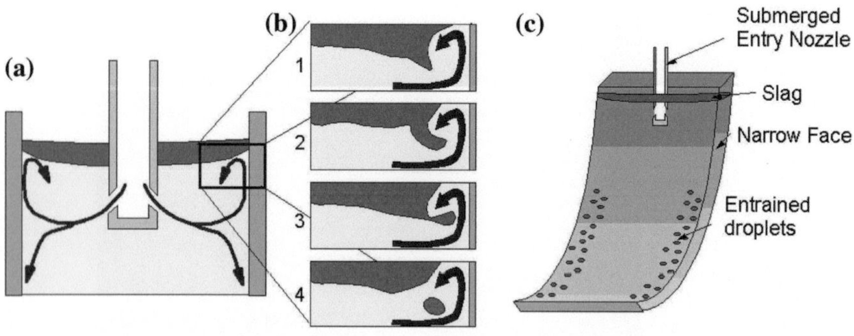

Fig. 11.77 Schematic drawings showing **a** the reversing flow formed in the mould, **b** the sequence of events during slag *necking and detachment* (**b**) location of entrapped slag on slab (Courtesy of RF Brooks, National Physical Laboratory)

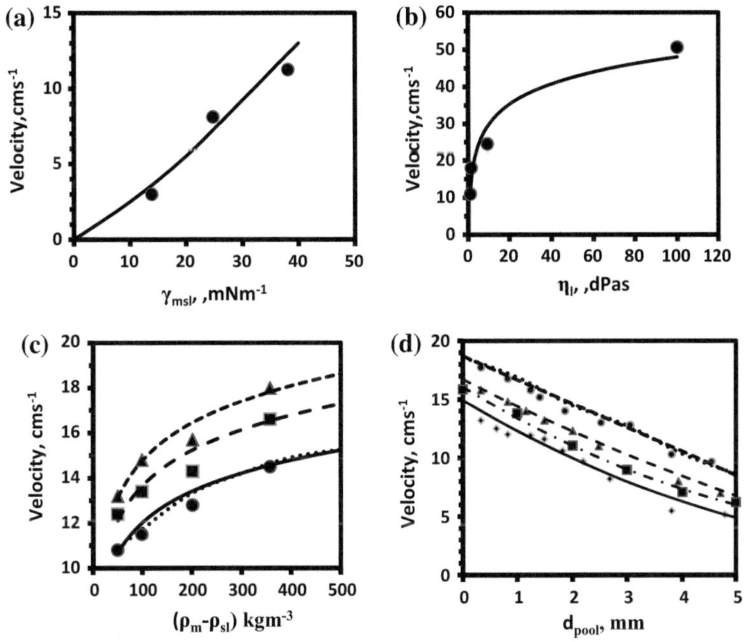

Fig. 11.78 Velocity at interface as a functions of **a** interfacial tension, **b** Slag viscosity, **c** differential density (Δρ = = $\rho_m - \rho_{sl}$) for slag pool depths of □ = 3 mm; ◇ = 8 mm; o = 1.3 mm; **d** slag pool depth; with $\Delta\rho = \rho_m - \rho_{sl}$ kg m^{-3} has the following values; ● = 1300; ▲ = 1200 ■ = 1100; ♦ = 1000 (permission granted, ISS/AIST, [198, 207])

result, sequentially, in meniscus instability, high vorticity in the flow and a rotating flow around the billet which causes slag entrapment [208]. Thus, billets were found to contain clusters of slag inclusions at periodic distances on the inner radius [208] (Fig. 11.79).

11.10.2.9 Meniscus Balding

This mechanism of slag entrapment occurs when an upward flow causes the "*balding*" of the mould powder and the creation of an "*eye*" in the powder layer which leads to reoxidation of the steel. A *double flow* system results in a bald patch being produced near the SEN and a *single roll* system in bald pitches forming near the narrow face. Entrapment by this mechanism can be minimised by using

(i) a deep powder bed so $d_{slag+powd} > h_{wave}$; the resulting decrease in q_{vert} will also help to reduce the size of the hook.
(ii) expanding mould powders reduce the tendency of spherical granules in the powder bed to run downhill [209].
(iii) EMBr to stabilise the meniscus and reduce the height of the standing wave [48].

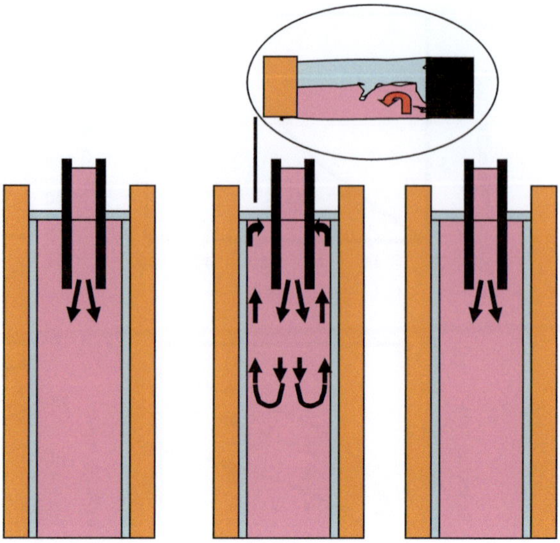

Fig. 11.79 Schematic drawings showing periodic formation of vortices and the formation of a back flow resulting in slag entrapment in billet casting. mould = *brown* *blue* = slag

11.10.2.10 Summary of Slag Entrapment Mechanisms

The following conclusions can be drawn from the review of slag entrapment mechanisms:

(i) Slag entrapment occurs by a number of different mechanisms.
(ii) The most potent mechanisms resulting in slag entrapment appear to be (a) slag crawling (aided by pressure differences across the SEN), (b) Karman vortices, (c) Kelvin–Helmholtz instabilities and (d) upward flows.
(iii) Most of the casting variables have contradictory effects on the various mechanisms but, overall, the slag entrapment can be reduced by (a) reducing the casting speed, (b) increasing interfacial tension and (c) increasing slag viscosity.
(iv) Slag entrapment can be reduced by using EMBr, EMS and EMC
(v) Slag entrapment can also be reduced by improving the thermal insulation of the powder bed (i.e. decreases q_{vert}) since this will have beneficial effect on slag entrapment associated with hook formation and meniscus balding and will help to reduce the length of the shell; however, it will also increase d_{pool} which tends to increase entrainment via the upward flow mechanism.

11.10.3 Gas Entrapment

Argon gas is fed through the SEN to minimise nozzle clogging, but it also helps to reduce the formation of hooks by increasing the vertical heat flux by increasing gas convection. Most of the Argon finishes up in the slag pool and powder layer but some of bubbles get incorporated into the metal flow [210]. Argon bubbling also aids the removal of inclusions from the metal by transporting them to the molten slag pool. However, Argon bubbling is the principal source of gas entrapment in the steel. Entrapped gas bubbles cause two types of defects in the cast product (i) "*pinholes*" and "*blow holes*" consisting of trapped gas bubbles like that shown in Fig. 11.61 and (ii) the defect known as "*pencil pipe*" (sometimes referred to as "*pencil blister*") which arises when gas bubbles carrying inclusions are captured (Fig. 11.81b). *Pinholes* consist of small argon bubbles. *Blowholes* usually contain CO (g) [46]; they are formed by dissolved oxygen content ($\underline{O}$) in the liquid steel gradually increasing as solidification proceeds and CO (g) bubbles are formed when the $\underline{O}$ concentration exceeds a certain, critical value. Atmospheric O_2 and N_2 can be drawn into the steel via an overturning wave in the *Standing Wave instability* and *Meniscus balding* mechanisms, and during periods of high mould level variations [211]. *Pencil pipe defects* are typically 1–2 mm wide and 12–300 mm long [212] and exhibit a raised surface ("*pencil*") after annealing (Fig. 11.80b).

Gas entrapment occurs principally through the formation of gas foams (Fig. 11.65), but bubbles can be swept into the metal flow where, subsequently, they are captured by the shell (Fig. 11.80a). Larger bubbles have more inertia than smaller bubbles and tend to emerge into the slag pool near the SEN, whereas, smaller bubbles emerge nearer the narrow-face walls. Consequently, there is a much greater risk of small bubbles being swept into the metal stream. Increased casting speeds have been reported to increase the number of small bubbles. Furthermore, there is a much greater risk of bubble entrainment with a *single roll flow* system since the *double roll* system gives the gas bubbles time to float to the slag pool whereas, the *single roll* system sweeps the bubbles straight down the mould (Fig. 11.80a). High Ar flow rates can cause a *double roll* system to transform ro a *single roll* system [173, 174]. The models used for determining bubble movement in the mould have been reviewed [210].

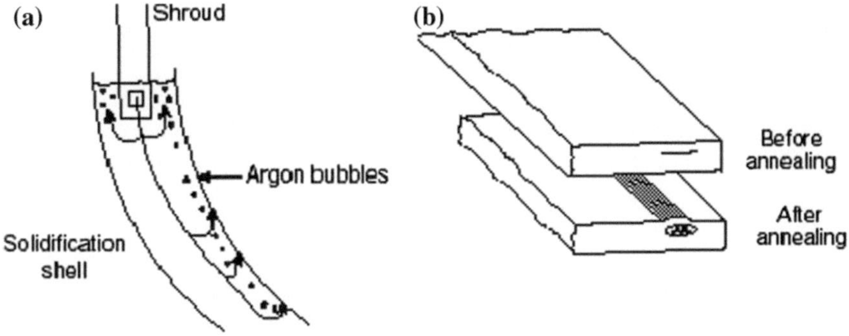

Fig. 11.80 Schematic drawings showing **a** how bubbles are trapped in steel [212] and **b** the formation of "*Pencil pipe*" defects [198] (permission granted, ISS/AIST, [198, 212])

11.10.3.1 Pinholes

The following factors affect the incidence of pinholes.

Argon Bubbling

The incidence of pinholes tends to increase with increasing Argon flow rate. However, excessive flow rates (>5 L min^{-1}) cause *double roll* flow system to transform to a *single roll* system [172–174, 207]. The aim is to keep the bubbles near the slag pool where they can be absorbed but *single roll* flows tend to move the bubble down the mould and away from the slag pool and should be avoided.

Reducing the Length of the Meniscus and Delayed Solidification

It has been proposed that bubble entrapment can be reduced by reducing the size of the solidified shell [122]. The principle of the proposed treatment is shown in Fig. 11.35 [122]. A shorter meniscus or shell can be obtained by (i) reducing the vertical heat flux or (ii) increasing the superheat (Fig. 11.81a). The vertical heat flux (q_{vert}) can be reduced by

- Increasing the depth of powder bed and decreasing the granule size.
- Using exothermic agents which reduce the thermal gradient in the bed (Fig. 11.81a).
- Decreasing Ar flow rates.

Delayed solidification, in which the shell solidifies further down the mould, results in a shorter solidified meniscus but also removes the shell away from the turbulent interface. Delayed solidification can be achieved with (i) meniscus-free casting [21, 22] (ii) Electromagnetic casting (EMC) [51] and (iii) a "*soft cooling*" mould powder with high crystallinity and a high-melting temperature which will create an insulating slag film in the meniscus region [38].

Use of Electromagnetic Devices

The pinch force generated in *Pulsative Electromagnetic casting* (EMC, for more details see Sect. 5.8.3) results, sequentially, in a thicker slag film, reduced horizontal heat flux and delayed shell solidification. This has been reported to lead to a 5–10-fold decrease in defects (inclusions and pinholes) [50, 51] and also provides the extra benefit of producing shallow oscillation marks.

Electromagnetic braking (EMBr or Flow control, for more details see Sect. 5.8.4) reduces (i) the velocity of the metal flow flowing from the SEN and (ii) the penetration depth of the metal flow into the mould which results in an increase in meniscus temperature [47, 48]. In addition, it has been reported to cause a 30% reduction in q_{vert} [124] leading to the formation of a short shell. It can be seen from

Fig. 11.81b that the use of Flow Control (EMBr) leads to significant reduction of pinholes [47, 48].

Electromagnetic stirring (EMS, for more details see Sect. 5.8.2) of liquid metal causes Saffman forces to be exerted on the gas bubbles and inclusions, the

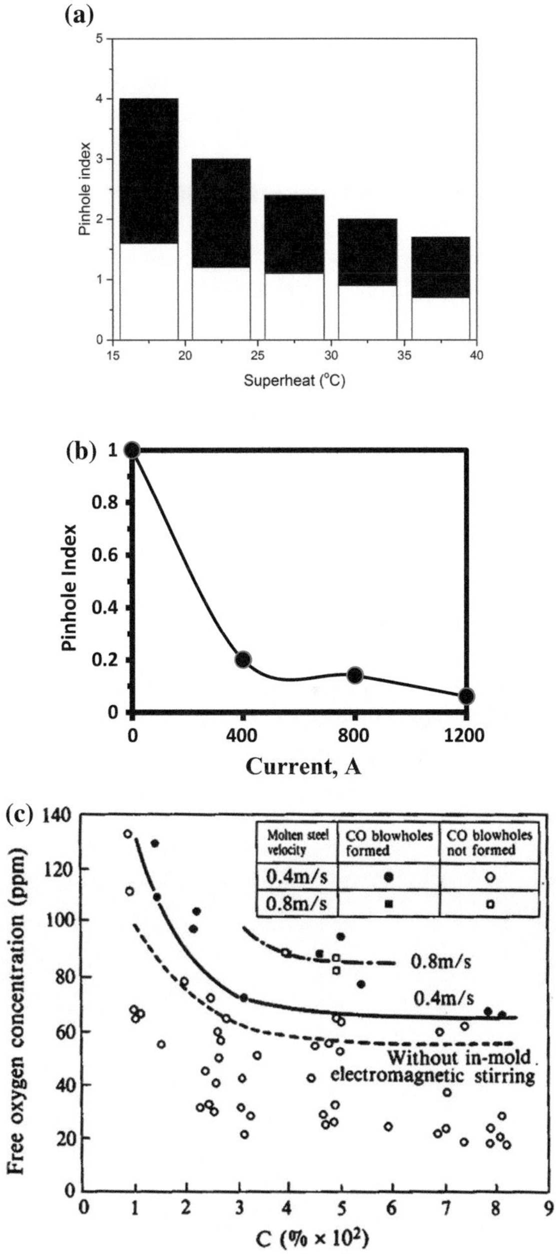

Fig. 11.81 Schematic drawings showing **a** effect of superheat and exothermic agents in the mould powder on the pinhole index [44]; conventional mould powder (■) and containing exothermic agents (□), **b** effect of Flow control (EMBr) on the pinhole index [47] and **c** Critical dissolved O concentration in steel versus C content of steel showing increased stirring velocity (- - = no EMS; ▬ = 0.4 ms^{-1}; dash-dot = 0.8 ms^{-1}) in EMS reduces CO blowhole formation (●, ■ = CO blowholes, o, □ = no CO blowholes) [46] (permission granted, **a** re-drawn using data from [44], **b** EPD Sci., re-drawn [47], **c** NSSM Corp.)

magnitude of the force increasing with increasing bubble/inclusion size [50, 213]. Bubble/inclusion separation occurs when the Saffman velocity exceeds the velocity of the solidification front. EMS is successful in removing bubbles and inclusions with particle diameters, >100 μm) [50]. Saffman forces easily remove bubbles with $D_{bubble} > 1$ mm and will even remove bubbles with $D_{bubble} < 0.4$ mm with steels with lower S contents (higher interfacial tension) [213]. EMS also reduces the level of CO *blowhole*s formed when casting Al-killed steels [46]. The metal flow created by EMS circulates the liquid metal at the solidification front which reduces the build up of the $\underline{O}$ concentration in the liquid steel in this region, thereby, suppressing CO blowhole formation (Fig. 11.81c).

Metal Flow Velocity and Bubble Size

Hanao et al. [214] found that the index of pinholes decreased as the molten steel flow velocity increased and tended to be slightly higher on the inner radius side than on the outer radius side (Fig. 11.82a). This is a surprising result since increased casting speeds tend to provide smaller bubbles. Hanao et al. [214] calculated $\Delta\gamma$ (given in Eq. 11.20) and found that it became increasingly negative as the casting speed increased. Negative $\Delta\gamma$ values indicate that the bubble trapped by the shell is unstable. This can be viewed as higher velocities liberate trapped gas bubbles. Their calculations also indicated that larger bubbles trapped by the shell were more unstable than smaller bubbles, i.e. bubble entrapment increases with decreasing bubble diameter.

Surface and Interface Tension

Hanao et al. [214] examined the various tensions acting on a gas bubble trapped against the solid shell and reported Eq. 11.20, where $\Delta\gamma$ = stability index of bubble

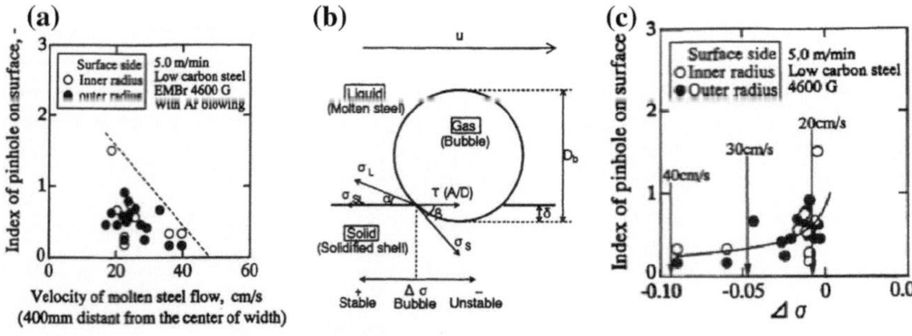

Fig. 11.82 Schematic diagrams showing **a** Index of pinholes as a function of metal flow velocity, **b** the tensions acting on a bubble trapped by the shell and **c** the pinhole index as a function of stability index, $\Delta\gamma$ (permission granted, ISS/AIST [214])

on the solid interface; γ_{LS} = interfacial tension between solid and liquid (steel), γ_L = surface tension of liquid (steel); γ_S = surface tension of solid (steel); α, β angles (°) made by tensions; τ = shear stress exerted by metal flow; δ = thickness of trapped part of bubble and D_{bub} = diameter of bubble and *A/D* = area of circular boundary line (Fig. 11.82b)

$$\Delta\gamma = \gamma_{LS} + \gamma_L \cos\alpha - \gamma_S \cos\beta - \tau(A/D_{bub}). \tag{11.20}$$

It can be seen from Fig. 11.82c that that the index of pinholes increases as the stability index, $\Delta\gamma$, increases [214].

Summary of Ways to Reduce Pinholes

The incidence of pinholes can be reduced by

- Optimising the casting speed and Ar flow to ensure that a *double roll*, flow system is established.
- Decreasing the vertical heat flux by (i) increasing depth of powder bed (ii) incorporating exothermic agents into powder bed (iii) increasing metal superheat.
- Delaying solidification by using (a) Meniscus-free casting (b) EMC or (c) using a high basicity mould slag (with high values of f_{crys} and $T_{br)}$ to reduce the horizontal heat flux.
- Using EMBr to increase the meniscus temperature and decrease the penetration depth of the metal flow or, EMS to remove pinholes and EMC to delay solidification.
- Increasing the casting speed which renders bubbles, captured by the shell, more unstable.
- Reducing the S content of the steel decreases the number of gas bubbles in the steel [213].
- Increasing the vertical section of the caster by 2.5 m which allows more time for pinholes to float out [215].
- Minimising mould level variations [211].

11.10.3.2 Pencil Pipe

The source of pencil pipe defects are small gas bubbles or gas foam swept up in the metal flow (Fig. 11.80a); a typical gas bubble is shown in Fig. 11.61. Gas bubbles (ca. 0.5 mm diam.) get captured at the solidification front, particularly on the inner radius (loose side) (Fig. 11.80a). The bubbles elongate on rolling where the inclusions (attached to the bubble) prevent them from being welded shut during rolling [212]. These defects are usually prevalent in steels with low yield strength (i.e. Ti- or Nb-stabilised LC and ULC grades) [212]; this allows the bubble to expand during annealing, especially if carried out at high temperature [198]. Pencil pipe defects tend to concentrated in a band around the 1/8th point and are prevalent in wide slabs (>1.8 m).

The effect of casting variables (e.g. casting speed, mould width, biased flow, metal level variations, etc.) on the incidence of pencil pipe defects, revealed that there was a strong correlation between the defects and steel throughput, the defects increasing with increasing throughput [212]. All other factors had little, if any, effect on the incidence of pencil pipe defects. Thus, the best way to reduce pencil pipe defects is to reduce the steel throughput [212] but this was found to result in a slightly higher incidence of *slivers* [212].

11.10.4 Inclusion Capture, Sliver Formation

There are two principal types of inclusions, namely, (i) those arising from mould slag entrapment and (ii) those (like Al_2O_3 or TiN) which originate from steelmaking reactions or from slag/metal reactions (e.g. Eq. 11.21); these occur in the tundish and the casting mould.

$$3\,SiO_{2\,sl} + 4\underline{Al} = 2\,Al_2O_{3\,sl} + 3\underline{Si} \tag{11.21}$$

Non-metallic inclusions, such as Al_2O_3 or TiN, reduce the mechanical strength of steels [216]. Furthermore, the mechanical strength decreases with increasing inclusion size [216]. Consequently, considerable efforts are made to reduce the number of inclusions in the steel. Inclusions tend to agglomerate through turbulent collisions and these agglomerated particles can restrict the flow of molten slag into the channel between shell and the solid slag film (adhering to the mould) (Fig. 11.83a [76, 77]). Large inclusions have a higher probability of forming clusters by collisions [217]. Agglomeration is encouraged by the fact that the solubilities of TiN and ZrO_2 in mould slag are both low. It can be seen from Fig. 11.83a that these agglomerates contain both solidified δ-*ferrite* and some gas bubbles; these give rise to *slivers* and *scums* in the product. *Slivers* are surface laminations and in thin strip products, the *slivers* consist of a tiny metal film which has been, either completely, or partially, torn away from the strip surface [218]. Skin laminations can be seen in Fig. 11.83b; inclusions in laminations were identified as TiN, mould powder and reaction products of mould slag and solute elements in steel [78]. *Slivers* originate from several sources, (inclusions) but also in association with bubbles, surface cracks and internal cracks [213].

Cracks are not an issue in ULC steels because of the good ductility at high temperatures [213]. *Slivers* associated with FeO have been reported to occur (i) from FeO carried over from the ladle [211] and (ii) from scale formed on the surface of the steel [219].

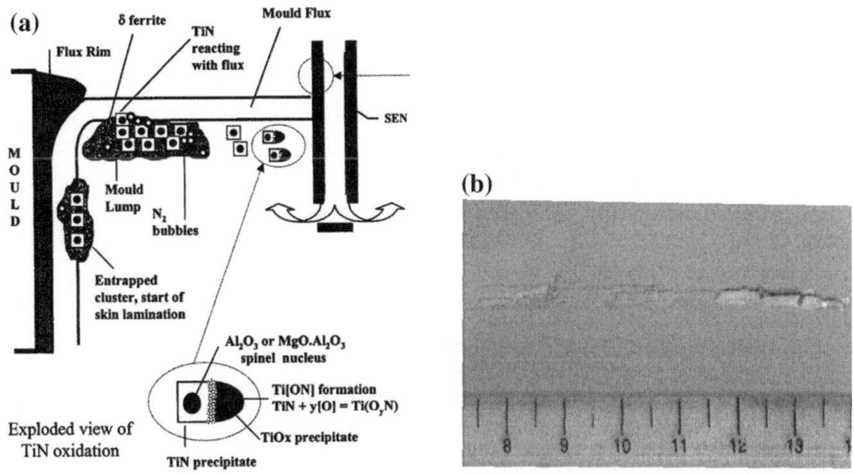

Fig. 11.83 Schematic diagrams showing **a** the formation of agglomerates (of TiN in Ti-stabilised stainless steel) [76–78] and **b** photograph of a skin lamination on the surface of Ti-stabilised stainless steel [78] (permission granted, ISS/AIST, [78])

11.10.4.1 Reducing Inclusion Levels in Steel

Great effort is made to reduce the number and size of inclusions in the steel. This is usually achieved by transporting the inclusions to the liquid slag phase (of the ladle, tundish and casting mould) where they can dissolve. This removal to the slag phase is achieved by the use of

(i) Flotation of the inclusions which is related to the density difference ($\rho_{steel} - \rho_{incl}$); see Sect. 11.10.3.2 below for further details.
(ii) Attachment to gas bubbles [210]; see Sect. 11.10.3.2.
(iii) Saffman forces (generated in EMS) which is effective in removing inclusions sized >100 μm [50, 215].

In addition, inclusion levels are reduced by the following:

(i) *Reducing the size of the solidified meniscus shell* (or hook) as shown in Fig. 11.35 [122] and it can be seen from Fig. 11.85 a that the inclusion levels decrease with reducing hook size [48]; this is achieved by (a) increasing superheat or meniscus temperature (b) by reducing the vertical heat flux by creating a deeper powder bed and using a powder containing exothermic agents and a small granule size.
(ii) *Maximising the time for flotation* by (a) adjusting the SEN immersion depth, port angle and Ar flow rate to ensure that a *double roll flow* system (Fig. 11.73) is established which keeps the inclusion close to the slag layer, in contrast to *single and asymmetric flows* which tend to carry inclusions away from the slag pool and (b) increase the vertical section of the caster (by 2.5 m) to aid flotation of inclusions [215].

(iii) *Delaying solidification* to a position further down the mould which can be achieved using meniscus-free casting, EMC or by forming a slag film with high crystallinity and T_{br}.
(iv) Increasing the slag/*metal interfacial tension* has been reported to reduce the number of inclusions in the steel [10, 220, 221]; the number of *slivers*, associated with gas bubbles, was found to decrease with decreasing S content of the steel (i.e. increasing γ_{msl}) [213].
(v) *Increasing slag viscosity* was found to reduce the number of mould-slag-related inclusions by ca. 20% [126].
(vi) *Casting speeds* were found to have little effect on the number of inclusions but abrupt changes in casting speed were found to increase the number of inclusions [222].
(vii) The amount of inclusions trapped increases as (i) the *size of the inclusion* increases (D_{incl} ↑) and (ii) the *size of the bubble* decreases (D_{bubble} ↓) [221].
(viii) Using *mould slags with high basicity* since this reduces the concentration of more reducible oxides (e.g. SiO_2) which can react with Al and Ti in the steel [223, 224].

11.10.4.2 Flotation of Inclusions

The velocity of an inclusion (V_I) rising in liquid steel is given by Stokes' Law (Eq. 11.22) where g is the gravitational constant, r is the radius of the inclusion (in m), η_M is the viscosity of the metal (in Pas) and the subscripts I, M, and G represent the inclusion, metal and gas, respectively.

$$V_I = 2\,(\rho_I - \rho_M) \cdot g \cdot r_I^2 / 9\eta_M \tag{11.22}$$

Thus, large particles will float quicker than small particles and agglomeration of inclusions will assist flotation.

11.10.4.3 Flotation by Attachment to Gas Bubbles

Consider a gas bubble approaching an inclusion in the liquid metal. For an inclusion to be taken up by the bubble, new interfaces must be formed. Work must be done to account for the change to the interfacial tension. This involves the eradication of the inclusion/metal (IM) interface and the creation of two new interfaces [i.e. metal/ gas (MG) and inclusion/gas (IG)].

The interfacial relationships are given by Young's Equation (Eq. 11.23) which was derived from a balance of forces (Fig. 11.84) where θ is the contact angle between inclusion and gas.

$$0 = \gamma_{IM} + \gamma_{MG} \cdot \cos\theta - \gamma_{IG}. \tag{11.23}$$

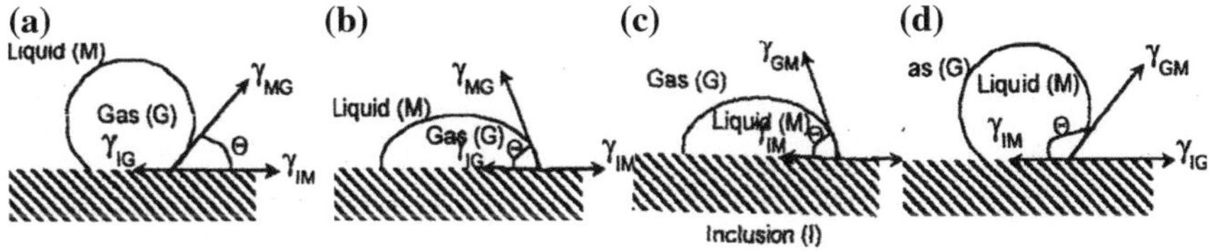

Fig. 11.84 Schematic drawings showing **a**, **b** bubble contacting inclusion for **a** wetting and **b** non-wetting conditions and **c**, **d** sessile drop experiments for, **c** wetting and **d** non-wetting conditions (permission granted, Verlag Stahleisen [225])

The work of adhesion (W_A) represents the work done and is defined in Eq. 11.24.

$$W_A = \gamma_{MG} + \gamma_{IG} - \gamma_{IM} = \gamma_{MG}(1 + \cos\theta) \tag{11.24}$$

For good flotation, it is necessary that the flotation coefficient (Δ, defined in Eq. 11.25) should be both positive and have a high value.

$$\Delta = \gamma_{MG} + \gamma_{IM} - \gamma_{IG} = \gamma_{MG}(1 - \cos\theta) \tag{11.25}$$

Thus, flotation is promoted by high values of both γ_{MG} (i.e. low S content of steel) and the contact angle, θ. At high temperatures, poor wettability (θ > 90°) is frequently associated with low reactivity; thus, a high contact angle is obtained when there is little reactivity between inclusion and metal.

Once the gas bubble has reached the metal/slag interface it is necessary for it to travel through the interface and on into the slag phase. The spreading coefficient (S*) is the measure of the ability of the liquid (metal or slag) to spread across the solid; it is defined by Eq. 11.26.

$$S^* = \gamma_{IG} - \gamma_{MG} - \gamma_{IG} = \gamma_{MG}(\cos\theta - 1) \tag{11.26}$$

The spreading increases as S* becomes more positive and is favoured by low values of (γ_{MG}) and θ, i.e. when the S content of the metal is high and there is significant reactivity between inclusion and metal.

It is also important that the inclusion should not be entrained and it has been shown [226] and this is favoured when ΔG in Eq. 11.27 is negative. These conditions are favoured by (i) high values of the inclusion surface tension (γ_{IG}) and metal surface tension, (γ_{MG}) (i.e. low S content) and (ii) low interfacial tension, γ_{MI}, (i.e. no adsorption layer or chemical reaction at the interface). Emergence of the inclusion is favourable when $\gamma_{MG} > \gamma_{IM,}$ this is usually the case for oxide inclusions in steels

$$\Delta G = \gamma_{IG} + \gamma_{MG} - \gamma_{IM} \tag{11.27}$$

11.10.4.4 Electromagnetic Devices Used to Reduce Inclusion Levels

EMS, EMBr and EMC are all effective in reducing the inclusion levels in steel. However, each method achieves inclusion reduction in its own individual way.

The Saffman forces induced on the inclusion (by EMS) suppress its entrapment; if the velocity of the Saffmann force exceeds the velocity of the solidification front then the inclusion will be washed away. Saffmann forces increase with increasing inclusion size, so EMS is very effective in removing inclusions >100 μm [50]. The application of EMS has been reported to give two–threefold [215, 227] reductions in alumina clusters (Fig. 11.85b). It can be seen from Fig. 11.86 that for a solidification velocity of 1.2 mm s^{-1}, inclusions of >100 μm will be removed when the steel velocity exceeds 0.3 m s^{-1} [50]. Thus, EMS is an effective method for removing larger inclusions.

Flow control (i.e. EMBr) causes (i) a 5–10 °C increase in meniscus temperature, (ii) a large decrease in vertical heat transfer and (iii) the suppression of turbulence at the slag/metal interface. All of these lead to a shorter meniscus shell and

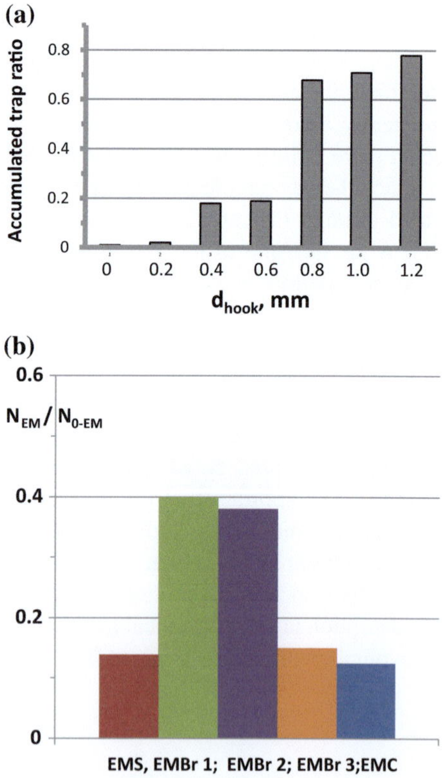

Fig. 11.85 **a** Trap ratio = (number of inclusions trapped under hook/total number of inclusions) as a function of depth of meniscus hook (Drawn using data from [48]), **b** The ratio (N_{EM}/N_{0-EM}), i.e. the ratio of (inclusions where EM devices used divided by inclusions where no EM was used) from the left, EMS (■), EMBr (■;■;■) and EMC (■) (Drawn using data from [17, 33, 48])

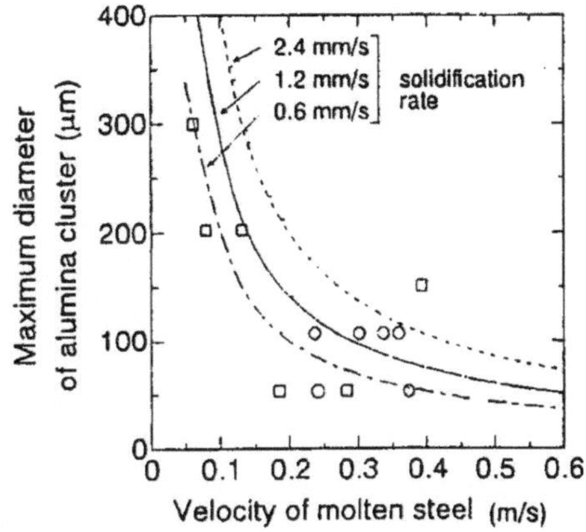

Fig. 11.86 Maximum diameter of inclusion cluster as a function of metal flow velocity for three different solidification rates (with units of 10^{-3} m s^{-1}) *solid line* 1.2; *dotted line* 2.4; *dash-dot line* 0.6 (permission granted, ISIJ, [50])

subsequently, to lower levels of inclusion entrapment. The use of EMBr has been reported to bring about significant reductions in inclusion levels, (e.g. reductions of fivefold [47], 2–3 fold [48] and twofold [13]) shown in Fig. 11.85b.

In EMC, the pinch force generated results in a thicker slag film with a consequent decrease in horizontal heat flux (q_{hor}). This, in turn, results in a short meniscus shell which is displaced a distance from the turbulent, interface area. The use of EMC has been reported to bring fivefold [50, 51] reduction in the inclusion levels (Fig. 11.85b).

11.11 Formation of Scales

Scales are formed by the oxidation of the steel surface [228–239]. Scaling tends to occur below the secondary cooling zone and results in the following:

- Yield loss.
- Reduction of equipment lifetime.
- Entrapped scales or scums which form *slivers* on rolling.

Scales can be formed at various stages of the process and scale formation is usually classified as follows:

- *Primary scale formation* occurs in the secondary cooling zone and scales (up to 2 mm thick) are usually removed [236].
- *Secondary scale formation* occurs during the re-heating and rolling stages and the scale has a thickness of 100 μm [237].
- *Tertiary* scale formation occurs, immediately before, and during, the final strip rolling process and can lead to wear in the rolls.

11.11.1 Factors Affecting Scale Formation

Si-containing steels are difficult to descale because low-melting, *fayalite* is formed on the steel surface and, subsequently, penetrates into the grain boundaries

The principal factors affecting scale formation are as follows:

- Surface temperature of the strand.
- Partial pressure of oxygen (p_{O2}) in the atmosphere (containing H_2O, air, furnace gas).
- The copper content of the steel, when this is >0.2%, it exceeds the Cu solubility level in *austenite* and is preferentially oxidised to form a low-melting, liquid phases which penetrate into the grain boundaries and which lead to cracks [149].
- Nickel additions are usually made to minimise the effect of copper since Ni increases the solubility of Cu in *austenite* and favours Cu-occlusion in the scale [149],
- Other elements (Sn and Sb), in addition to Cu, promote the formation of the oxidised, low-melting liquid; the Cu equivalent is used as a measure of the propensity to form liquid ($Cu_{eq} = \%Cu + n(\%Sn + \%Sb) - \%Ni$), where *n* has a value between 6 and 8). [149].
- The carbon content of the steel (scaling decreases as %C in the steel increases).
- Certain elements (e.g. Al and Cr) form dense oxides which lower the diffusion rates.

There is anecdotal evidence that certain mould slags tend to produce more scale than other slags. This is thought to be due to the fact that certain slag constituents (e.g. Na_2O) reduce the surface tension and cause the slag to wet the steel and hence, adhere strongly to the steel. Wetting is defined by the contact angle (θ) which is the angle between γ_{GM} and γ_{IM} (in Figs. 11.84c, d) and a liquid is considered to *wet* the solid when $\theta < 90°$ (Fig. 11.84c) and is *non-wetting* when $\theta > 90°$ (Fig. 11.84d).

Nakato et al. [219] calculated the adhesion energy (γ_{ad}) from the contact angle (θ), with γ_{ad} increasing as θ decreased (i.e. as wetting of steel (by slag) improves); experiments were carried out with two mould fluxes, A and B, with three different steels. It can be seen from Fig. 11.87a, b that:

- $\gamma_{ad}^{A} > \gamma_{ad}^{B}$ and γ_{ad} decreases as the S content of the steel decreases ($\gamma_m \uparrow$).
- Mould slag A (with high Na_2O%) wets the steel more than mould slag B (with low Na_2O%), i.e. $\theta^A < \theta^B$, which suggests θ decreases with increasing Na_2O%.

An analysis of the number and size of *sliver* defects in cold-rolled sheets showed that the number of macro inclusions was greater for slag A than for slag B (Fig. 11.88).

The constituents with lowest surface tension in mould slag are $B_2O_3 < K_2O < Na_2O < Li_2O$ and these oxides would be expected to promote wetting. The effect of Na_2O on the interfacial tension (γ_{msl}) can be clearly seen in Fig. 11.89.

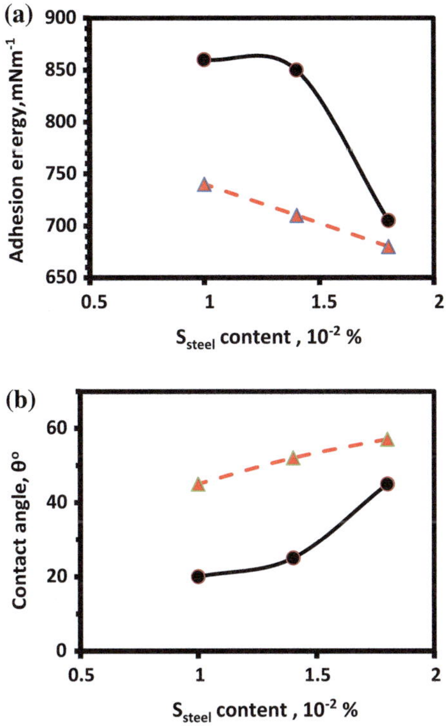

Fig. 11.87 Values of the **a** adhesion energy and **b** the contact angle for two mould fluxes used to cast LCAK and two ULC steels; powder A (=●) and B (▲) showing the effect of Na_2O% (15.2 and 0.7%, respectively) in slag and S content of steel on these properties (drawn using data from [219]); Composition %:powders A(B); SiO_2, 30 (31.4), CaO, 30 (31.1), Al_2O_3 7.5 (5.4); Na_2O 15.2 (0.7); C_{tot} 2.9, (5); Steels LCAK, (ULCAK1) {ULCAK2}; C, 0.041 (0.019); Mn, 0.023 (0.021) {0.021}; P, 0.018 (0.019) {0.015} S, 0.017 (0.013) {0.010}; Al, 0.024 (0.045) {0.065}; Ti, -, (0.032) {0.063}

Fig. 11.88 Diagram showing effect of mould powders A and B on the numbers and sizes of macro inclusions in steel when casting with mould slags A (■) and B (■); Re-drawn from data in [219]; Sliver defects with and {without} conditioning; powder A. 40 {100}; powder B, 0.08, {0.09}

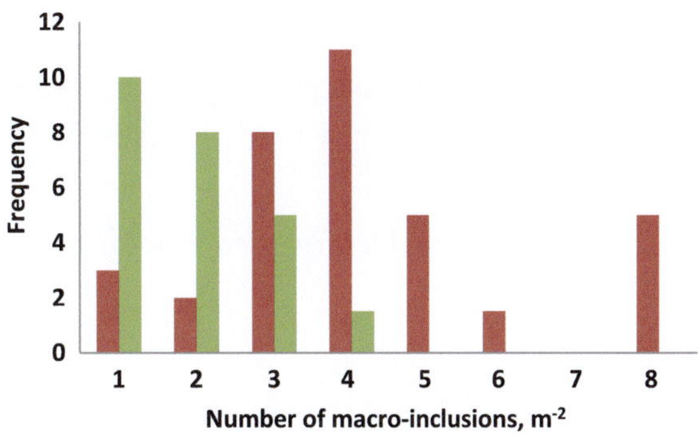

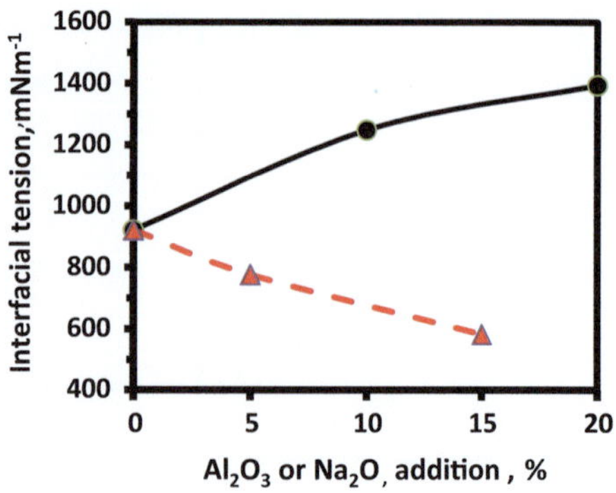

Fig. 11.89 The effect of additions of Al_2O_3 (●) and Na_2O (▲) on the interfacial tension (γ_{msl}) between molten steel and mould slag (permission granted, ISS/AIST, re-drawn from [198, 207])

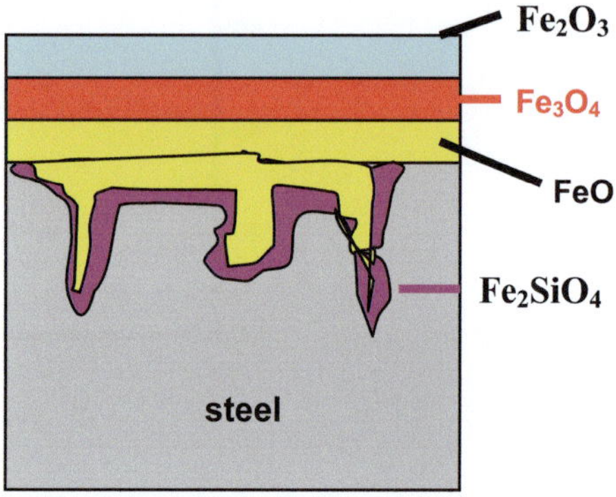

Fig. 11.90 Schematic diagram showing the location of various iron oxides and the penetration of *fayalite* (Fe_2SiO_4 = *purple*) down the grain boundaries

11.11.2 Causes, Mechanisms

Oxidation of the steel surface by oxygen and water vapour in the atmosphere causes the formation of a surface coating of iron oxides. In Si-steels, the diffusion of Fe^{2+} in the Fe_{1-x} O phase (Fig. 11.90), results in the formation of Fe_2SiO_4 (*fayalite*) with a low-melting point, 1177 °C. Molten *fayalite* penetrates into the steel via the grain boundaries and, on cooling, provides "*handholds*" for the scale to adhere to the steel, making it difficult to remove. Mould fluxes, adhering to the steel surface, are another source of SiO_2 and promote the formation of Fe_2SiO_4.

11.11.3 Ways of Dealing with Scaling

1. Use an inert atmosphere.
2. Ensure that the copper and tin contents of steel are low or treat with same mass of Ni to remove the copper from the grain boundaries.
3. Use mould slags with lower K_2O and Na_2O contents to reduce adhesion to the steel surface.

11.12 Carbon Pick-up

11.12.1 Factors Affecting Carbon Pick-up

Carbon pick-up by the steel is a serious problem, especially when casting ULC steel grades [126, 240–243]. There are several potential sources of carbon, namely:

- The bed of powder in the mould and core samples taken from the mould frequently exhibit a carbon-rich layer at the top of the liquid pool which is due to unreacted carbon particles floating at the top of the liquid pool.
- The slag rim which is a mixture of unreacted mould powder and casting slag which results in the formation of "*amorphous graphite*" in the slag rim [243].
- Any broken pieces of Carbon-oxide refractory which is swept into the slag pool.

Carbon pick-up has been found to increase as

- The carbon content of the powder increases (as shown in Fig. 11.91a) [126, 241, 242].
- Mould level variations increase [162].
- The depth of the slag pool decreases (Fig. 11.91b) [126, 219]; these results could also explain Fig. 11.91a because the slag depth is connected to the melting rate, which, in turn, is inversely related to the carbon content in the mould powder (C% $\uparrow \rightarrow Q_{MR} \downarrow \rightarrow d_{pool} \downarrow$).
- The particle size of the carbon particles present, since smaller particles (with a high surface area to mass) tend to combust quicker than larger particles; this results in lower carbon concentrations in the lower bed and the slag pool [244]. Note: smaller carbon particles also slow the melting rate and thus, less carbon is required to obtain the required melting rate.

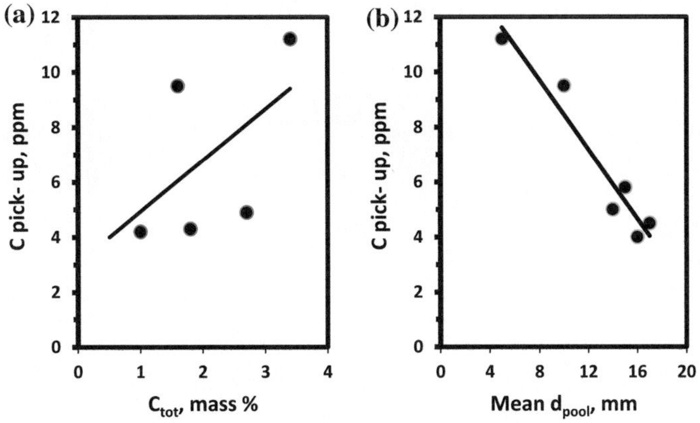

Fig. 11.91 Carbon pick-up by the steel as functions of **a** the initial carbon content of the casting powder and **b** the depth of the slag pool (permission granted, ISS/AIST, re-drawn from [126])

11.12.2 Causes, Mechanisms

Two mechanisms have been proposed for the carbon pick-up:

- The *Painting mechanism* in which the slag rim is considered to be the major source of carbon and where the carbon in the rim "*paints*" carbon onto the shell as the rim moves up and down as shown in Fig. 11.92a, b [242, 245].
- In the alternative mechanism, the metal dissolves carbon because the flow creates a standing wave which allows contact with either the powder bed or the carbon-rich layer of the slag pool and this carbon dissolves in the metal (Fig. 11.92c) [242, 246]. Thus, carbon pick-up by this mechanism would be expected to increase with increasing casting speed since this would lead to taller, standing waves.

High casting speeds lead to the formation of both standing waves and Karmann vortices (Figs. 11.69 and 11.68, respectively). It is probable that these vortices could lead to carbon pick-up by a similar mechanism to that responsible for slag entrapment.

It is apparent that higher carbon contents in the powder will tend to increase (i) the amount of carbon in the slag rim (Painting mechanism), (ii) the amount of carbon in the lower section of the powder bed and (iii) the amount of carbon floating in the slag pool (alternative mechanism). It has been reported that the slag rim is built up by the painting mechanism [247].

The slag pool depth affects both mechanisms. A deep slag pool reduces the area of contact between carbon and the shell in both mechanisms.

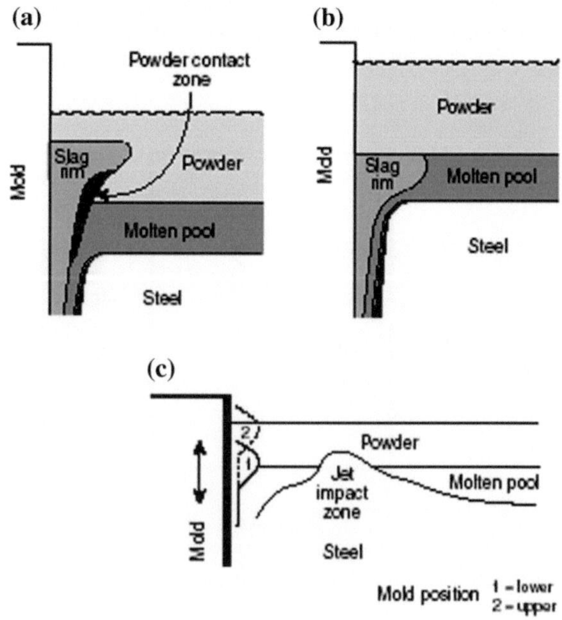

Fig. 11.92 Mechanisms proposed for the recarburisation of the steel in the continuous casting mould **a**, **b** the painting mechanism, **c** the alternative mechanism arising from the standing wave (Permission granted, EPD Sci. [242])

11.12.3 Ways of Dealing with Carbon Pick-up

11.12.3.1 Reducing the Carbon Concentration in the Lower Bed and the Slag Pool

This can be achieved by

- decreasing the overall, carbon content of the casting powder; this should be carried out by reducing the carbon content whilst maintaining the melting rate at the same level (by using smaller carbon particle sizes, e.g. carbon black). However, it should be noted that small carbon particles also combust more rapidly.
- by using fluxes containing MnO to promote the oxidation of carbon particles in the lower mould (but elements such as Al and Ti in steel will react with the MnO to give more oxide inclusions).

11.12.3.2 Increase the Depth of the Slag Pool

By using the methods summarised in Sect. 4.4, namely:

Increase the Melting Rate of the Powder

This can be achieved by (i) reducing the carbon content of the powder, (ii) increasing the particle size of carbon particles in the powder and (iii) increasing the vertical heat transfer (note introducing EMBr would tend to reduce both the metal flow turbulence and the vertical heat transfer).

Increase the Thermal Insulation of the Powder Bed

This can be achieved by (i) increasing the depth of the powder bed, (ii) by using casting powders with smaller granule size or (iii) by introducing exothermic agents into the casting powder.

11.12.3.3 Reduce the Stroke Length and Improve Mould Level Control

This seeks to minimise the area of contact between the carbon-rich zones of the powder bed and slag pool and the shell (Painting mechanism).

11.12.3.4 Reduce Metal Flow Rate to Decrease the Height of the Standing Wave

The height of the standing wave increases with increasing metal flow rate (i.e. increasing turbulence).Where high production (or casting) rates are needed the use of EMBr is usually introduced to minimise the effects of metal flow turbulence.

11.12.3.5 Replace Carbon in Powders with Carbides or Nitrides

The carbon in the mould powder is there to control the melting rate. This task can be carried out by other materials providing they are non-wetting to molten casting powder and combust with air. Casting powders, where the carbon has been replaced with boron nitride, silicon nitride or silicon carbide, have been produced commercially [242, 243, 248]. Most of these powders have proved sensitive to sintering and resulted in the formation of large rims and consequently, have not been widely used.

It should be noted that some of the factors above appear to be acting in a contradictory manner, for instance, decreasing the carbon particle size will reduce, sequentially, the melting rate, the pool depth and increase the carbon pick-up. However, decreasing particle size will also lead to faster combustion and produce lower carbon contents in the bed and pool and thus, reduced carbon pick-up. It would appear that the latter effect is dominant [244]. In the same way, EMBr leads to reduced pool depth and greater carbon pick-up but would result in a lower standing wave and hence a lower carbon pick-up.

11.13 SEN Erosion

Most SENs are fabricated from oxide/graphite refractories (such as Al_2O_3/C, ZrO_2/C, etc.) with inserts of ZrO_2 or ZrO_2/C in the region subjected to extensive wear which is usually referred to as the Z-band. The principal area of erosion lies between the region of the SEN in contact with the slag pool and to the impact zone of the circulating metal flow. The eroded region consists of a deep groove in the SEN. It is usual practice to alter the immersion depth of the SEN during casting to even out the wear in the SEN. Wear can also occur in both the interior of the SEN [49] and in the Z–band.

Erosion rate studies are usually carried out in simulation experiments in which the refractory sample is either immersed or rotated in the liquid (slag and steel) for a known time. The refractory sample is sometimes covered by a mould powder to minimise oxidation of the graphite (Fig. 11.93a). It is important to exclude air during these experiments since oxidation of the carbon in the SEN can have a serious effect on the erosion rates. The erosion rate is measured from either the average depth of the groove or by estimating the volume of the eroded area (i.e. the groove) (Fig. 11.93b) in a known time period.

There are a number of factors affecting SEN erosion:

- The slag phase will dissolve the oxide particles and the molten steel will dissolve the graphite particles.
- The driving force for both of these processes is the concentration difference ($C_{sat} - C_0$) where C is the concentration and the subscripts, *sat* and *0*, denote the saturation limits and the current value, respectively.
- The saturation limits for various oxides in mould slags vary considerably (e.g. Al_2O_3 is ca. 40%, TiO_2 is ca. 10% and ZrO_2 ca. 2%).
- These saturation limits also affect the kinetics of dissolution, for example in low C steels (where $C_{sat} - C_0$ is large) the slowest (rate-determining) step is the dissolution of oxide whereas, in high-C steels (where $C_{sat} - C_0$ is small) the rate-determining step is the dissolution of graphite by the steel [250, 251].
- The velocities of the flow of liquid metal and the concomitant flow of slag on the SEN would also be expected to be significant factors affecting the erosion rate.
- The SEN erosion rate was found to be greater when using EMBr that when not using EMBr [48]; this was attributed to the better mould level control obtained with EMBr which thus reduced the erosion zone and hence increased the depth of erosion [48]. The increased meniscus temperature obtained with EMBr may also be factor. EMS has also been reported to increase SEN erosion rates but this was not attributed to the swirling flow produced by EMS but to slag crawling and Chemical attack of the SEN by the entrained slag [49].

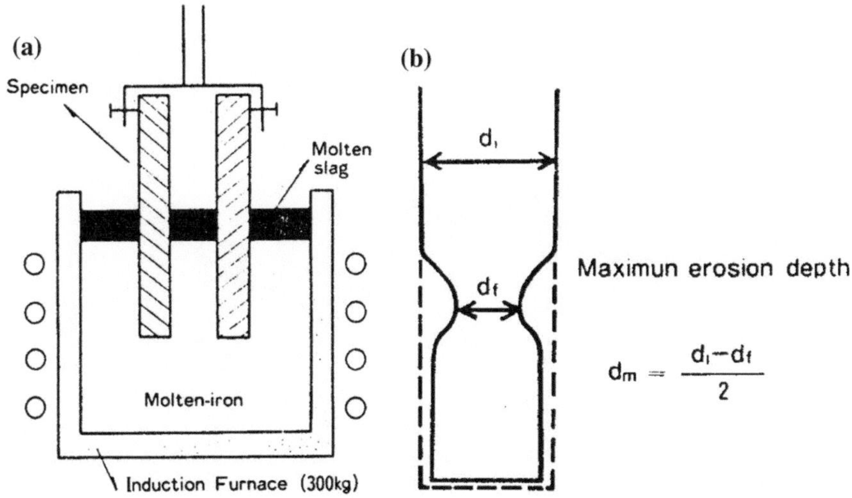

Fig. 11.93 Schematic drawings showing **a** dipping test and **b** measurement of volume of sample eroded (permission granted, ISIJ, [249])

11.13.1 Factors Affecting SEN Erosion Rates

Nakamura et al. [249] and Tsukamoto [252] studied the erosion of ZrO_2/C and $A_{12}O_3$/C refractories, respectively, in simulation tests and found that the erosion rate increased as

- The mould slag viscosity decreased (or as the fluidity increased) and shown in Fig. 11.94a [249, 252]. Plant data also indicate that erosion rates increase as the mould slag viscosity decreases [81].
- The fluoride content of the mould slag increased (Fig. 11.94b) [81, 249, 252]. Several workers have reported that SEN erosion rates were halved when the conventional powder was replaced with a F-free powder [253, 254].
- In the hierarchy of erosion rates for different SEN materials, Baddleyite > MgO-stabilised ZrO_2 > CaO-stabilised ZrO_2 > Yttria-stabilised ZrO_2 [252] (Fig. 11.94c).

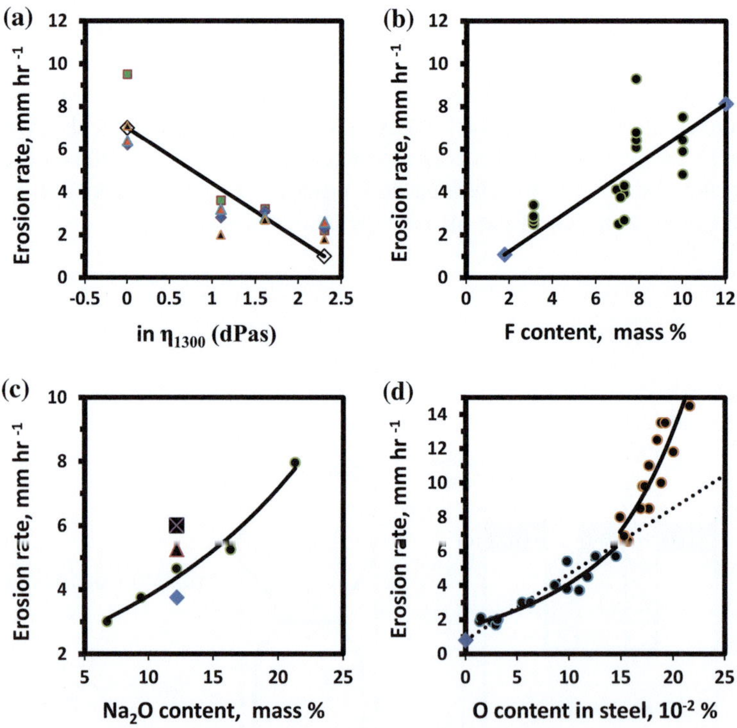

Fig. 11.94 SEN erosion rates of ZrO_2/C refractory as a function of **a** ln $\eta_{1300°C}$ (d Pas) using four mould powders [249], **b** F content of mould slag [249] and **c** Na_2O content of mould slag showing the effect of the ZrO_2 stabiliser (● = stabiliser CaO; *square* = baddelyite; ▲ = MgO; *diamond* = Y_2O_3) on the refractory [252] and **d** as a function of O content of the steel [252] (permission granted, ISIJ, [249, 252]; re-drawn)

- The basicity, CaO/SiO_2 ratio, of the mould slag increased (i.e. increasing fluidity) [249, 252].
- The C content of the refractory increased (or the Al_2O_3 or ZrO_2 content decreased) [252].
- As the *soluble* Oxygen in the steel increases [252] (Fig. 11.94d).
- As MnO content of the slag increases [252].
- As Na_2O content of the slag increases [252] (Fig. 11.94c).

It should be noted that ZrO_2 undergoes a phase transition at high temperature and consequently, ZrO_2 is usually stabilised by addition of CaO, MgO orY_2O_3. It has been reported that fluorine and SiO_2 in the mould slag tend to leach out the stabiliser (e.g. CaO) and hence destabilise the ZrO_2 [81]. SEN erosion has also been found to increase with increasing metal flow velocity.

11.13.2 Causes, Mechanisms

The erosion of refractories is common to many high-temperature processes and is frequently concentrated in the interfacial region in contact with both slag and metal. Such erosion is often referred to as "*slag-line attack*". In traditional oxide refractories, slag *wets* the oxide refractory and forms a meniscus (Fig. 11.95, where the refractory is SiO_2) and the slag, subsequently, dissolves the oxide. If the refractory oxide causes a reduction in surface tension, the surface tension of the slag at the point A will be lower than that at point B (i.e. $\gamma_A < \gamma_B$). This surface tension gradient results in a Marangoni flow (which is always in the direction of *low* γ *to high* γ, i.e. along liquid surface from A to B). This creates a clockwise vortex which continually erodes the refractory. If the dissolution of oxide increases the surface tension of the slag then $\gamma_A > \gamma_B$ and the Marangoni flow will be in the direction of B to A resulting in the formation of a vortex moving with an anticlockwise motion.

Several investigators have suggested that Marangoni flow is involved in refractory erosion [250, 251, 255]. Hauck and Potschke [255] proposed that the erosion takes place in two steps:

- Dissolution of carbon in the molten steel at the metal surface,
- Dissolution of the carbon in the slag (this would seem improbable since the saturated limit for C in most slags is ca. 0.2%).

It was suggested that these steps resulted in the changes in surface tension, and the Marangoni forces caused the formation of two vortices, at the metal /slag and slag/air interfaces. It was postulated that these two erosion zones gradually expanded with time and eventually, merged.

In an alternative mechanism based on erosion rate measurements with Al_2O_3/C and ZrO_2/C refractories [256], it was proposed that oxide dissolution into the slag was the rate-determining step; C-pick-up by the metal was considered as a

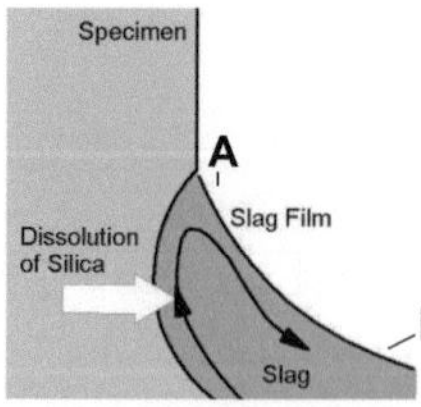

Fig. 11.95 Marangoni convection of slag film in the local corrosion zone of a SiO_2(s)–(FeO–SiO_2) slag system showing typical "slag-line" attack (permission granted, IOM/Taylor & Francis, [250])

qualitative measure of refractory attack [256]. Their results supported the view that ZrO_2 increased the viscosity of the slag and created an inert protective layer [256].

Mukai [250, 251] proposed another mechanism to explain the erosion of oxide/C refractories. The molten metal dissolves carbon and the slag dissolves oxides. At high temperatures, most reactions (like dissolution) are accompanied by "*reactive wetting*". Mukai [250, 251] proposed that refractory erosion of oxide/C refractories involved a cyclical two-stage process. Starting with the situation where the dissolution of carbon by the metal has left the refractory surface predominantly covered by oxide (Fig. 11.96a). Since oxide is non-wetting to most metals but reacts with slags, the interfacial conditions favour the covering of the surface by slag, so the metal retreats and the refractory surface is covered by slag. The slag then dissolves the oxide particles in the refractory until graphite particles mainly occupy the surface sites (Fig. 11.96b). Since slag is non-wetting on graphite, the interfacial conditions now favour the covering of the refractory by metal, so the slag retreats and metal covers the refractory. Then graphite dissolves in the metal until the surface is predominantly covered by oxide and the whole process starts over again. Thus, dissolution is a cyclical process.

The above mechanisms apply to steady state conditions. However, in continuous casting the metal flow is both vigorous and fluctuating and the metal flow causes drag forces in the slag pool. The velocities of these flows would be expected to make a significant contribution to the erosion process.

11.13.3 Ways of Dealing with SEN Erosion

11.13.3.1 Reduce the Metal Flow Viscosity and the Concomitant Flow in the Slag Pool

This can be achieved in the following ways:

- By optimising the SEN immersion depth and port design to minimise the metal flow velocity and turbulence for a given casting speed [49].

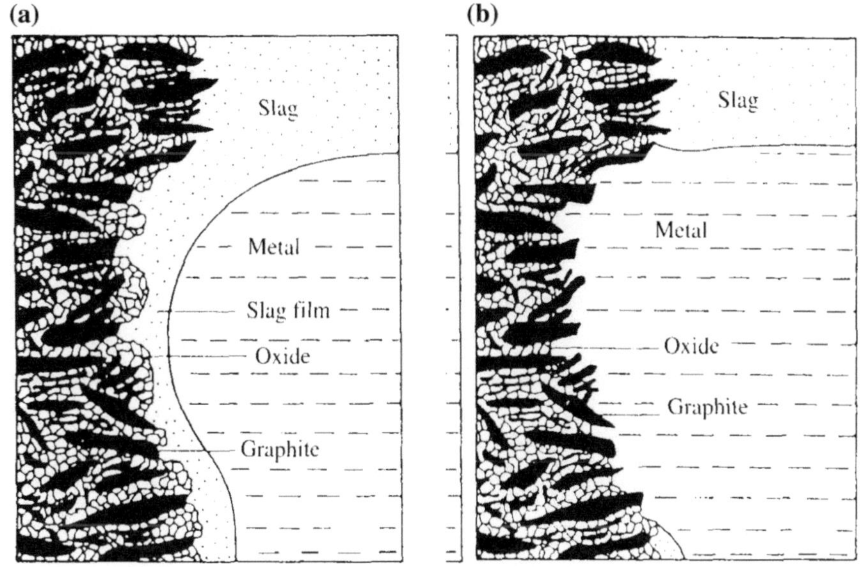

Fig. 11.96 Schematic diagrams illustrating the proposed mechanism for oxide/C refractories where refractory surface is principally covered by **a** Oxide (*white*) and **b** Graphite (*black*) (permission granted, IOM/Taylor & Francis, [250])

- By applying EMBr or EMS to reduce the metal flow velocity; however, plant data seem to indicate that EMBr and EMS seem to lead to slightly higher erosion rates.
- Use a high viscosity mould powder to restrict the flow velocity in the slag pool (although this will also reduce powder consumption which is not desirable).

11.13.3.2 Reduce the Driving Force for Dissolution of Refractory Oxide

Use a mould powder containing 2% ZrO_2 to minimise the driving force for ZrO_2 dissolution ($C_{sat} - C_0$) since C_{sat} for most mould slags is about 2% ZrO_2 and thus ($C_{sa} - C_0$) ≈ 0.

11.13.3.3 Reduce the Fluoride Content of the Mould Flux

It was reported that the replacement of a conventional powder with a fluoride-free powder reduced the rate of SEN erosion [253] to half of that for a conventional, F-containing, slag.

11.13.3.4 Increasing the Oxide Content of the SEN Refractory

Increasing the ZrO_2 content of the refractory reduces the rate of SEN erosion [249, 252, 255, 256].

11.14 Fluorine Emissions

Casting powders consist of various oxides and most contain fluorides. These oxyfluorides are basically unstable on heating since there are reactions between oxides and fluorides at higher temperatures producing gaseous fluorides or HF (g). These fluoride emissions pose several problems since

- They present a potential health hazard to plant operators [257].
- They change the composition of the mould powder (note these changes tend to be much larger in laboratory experiments than in plant operation since the ratio [surface area/mass (or volume)] is much larger in the laboratory experiments).
- The acidic emissions can cause corrosion to the plant.
- The F-emissions acidify the secondary cooling water which too can cause plant corrosion [258, 259] and alkaline agents are added to the cooling water.

11.14.1 Factors Affecting Fluoride Emissions

When mould powders are heated they undergo a series of reactions occurring between the components of the powders producing, for example [257]:

$$Na_2O + CaF_2 = CaO + 2NaF\,(g) \tag{11.28}$$

$$SiO_2 + 2CaF_2 = 2CaO + SiF_4\,(g) \tag{11.29}$$

$$Al_2O_3 + 3\,CaF_2 = 3CaO + 2AlF_3\,(g). \tag{11.30}$$

These gaseous fluorides react with any moisture present in the atmosphere to form

HF (g) for example:

$$H_2O\,(g) + 2NaF\,(g) = CaO + 2HF\,(g). \tag{11.33}$$

The various gaseous fluorides emitted have been investigated by several workers [257–261]. It was noted that NaF(g) and KF(g) were emitted at temperatures above

600 °C and the pressure of these fluorides increases with increasing temperature [207]. Thermodynamic calculations indicate that the vapour pressure of SiF_4(g) at high temperature can exceed 0.5 atm [207, 260–262]. The kinetics of these fluoride emissions has also been studied [263].

It has also been observed that fluoride emissions increased as

- The fluoride content of the mould slag increased. [257, 264].
- The CaO and MgO content decreased. [257, 264].
- The SiO_2, Na_2O and TiO_2 contents increased [264].
- As CaF_2 was replaced with *cryolite* (Na_3AlF_6) (causing a 10-fold increase in emissions) [257].
- As the basicity decreased [257].

All of the above findings are consistent with recent observations on the effect of CaF_2 on slag structure, namely, that F^- ions tend to bond preferentially with cations of high-field strength (z/r^2) like Mg^{2+} and Ca^{2+} and do not tend to bond with other components in the slag, e.g. Si^{4+}. High p_{SiF4} values arise because of the interaction of SiO_2 (which is present in large amounts) with the fluorides present. Prefusing the slag reduces the chemical activities of the various fluorides in the mould slag and thereby reduces the partial pressures of the gaseous fluorides.

11.14.2 Ways of Dealing with Fluoride Emissions

Fluoride emissions can be reduced by

- Reducing the amount of fluorine in the slag.
- Using F-free slags which would have the additional benefit of reducing SEN-wear [253, 254].
- Using a pre-fused casting powder which reduces the chemical activity of the CaF_2 (a_{CaF2}) and hence, reduces the vapour pressures of fluoride species [257].
- Use an alkaline additive to the cooling water to reduce its acidification [258].

11.15 Summary

The causes of the various defects and process problems, the strategies adopted and suggested remedial treatments are all summarised in Table 11.6 for each type of defect.

Table 11.6 Summary of the causes, strategies and treatments used to overcome defects and process problems in continuous casting. *TSL* thermal shrinkage coefficient

Defect/problem	Cause	Strategy	Treatment
Longitudinal cracking [6–10, 12–16]	4% difference in TSL for ∂-Fe and γ-Fe especially, MC steels→ Variations in shell thickness → "hoop stresses" → relieved by cracking	Produce thin uniform shell by reducing q_{hor} in meniscus region, i.e. create a thick (with high T_{br}) crystalline slag film	Reduce q_{hor} by creating a thick, crystalline slag film –use flux with high T_{br} and (C/S) $\geq$ 1.3
Longitudinal corner cracking [42, 60–64]	Local variations in shell thickness 1. Billets: over-cooled corners 2. Slabs: Metal flow causes shell thinning in corners [50]	Produce thin uniform shell	1. Use flux with high T_{br} and forms crystalline slag film 2. Redesign SEN ports use lower casting speed
Sticker breakouts [67, 68, 70, 72, 73, 81, 83, 85, 86, 90]	Various causes all involve forming of a thin, weak shell 1. Large agglomerates blocking slag infiltration plus Carbon pick-up → low-melting shell—does not freeze in t_p 2. H_2 in slag film [60]	Produce thick stronger shell by increasing q_{hor}	Use a glassy slag with low T_{br} with (C/S) $\leq$ 0.9 Ensure % H_2O < 0.5% and no water leaks
Oscillation marks (OMs) [43, 55, 103, 107, 109, 111, 112] [105, 113]	Natural event due to tide changes in slag flow. (Sect. 5.3)	Decrease d_{OM} by 1. Reducing t_n; 2. increase $d_{rim/tip}$	1. Reduce t_n by $\uparrow f$; $\uparrow V_c$:$\downarrow s$ Increase slag viscosity 2. Delay solidification by $\uparrow$superheat
Transverse cracks [45, 139, 145–147, 149]	1. Due to incorrect taper for steel 2. Associated with deep OM's	Ensure good powder consumption [29] Reducing q_{hor} in meniscus region, i.e. create a thick (with high T_{br}) crystalline slag film	Ensure correct taper for steel [29] Reduce d_{OM} → $\downarrow t_n$ $\downarrow s$
Star cracking [156–158]	Loss of liquid slag in lower mould → spalling of solid slag film and variable heat flux [51]	Create liquid slag film through mould	1. Increase casting speed and superheat 2. Decrease water flow rate
Depressions 1. Longitudinal [129–131] 2. Transverse [160]	Poor mould level control + formation of latge slag rim → overflow and fracture and capture of rim	Allow longer time before replacing starter flux with conventional flux [111]	Improve mould level control and use a lower viscosity flux

(continued)

Table 11.6 (continued)

Defect/problem	Cause	Strategy	Treatment
Slag entrapment [163–166, 179, 183, 191, 197, 198] Slivers [78, 211, 213, 219–221] Gas entrapment *Pinholes* (LC, ULC steels) [46, 50, 51, 213, 214] "*Pencil pipe*" [198, 212]	1. Slag entrapment Turbulence in metal and slag flows Al_2O_3 formed by reaction of Al in steel with FeO in slag-trapped by shell Gas trapped by meniscus-metal flow takes bubble too far down-inclusion attaches to bubble and is not welded shut in rolling	a. Decrease metal flow velocity (e.g. by EMBr) b. Increase slag viscosity Minimise slag carry-over from ladle Reduce the length of meniscus hook [28] Optimise metal flow Delay solidification	Optimise SEN depth, port design and Ar flow rate Use EMBr Minimise slag carry-over and decrease throughput Thicker bed, exothermic agents-Optimise SEN depth, port design and Ar flow rate, EMBr
SEN erosion [249, 250, 252, 253]	ZrO_2 has 2% solubility in slag; C dissolves in steel Metal (& slag) flow impact on SEN	Possible to: Saturate slag with 2% ZrO_2 Increase viscosity of slag	Reduce metal and slag flows —Use F-free powders
Carbon pick-up Especially in LC, ULC grades [242–244, 247, 248]	1. Contaminated by C in slag rim 2. Steel poking through into powder bed	1. Minimise standing wave 2. Keep a deep slag pool	1. Reduce C in flux 2. Replace C partially with SiN
Fluoride emissions [207, 257, 258, 263, 264]	Reactions of oxides and fluorides at higher temperatures	1. Reduce F-in powder 2. Use F-free powders 3. Use pefused powders	Add alkaline additive to cooling water

References

1. NS Hunter, JD Madill, DJ Coones, PN Hewitt, D Stewart, Steel Technol. Intl., **1996**, 171, (1996).
2. D Stewart, PN Hewitt, L Peeters, Proc. 79th Steelmaking Conf. 1996, (ISS, Warrendale, PA, 1996) p. 207.
3. Y Maehara, K Yasumoto, Y Sugitani, K Gunji, Trans. ISIJ, **25**, 1045, (1985).
4. E Schmidtmann, L Pluegel, Archiv. f. Eisenhuttenw., **51**, 49, (1980).
5. M Wolf, Proc. METEC Congress '94, 2nd Europ. Conf. Continuous Casting, Dusseldorf, 1994, (VDEh, Dusseldorf, 1994) vol. 1 p. 78 see also Proc. 81st Steelmaking Conf., 1998, (ISS, Warrendale, PA, 1998) p. 53.
6. S Hiraki, K Nakajima, T Murakami, T Kanazawa, Proc. 77th Steelmaking Conf., Chicago, 1994, (ISS, Warrendale, PA, 1994) p. 397.
7. M Kawamoto, T Kanazaka, S Hiraki, S Kumakura, Proc. 5th Intl. Conf. Molten slags, salts and fluxes, Sydney, 1997, (ISS, Warrendale, PA, 1997) p. 777.
8. D Springorum, Steel Times, **1969** (Nov), 727, (1969).
9. K. Nakajima, S Hiraki, T Kanazawa, T Murakami, CAMP-ISIJ, **5**, 1221, (1992).

10. KC Mills, AB Fox, R P Thackray, Z Li, Proc. 7th Intl. Conf. Molten slags, fluxes and salts, Capetown, 2004, (SAIMM, Johannesburg, 2004) p. 713.
11. H Uchiyama, Paper prepared for AISI Technical Committee on Strand casting, Sept., 1985.
12. HP Narzt, R Martinelli, C Furst, G Xia, H Pressinger, Proc. 3rd Europ. Conf. Continuous casting, Madrid, 1998, (UNESID, Madrid, 1998) p. 747.
13. SG Kollberg, HR Hackl, PJ Hanley, Iron and Steel Engineer, **73**, (7), 24, (1996).
14. S Umeda, Y Tsukaguchi, H Mikai, Y Hitomi, M Kawamoto, CAMP- ISIJ, **7**, 302, (1994).
15. CA Dacker, T Sohlgren, Steel Research Intl., **81**, 899, (2010).
16. K Brimacombe, F Weinberg, EB Hawbolt, Met. Trans. B, **10B**, 279, (1979).
17. RJ Gray, A Perkins, B Walker, Proc. Intl. Conf. on Soldification, Sheffield, 1977, (Metals Soc., London, 1977) p. 300.
18. T Narita, T Nozaki, T Mori, H Yasunaka, T Fujimoto, T Ohnishi, Trans. ISIJ, **21**, B 159, (1981).
19. Y Ito, I Shimoda, M Suzuki, Y Miki, Proc. 4th Intl. Congress Science, Technol. Steelmaking, Gifu City, 2008, (ISIJ, Tokyo, 2008) p. 698.
20. MJ Lu, JH Chen, DH Tsai, CH Huang, Proc. 78th Steelmaking Conf., 1995, (ISS, Warrendale, PA, 1995) p. 359.
21. C Bertolletti, P Courbe, JM Jolivet, PP Naveau, A Oper, E Perrin, C Salaris, AL Spierings, E Wiesseldinger, Proc. 3rd Europ. Conf. Continuous Casting, Madrid, 1998, (UNESID, Madrid, 1998) p. 65.
22. P Courbe, JM Jolivet, G Lesoult, E Perrin, P Nilles, Steel Research, **70** (8/9), 338, (1999).
23. M M'Hamdi. G Lesoult, E Perrin, JM Jolivet, ISIJ Intl., **36**, S 197, (1996).
24. K Nakai, T Sakashita, M Hashio, M Kawasaki, K Nakajima, Y Sugitani, Tetsu-to-Hagane, **73**, (3), 498, (1987).
25. Y Sugitani, M Nakamura, M Okuda, M Kawasaki, K Nakajima,Trans. ISIJ, **25**, B 91, (1985).
26. JW Cho, HT Jeong, Met. Mater. Trans. B, **44B**, 146, (2013).
27. H Murukami, M Suzuki, T Kitagawa, S Miyahawa, Tetsu-to- Hagane, **78**, 105, (1992).
28. T Mochida, S Kohriyama, S Itoyama, T Yamashita, M Suzuki, Y Kishimoto, CAMP-ISIJ, **20**, 851, (2007).
29. K Ichikawa, A Morita, Y Kawabe, Shinagawa Tech. Report, **36**, 283, (1993).
30. K Watanabe, M Suzuki, K Murakumi, H Kondo, A Miyamoto, T Shiomi, Tetsu-to-Hagane, **83**, 115, (1997).
31. V Guyot, JF Martin, A Ruelle, A d'Anselme, JP Radot, M Bobadilla, JV Lavant, JN Pontoire, ISIJ Intl., **36**, S 227, (1996).
32. A Yamauchi, S Itoyama, Y Kishimoto, H Tozawa, K Sorimachi, ISIJ Intl., **42**, 1094, (2002).
33. T Watanabe, CAMP- ISIJ, **14**, 153, (2001).
34. A Cristallini, R Langella, A Polk, Proc. 3rd Europ. Conf. Continuous Casting, Madrid, 1998, (UNESID, Madrid, 1998) p. 589.
35. A Delhalle, M Larrecq, J Petegnief, JP Radot, 4th Intl Conf. Continuous Casting, Brussels, 1988, (VDEh, Dusseldorf, 1988) Preprints 1, p. 38.
36. T Tsai, H Yasunaka, M Kogita, T Inoue, K Kan, CAMP-ISIJ, **7**, 1149, (1994).
37. M Ozgu, MB Perrini GP Buckley, Proc. 79th Steelmaking Conf., 1996, (ISS Warrendale, PA, 1996) p. 195.
38. C-Å Däcker, A Salwén, P Andersson, C Eggertsson, Proc. 7th Europ. Continuous Casting Conf., Düsseldorf, 2011, (VDEh, Dusseldorf, 2011). Session 12.
39. M Kawamoto, T. M u r a k a m i, M. H a n a o, H. K i k u c h i, T. Wa t a n a b e, Proc. 6th Intl. Conf. Molten slags, fluxes and salts, Stockholm, paper 146, (2000).
40. M Kawamoto, Paper presented at Intl. workshop on " *Thermophysical Data for development of mathematical models of solidification*", Gifu City, Japan, (1995).
41. M Suzuki, S Miyahara, T Kitagawa, S Uchida, K Okim, Tetsu- to Hagane, **78**, 113, (1992).
42. K Sorimachi M Kuga, K Marumoto, T Koshikawa, K Hamagami, H Kitaoka, Trans. ISIJ, **22** (2), B 17, (1982).
43. C-Å Däcker, T. Sohlgren, Proc. 5th Europ. Cont. Casting Conf., Nice, 2005, pp. 10–17.

44. K Tsukaguchi, S Ura, A Shiraishi, Y Hitomi, T Nagahata, Steel Technol. Intl., **6**, 175, (1995/6).
45. S Ogibayashi, Paper Intl. Workshop on " *Thermophysical Data for development of mathematical models of solidification*", Gifu City, Japan, (1995).
46. E Takeuchi, H Harada, H Tanaka, T Ishii, T Toh, M Zeze, M Hojo, K Shigematsu, Nippon Steel Technical Report**, 61**, 29, (1994).
47. M Washio, M Sugizawa, S Moriwaki, K Kariya, S Idogawa, S Takeuchi, Revue de Metallurgie, CIT, **90**, (April), 507, (1993).
48. H Take, H Osanai, J Hasunuma, T Yamamoto, H Bada, H Tozawa, Proc. Conf. "Quality of ordinary steel in Iron- and steel-making", Bangkok, (1994) Session 3, Paper 1.
49. E Favre, S Kunstreich, MC Nove, D Rotelec, W Courths, E Korte, Proc. 3rd Intl Conf. Continuous Casting, Madrid, 1998, (UNESID, Madrid, 1998) p. 595.
50. T Matsumiya, ISIJ Intl., **46**, 1800, (2006).
51. M Tani, T Toh, K Umetsu, K Tanaka, M Zeze, K Tsunenari, K Hayashi, S Fukunaga, Nippon Steel Technical Report, **104**, 62, (2013).
52. M Hanao, M Kawamoto, M Hara, T Murakami, H Kikuchi, A Yamanaka, Proc. 5th Europ. Conf. Cont, Casting, Nice, 2005 (La Revue Metall., Paris, 2005), see also Tetsu-to Hagane, **88**, (1), 23, (2002).
53. D Gottlelf, P Andrezjewski, E Julius, H Haubrich, Proc. 3rd Europ. Conf. Continuous Casting, Madrid, 1998, (UNESID, Madrid, 1998) p. 825.
54. P E Ramirez-Lopez, P.D. Lee, K.C. Mills, ISIJ Intl., **50**, (3), 425, (2010).
55. PE Ramirez-Lopez, KC Mills, PD Lee, B Santillana, Met. Mater. Trans., B, **43B**, 109 (2012).
56. SN Singh, KE Blazek, J. Metals, **26**, (10), 17, (1974).
57. C-Å Däcker, T Sohlgren, Steel Research Intl., **10**, 899, (2010).
58. AS Normanton, PN Hewitt, NS Hunter, B Harris, D Scoones, Proc. 4th Europ. Conf. Continuous casting, Birmingham, 2002, (IOM, London, 2002) p. 561.
59. T Emi, Proc. 74th Steelmaking Conf., Washington, DC, 1991 (ISS, Warrendale, PA, 1991), p. 623.
60. JK Park, BG Thomas, IV Samarasekera, Ironmaking and Steelmaking, **29**, 359, (2002).
61. H Tai, M Morashita, T Miyake, Proc. 3rd Europ. Conf. Continuous Casting, Madrid, 1998, (UNESID, Madrid, 1998) p. 447.
62. T Miyake, K Nakayama, M Moroshita, H Tai, Kobelco Technol. Review, **21**, (Apr), 7, (1998).
63. E Kivela, Revue de Metallurgie, CIT, **91**, (Jan), 115, (1994).
64. CS Li, BG Thomas, Proc. ISS Tech Conf., Indianapolis, 2003, (ISS, Warrendale, PA, 2003) p. 685 see also Met. Mater. Trans. **35B**, 1151, (2004).
65. R Nishimachi, Y Ogura, SEAISI Quarterly, **25**, (Oct.), 44, (1996).
66. M Miyazaki, CAMP- ISIJ, **20**, 866, (2007).
67. A. Tsuneoka, W. Ohashi, S. Ishitobi, Proc. 68th Steelmaking Conf., 1985, (ISS, Warrendale, PA, 1985) p. 3.
68. Y Liu, XD Wang, M Yao, AB Zhang, H Ma, Z Wang, JC Ma, X Wang, GQ Shi, Ironmaking and Steelmaking, **41**, 748, (2014).
69. MM Wolf, Iron and Steelmaker, **23**, (2), 47, (1996).
70. KC Mills, TJ Billany, AS Normanton, B Walker, P Grieveson, Ironmaking and Steelmaking, **18**, 253, (1991).
71. J Kromhout, RS Schimmel, Proc. 8th Europ. Conf. Continuous casting, Graz, Austria, 2014, (Austrian Metals Soc., 2004).
72. M Hanao, Y Tsukaguchi, M Kawamoto, Proc. 4th Intl. Cong. Sci. Technol. Steelmaking, Gifu, p. 694 (ISIJ, Tokyo. 2008).
73. YA Meng, BG Thomas, Met. Mater. Trans. B, **34B**, 707, (2003).
74. JA Kromhout, PhD Thesis "*Mould powders for the high speed continuous casting of steel*" Univ. of Delft, (2011).
75. Q Wang Y. Lu, S. He, K.C. Mills, Z.S. Li, Ironmaking and Steelmaking, **38**, 297, (2011).

76. T Mukongo, *Effect of titanium pick up on mould flux viscosity in continuous casting of Ti-stabilised steel.* MSc Thesis, Univ. Pretoria: (2003) see also " Proc. ISSTech. Conf., 2003, (ISS, Warrendale, PA, 2003) p. 675.
77. T Mukongo, C Pistorius, A Garbers-Craig, Ironmaking Steelmaking, **31**, 135, (2004).
78. RC Nunnington, N Sutcliffe, Proc. 59th Electric Furn. Conf., Pheonix, AZ, Nov. (2001), (AIST, Warrendale, PA) p. 361.
79. T Kishi, H Takeuchi, M Yamamiya, H Tsuboi, T Nakano, T Ando, Nippon Steel Technical Report, **34**, 11, (1987).
80. RJ Phillips, WF Salem, WR Emling, HD Baker, Proc. 73rd Steelmaking Conf., 1990, (ISS, Warrendale, PA, 1990) p. 247.
81. DW Bruce, NS Hunter, Proc. 2nd Europ. Conf. Continuous Casting, Dusseldorf, 1994 (VDEh, Dusseldorf, 1994) p. 156.
82. K Tokiwa *et al.*, Tetsu-to-Hagane, **69**, S 1034, (1983).
83. T. Mizukami, K Kawakami, S Miyahara, M Suzuki, T Kitagawa, O Terada, Trans. ISIJ, **25**, B 300, (1985).
84. T. Mizukami, A Ozeki, A Kurabayashi, N Hsebe, S Uchida, T Kitagawa, Trans ISIJ, **25**, B 301, (1985).
85. M Mukai, K Yamaguchi, S Ogibayashi, Trans ISIJ, **26,** (4), B 163, (1986).
86. JN Pontoire, JP Radot, V Delvaux, E Dehaussy, Revue de Metallurgie, CIT, **93**, 1237, (1996).
87. Y Ueshima, T Mizoguchi, T Kajitani: Proc. 9th Intl. Conf. Molten slags, fluxes and salts (Molten 12), Beijing, 2012, (Chinese Metals Soc., Beijing, 2012).
88. H Yamamura, T Kajitani, J Nakashima, M Yamasaki, S Mineta, Nippon Steel Technical Report, **104**, 54, (2013), see also T. Kajitani et al. ISIJ Intl. **48**, 1215, (2008).
89. S Ogibayashi, K Yamaguchi, T Mukat, T Takahashi, Y Mimura, K Koyama, Y Nagano, T Nagano., Nippon Steel Technical Rept, **34**, 1, (1987).
90. K Sorimachi, M Kuga, H Nishikawa, K Marumoto, H Nakato, Trans ISIJ, **22** (4), B 435, (1982).
91. H Nakato, T Sakuraya, T Nozaki, T Emi, H Nikoshawa, Proc. 69 th Steelmaking Conf., Washington, DC, 1986, p. 23.
92. H Kyoden, T Diohara, O Nomura, *Mould Powders for Continuous Casting and Bottom Pour Teeming* (ISS, Warrendale, PA, 1987) p. 45.
93. P Zasawoski, DJ Sosinski, Proc. 73rd Steelmaking Conf., 1990, (ISS, Warrendale, PA, 1990) p. 253.
94. SY Kim, J Choi, Proc. 4th Europ. Conf. Continuous casting, Birmingham, 2002, (IOM, London, 2002) p. 594.
95. XD Wang, M Yao, X Chen, ISIJ Intl., **48**, 1047, (2008).
96. CH Moon, DM Lee, SC Moon, HD Park, ISIJ Intl., **48**, 48, (2008).
97. S Tanaka, H Misumi, S Mizoguchi, H Horiguchi, Trans. ISIJ, **21**, B 350, (1981).
98. H Takeuchi, S Matsumara, R Hidaka, Y Nagano, Y Suzuki, Trans. ISIJ, **22**, B 16, (1982).
99. H Jacobi, Arch. Eisenhüttenwesen, **47**, 441, (1976).
100. M Tomono, W Kurz, W Heinemann, Met. Trans. B, **12B**, 409, (1981).
101. K.C. Mills, P.E. Ramirez-Lopez, P.D. Lee, Sano Symp., Tokyo, 2008, (Univ. Tokyo, 2008) p. 158.
102. C-Å Däcker, Research Report "*Case study-mould powder*" IM -2003-534, Swerea KIMAB, Stockholm, 2003.
103. PD Lee, R Ramirez-Lopez, KC Mills, B Santillana: Ironmaking and Steelmaking, **39**, 244, (2012).
104. E Takeuchi, J K Brimacombe, Met. Trans. B, **15B**, 493, (1984).
105. R. Aigner, H. Steinruck, Eurotherm 82, Numerical Heat Transfer, ed. R.A. Białecki and A. Nowak, Gliwice–Cracow, Poland, 2005.
106. JM Hill, YH Wu, B Witwatanapataphee, J. Eng. Math., **36**, 311, (1999).
107. S Itoyama, CAMP-ISIJ, **5**, 1225, (1992).
108. S Itoyama, Curr. Adv. Mater. Proc. (Japan), CAMP-ISIJ, **5**, 1225, (1992).

109. S Itoyama, H Tozawa, K Sorimachi, Proc. 77 th Steelmaking Conf. (ISS, Warrendale, PA, 1994) p. 365.
110. T Fastner, C Furst, HP Narzt, G Xia, G Zuba, Proc. 3 rd Intl. Conf. Continuous Casting, Madrid, 1998, (UNESID, Madrid,1998) p. 791.
111. K Schwerdtfeger, Hong Sha, Met. Trans. B, **31B**, 813, (2000).
112. S Nebeshima, Y Itoh, H Tozawa, H Nakato, K Sorimachi, Proc. 4th Intl. Conf. Solidification Processing, Sheffield, 1997 (Sheffield Univ., 1997) p. 10.
113. H.J. Shin, S.H. Kim, B.G. Thomas, G.G. Lee, J.M. Park, J. Sengupta, ISIJ Intl., **46**, 1635, (2006).
114. K Shimizu, AW Cramb, Iron and Steel Maker, **2002**, (6), 43, (2002).
115. Y Murunaka *et al,* 102nd ISIJ Meeting, November 1981, S 905.
116. OD Kwon, J Choi, IR Lee, JW Kim, KH Moon, YK Shin, Proc. 74th Steelmaking Conf., 1991, (ISS, Warrendale, PA, 1991) p. 561.
117. RB Mahapatra, K Brimacombe, IV Samarasekera. Met. Trans. B, **22 B**, 875, (1991).
118. G.G. Lee, H. Shin, S.-H. Kim, S.-K. Kim, W.-Y. Choi, B.G. Thomas, Ironmaking Steelmaking, **36** (1), 40, (2009).
119. H Steinruch, C Rudischer, W Schneider, Proc. Conf. Modelling Casting, Welding and Advanced Solidification processes VIII (MCWASP) (Minerals, Metals and Materials Soc., 1998).
120. H Yasunaka, T Mori, H Nakata, T Mori, F Kamai, S Narada, Trans. ISIJ, **25**, B90, (1985).
121. M Suzuki, T Kitagawa, S Uchida, K Ozawa, T Mori, Trans. ISIJ, **26**, B164, (1986).
122. F Neumann, J Neal, MA Pedroza, AH Castillejo, FA Acostra, Proc. 79th Steelmaking Conf. (ISS, Warrendale, PA., 1996) p. 249.
123. D W van der Plas, C Platvoet, B Diesesme, JP Radot, JM Galpin, Proc. 2nd Europ. Conf. Continuous casting, Dusseldorf, 1994, Metec Congress'94 (VDEh, Dusseldorf, 1994) p. 109.
124. R Koldewijn Unpublished Corus Internal Rept (2007) cited in KC Mills, J Kromhout, A Hamoen, R Boom, Proc. Admet Conf, Dnipetrovsk, 2007, (Natl. Metall. Acad. Ukr., Dnipropetrovsk) vol 2, p. 174.
125. T. Toh, E. Takeuchi, M. Hojo, H. Kawai, S. Matsumura, ISIJ Intl, **37**, 1112, (1997).
126. G. Skoczylas, Proc. 79th Steelmaking Conf., 1996, (ISS, Warrendale, PA, 1996) p. 269.
127. Y Fukuda, H Kawai, M Okimori, M Hojo, S Tanaka, Proc. 5th Intl. Conf. Slags, Fluxes and molten slags, Sydney, 1997, (ISS, Warrendale, PA., 1997) p. 791.
128. H Fredrikson, J Elfsberg, J Scand. Metall., **31**, 292, (2002).
129. JW Kim, S. K. Kim, D. S. Kim, Y. D. Lee, J. I. Eum, E. S. Lee, Proc. 78th Steelmaking Conf.,1995, (ISS, Warrendale, PA, 1995) p. 333.
130. J Madill, Proc. Workshop on *Steelmaking, Steel cleanliness and mould thermal monitoring,* Brussels, 1993, (EC Sci. Tech. Comm. Publ., Luxembourg, 1994) p. 104.
131. M Jenkins, BG Thomas, WC Chen, RB Mahapatra, Proc. 77th Steelmaking Conf., 1994, (ISS, Warrendale, PA. 1994) p. 337.
132. Jae Eung. CAMP-ISIJ, **8**, 218, (2003).
133. P. Ackermann, Ph.D. Dissertation, Ecole Polytechnique Federale de Lausanne, Lausanne, (1983).
134. T Emi, H Nakato, Y Iida, K Emoto, R Tachibana, T Imai, H Bada, Proc. 61st NOH-BOS Conf., 1978, (ISS, Warrendale, PA, 1978) p. 350.
135. H. Tomono, Ph.D. Dissertation, Ecole Poly. Federale, d'Lausanne, Lausanne, 1979.
136. H Tomono, H Ackermann, W Kurz, W Heinemann, Solidification Technology in the foundry and Cast house, Warwick, 1983 (The Metals Soc., London, 1983) Book 273 p. 524.
137. M Nakato, Tetsu-to Hagane, **61** (12), 131, (1975).
138. P Riboud and M Larrecq, Proc. 62nd NOH-BOS Conf., Detroit, 1979, (ISS, Warrendale, PA., 1979) p. 78.
139. A.S. Normanton, V. Ludlow, M. Hecht, B. Harris, C-Å. Däcker, A. Di. Donato, T. Sohlgren, ECSC-Report 7210.PR/273, 2005, (Europ. Comm. Sci. Tech. Publ., Luxembourg, 2005).

140. J Sengupta, BG Thomas, HJ Shin, GG Lee, SJ Kim, Met. Mater. Trans. A, **37A**, 1597, (2006).
141. J. Sengupta, B.G. Thomas, JOM, **58**, 16, (2006).
142. PE Ramirez-Lopez, KC Mills, PD Lee, B Santillana, Proc. 7th Europ., Conf. Continuous Casting, 2011, Metec Congress, Dusseldorf, 2014 (VDEh, Dusseldorf, 2011).
143. BG Thomas, J Sengupta, C Ojeda, Proc. 2nd Baosteel Biennial Conf., 2006, Shanghai, vol. 1, (2006), p. 112.
144. C. Ojeda, J. Sengupta, B.G. Thomas, J. Barco, J.L. Aruna, Pr*oc.* AIST Tech., 2006, vol. 1 (ISS Warrendale, PA) p. 1017.
145. C. Offerman, C-Å. Däcker, C. Enström, Scand. J Metallurgy, **10**, 115, (1981).
146. M.M. Wolf, Continuous Casting, Volume 9, Iron & Steel Society, Warrendale, PA, p. 285 (1997).
147. A.M. El-Wazri, E.Es-Sadiqi, S. Yue, Proc. 39th MCWSP Conf., ISS, XXXV, 1998, p. 721.
148. H. Bruce, Swerea KIMAB Report, IM-2002-530 (2002).
149. Z Buzek, E Mazancova, Z Jonsta, F Mazanek, www.metal2013.com/files/proceedings.
150. C Houpert, V Lautari, JM Jolivet, M Guttmann, JP Birat, M Jallon, M Contente, Revue. de Metall., **94**, 1369, (1997).
151. M R Ozgu, B Kocatulum, Iron and Steelmaker, **21,** (5), 77, (1994).
152. J Nieto, S Morales, R Garcia, G Campos, Proc. 3rd Intl. Conf. Cont. Casting, Madrid, 1998, (UNESID, Madrid, 1998) p 485.
153. G Bocher, U Hoffmann, P Muller, Proc. 2nd Europ. Conf. Continuous casting, Dusseldorf, 1994, Metec Congress'94 (VDEh, Dusseldorf, 1994) p. 103.
154. L. Schmidt, A. Josefsson, Scand. J. Metallurgy, **10**, (1981).
155. B. Mintz, R. S. Yue, J.J. Jonas, Intl. Materials Review, **36**, 5, (1991).
156. A Wyckaert, Ironmaking and Steelmaking, **4,** (6), 340, (1977).
157. K Relander, Jernkonorets Ann., **155**, (9), 565, (1971).
158. TJ Billany, AS Normanton, KC Mills, P Grieveson, Ironmaking and Steelmaking, **18**, 403, (1991).
159. RJ O'Malley, J Neal, Proc. Intl. Conf. New Developments in Metallurgical Process Technol., Dusseldorf, METEC Congress, 1999, (VDEh, Dusseldorf, 1999) p. 73.
160. M S Jenkins, BG Thomas, Proc. 80th Steelmaking Conf., Chicago, IL, 1997, (ISS, Warrendale, PA, 1997) p. 285.
161. BG Thomas, A Moitra, R McDavid. Process Technol. Conf., **13**, 1996, (ISS, Warrendale, PA, 1996) p. 143.
162. A Kusano, N Sato, M Okimori, S Fukunaga, K Nishihara, M Sato, Y Minigawa, Proc. 74th Steelmaking Conf., 1991, (ISS, Warrendale, PA, 1991) p. 111.
163. LC Hibbeler, BG Thomas, Proc. AISTech Conf. 2010, Pittsburgh (ISS, Warrendale. PA, 2010) p. 1215.
164. LC Hibbeler, BG Thomas, Iron and Steel Technology, **2013,** (Oct), 121, (2013).
165. T. Teshima, M. Osame, K. Okimoto, Y. Nimura, Proc. 71st Steelmaking Conf., 1988, (ISS, Warrendale, PA, 1988), p. 111.
166. J. Kubota, K. Okimoto, A. Shirayama, H. Murakami, Proc. 74th Steelmaking Conf., 1991, (ISS, Warrendale, PA, 1991) p. 233.
167. H. Nakato, K. Saito, Y. Oguchi, N. Namura, K. Sorimachi, Proc. 70th Steelmaking Conf., 1987, (ISS, Warrendale, PA, 1987) p. 427.
168. Z. Wang, K. Mukai, Z. Ma, M. Nishi, H. Tsukamoto, F. Shi, ISIJ Intl, **39**, 795, (1999).
169. M Cervantes, H Gustavsson, Proc. 3rd Europ Conf. Continuous casting, Madrid, 1998, (UNESID, Madrid, 1998) p. 202.
170. A. Matsushita, K. Isogami, M. Temma, T. Ninomiya, K. Tsutsumi, Trans. ISIJ, **28**, 531, (1988).
171. B.S. Moghaddam, E. Steinmetz, P. R. Scheller, Proc. 3rd Intl. Congress Sci. Technol. Steelmaking, 2005, (AIST, Warrendale, PA, 2005) p. 911.
172. G.A. Panaras, A. Theodorakakos, G. Bergeles, Met. Mater. Trans, **29B**, 1117, (1998).

173. PE Ramirez- Lopez, PN Jalali, J Bjorkvall, U Sjostrom, C Nilsson, ISIJ Intl. 54, 342. (2014) and Proc. 8th Europ. Conf. Continuous Casting, Graz, Austria, 2014 (ASMET, Vienna, 2014).
174. E van Vliet, DW van der Plas, SP Carless, A A Kamperman, AE Westendorp: Proc. 7th Europ Conf. Cont. Casting, Dusseldorf, 2011, (VDEh, Dusseldorf, 2011) Session 4.
175. S. Yamashita, M. Iguchi, ISIJ Intl., **41**, 1529, (2001).
176. T. Watanabe, M. Iguchi, ISIJ Intl., **49**, 182, (2009).
177. R.M. McDavid, B.G. Thomas, Met. Mater. Trans., **27B**, 672, (1996).
178. B Zhao, SP Vanka, BG Thomas, Intl. J Heat Fluid Flow, **26**, 105, (2005).
179. Q. He, G. Evans, R. Serje, T. Jaques, Proc. AISTech, 2009, (AIST, Warrendale, PA, 2009), No. 2, p. 57.
180. H. Bai, B.G. Thomas, Met. Mater. Trans, **32B**, 707, (2001),
181. G.-G. Lee, H.-J. Shin, B.G. Thomas, S.-H. Kim, Proc. AISTech, 2008, (AIST, Warrendale, PA, 2008), No. 2, p. 63.
182. B. Li, T. Okane, T. Umeda, Met. Mater. Trans.; **32B**, 1053, (2001).
183. B. Li, F. Tsukihashi, ISIJ Intl., **45**, 30, (2005).
184. N. Kasai, M. Iguchi, ISIJ Intl., **47**, 982, (2007).
185. B. Li, T. Okane, T. Umeda, Met. Mater. Trans, **32B**, 1053, (2001).
186. M. Gebhard, Q.L. He, J. Herbertson, Proc. 76th Steelmaking Conf., (ISS, Warrendale, PA, 1993), p. 441.
187. Y.H. Wang, Proc. 73rd Steelmaking Conf., 1990, (ISS, Warrendale, PA, 1990), p. 473.
188. D. Gupta, A.K. Lahiri, Met. Mater. Trans, **25B**, 227, (1994).
189. K Tsutsumi, K Watanabe, M Suzuki, M Nakada, T Shiomi, Proc. 7th Intl. Conf. Molten slags, fluxes and salts, Cape Town, 2004, (SAIMM, Johannesburg, 2004) p. 803.
190. Q. He, ISIJ Intl., **33**, 343, (1993).
191. A. Theodorakakos, G. Bergeles, Met. Mater. Trans, **29B**, 1321, (1998).
192. D. Gupta, A.K. Lahiri, Met. Mater. Trans, **27B**, 695, (1996).
193. T. Honeyands, J. Herbertson, Steel Research Intl, **66**, 287, (1995).
194. J. Anagnostopoulos, G. Bergeles, Met. Mater. Trans, **30B**, 1095, (1999).
195. M. Kamal, Y. Sahai, Steel Research Intl., **76**, 44, (2005).
196. R. Chaudhary, B.T. Rietow, B.G. Thomas, Proc. 2009 Mater. Sci. and Technol. Conf., (AIST/TMS, Warrendale, PA., 2009), p. 1090.
197. M. Harman, A.W. Cramb, Proc. 79th Steelmaking Conf. 1996, (ISS, Warrendale, PA, 1996), p. 773.
198. W Emling, T A Waugaman, SL Feldbauer, AW Cramb, Proc. 77th Steelmaking Conf. (ISS, Warrendale, PA., 1994) p. 371.
199. K. Tsukaguchi, S Ura, A Shiraishi, Y Hitomi T Nagahata, Steel Technol. Intl., **16**, 175, (1995/6).
200. J Yoshida, T Ohmi, M Iguchi, ISIJ Intl., **45**, 1160, (2005).
201. K Watanabe, K Tsutsumi, M Suzuki, M Nakada, T Shiomi, ISIJ Intl,**49**, 1161, (2009).
202. P.E. Ramirez-Lopez, PhD Thesis, Modelling shell and oscillation mark formation during continuous casting via explicit incorporation of slag infiltration, Imperial College, London (2010)
203. T. Funada, D.D. Joseph, J. Fluid Mechanics, **445**, 263, (2001).
204. L.M. Milne-Thomson, *Theoretical Hydrodynamics*, Macmillan Press, London, (1968).
205. S. Chandrasekhar, *Hydrodynamic and Hydromagnetic Stability*, (Clarendon Press, Oxford,,1961), p. 481.
206. P. Cha, J. Yoon, Met. Mater. Trans. B, **31B**, 317, (2000).
207. S Feldbauer, I Jimbo, A Sharan, K Shimizu, W King, J Stepanek, J Harman, AW Cramb, Proc. 78 th Steelmaking Conf., 1995, (ISS, Warrendale, PA, 1995), p. 655.
208. E Torres- Alonso, RD Morales, S Hernandez-Garcia, J Palafox- Ramos, Met. Mater. Trans. B, **39B**, 840, (2008).
209. S Diehl, JA Moore, RJ Phillips, Proc. 78th Steelmaking Conf., Nashville, TN, 1995 (ISS Warrendale, PA, 1995), p. 351.

210. BG Thomas, LF Zhang, ISIJ Intl., **41**, 1181, (2001).
211. S Chakraborty, W Hill, Proc. 77th Steelmaking Conf., 1994, (ISS, Warrendale, PA, 1994) p. 389.
212. J Knoepke, M Hubbard, J Kelly, R Kittridge, J Lucas, Proc. 77th Steelmaking Conf., 1994, (ISS, Warrendale, PA., 1994) p. 381.
213. T Miyake, M Morishita, H Nakata, M Kokita, ISIJ Intl., **46**, 1817, (2006).
214. M Hanao, M Kawamoto, H Mizukami, K Hanazaki, Ironmaker Steelmaker, **27**, (Nov.), 55, (2000).
215. S Shima, Y Maruki, T Toh, K Tsunenari, Proc. ISS Tech. 2003 Conf. (ISS, Warrendale, PA, 2003) p. 721.
216. L Zhang, BG Thomas, Proc. 24th National Steelmaking Conf., Morelia, Mich. Mexico, p. 138.
217. H Lei, Y Zhao, DQ Geng, ISIJ Intl., **54**, 1629, (2014).
218. PV Riboud, C Gatellier, H Gaye, JN Pontoire, P Rocabois, ISIJ Intl., **36**, 522, (1996).
219. H Nakato, S Takeuchi, T Fujii, T Nozaki, N Washio, Proc. 74th Steelmaking Conf., Washington, DC,1991, (ISS, Warrendale, PA, 1991) p. 639.
220. M Hanao, M Kawamoto, ISIJ Intl., **48**, 1210, (2008).
221. P Wei, K Uemura, S Koyama, Tetsu-to- Hagane, **78**, 1361, (1992).
222. QY Zhang, L Wang, XH Wang, ISIJ Intl., **46**, 1421, (2006).
223. M Hanao, CAMP- ISIJ, **19**, 803 (2006).
224. M Hanao, CAMP-ISIJ, **21**, 822, (2008).
225. *Slag Atlas*, (Stahl u Eisen, VDEh, Dusseldorf, 1995).
226. P Kozakevitch, LD Lucas, Revue Metallurgie, **68a**, 589, (1968).
227. E Takeuchi, JOM, **1995**, (May), 42, (1995).
228. MM Wolf, Ironmaker & Steelmaker, **2000**, (Jan), 22, (2000).
229. MM Wolf, Ironmaker & Steelmaker, **2000**, (Feb), 65, (2000).
230. MM Wolf, Ironmaker & Steelmaker, **2000**, (March), 69, (2000).
231. MM Wolf, Ironmaker & Steelmaker, **2000,** (April), 58, (2000).
232. MM Wolf, Ironmaker & Steelmaker, **2000,** (May), 78, (2000).
233. MM Wolf, Ironmaker & Steelmaker, **2000**, (June), 22, (2000).
234. MM Wolf, Ironmaker & Steelmaker, **2000**, (July), 63, (2000).
235. MM Wolf, Ironmaker & Steelmaker, **2000,** (Aug.), 75, (2000).
236. MM Wolf, Ironmaker & Steelmaker, **2000**, (Sept.), 90, (2000).
237. MM Wolf, Ironmaker & Steelmaker, **2000**, (Oct.), 114, (2000).
238. MM Wolf, Ironmaker & Steelmaker, **2000**, (Nov.), 67, (2000).
239. MM Wolf, Ironmaker & Steelmaker, **2000**, (Dec.). 45, (2000).
240. S Suresh, RC Sinha, JB Singh, A Narayan, MS Sadhu, Tata Search, **2006**, 291, (2006).
241. C Knocke, R Berger, W Pluschkell, Stahl u Eisen, **112**, 81, (1992).
242. C Lefebvre, JP Radot, JN Pontoire, Y Roux, Revue de Metallurgie, CIT, **94**, 489, (1997).
243. JJ Macho, G Hecko, B Golinmowski, M Frazee, Preprints 33rd McMaster Symp. Iron and Steelmaking, Hamilton, Ont., 2005, (McMaster Univ. Hamilton, 2005) p. 131.
244. P Valentin, C Brucke, K Harste, J Potschke, Steel Res. Intl., **74**, 139, (2003).
245. Y Tanizawa, CAMP-ISIJ, **4**, 257, (1991).
246. K Yamaguchi, CAMP-ISIJ, **3**, 181, (1990).
247. C Perrot, JN Pontoire, C Marchionni, MR Ridolfi, LF Sancho, Proc. 5th Europ. Conf. Continuous Casting. Nice, 2005 (La Revue Metall., Paris, 2005) p. 36.
248. B. Debiesme, J Radot, D Coulombet, C Lefevbre, Y Roux, C Demarval, US Patent No 5, 876,482 (1999) and US Patent No 6, 328, 781 (2001).
249. Y Nakamura, T Ando, K Kurata, M Ikeda, Trans. ISIJ, **26**, 1052, (1986).
250. K Mukai, *Marangoni and Interfacial Phenomena in Materials Processing* ed. ED Hondros, M McLean and KC Mills, (IOM, London, 1998) p. 201.
251. K Mukai *et a*l, Taikabutsu (Refractories), **42**, 710, (1990).
252. N Tsukamoto, Taikabutsu (Refractories), **44**, (5), 270, (1992).

253. MC Bezerra, CA Valadares, IP Rocha, JP Bulota, MC Carboni, IL Scripnic, CR Santos, K Mills, D Lever, Proc. 37th Steelmaking Seminar, Porto Allegre, RS, Brazil, 2005) (ABM, Sao Paulo, 2005) p. 190. http//www.carboox.com/pdf/fluxante_2007.pdf.
254. Q Wang, JH Chi, B Xie, Y He, B Zhu, W Chen, Proc. Conf. Continuous Casting in Developing Countries, Beijing, 1993 (SEAISI, 1993) p. 842.
255. F Hauck, J Potschke, Arch f. Eisenfuttenw. **53**, 133, (1982).
256. AF Dick, S Zarrug, RF Pomfret, KS Coley. Proc. 4th Intl Conf. Molten slags and fluxes, Sendai, (1992) (ISIJ, Tokyo, 1992) p. 380.
257. AI Zaitsev, AV Leites, AD Litvina, B Mogutov, Steel Research, **65**, 368, (1994).
258. DE Sturgill. Proc. 79th Steelmaking Conf., Pittsbugh, PA, 1996, (ISS Warrendale, PA) p. 265.
259. IR Lee, JW Kim, J Choi, D Kwon, YK Shin, Proc. Conf. on Continuous casting in Developing countries, Beijing, China, (1993), p. 814.
260. M Shinmei, T Machida, Met. Trans., **4**, 1996, (1973).
261. K Shimizu, T Suzuki, I Jimbo, AW Cramb, Proc. 79th Steelmaking Conf., Pittsburgh, 1996, (ISS, Warrendale, PA, 1996) p. 723.
262. NN Viswanathan, F Shahbazian, S Du, S Seetharaman, Steel Research, **70**, 53, (1999).
263. M Persson, S Seetharaman, S Seetharaman, ISIJ Intl., **47**, 1711, (2007).
264. H Heimbach, K Schulz, J Markardt, HJ Ehrenberg, Stahl u Eisen, **117**, 105, (1997).

Correction to: The Casting Powders Book

Correction to:
K.C. Mills and C.-Å. Däcker, *The Casting Powders Book*,
https://doi.org/10.1007/978-3-319-53616-3

Within this book, minor changes to some of the figure permissions and the readability of the chapters have been made and several figures have been replaced with new versions.

The updated version of the book can be found at
https://doi.org/10.1007/978-3-319-53616-3

K.C. Mills and C.-Å. Däcker, *The Casting Powders Book*,
https://doi.org/10.1007/978-3-319-53616-3_12

MIX
Papier aus verantwortungsvollen Quellen
Paper from responsible sources
FSC® C105338

If you have any concerns about our products,
you can contact us on
ProductSafety@springernature.com

In case Publisher is established outside the EU,
the EU authorized representative is:
Springer Nature Customer Service Center GmbH
Europaplatz 3, 69115 Heidelberg, Germany

Printed by Libri Plureos GmbH
in Hamburg, Germany